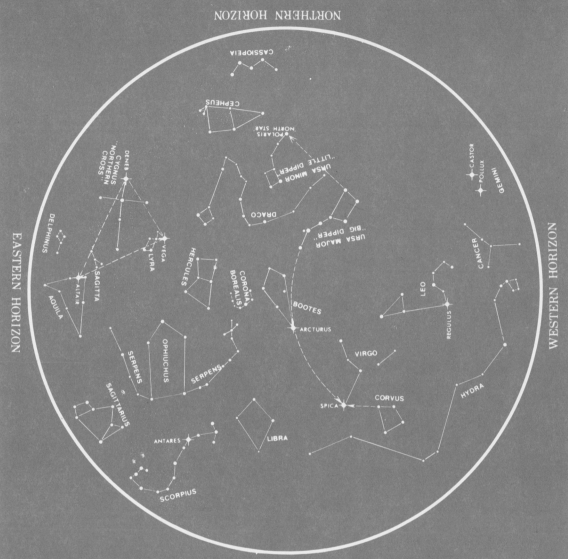

THE NIGHT SKY IN JUNE

Latitude of chart is 34° N, but it is practical throughout the continental United States.

To use: Hold chart vertically and turn it so the direction you are facing shows at the bottom.

Chart time (Local Standard):
10 p.m. First of month
9 p.m. Middle of month
8 p.m. Last of month

ESSENTIALS OF THE DYNAMIC UNIVERSE
AN INTRODUCTION TO ASTRONOMY

THEODORE P. SNOW
University of Colorado at Boulder

ESSENTIALS OF
THE DYNAMIC UNIVERSE
AN INTRODUCTION TO ASTRONOMY

WEST PUBLISHING COMPANY
St. Paul   New York   San Francisco   Los Angeles

COPYRIGHT © 1984 By WEST PUBLISHING CO.
50 West Kellogg Boulevard
P.O. Box 3526
St. Paul, Minnesota 55164

Library of Congress Cataloging in Publication Data

Snow, Theodore P. (Theodore Peck)
The essentials of the dynamic universe.

Includes bibliographies and index.
1. Astronomy. I. Title.
QB43.2.S663 1984   520   83-23560
ISBN 0-314-77798-9
1st Reprint—1985

Composition: Clarinda Company
Copy editing: Deborah Annan
Text Design: Janet Bollow
Text illustrations: Henry Taly Design

Cover Photo

Front Top: Association of Universities for Research in Astronomy, Inc., Kitt Peak National Observatory.

Front Bottom: NASA.

Back: Association of Universities for Research in Astronomy, Inc., the Kitt Peak National Obervatory.

Chapter opening credits

Chapter 1: Anglo-Australian Telescope Board. **Chapter 2:** NASA. **Chapter 3:** The Granger Collection. **Chapter 4:** NASA. **Chapter 5:** NASA. **Chapter 6:** NASA. **Chapter 7:** NASA. **Chapter 8:** NASA. **Chapter 9:** Lowell Observatory. **Chapter 10:** The Granger Collection. **Chapter 11:** High Altitude Observatory, National Center for Atmospheric Research, sponsored by the National Science Foundation. **Chapter 12:** © West Publishing. **Chapter 13:** Harvard College Observatory. **Chapter 14:** © West Publishing. **Chapter 15:** U.S. Naval Observatory photograph. **Chapter 16:** Lick Observatory. **Chapter 17:** R. E. Royer. **Chapter 18:** Barnard Collection. **Chapter 19:** © West Publishing. **Chapter 20:** Mount Palomar Observatory, California Institute of Technology. **Chapter 21:** Bell Laboratories. **Chapter 22:** Mount Palomar Observatory, California Institute of Technology. **Chapter 23:** © West Publishing. **Chapter 24:** NASA.

Color Photo Credits

V The Granger Collection. **VI** University of Hawaii. **VII** NASA photograph. **VIII** NASA photograph. **IX** © 1979 Anglo-Australian Telescope Board. **X** Photo by F. Espenak. **XI** James D. Wray, University of Texas. **XII** NASA photograph. **Color Plate 1.** *upper:* Photograph by Patrick Wiggins/Hansen Planetarium; *lower:* Sommers-Baush Observatory, University of Colorado, Photography by J. Kloeppel. **Color Plate 2.** *upper:* © West Publishing Company; *lower:* © Association of Universities for Research in Astronomy, Inc., the Kitt Peak National Observatory. **Color Plate 3.** *upper:* The National Radio Astronomy Observatory, operated by Associated Universities, Inc. under contract with the National Science Foundation; *lower left:* Lick Observatory photograph.; *lower center:* © Association of Universities for Reseach in Astronomy, Inc., the Cerro Tololo Inter-American Observatory; *lower right:* MMTO: University of Arizona and Smithsonian Institution. **Color Plate 4.** *upper left, upper right, and lower:* NASA photographs. **Color Plate 5.** *upper left, upper right, and lower:* NASA photographs. **Color Plate 6.** *upper, center, and lower:* M. Kobrick, JPL. **Color Plate 7.** *upper:* TASS from SOVFOTO; *lower left:* NASA photograph; *lower right:* Laboratory for Atmospheric and Space Physics, University of Colorado, sponsored by NASA. **Color Plate 8.** *upper, lower left, and lower right:* NASA photographs. **Color Plate 9.** *upper lower left, and lower right:* NASA photographs. **Color Plate 10.** *upper, lower left, and lower right:* NASA photographs. **Color Plate 11.** *lower, upper left, and upper right:* NASA photographs. **Color Plate 12.** *upper and lower left:* NASA photographs; *lower right:* Laboratory for Atmospheric and Space Physics, University of Colorado, sponsored by NASA. **Color Plate 13.** *upper left and lower:* NASA photograph; *upper right:* © F. Espanek. **Color Plate 14.** *upper left:* Solar Maximum Mission, High Altitude Observatory National Center for Atmospheric Research, sponsored by NASA; *upper right:* Laboratory for Atmospheric and Space Physics, University of Colorado sponsored by NASA; *lower:* NASA photograph. **Color Plate 15.** *lower left:* © Association of Universities for Research in Astronomy, the Kitt Peak National Observatory; *lower right:* Harvard-Smithsonian Center for Astrophysics; *upper:* © 1961, California Institute of Technology. **Color Plate 16.** *upper left:* © 1981 Anglo-Australian Telescope Board; *upper right:* Lick Observatory photograph. *lower left:* N. R. Walborn; *lower right:* K. Davidson, University of Minnesota. **Color Plate 17.** *upper:* © 1981 Anglo-Australian Telescope Board; *lower:* R. D. Gehrz, J. Hackwell, and G. Grasdalen, University of Wyoming. **Color Plate 18.** *upper left and upper right:* © 1980 and 1981 Anglo-Australian Telescope Board; *lower:* R. J. Dufour, Rice University. **Color Plate 19.** *lower left:* Anglo-Australian Telescope Board; *lower right:* T. R. Gull, Goddard Space Flight Center. *upper:* Lick Observatory photograph. **Color Plate 20.** *upper:* J. R. Dickel; *lower:* NASA photograph. **Color Plate 21.** *upper left and lower:* R. J. Dufour, Rice University; *upper right:* © California Institute of Technology. **Color Plate 22.** *upper left and upper right:* © 1980, Anglo-Australian Telescope Board; *lower right:* J. D. Wray, University of Texas; *lower left:* JPL, Photo by H. Arp, processing, Jean J. Lorre, Imaging Processing Lab-

Continued on page 487.

CONTENTS

3 Messages From the Cosmos: Light and Telescopes 44

Introduction to Section II
The Solar System 71

4 The Earth and Its Companion 73

The Outer Planets   163

Space Debris   177

The Sun   194

12 Adding It Up: Formation of the Solar System 212

SECTION III

Introduction to Section III
The Stars 229

13 Observations and Basic Properties of Stars 231

14 Stellar Structure: What Makes A Star Run? 252

15 Life Stories of Stars 267

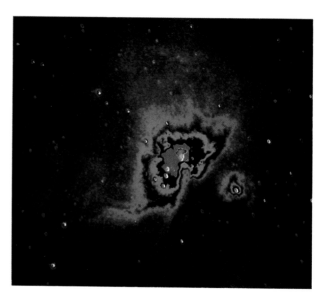

16 Stellar Remnants 290

SECTION V

Introduction to Section V.
Extragalactic Astronomy 355

20 Galaxies Upon Galaxies 357

21 Universal Expansion and the Cosmic Background 380

22 Peculiar Galaxies, Explosive Nuclei, And Quasars 394

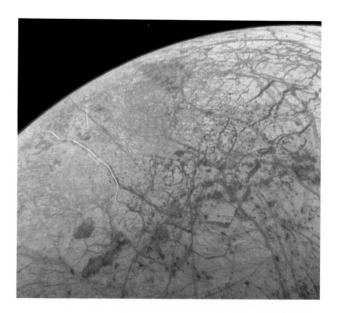

23 Cosmology: Past, Present, and Future Of The Universe 411

SECTION VI

Introduction to Section VI Life In The Universe 431

24 The Chances of Companionship 433

PREFACE

To study astronomy is, in a sense, the quintessential human endeavor. What distinguishes us from lower creatures, if not our curiosity, our compulsion to explore and discover? And what exemplifies this compulsion more than the study of the universe?

We probe the heavens (and the earth) with all possible means, and we do it for no other reason than to learn whatever there is to be known. Astronomy has produced many important and useful by-products, of course, and could be (and often is) justified solely on that basis. That is not the real motivation for astronomy, however.

This textbook represents an attempt by an astronomer to share both the knowledge and the intellectual gratification of our science. There is considerable beauty in the universe, for the eye and the mind to behold. Just as it is visually stimulating to gaze at a great glowing nebula or a colorful moon, it is pleasing to the intellect to grasp a new understanding of one of the grand themes of the cosmos. It is hoped that the reader of this book will gain by doing both.

The *Essentials of the Dynamic Universe* is based on the earlier hardback version, *The Dynamic Universe*. The new version is a distillation of the old, aimed at general astronomy courses that survey the entire subject in one semester or quarter. The reduction in length was achieved by merging chapters here and there, while retaining the original organization. Little that is essential to the study of astronomy was dropped; the major deletion was in the discussion of the history of astronomy (although a sketch of the principal people and events remains). Further economy was accomplished by tightening up the descriptions throughout the text, especially in chapters that were created by combining pairs of chapters from the *Dynamic Universe*.

The entire text has been kept current by revisions and additions made in response to new developments. For example, a variety of new discoveries made by the *Infrared Astronomical Satellite,* including those announced in November 1983, are included. The chapter on cosmology (chapter 23) has been completely revised, in view of the new inflationary-universe theory, which alters the viewpoint of astronomers on the im-

portant issue of whether the universe is open or closed. New evidence that the universe itself may not be homogeneous is now described. Discussions of the grand-unified theory, which relates the fundamental forces of nature to each other and to the origin and evolution of the universe, are included in this chapter. A great deal of revision has occurred in both the details (planetary and satellite data) and the overall understanding of the solar system, and these updates are included both in the section on the solar system and in the appendixes.

The text is arranged in a traditional manner, beginning with discussions of the nighttime sky, basic physical principles, and the basic tools of astronomy. A thorough discussion of the solar system follows, and then there are sections on the stars, our galaxy, extragalactic astronomy, and life in the universe. There are alternative logical sequences through the material: one may follow the introductory three chapters immediately with the stellar, galactic, and extragalactic sections, for example; or take the solar system and then the closing chapter on life.

There is a brief introduction for each of the six major sections, and each is followed by a guest editorial written especially for this book. These essays contain the personal viewpoints of prominent scientists on important current mysteries or controversies in astronomy, and lend to the reader additional excitement and understanding of how research in our science proceeds.

Each chapter begins with an outline of topics to be covered, and ends with a brief Perspective, placing the chapter in context with those to come. Within the chapters are Astronomical Insights, brief essays on topics related to the main text, but set apart because they contain anecdotal material, somewhat more technical information than appropriate for the general text, historical background, or additional insights that may apply to other areas of astronomy as well as to the current chapter. At the end of each chapter is a summary of the principal points, an assortment of review questions, and a list of supplementary readings.

Extensive appendices at the back of the book provide basic data on a variety of topics, such as the elements, telescopes, the planets and satellites, bright stars, the constellations, interstellar molecules, and galaxies of the Local Group. Other appendixes contain detailed explanations of certain concepts that are necessarily treated only qualitatively in the text. These include temperature scales, a logarithmic treatment of the magnitude system, a comprehensive discussion of radiation laws, summaries of important nuclear-reaction sequences, and a discussion of the relativistic Doppler effect.

Ancillary materials for this text include an Instructor's Manual, written by Stephen J. Shawl of the University of Kansas, and a Study Guide, authored by Catharine D. Garmany and the undersigned, of the University of Colorado. The Instructor's Manual contains helpful discussions of strategies in teaching, provides a large number of possible exam questions (with answers), and gives complete answers to all the problems from the main text. The Study Guide, intended to help the student make the most of the text, contains brief chapter summaries, lists of key words and phrases, self-tests, answers to selected (usually the more difficult) problems from the text, and complete bibliographies of articles on relevant topics, taken from a wide assortment of popular magazines and journals. In addition to the Instructor's Manual and the Study Guide, another aid to teaching from the *Essentials of the Dynamic Universe* is available to large adopters: a set of transparencies for use with overhead projectors, showing a number of useful diagrams and charts from the text.

The artwork in the *Essentials of the Dynamic Universe* is both extensive and of great value in illustrating the text material. The line drawings are intended to show, often schematically, the relationships of major concepts, and the photographic reproductions demonstrate the true appearance of the skies and the tools with which they are probed. The color-plate sections add beauty, helping the reader gain an aesthetic appreciation of astronomy, to complement the intellectual one to be derived from reading the text.

Very little of this book could have been created without the benefit of interaction with other people. It is a pleasure to acknowledge this assistance (with apologies to those inadvertently omitted).

As in the case of the *Dynamic Universe,* this book owes much to the involvement of three key people. The most important was the support, enthusiasm, and patience of my wife, Connie, who also did heroic work as manuscript typist on the large edition, from which this one is derived. The editor who provoked me into undertaking the original project, and then guided the development of this one, was Denise Simon. When it got to the hard part, where a manuscript with nothing but text is somehow translated into a book, with illustrations, appendixes, captions, and a cover, production editor John Orr once again made it all come together.

A large number of scientists in our field helped in various indispensable ways, from allowing me to pick their brains, to reviewing portions of the manuscript, to providing illustrations and photographs. I am especially grateful to several of my colleagues at the University of Colorado, who, unable to escape me, were unfailingly generous with their time and energy whenever I sought their assistance. Although these people were called upon less often than during the preparation of the first edition, there are still a few who ought to be singled out. These include J. M. Shull, L. W. Esposito, C. D. Garmany, R. A. West, G. Rottman, and G. Lawrence.

Others around the country also assisted with advice, illustrations, or both. Individual photo and illustration credits are listed elsewhere, but it is a pleasure to make specific mention here of those who were especially helpful. The following list includes people who provided materials for the *Dynamic Universe,* and therefore for this edition as well; a few who were accidentally omitted from the acknowledgments in the first edition; and several who have helped with new information and illustrations for the *Essentials of the Dynamic Universe.* These include, more or less in the same order as their areas of expertise appear in the book, P. M. Routly, U.S. Naval Observatory; D. F. Malin, Anglo-Australian Observatory; E. C. Krupp, Griffith Observatory; J. Eddy, High Altitude Observatory; O. Gingerich, Harvard College Observatory; G. Emerson, E. E. Barnard Observatory and Ball Aerospace Corp.; A. N. Witt, University of Toledo; R. D. Gehrz, University of Wyoming; H. Masursky, U.S. Geological Survey; M. Kobrick, NASA JPL; P. Thomas, Cornell University; G. Newkirk, L. House, and J. K. Watson, High Altitude Observatory; T. Lee, Carnegie Institute of Washington; M. H. Liller, Harvard College Observatory; S. J. Shawl, University of Kansas; R. L. Kurucz, Harvard-Smithsonian Center for Astrophysics; J. C. Wheeler, University of Texas; P. M. Flower, Clemson University; I. Iben, University of Illinois; G. A. Wegner, Dartmouth College; J. Heckathorn, Computer Sciences Corporation; N. Walborn, NASA: Goddard Space Flight

Center; C. Heiles, University of California, Berkeley; P. Thaddeus, Goddard Institute for Space Studies; A. F. J. Moffat, Université de Montreal; B. J. Bok, University of Arizona; J. J. Lorre, NASA JPL; F. N. Owen, National Radio Astronomy Observatory; M. Pettini, Royal Greenwich Observatory; R. J. Dufour, Rice University; P. J. E. Peebles, Princeton University; and F. D. Drake, Cornell University.

A special debt is owed to those who wrote the guest editorials, for adding their thoughts and visions to my own less elegant discussions. Much of the excitement of astronomy lies in the pursuit of new revelations beyond the scope of current knowledge, and the essays contributed by leaders in this pursuit help immeasurably to impart this excitement to the reader. The essayists for this edition are John N. Bahcall, Institute for Advanced Study; David N. Morrison, University of Hawaii; Andrea K. Dupree, Center for Astrophysics; Ben N. Zuckerman, UCLA; Martin Rees, Cambridge University; and Gerard K. O'Neill, Princeton University and the Space Studies Institute.

Reviewers of the manuscript, at various stages, were:

Steve Lattanzio
Orange Coast College

David Thieson
University of Maryland

Stephen Shawl
University of Kansas

For all of these people, and to the students whose responses to my teaching philosophies have also helped shape this book, I am grateful. With their continued input, I trust that this book will continue to evolve, as does the dynamic universe.

Theodore P. Snow
November, 1983

for Mac and Tyler

THE NIGHTTIME SKY AND THE TOOLS OF ASTRONOMY

1

Introduction to Section I

We begin our study of astronomy with an introduction to the nighttime sky and a brief overview of early astronomical developments. This will provide us with an immediate understanding of many of the phenomena that can be seen and appreciated with only our eyes as observing equipment, and it will arm us with all the knowledge our ancestors had as they attempted to develop a successful picture of the cosmos and their place in it. We will then see how our ancestors fared as they sought to unravel the true nature of the cosmos. We will concentrate our historical discussions on the civilizations, primarily the Greek, that arose on the shores of the Mediterranean, for it was here that the foundations of modern astronomy were laid.

In the second chapter we will discuss the major developments of the Renaissance, when fresh ideas arose in astronomy, as in all forms of human endeavor. We will learn to appreciate the awesome breakthroughs made by giants such as Copernicus, Brahe, Kepler, Galileo, and Newton, who led the way toward a correct understanding of the universe and the earth's position within it. We move on to the laws of physics that govern such diverse phenomena as planetary orbits, the motions of molecules in a gas, and tides on the earth and other bodies.

We would know nothing of the external universe were it not for the light that reaches us from faraway objects, and chapter 3 describes the nature of light and the way we decipher its messages. We will find that an amazing variety of information can be derived from the spectra of objects like planets and stars. Things once considered forever beyond our grasp are now routinely measured, and in this chapter we will learn how this is done. Chapter 3 also includes a description of telescopes and their principles, and how they are used to measure light in all portions of the spectrum.

LEARNING ABOUT THE NIGHTTIME SKY

Learning Goals

To sit outdoors on a clear night is a special experience, and few of us can avoid being forced into a contemplative mood by the grandeur of the sky. In this chapter we will interpret the sights that face us on a clear night, and we will learn how the first understandings of these phenomena developed historically.

When we look at the sky, we do not see it in three dimensions, because there are no obvious clues to tell us the distances to the objects we see. This fact led, long ago, to the concept of the **celestial sphere,** in which the stars and other objects in the sky are said to lie on the surface of a sphere that is centered on the earth. Although we no longer think of this as literal truth, it is still a convenient device for discussing and describing the heavens. Positions of objects on the celestial sphere are measured in angular units, because without knowledge of how far away things are, we have no means of determining their actual separations in any true distance units such as kilometers.

The Rhythms of the Cosmos: Daily, Monthly, and Annual Motions

It soon becomes apparent to any observer of the heavens that there are cycles of motions ranging from those which occur daily to those which define the year. These cycles have had a profound influence on us and on other life forms on the earth. Some of the cycles are caused by the earth's rotation, others by the motions of the moon or planets, and still others by the earth's orbital motion about the sun.

Daily Motion

The most obvious of the many motions that affect our view of the universe are the daily cycles of all celestial objects, resulting from the rotation of the earth. The earth spins on its axis in twenty-four hours, so we, on its surface, see a continuously changing view of the heavens. We see the sun rise and set, along with the moon, the planets, and most of the stars. Even though we understand that these daily, or **diurnal,** motions are the result of the earth's spin, we still refer to them as though the objects themselves were moving.

The rotation of the earth forms the basis of our timekeeping system, since the length of a day is a natural unit of time on which to base our lives; one to which, indeed, nearly all earthly species have adapted. The day is divided into twenty-four hours, each containing sixty minutes, and each minute consisting of sixty seconds. These divisions are based on the numbering system developed several thousand years ago, largely by the Babylonians.

Careful observation shows that the sun and the moon take longer to complete their daily cycles than do the stars (Fig. 1.1). Each has its own motion that carries it in the same direction as the earth's rotation, opposite to the daily rising and setting of the stars. The sun and the moon therefore "get ahead" of the stars a little each day, so that it takes a little longer for each to return to the same position, from our point of view. Thus, the sun rises about four minutes later each day, in comparison to the stars, and the moon rises almost an hour later each day. The planets also move, but so slowly that their diurnal motions are not easily distinguished from those of the stars.

Our "official" day, the one that forms the basis of our timekeeping system, is based on the **solar day** rather than the **sidereal day,** which is the true rotation period of the earth (the term *sidereal* means "with respect to the stars"). Because the earth's orbital speed is not precisely constant, the length of the solar day varies a little throughout the year. It would be inconvenient to allow our hour, minute, and second to vary

FIGURE 1.1. *The Contrast Between Solar and Sidereal Days.* The arrow indicates the overhead direction from a fixed point on the earth. From noon one day (left), it takes one sidereal day for the arrow to point again in the same direction, as seen by a distant observer. Because the earth has moved, however, it will be about four minutes later when the arrow points directly at the sun again; hence the solar day is nearly four minutes longer than the sidereal day.

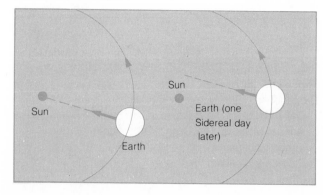

along with it, so the average length of the solar day, called the **mean solar day,** has been adopted as our timekeeping standard. The mean solar day is 3^m56^s longer than the sidereal day.

Special telescopes called **transit telescopes** (Fig. 1.2), which measure precisely the moment when a chosen star passes overhead and therefore can determine the earth's rotation period, have been used for time-keeping. Today's official time standards are based on the vibration frequencies of certain kinds of atoms, however, with reference to stellar transit observations to make sure that the official time does not get out of synchronization with the stars.

The earth's rotation causes the stars to travel through the sky in a daily cycle (Fig. 1.3). Our view of these motions depends on how far north or south of the equator we are: if we are north of the equator, then stars near the north pole's projection onto the sky (that is, stars near the **north celestial pole**), appear to circle the pole, without setting. If our latitude is 40° N,

FIGURE 1.2. *A Transit Telescope.*
Such a device points straight up. It is used to record the times when certain reference stars pass over the meridian and is therefore helpful in measuring the sidereal day.

FIGURE 1.3. *Star Trails Illustrating the Earth's Rotation.*
The circular trails are created by stars near the south celestial pole, which completed about half a circle during this all-night exposure.

for example, then all the stars within 40° of the north celestial pole stay up all night, whereas stars farther than 40° from the pole rise and set each day. At the same time, for those of us living in the northern hemisphere, a large region of the southern sky is forever beyond our view. The constellations we see vary as we travel north or south, a fact not lost on early astronomers, who concluded from this that the earth is round.

The system of latitude and longitude used for measuring positions on the earth's surface has provided the basis for the most commonly used system of measuring positions on the sky (Fig. 1.4). In the **equatorial coordinate system,** a star's angular distance north or south of the **celestial equator** (the projection of the earth's equator onto the sky) is called its **declination,** and its position in the east-west direction is its **right ascension.** Right ascension, measured from a fixed direction on the sky, is expressed in units of hours,

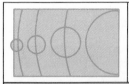

Astronomical Insight 1.1

WHAT IS AN ASTRONOMER ANYWAY?

Throughout this text we will be referring to astronomers, scientists, astrophysicists, and physicists. In a book that endeavors to summarize all that we know about the universe and its contents, we can hardly omit a description of the people who devote their time to developing this knowledge.

There are thousands in the United States alone who study astronomy, either as a vocation or an avocation. Representative of the latter are the amateur astronomers who, often on their own, but in many cases through local and even nationwide organizations, engage in a variety of astronomical activities. Telescope making, astro-photography, long-term monitoring of variable stars, public programming, and just plain star-gazing are included. If you wish to join such a group, to get advice on buying a telescope, or to learn techniques such as photography of celestial objects, your best bet is to get in touch with an amateur astronomy group. Local clubs exist in most major cities, and regional associations are

everywhere. These groups may be difficult to find in the telephone book, but a telescope shop or planetarium is bound to know whom to contact.

The amateur astronomers in the United States outnumber the professionals. There are some 3,500 members of the American Astronomical Society, the principal professional astronomy organization in the United States. These people, as a rule, make up a limited number of categories: those teaching and doing research at colleges and universities; those doing research work at government-sponsored institutions, such as the national observatories and federal laboratories, like the National Aeronautics and Space Administration (NASA); and those performing research and related engineering functions in private industry, most often with companies involved in aerospace activities.

About 8 percent of the members of the American Astronomical Society are women. This is certainly a low percentage, but a recent study has indicated

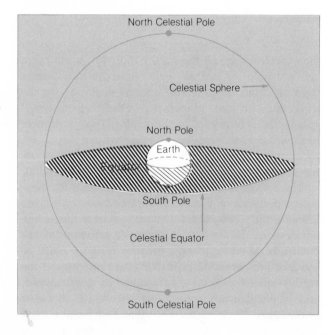

FIGURE 1.4. *The Celestial Sphere.*
For practical reasons, the sky can be imagined as a sphere centered on the earth. The earth's poles and equator, projected onto the celestial sphere, provide the basis for measuring star positions.

minutes, and seconds of time, because it is convenient to specify positions on the sky according to the time they cross the **meridian,** the north-south line that passes overhead. Thus a star's equatorial coordinates might be declination $+27°14'54''$ and right ascension $6^h12^m8^s$ (the plus sign indicates declinations north of the equator, which has declination $0°$, and a minus sign indicates southern declinations).

There is one complication with using the equatorial coordinate system: the earth slowly wobbles on its axis, in a motion called **precession,** and this causes the celestial equator to move slowly through the sky (Fig. 1.5). It takes some 26,000 years for the earth to complete one cycle of this motion, which is exactly like the

that there is little discrimination in the hiring of professional astronomers. The small number of women in astronomy thus reflects the low number who choose to enter the field at the graduate level, so it is at this level that improvement must be made.

Most of the funding for research in astronomy, even for those not working directly for federal laboratories and observatories, comes from the government. A significant function of an astronomer on a university faculty is to write proposals, usually to the National Science Foundation or to NASA, for support of the research programs he wishes to carry out.

The terms *astronomer* and *astrophysicist* have come to mean pretty much the same thing in modern usage, although historically, there was a difference. An astronomer was one who studied the skies, gathering data, but doing relatively little interpretation; an astrophysicist was primarily interested in understanding the physical nature of the universe, and therefore carried out comprehensive analyses of astronomical data, or did theoretical work, in both cases applying the laws of physics to phenomena in the heavens. Nearly all modern astronomers do astrophysics, however, to varying degrees ranging

from the observational astronomer at one end of the spectrum to the pure theorist at the other. The two terms for people engaged in these pursuits are therefore used interchangeably today. Many modern astronomers call upon the fields of engineering (for instrument development), chemistry (in studying planetary and stellar atmospheres and the interstellar medium), geophysics (in probing interior conditions in planets and other solid bodies), and possibly even biology, but always with an underlying foundation in physics.

If you want to become a professional astronomer, you should be aware from the outset that the field is small and job opportunities are both limited and often subject to the vagaries of federal funding. If you persist, the best course is to study physics, at least through the undergraduate level, and then plan to attend graduate school in an astronomy department or in physics (the latter option may make you a bit more versatile, and would not be a handicap when you entered astronomy later). On the bright side, demographic studies have shown that there may be a shortage of astronomers for a period beginning in the late 1980s and extending through the rest of this century. Perhaps the timing will be right for you.

FIGURE 1.5. *Precession.*
The earth's axis is tilted 23½° with respect to its orbital plane, and it wobbles on its axis, so that an extension of it describes a conical pattern (left) in a time of about 26,000 years. The north celestial pole therefore follows a circular path on the sky (right).

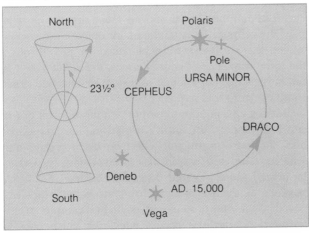

wobbling of a spinning play top or gyroscope. The motion of our coordinate system is therefore very gradual, causing star positions to shift extremely slowly. Despite the small magnitude of the effect, it was noticed more than two thousand years ago, and modern astronomers must allow for it when planning observations.

Annual Motions: The Seasons

We turn our attention now to celestial phenomena that are caused by the earth's motion as it orbits the sun. One aspect of this, the daily eastward motion of the sun with respect to the stars, has already been mentioned, in connection with the difference between the solar and sidereal day. It must be emphasized that this is only an apparent motion, caused by our changing angle of view as we move with the earth in its orbit. As the earth moves about the sun, it travels in a fixed plane, so that the apparent path of the sun through the constellations is the same each year. The apparent path of

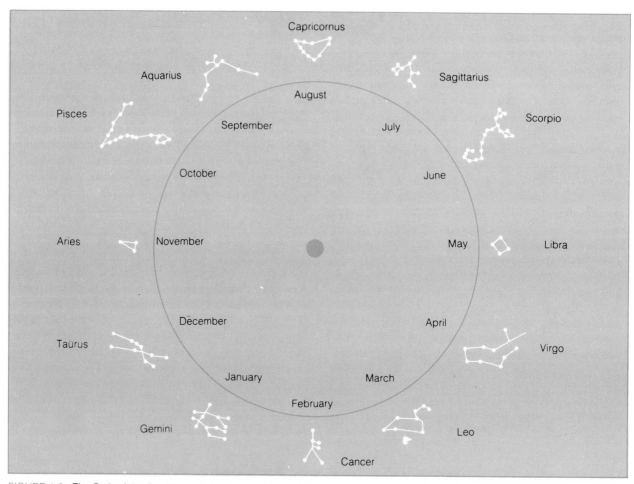

FIGURE 1.6. *The Path of the Sun Through the Constellations of the Zodiac.* The dates refer to the position of the earth each month. To see which constellation the sun is in during a given month, imagine a line drawn from the earth's position through the sun; that line will extend to the sun's constellation. For example, in March the sun is in Aquarius.

the sun is called the **ecliptic,** and the sequence of constellations through which it passes is called the **zodiac** (Fig. 1.6). There are twelve principal constellations of the zodiac, identified since antiquity, and once thought by astronomers to have significance in our daily lives. To this day, an astrologer will tell you that your fate is influenced by the sign under which you were born. This is a very old idea; astronomers have refined their understanding of the heavens in the last two thousand years, even if astrologers have not. There is, in fact, no evidence to support the claims of astrologers. Ironically, because of precession, the sign associated with a particular date is no longer the same as the constellation the sun is passing through on that date.

It happens that the orbital planes of the other plan-

ets and the moon are closely aligned with that of the earth, so that the planets and the moon are always seen near the ecliptic. Thus all the major objects in the solar system that can be seen by the naked eye pass through the same sequence of constellations, the zodiac, so it is not surprising that ancient astronomers attached great significance to this sequence.

Besides causing the apparent annual motions of the sun and planets, the earth's orbital motion has a second, far more significant, effect on us; it creates our seasons. The earth's spin axis is tilted with respect to its orbital plane, so that during the course of a year, portions of the earth away from the equator are exposed to varying amounts of daylight. Summer in the northern hemisphere occurs when the north pole is tipped toward the sun; winter occurs during the opposite part of the earth's orbit, as the pole, which remains fixed in orientation, is tilted away from the sun. The tremendous seasonal variations in climate at intermediate latitudes are caused by a combination of two effects: (1) the length of the day varies, so that in summer, for example, the sun has more time to heat the earth's surface; and (2) the angle at which the sun's rays strike the ground is more nearly perpendicular in the summer, so that the sun's intensity is much greater, heating the surface more efficiently.

The earth's axis is tilted $23\frac{1}{2}°$ from the perpendicular to the orbital plane. Therefore during the year the sun, as seen from the earth's surface, can appear directly overhead as far north and south of the equator as $23\frac{1}{2}$, defining a region called the **tropical zone** (Fig. 1.7). For those of us who live outside the tropics, the sun can never be directly overhead. When the north pole is tilted toward the sun, an occasion occurring near June 22 and called the **summer solstice,** the sun passes directly overhead at $23\frac{1}{2}°$ north latitude.

FIGURE 1.7. *The Definition of Latitude Zones on the Earth.* As shown, at summer solstice the sun is overhead at $23\frac{1}{2}°$ north latitude, its northernmost point. This defines the Tropic of Cancer, the northern limit of the tropical zone. At the same time, the entire area within $23\frac{1}{2}°$ of the north pole is in daylight throughout the earth's rotation, and the boundary of this region is the Arctic Circle. Similarly, the Antarctic Circle receives no sunlight at all during a complete rotation of the earth. Six months later, the sun is overhead at the Tropic of Capricorn.

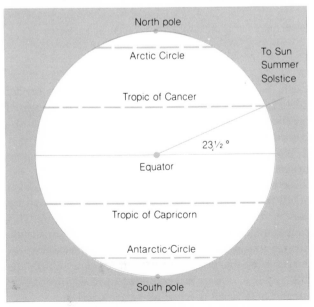

During this solstice, sunlight covers the entire north polar region to a latitude as far as $23\frac{1}{2}°$ south of the pole. This defines the **artic circle,** and at the time of the solstice the entire circle has daylight for all twenty-four hours of the earth's rotation. At the pole itself there is constant daylight for six months. At the **winter solstice,** when the south pole is pointed most closely in the direction of the sun, the sun's midday height above the horizon, as viewed from the northern hemisphere, is the lowest of the year.

If we follow the sun's motion north and south of the equator throughout the year, we find that it follows a graceful curve as it traverses its range from $+23\frac{1}{2}°$ (north) declination to $-23\frac{1}{2}°$ (south) declination (Fig. 1.8). The sun crosses the equator twice in its yearly excursion, at the times when the earth's north pole is pointed in a direction $90°$ from the earth-sun line. At these times the lengths of day and night in both hemispheres are equal, and these occasions are referred to as the **vernal** (spring) and **autumnal** (fall) **equinoxes,** taking place on about March 21 and September 23, respectively. The direction of the sun at the time of the vernal equinox coincides exactly with the direction of 0^h right ascension (the definition of this direction is that it lies along the line of intersection of the earth's equatorial plane and the orbital plane).

Lunar Motions

The most prominent object in the nighttime sky is the moon. The moon orbits the earth, its orbital plane nearly coinciding with that of the ecliptic. During each

FIGURE 1.8. *The Path of the Sun Through the Sky.* Because of the earth's orbital motion and the tilt of the earth's axis, the sun's annual path through the sky has the shape illustrated here.

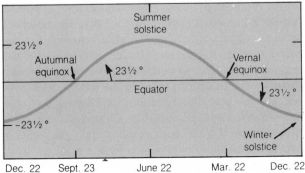

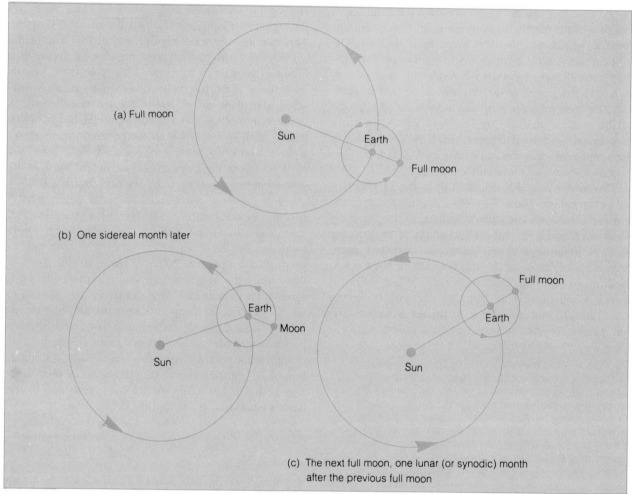

FIGURE 1.9. *Sidereal and Synodic Periods of the Moon.* Because the earth moves in its orbit while the moon orbits it, the moon goes through more than one full circle (as seen by an outside observer) to go from one full moon to the next. Hence the lunar (or synodic) month is about two days longer than the moon's sidereal period.

circuit that the moon makes around the earth, we see its full cycle of phases. The true orbital period is called the **sidereal period,** because it is the time required for the moon to go around the earth, as seen in a fixed reference frame with respect to the stars. The observed cycle of phases, which is the time required for the moon to return to a given alignment with respect to the sun, is called the **synodic period** or the **lunar month.** The difference between the synodic and sidereal periods is akin to the distinction between the solar and sidereal days, discussed in the preceding section. In both cases it is the earth's motion about the sun that lengthens the time it takes to complete a full cycle as we see it (Fig. 1.9).

The moon always keeps the same side facing the earth, because its rotation period is equal to its orbital period. This is caused by gravitational forces exerted on the moon by the earth. (These forces will be discussed more fully in chapters 2 and 4.)

Since the moon does not emit light of its own, but instead shines by reflected sunlight, we easily see only those portions of its surface that are sunlit. As the moon orbits the earth and our viewing angle changes, we see varying fractions of the daylit half of the moon. This causes the moon's apparent shape to change drastically during the month, the sequence of shapes being referred to as the **phases** of the moon (Fig. 1.10). The full cycle of phases is completed during one synodic period, or lunar month, of about 29½ days.

The extremes of the cycle are represented by the **full**

moon, occurring when it is directly opposite the sun, so that we see its entire sunlit hemisphere; and the **new moon,** when it is between the earth and the sun, with its dark side facing us. The full moon is on the meridian at local midnight, whereas the new moon is on the meridian at local noon. The new moon cannot be observed, because only the dark side faces us and because it is so close to the sun in the sky.

Just as we speak of the moon's phases, which really refer to its apparent shape as seen from the earth, we can also speak of its **configurations,** which describe its position with respect to the earth-sun direction. For example, when a full moon occurs, the moon is at **opposition,** meaning it is in the direction opposite that

of the sun. A new moon occurs at **conjunction,** when the moon lies in the same direction as the sun. We can follow the moon through its phases as we trace its configurations, beginning with the new moon. During the first week after conjunction, as the moon moves toward **quadrature** (the position 90° from the earth-sun line), it appears to us to have a crescent shape that grows in thickness each night. This is called the **waxing crescent** phase, and when the moon reaches quadrature, so that we see exactly one-half of the sunlit hemisphere, it has reached the phase called **first quarter.** For the next week, as the moon goes from quadrature toward opposition, and we see more and more of the daylit side, its phase is said to be **waxing**

FIGURE 1.10. *Lunar Phases and Configurations.* As the moon orbits the earth, we see varying portions of its sunlit side. The phases sketched here (outside the circle representing the moon's orbit) show the moon as it appears to an observer in the northern hemisphere.

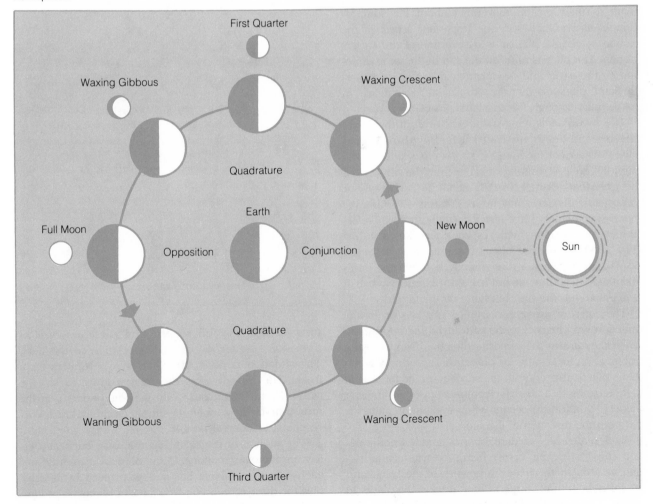

gibbous. After the full moon, as it again approaches quadrature, the phase is **waning gibbous,** and this time at quadrature, the phase is called **third quarter.** We can distinguish first quarter from third quarter by noting the time of night: if the moon is already up when the sun sets, it is first quarter, but if it does not rise until midnight, it is third quarter. After third quarter, as the moon moves closer to the sun, we see a diminishing slice of its sunlit side, and it is in the **waning crescent** phase.

Planetary Motions

Like the sun and the moon, the planets, too, move with respect to the background stars. It was this fact that lent the planets their generic name, since *planet* is the Greek word for *wanderer.*

The observed motions of the planet are primarily a result of their orbital movement about the sun, although, as we shall see, one important aspect of the motion of certain planets is a reflection of the earth's motion. The planets all orbit the sun in the same direction as the earth and, as mentioned earlier, in nearly the same plane, so that they appear to move nearly in the ecliptic, through the constellations of the zodiac.

The two planets lying within the orbit of the earth, Mercury and Venus, are called **inferior planets,** and can never appear far from the sun in our sky. For Mercury, the greatest angular distance from the sun (called the **greatest elongation**) is about 28°, whereas for Venus, this distance may be as great as 47°. Like the moon, the planets have specific configurations, referring to their positions with respect to the sun-earth line (Fig. 1.11). An inferior planet is said to be at **inferior conjunction** when it lies directly between the earth and the sun, and at **superior conjunction** when it is aligned with the sun, but lying on the far side.

The outer, or **superior,** planets can be seen in any direction with respect to the sun, including opposition, when they are in the opposite direction from the sun. Conjunction for a superior planet can occur only when the planet is aligned with the sun but on the far side of it (analogous to a superior conjunction for an inferior planet). Quadrature occurs when a superior planet is 90° from the direction of the sun.

Each planet has a sidereal period and a synodic period (Fig. 1.12), the former being the true orbital period as seen in the fixed framework of the stars, and the latter being the length of time it takes the planet to

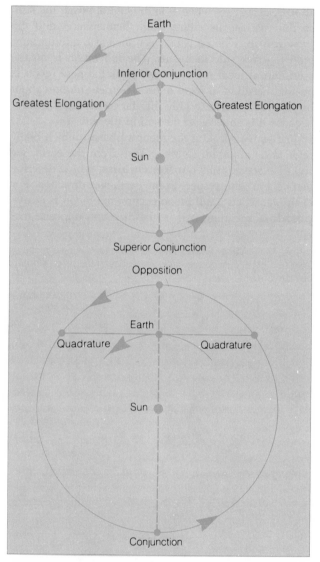

FIGURE 1.11. *Planetary Configurations.*

pass through the full sequence of configurations, as from one conjunction or one opposition to the next. The situation is much like that of two runners on a track. The time it takes the faster runner to lap the slower one is analogous to the synodic period, and the time it takes the runners to simply circle the track corresponds to sidereal period.

The motion of the earth has one very important effect on planetary motions. Going outward from the sun, each successive planet has a slower speed in its orbit (see the discussion of Kepler's laws of planetary motion

Astronomical Insight 1.2
ASTRONOMY AND ASTROLOGY

There is an unfortunate tendency in modern society to confuse astrology and astronomy, or, worse yet, to consider one a legitimate alternative to the other.

Astrology, the pseudoscience based on the belief that our lives are influenced by the configurations of heavenly bodies, arose at a primitive stage in the development of civilization, at a time when the earth was thought to be a flat disk under the dome of the heavens. Although ancient Greek astronomers did much to raise the study of the heavens to a scientific level, during the subsequent Dark Ages astrology and the governing of human lives by the stars became once again the primary basis for study of the heavens. This trend changed radically with the Renaissance, when the true nature of the heavens and the motions of astronomical objects were untangled. It is unfortunately true, however, that a segment of the populace has continued to profess a faith in astrology.

One of the basic lessons we have learned about the universe is that it is easy to make mistakes unless we are careful to be objective, accept only conclusions that can be verified by repeated observations or experiments, and make predictions that can be tested. Astrology, utterly and abysmally, fails to meet these criteria. By performing statistical analyses of people born under different signs, researchers have made serious attempts to test astrological lore, with no trace of a correlation ever being seen.

There certainly are many phenomena that defy the understanding of modern science, and many of them deserve more attention than they are getting. Astrology is not one of them. As a phenomenon, it is worth of study only by the fields of sociology and psychology, for its effects do not exist in the physical universe. It may be interesting party talk to compare astrological signs, but it would be good also to keep in mind the difference between objective reality and subjective impressions. Failure to do so would be failure to understand what science is.

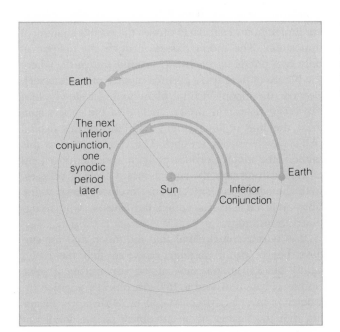

FIGURE 1.12. *The Synodic Period of an Inferior Planet.* The inner planets travel faster than the earth in their orbits, and therefore "lap" the earth, much as a fast runner laps a slower one on a track. This illustration shows roughly the situation for Mercury, which has a synodic period of about 116 days, or about one-third of a year.

in chapter 2). This means that the earth, moving faster than the superior planets, periodically passes each of them (this occurs once every synodic period). As the earth overtakes one of the superior planets, there is an interval of time during which our line of sight to that planet sweeps backward with respect to the background stars, making it appear that the planet is moving backwards (Fig. 1.13). The same thing happens when we, in a rapidly moving automobile, pass a slowly moving vehicle: for a brief moment, the other vehicle appears to move backward with respect to the fixed background. This apparent backward movement, called **retrograde motion,** was thought in ancient

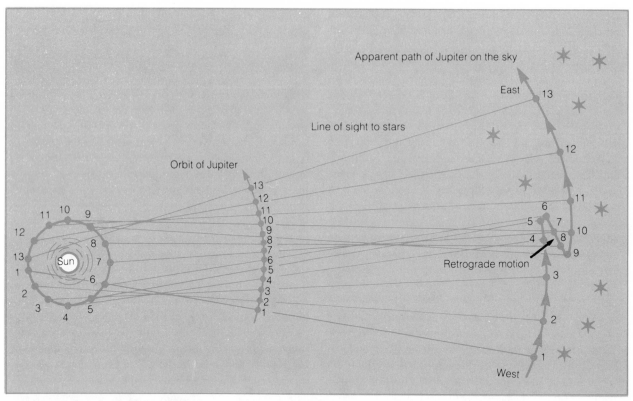

FIGURE 1.13. *Retrograde Motion*. As the faster-moving earth overtakes a superior planet in its orbit, the planet temporarily appears to move backward with respect to the fixed stars. This sketch illustrates the modern explanation of something that took ancient astronomers a long time to correctly understand.

times to be a real motion of the superior planets, rather than a reflection of the earth's motion. This idea very much complicated many of the early cosmologies, as we shall see later in this chapter. The inferior planets also undergo retrograde motion, although it is not so easily observed.

Eclipses

From the discussion in the preceding sections, it may seem that the moon should pass directly in front of the sun on each trip around the earth, and through the earth's shadow at each opposition, producing alternating solar and lunar eclipses at two-week intervals, but this is obviously not the case. The reason is that the moon's orbital plane does not lie exactly in the ecliptic, but is tilted by about 5°. Therefore the moon usually

passes just above or below the sun as it goes through conjunction, and similarly misses the earth's shadow at opposition. The moon passes through the ecliptic at only two points on each trip around the earth, where the planes of the earth's and the moon's orbits intersect. Because the moon's orbital plane wobbles slowly, in a precessional motion similar to that of the earth's spin axis, the line of intersection with the earth's orbital plane slowly moves around. The combination of this motion, the moon's orbital motion, and the movement of the earth around the sun creates a cycle of eclipses, with the same pattern recurring every eighteen years. This cycle of eclipses, called the **Saros,** was recognized in antiquity.

It is purely coincidental that the moon and the sun have nearly equal angular sizes, so that the moon neatly blocks out the disk of the sun during a solar eclipse (Figs. 1.14 and 1.15). An object's angular diameter is inversely proportional to its distance, meaning that the farther away an object is, the smaller it looks. The sun is much larger than the moon, but is also much more distant. The two objects have the same angular diameter because the ratio of the sun's diameter to that of the moon just happens to be compensated by the ratio of the sun's distance to that of the moon.

If a total solar eclipse occurs at the time when the moon (in its slightly noncircular orbit) is farthest from the earth, it does not quite block all of the sun's disk, but rather, leaves an outer ring of the sun visible. This is called an **annular** eclipse. Because a total (or annular) solar eclipse requires a precise alignment of sun and moon, an eclipse will appear total only along a well-defined, narrow path on the earth's surface (Fig. 1.16). There is a wider zone outside of that where the moon appears to block only a portion of the sun's disk; people in this zone see a partial solar eclipse.

During a lunar eclipse, when the moon passes through the earth's shadow, observers everywhere on earth see the same portion of the moon eclipsed. If the entire moon passes through the **umbra,** the dark inner portion of the earth's shadow, then the eclipse is total, as no part of the moon's surface is exposed to direct sunlight. If only a portion of the moon passes through the umbra, then we see a partial eclipse. A penumbral eclipse occurs when all or part of the moon passes

FIGURE 1.14. *A Total Solar Eclipse.*
This occurs when the moon entirely blocks our view of the sun.

FIGURE 1.15. *A Solar-Eclipse Sequence.*
This series of photographs illustrates the moon's progression across the disk of the sun during a total solar eclipse.

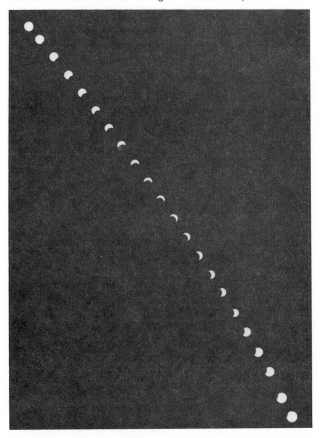

FIGURE 1.16. *The Moon's Shadow During a Solar Eclipse.*
This sequence shows the path of the moon's shadow across the surface of the earth during a solar eclipse. The eclipse appeared total only to observers on the earth who were located directly in the center of the shadow's path (that is, in the umbra).

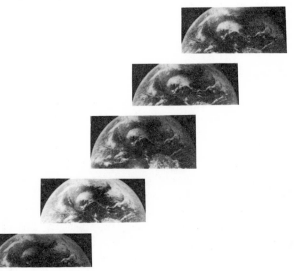

through the **penumbra,** the lighter outer portion of the earth's shadow. Such eclipses are not readily noticeable by the eye, however.

Historical Developments

Now that we have discussed most of the phenomena that can be observed without telescopes, and thus are aware of nearly all the data available to ancient watchers of the sky, it is appropriate to review how we come to understand these phenomena.

It is likely that people were preoccupied with the heavens from the time they first became aware of their environment. The speculative mood that we, in these modern times, can conjure up only by disregarding our daily pressures and escaping into the countryside on a clear night to look at the stars, must have dominated the nighttimes of the earliest cultures. In the following discussion, the emphasis is on developments that oc-

curred around the shores of the Mediterranean, because these developments laid the foundation for the modern understanding of the universe. Parallel developments occurred in many other parts of the world, but in most cases either reached dead ends or were eventually absorbed into the Western scientific culture.

The Earliest Astronomy

The earliest records of astronomical lore have been found in the region east of the Mediterranean now known as Iraq (Fig. 1.17), where the Babylonian culture flourished for many centuries, beginning around 3000 B.C. The Babylonians developed an accurate knowledge of the length of the year (establishing the basis for the modern twelve-month calendar), and bequeathed us our timekeeping and angular measures that are based on the number sixty.

In parallel with developments in Babylonia, an early Greek civilization arose on the shores of the Mediterranean, at some unknown time coming in contact with the Babylonian culture. The ancient Greek traditions,

FIGURE 1.17. *The Ancient Mediterranean.* This map shows the locations of many of the sites mentioned in the text.

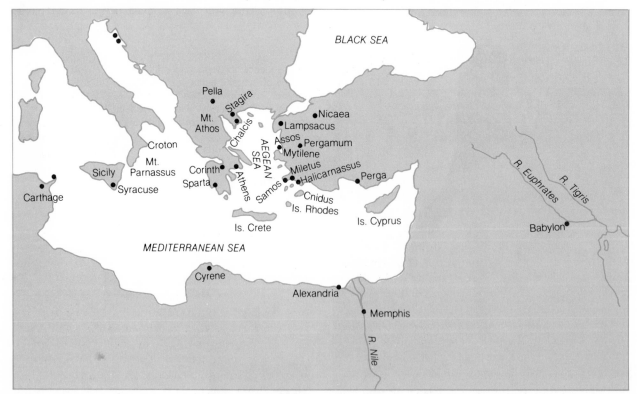

known to us largely through the writings of Homer (Fig. 1.18), gave birth to our modern constellation names and other astronomical lore, and laid the foundation for the first rational scientific inquiry. The foundations of modern science were established by the philosophers of Greece. (Table 1.1).

The first formal scientific thought is associated with the philosopher **Thales** (624–547 B.C.) and his followers, who formed the so-called Ionian school of thought. The principal contribution of Thales was the idea that rational inquiry can lead to *understanding* the universe, going beyond describing it. The Ionians developed a primitive **cosmology,** or theory of the universe, in which all the basic elements of the universe were formed from water, the primeval substance.

FIGURE 1.18. *Homer's Odyssey and Iliad.*
In these two epic poems, Homer described the much more ancient legends of astronomical lore from the civiliation that had flourished on the isle of Crete.

TABLE 1.1 Ancient Greek Achievements

Date	Philosopher	Discovery or achievement
c.900–800 B.C.	Homer	*Iliad* and *Odyssey;* summaries of legends
c.624–547 B.C.	Thales	Rational inquiry leads to knowledge
c.611–546 B.C.	Anaximander	Universal medium; Primitive cosmology
c.570–500 B.C.	Pythagoras	Mathematical representation; round earth
c.500–400 B.C.	Philolaus	Earth orbits central fire
c.500–428 B.C.	Anaxagoras	Moon reflects sunlight; eclipses explained
c.428–347 B.C.	Plato	Material world imperfect; learn by reason
c.408–356 B.C.	Eudoxus	First mathematical cosmology
c.384–322 B.C.	Aristotle	Concept of physical laws; round earth
c.310–233 B.C.	Aristarchus	Size of sun; first heliocentric theory
c.273–? B.C.	Eratosthenes	Size of earth
c.265–190 B.C.	Apollonius	Introduction of the epicycle
c.200–100 B.C.	Hipparchus	Many astronomical firsts; epicyclic theory
A.D. c.100–200	Ptolemy	*Almagest;* elaborate epicycle model

Another major school of thought was developed by **Pythagoras** (c. 570–500 B.C.) and his followers, who believed that natural phenomena could be described mathematically, a belief that is at the very heart of all modern science. Pythagoras is credited with being the first to assert that the earth is round, and that all heavenly bodies move in circles, ideas that never thereafter lost favor in ancient times.

Plato and Aristotle

One of the most influential people in the development of Greek philosophy was **Plato** (c. 428–374 B.C.), who established an academy near Athens in about the year 387 B.C., where he taught his ideas of natural philosophy (Fig. 1.19). His fundamental precept was that what we see of the material world is only an imperfect representation of the ideal creation. The corollary of this doctrine was that we can learn more about the universe by reason than by observation, since, after all, observation can give us only an incomplete picture. Hence Plato's ideas of the universe, described in his

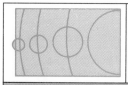

Astronomical Insight 1.3

THE MYSTERY OF STONEHENGE

By now we are aware that some form of astronomical influence affected nearly every ancient culture. The timing of major developments seems to have been surprisingly similar in most parts of the world. The earliest records of accurate positional measurements and the establishment of calendars often date from the second millennium B.C., with sufficient sophistication for predictions of eclipses and planetary motions not developing until the last few centuries B.C.

It is most impressive, then, to consider the stone circle monuments of the British Isles, for they predate recorded astronomical achievements in the rest of the world, yet they show that their builders had substantial knowledge of astronomical skills, possibly including the prediction of eclipses. Of the nine hundred or so of these monuments, by far the best known is Stonehenge, located on Salisbury Plain, about seventy miles west of London. This is a large and complex monument characterized by immense upright stones arranged in various circular patterns, in some instances with horizontal lintels connecting the vertical members.

Little is known about the builders of Stonehenge. The monument was built in three stages, the first commencing about 2800 B.C., and the most recent being completed by 2075 B.C. The sophisticated astronomical principles that guided the construction were incorporated from the first, even though it appears that the people inhabiting the region changed from the original Stone Age civilization to a new group, called the Beaker People, who arrived from what is now the Netherlands in approximately 2500 B.C. Subsequent stages were built by these later arrivals on the scene, who apparently embraced the logic that had led to the original construction.

It is not known how the stones, which weigh up to fifty tons, were moved to Salisbury Plain. A number of them came from Wales, possibly by sea, and many are from a region some twenty miles distant. The monument is now in a state of partial ruin and disarray, with some stones evidently missing. Besides the large stones, there are a number of holes that may have once contained wooden posts used to mark certain directions. Even today new parts of Stonehenge are found occasionally, in the form of postholes or distant mounds that may have been used to mark significant alignments.

There is no question that Stonehenge was laid out according to astronomical principles. Substantial controversy exists, however, concerning the details. Despite a great deal of publicity in the 1960s about interpretations of Stonehenge as an early computer used to predict eclipses, there is today little agreement on what is the correct interpretation.

One axis of Stonehenge, delineated by the sight line to the Heelstone from the center of the monument, aligns with the position of sunrise at the summer solstice, although, oddly enough, because of precession the alignment is better now than it was in 2800 B.C., when the first stage was built. (This has led at least one student of Stonehenge to hypothesize that this axis was meant to signify the midwinter rising of the full moon, with which the alignment was better.) A circle of fifty-six holes around the center of Stonehenge has also attracted a great deal of attention; some people think that these holes could have been used to predict eclipses, although this appears unlikely.

Whatever the correct interpretation of Stonehenge, the fact remains that extensive knowledge of astronomical phenomena was available to its builders as early as 2800 B.C., several centuries before other cultures such as the Babylonians are known to have developed such a high level of sophistication. It is unfortunate that we do not know more about the remarkable people who built it.

FIGURE 1.19. *Plato*.

seek the center by falling straight down (it was another premise of his that falling objects were following their natural inclination to reach the center of the universe); (2) the view of the constellations changes as one travels north or south; and (3) during lunar eclipses it could be seen that the shadow of the earth is curved (Fig. 1.20). By relating his theories to observation in this way, Aristotle broke with the tradition of Plato to some extent, although, as we have seen, he still approached the problem in the same manner, letting reason rather than observation guide the way.

Other tenets of Aristotle included the conclusion that the universe was finite in size (this led to his belief that the heavenly bodies could follow only circular motions, because otherwise they might encounter the edge of the universe), and that the heavenly bodies were made of a fifth fundamental substance, which he called the aether.

The Later Greeks: Prelude to the Dark Ages

After Aristotle's period, the center of Greek scientific thought moved across the Mediterranean Sea to Alexandria (the capital city established in 332 B.C. by Alexander the Great), near the site of the present-day city

FIGURE 1.20. *Curvature of the Earth's Shadow on the Moon.* Only a spherical body can cast a circular shadow for all alignments of the sun, moon, and earth.

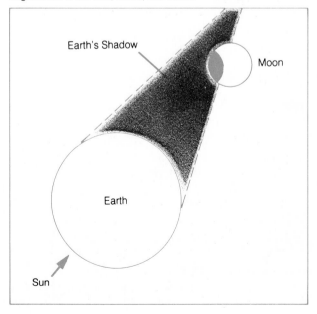

Republic, were based on certain idealized assumptions that he found reasonable. One of the most important of these was that all motions in the universe are perfectly circular, and that all astronomical bodies are spherical in shape. Thus, he adopted the Pythagorean view that the sun, moon, and all the planets moved in combinations of circular motions about the earth. Plato evidently thought of these objects as affixed to clear, ethereal spheres that rotated.

The most renowned student of Plato was **Aristotle** (c. 384–322 B.C.), who was the first to adopt physical laws and then to show why, in the context of those laws, the universe works as it does. He taught that circular motions are the only natural motions, and that the center of the earth is the center of the universe. He also believed that the world is composed of four elements: earth, air, fire, and water. He could demonstrate, in the context of his adopted physical laws, that the universe was spherical in shape, and that the earth was also spherical. He had three proofs of the latter: (1) only at the surface of a sphere do all falling objects

of Cairo. The first prominent astronomer of this era was **Aristarchus** (c. 310–230 B.C.), the first scientist to adopt the idea that the sun, not the earth, is at the center of the universe. His conclusion was based on geometrical arguments which showed that the sun was much larger than the earth, and thus was more likely to be at the center of the universe. The heliocentric hypothesis of Aristarchus failed to attract many followers at the time, largely because of a lack of concrete evidence that the earth was in motion, but also because of a general satisfaction with the Aristotelian view point, which had no recognized flaws.

By the third century B.C. the need for more precise mathematical models of the universe became apparent, as better observing techniques were developed. A mathematical concept that provided the needed precision while still preserving the precepts of Aristotle was the **epicycle,** a small circle on which a planet moves, whose center in turn orbits the earth following a larger circle called a **deferent** (Fig. 1.21). An important advantage of the epicycle was that it could explain the retrograde motions of the superior planets.

The epicyclic motions of the celestial bodies were refined further by **Hipparchus** (Fig. 1.22), who was active during the middle of the second century B.C. (very little is known about his life, even the dates of his

FIGURE 1.22. *A Fanciful Rendition of Hipparchus at His Observatory.*

FIGURE 1.21. *The Epicycle.*
It was realized that planetary motions could be represented by a combination of motions involving an epicycle, which carries a planet as it spins while orbiting the earth.

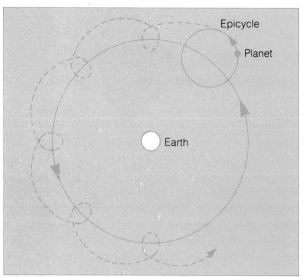

birth and death). Hipparchus, most of whose work was done at his observatory on the island of Rhodes, was one of the greatest astronomers of antiquity. Among his major contributions were: (1) the first use of trigonometry in astronomical work (in fact, he is largely credited with its invention, although many of the concepts were developed earlier); (2) the refinement of instruments for measuring star positions, along with the first known use of a celestial coordinate system akin to our modern equatorial coordinates, enabling him to compile a catalogue of some 850 stars; (3) the refinement of the methods of Aristarchus for measuring the relative sizes of the earth, moon, and sun; (4) the invention of the stellar magnitude system for estimating star brightnesses, a system still in use today (with minor modifications); and, perhaps most impressively, (5) the discovery of the precession of star positions, which he accomplished by comparing his observations with some that were made 160 years earlier.

It is worthwhile to consider why Hipparchus, with access to the work of Aristarchus (including the knowledge that the sun is much larger than the earth), did not adopt the heliocentric view. Apparently he was motivated to reject this idea for one very significant observational reason: he could not detect any apparent shifting of star positions during the course of the year, shifting that he realized should appear as a result of the changing point of view if the earth moved about the sun (Fig. 1.23). The lack of any detectable **stellar parallax,** as such an apparent motion is called, caused many to resist the heliocentric viewpoint for centuries to come.

After the great work of Hipparchus, almost three hundred years passed before any significant new astronomical developments occurred. Claudius Ptolemaeus, or simply **Ptolemy** (Fig. 1.24), lived in the middle of the second century A.D., when he undertook to summarize all the world's knowledge of astronomy. He did this with the publication of a treatise called the *Almagest,* which was based in part on the work of Hipparchus, but which also contained some new developments of his own. The thirteen books of the *Almagest* range in coverage from a summary of the observed motions of the planets to a detailed study of the motions of the sun and moon; from a description of the workings of all the astronomical instruments of the day to a reproduction of the star catalogue of Hipparchus. Most important, however, his treatise contained detailed models of the planetary motions (Fig. 1.25), based on

FIGURE 1.24. *Claudius Ptolemy.*

FIGURE 1.25. *The Cosmology of Ptolemy.*
To account for nonuniformities in the planetary motions, Ptolemy devised a complex epicyclic scheme. The deferent, the large circle on which the epicycle moves, is not centered on the earth, and has a rotation rate that is constant as seen from another displaced point called the equant. This meant that when observed from the earth, the planet appeared to move faster through the sky when on one side of its deferent than when on the other. Each planet had its own deferent and epicycle, with motion and offsets adjusted to reproduce the observations.

FIGURE 1.23. *Stellar Parallax.*
As the earth orbits the sun, our line of sight toward a nearby star varies, causing the star's position (with respect to more distant stars) to change. The change of position is greatly exaggerated in this sketch; in reality, even the largest stellar parallax displacements are too small to have been measured by ancient astronomers.

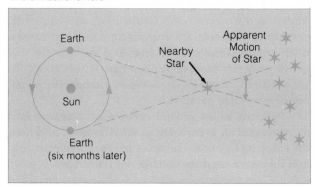

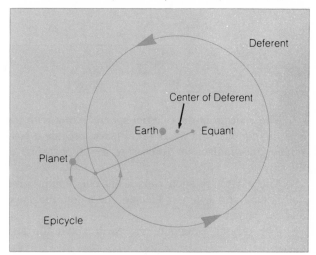

the epicyclic theory. Here Ptolemy made his greatest contribution, for these models were sufficiently accurate in their ability to predict the planetary positions that they were used for the next thousand years.

The history of Greek astronomy came to an end with the work of Ptolemy, as did most significant astronomical development in Europe and western Asia until the time of the Renaissance. There followed a period known as the Dark Ages, when western civilization was largely dormant. During this time Arab astronomers preserved many of the Greek traditions, so that the ancient teachings were still firmly entrenched centuries later, when the first stirrings of a new spirit of inquiry began to be felt in Europe.

PERSPECTIVE

We have now developed a modern picture of all the astronomical phenomena visible to the naked eye. There is much left for us to learn about the universe and the laws that govern it, but already we have available to us all the information upon which the first astronomers based their crude theories of cosmology. We have seen how the ancient philosophers interpreted these phenomena in developing the earliest concepts of the universe. In the next chapter we shall see how these foundations were built upon as a modern understanding of physics and astronomy arose.

SUMMARY

1. The concept of the celestial sphere, along with the celestial equator and the celestial poles, provide a convenient mechanism for discussing the appearance of the sky.

2. The earth's rotation causes diurnal motions: the daily rising and setting of the sun, the moon, the stars, and the planets.

3. The orbital motion of the earth about the sun causes annual motions such as the apparent motion of the sun through the constellations of the ecliptic, and is responsible, along with the tilt of the earth's axis, for our seasons.

4. The planetary orbits lie nearly in the same plane, so they are always seen along the ecliptic, in various configurations with respect to the sun-earth line. The planets go through temporary retrograde motion owing to the relative speed with which they pass or are passed by the earth.

5. The moon orbits the earth while the earth orbits the sun, and from the earth we see the various phases of the moon as it passes through different configurations.

6. Solar and lunar eclipses occur when the moon passes directly in front of the sun or through the earth's shadow, respectively. The occurrence of these alignments is affected by the tilt and precession of the moon's orbit.

7. The earliest recorded astronomical data were found in the region known as Babylonia, now Iraq.

8. The first rational inquiry and the earliest cosmological theories arose in the ancient Greek empire, on the shores of the Mediterranean.

9. From 570 B.C., Pythagoras and his followers developed the notion that the universe can be described by numbers, and adopted the belief that the earth is spherical.

10. Plato, and his pupil Aristotle, stated underlying principles that they thought governed the universe, and which could not be tested by observation. Their principles included the beliefs that the earth is the center of the universe and that all heavenly bodies are spherical.

11. In the third century B.C., Aristarchus used geometrical arguments to conclude that the sun, not the earth, is at the center of the universe.

12. Hipparchus, in the second century B.C., made precise observations, applied new mathematical techniques to astronomy, compiled a large star catalogue, developed a system for measuring star brightnesses, discovered precession, and developed the epicyclic theory of planetary motion. Hipparchus did not adopt the sun-centered cosmology, however, because he could not detect stellar parallax.

13. Ptolemy, in the second century A.D., summarized all astronomical knowledge in the *Almagest,* and used the epicyclic theory to construct tables of planetary motion that were used throughout the Dark Ages.

REVIEW QUESTIONS

1. Explain why it is necessary to have astronomical observatories in both the northern and southern hemispheres.

2. Imagine that you live at a latitude of 30°N. At the time of the winter solstice (on or about December 21), how far above the southern horizon would the sun rise at midday? How would your answer differ if you lived at 30°S latitude?

3. Imagine that the earth's rotation axis is perpendicular to the ecliptic, instead of being tilted 23½° away from perpendicular. What would be the length of the day at the time of the summer solstice for a person living at 40°N latitude?

4. The moon's synchronous rotation means that its spin and orbital periods are equal. Is it the synodic or the sidereal orbital period that is equal to the rotation period?

5. Would retrograde motion of the planets occur if each planet moved more rapidly than the next one closer in to the sun, rather than more slowly? If it would occur, would it be easily observed?

6. Lunar eclipses always occur at the same phase of the moon. Which phase is it? During which lunar phase do solar eclipses always take place?

7. The Babylonians adopted a 360-day year, divided evenly into 12 months. Since the year is actually 5¼ days longer than this, how often did they have to add an extra month, in order to make things come out evenly? How do you suppose they allowed for the remaining error that accumulated over many years?

8. Even the best measuring instruments at the time of Hipparchus could only measure angular separations or positions with an accuracy of several arcminutes. If you assume that the error or uncertainty of a single measurement was 40′, how many years of observation would be required in order for precession, which occurs at a rate of around 50″ per year, to be noticed?

How much precession had occurred over the 160-year period covered by the observational data available to Hipparchus?

9. Explain how the epicyclic theories of Hipparchus and Ptolemy satisfied the principles of Plato and Aristotle, yet in certain ways violated some of them as well.

10. Why did Hipparchus, who was aware of the work of Aristarchus, not accept the idea that the earth orbits the sun?

ADDITIONAL READINGS

There are a number of magazines for the sky-watcher, for example, *Mercury, Sky and Telescope, Astronomy,* and the *Griffith Observer,* that contain practical information, such as planetary positions and the seasonal appearance of the constellations. There also exist annually published handbooks with similar data. One of the most widely used is the *Observer's Handbook,* by Roy L. Bishop (Toronto: Royal Astronomical Society of Canada). Some practical exercises in astronomy can be found in books such as *Astronomy: A Self-Teaching Guide,* by Dinah L. Moché (New York: Wiley, 1981).

It is also possible to find readings on the history of astronomy. Listed here are some books relevant to this chapter, including material on ancient astronomical developments (for example, the astronomies of Asia and the Americas), that is not covered here. It is also useful to browse through professional journals that cover the history of astronomy. These include *Vistas in Astronomy* and the *Journal for the History of Astronomy,* both of which can be found in most science-oriented libraries.

Heath, T. L. 1969. *Greek astronomy.* New York: AMS.

Krupp, E. C. 1978. *In search of ancient astronomies.* Garden City, N.J.:Doubleday.

———. 1983. *Echoes of the ancient skies.* New York: Harper and Row.

Neugebauer, O. 1969. *The exact sciences in antiquity.* New York: Dover.

THE RENAISSANCE

Learning Goals

COPERNICUS: THE HELIOCENTRIC VIEW
TYCHO BRAHE: ADVANCED OBSERVATIONS
JOHANNES KEPLER: THE NATURE OF ORBITS
GALILEO: EXPERIMENTAL PHYSICS AND
    OBSERVATIONAL ASTRONOMY
ISAAC NEWTON: THE LAWS OF MOTION
Mass and Inertia
Force and Acceleration
Action and Reaction

THE LAW OF UNIVERSAL GRAVITATION
A MODERN VIEW OF ORBITS
The concept of Energy: Kinetic and Potential
Orbits Involving Two Bodies and a Center of Mass
Angular Momentum and Its Preservation
Kepler's Third Law Revised
Escape Velocity
TIDAL FORCES

2

The fifteenth century saw the beginnings of a reawakening of intellectual spirit in Europe. Some scientific studies began at the major universities, increased maritime explorations brought demands for better means of celestial navigation, and the art of printing was discovered, opening the way for widespread dissemination of information.

Major advances in all the sciences accompanied the new developments in other fields of human endeavor. In this chapter we discuss the principal achievements in Renaissance science that led to the development of modern astronomy. In doing so we discuss the accomplishments of five major figures: Copernicus, Brahe, Kepler, Galileo, and Newton.

Copernicus: The Heliocentric View Revisited

FIGURE 2.1. *Nicolaus Copernicus.*

Some nineteen years before the epic voyage of Columbus, Niklas Koppernigk (Fig. 2.1) was born in Toruń, in the northern part of Poland. As a young man he attended the university in Cracow, where his entrancement with Latin, the universal language of scholars, led him to change his surname to Copernicus. At Cracow he nourished an abiding interest in astronomy, becoming fully acquainted with the Aristotelian view as well as the Ptolemaic model of the planetary motions. During his subsequent years of education, he persisted in his studies of astronomy, and it is known that, by 1514, he had developed some doubts about the validity of the accepted system.

His reasons for doing so have been the subject of some uncertainty and misconception. It was long assumed that Copernicus was encouraged to adopt the sun-centered view of the universe because he recognized shortcomings in the geocentric model of Ptolemy. There is, however, no evidence of widespread dissatisfaction with the Ptolemaic system, nor any records indicating that Copernicus found serious inaccuracies in it. His reasons for adopting the heliocentric viewpoint were more subtle.

The basis for his conversion was primarily philosophical. The new system, in the mind of Copernicus, presented a pleasing and unifying model of the universe and its motions. Although he was, no doubt, en-

couraged by the climate of change and cultural revolution that was sweeping Europe with the advent of the Renaissance, he did not adopt the heliocentric view just to be different, nor did he do it to improve the accuracy or reduce the complexity of the accepted view. The Copernican model was, in fact, no more accurate in predicting planetary motions than was the Ptolemaic system, and it was just as complex. Copernicus adhered to the notion of perfect circular motions, and was obliged to include small epicycles in order to match the observed planetary orbits. Apparently one of the most pleasing aspects of the sun-centered system to Copernicus was the fact that the relative distances of the planets could be deduced (Fig. 2.2), and were found to have a certain regularity, the spacings between planets growing systematically with distance from the sun. Copernicus was also able to determine the relative speeds of the planets in their orbits, finding that each planet moves more slowly than the next one closer to ths sun.

Copernicus first circulated his ideas informally sometime before 1514, in a manuscript called *Commentariolis,* and it drew increasing attention over the next several years. The Church voiced no opposition, despite the fact that the ideas expressed in the work were in strong contradiction with commonly accepted

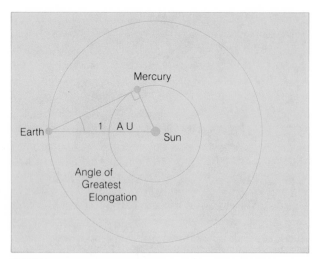

FIGURE 2.2. *The Method of Copernicus for Finding Relative Planetary Distances from the Sun.*
For an inferior planet such as Mercury or Venus, Copernicus knew the angle of greatest elongation, and was therefore able to reconstruct the triangle shown, providing him with the sun-Mercury (or sun-Venus) distance relative to the sun-earth distance (that is, the astronomical unit). For superior planets, similar but slightly more complicated considerations yielded the same information.

Church doctrine. Copernicus was apparently reluctant to publish his findings in a more formal way, for fear of raising controversy, and was continually rechecking his calculations. Publication of his findings in a book called *De Revolutionibus Orbium Coelestium (On the Revolution of the Celestial Sphere),* or simply *De Revolutionibus,* finally took place in 1543, when Copernicus was near his death. He did not live to see the profound impact of his work.

Tycho Brahe: Advanced Observations and a Return to a Stationary Earth

Although Copernicus had contributed little observationally, and had not demanded a close match between theory and observation as long as general agreement was found, there was, in fact, a strong need for improved precision in astronomical measurements. The next influential character in the historical sequence filled this need; if he had not, further progress would have been seriously delayed, as would have the eventual acceptance of the Copernican doctrine.

Tycho Brahe (Fig. 2.3) was born in 1546, some three years after the death of Copernicus, in the extreme southern portion of modern Sweden (the region was at the time part of Denmark). Of noble descent, Tycho spent his youth in comfortable surroundings, and was well educated, first at Copenhagen University, then at Leipzig, where he insisted on studying mathematics and astronomy despite his family's wish that he pursue a law career. As a result of some notable observations and their interpretation, Tycho eventually developed a strong reputation as an astronomer (and astrologer; the distinction was still scarcely recognized), and attracted the attention of the Danish King, Frederick II. In 1575 the king ceded to Tycho the island of Hveen, about fourteen miles north of Copenhagen, along with enough servants and financial assistance to allow him to build and maintain his own observatory.

Tycho had, even before this time, shown an acute interest in astronomical instruments, and with the grant to build his observatory, this interest bore fruit. He devised a variety of instruments (Fig. 2.4) which, although they really did not encompass any new prin-

FIGURE 2.3. *Tycho Brahe.*

FIGURE 2.4. *The Great Mural Quadrant at Hveen.*
The quadrant was an instrument used to measure the angular
positions of stars and planets with respect to the horizon.

ciples, were capable of more accurate readings than
any before his time.

His observational contribution consisted of the unprecedented accuracy of his data, and the completeness
of his records. Until his time, it had been the general
practice of astronomers to record the positions of the
planets only at notable points in their travels, such as
when a superior planet comes to a halt just before beginning retrograde motion. Tycho made much more
systematic observations, recording planetary positions
at times other than just the significant turning points in
their motions. He also made multiple observations in
many instances, allowing the results to be averaged together to improve their accuracy. Tycho himself did not
attempt any extensive analysis of his data, but the vast
collection of measurements that he gathered over the
years contained the information needed to reveal the
basis of the planetary motions.

Tycho was unable to accept the heliocentric view,
primarily because he could find no evidence that the
earth was moving. He tried and failed to detect stellar
parallax, which he supposed he should see with his
accurate observations if the earth really moved. Furthermore, as a strict Protestant, he found it philosoph-
ically difficult to accept a moving earth, when the
Scriptures stated that the earth is fixed at the center of
the universe. On the other hand, he realized that the
Copernican system had advantages of mathematical
simplicity over the Ptolemaic model, and in the end he
was ingenious enough to devise a model that satisfied
all his criteria. He imagined that the earth was fixed
with the sun orbiting it, but that all the other planets
orbited the sun (Fig. 2.5). Mathematically, this is
equivalent to the Copernican system in terms of accounting for the motions of the planets as seen from the
earth. The idea never received much acceptance, however, and Tycho is remembered primarily for his fine
observations.

Tycho Brahe died in 1599. The task of seeking the
secrets contained in Tycho's data was left to those who
followed him, particularly a young astronomer named
Johannes Kepler.

Johannes Kepler and the Laws of Planetary Motion

Until now it has been possible to follow developments
in a straight sequence, but here we must begin to cover
events that occurred at nearly the same time, although
in different locations. To complete the thread begun
with our discussion of Tycho Brahe, we now turn to
Johannes Kepler (Fig. 2.6), who worked briefly at the
side of Tycho, and then spent many years analyzing the
great wealth of observational data that Tycho had accumulated. We must keep in mind, however, that during the same period a scientist and philosopher named
Galileo Galilei was at work to the south, in Italy, and
that the two were aware of each other's accomplishments.

Born in 1571, Kepler was a sickly youngster, seemingly headed for a career in theology. While attending
university, however, he encountered a professor of astronomy who inspired in him a strong interest in the
Copernican system, which he adopted wholeheartedly.

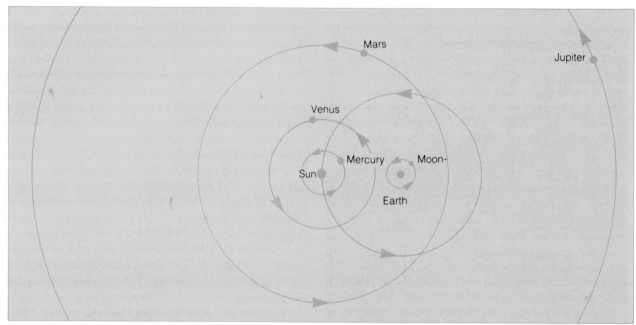

FIGURE 2.5. *Tycho's Model of the Universe.* The earth was held fixed, and the sun was thought to orbit the earth. The planets in turn were envisioned as orbiting the sun. This system was never worked out in mathematical detail, but it successfully preserved both the advantages of the Copernican system and the spirit of the ancient teachings. Most significant, it accounted for the lack of observed stellar parallax, since the earth was fixed.

FIGURE 2.6. *Johannes Kepler.*

Kepler thereafter devoted his life to seeking the underlying harmony of the cosmos. He put forth great effort in searching for simple numerical relationships among the planets, and in doing so embarked on several false turnings and erroneous paths. He was a true scientist, however, in that he always was willing to discard his ideas if the data did not support them.

Kepler went to work as Tycho's assistant in 1598, and a year later found himself the beneficiary of the massive collection of data left behind by Tycho's death. Kepler's main mission, owing to both Tycho's wishes and his own interests, was to develop a refined understanding of the planetary motions and to upgrade the tables used to predict their positions. His first work involved the planet Mars, for which the data were particularly extensive, and whose motions were among the most difficult to explain in the established Ptolemaic system (and, in fact, in the Copernican system, with its requirement of only circular motions). By a very complex process, Kepler was able to separate the effects of the earth's motion from those of Mars, so that he could map out the path that Mars followed with respect to the sun.

By 1604 Kepler had determined that the orbit of Mars was some kind of oval, and further experimentation revealed that it was fitted precisely by a simple geometric figure called an **ellipse** (Fig. 2.7). This is a closed curve defined by a fixed total distance from two points called **foci,** and indeed Kepler found that the sun was at the precise location of one focus of the ellipse. This discovery was later generalized by Kepler to apply to all the planets, and it became known as Kepler's first law of planetary motion: the orbit of each planet is an ellipse, with the sun at one focus.

Further analysis of the motion of Mars revealed another characteristic, having to do with the fact that the planet moves fastest in its orbit when it is nearest the sun, and slowest when it is on the opposite side of its orbit, farthest from the sun. Kepler's second mathematical discovery was that a line connecting Mars to the sun sweeps out equal areas of space in equal intervals of time (Fig. 2.7). Again, Kepler later stated that this law applies to all the planets.

Kepler's results concerning the orbit of Mars appeared in 1609 in a work entitled *The New Astronomy: Commentaries on the Motions of Mars.* The book received a great deal of attention. In 1619 Kepler's book called *The Harmony of the World* was published, in which is reported his discovery of the simple relationship between the orbital periods of the planets and their average distances from the sun. Now known as Kepler's third law, or simply the Harmonic law, it states that the square of the period of a planet is proportional to the cube of the semimajor axis (which is half of the long axis of an ellipse). In other words, this says that $P^2 = a^3$, where P is the sidereal period of a planet in years, and a is the semimajor axis in terms of the sun-earth distance (that is, in terms of the **astronomical unit** or **AU**).

In another major work called the *Epitome of the Copernican Astronomy,* published in parts in 1618, 1620, and 1621, Kepler presented a summary of the state of astronomy at that time, including Galileo's discoveries. In this book Kepler generalized his laws, explicitly stating that all the planets behaved similarly to Mars, something that had clearly been his belief all along.

By the time the *Epitome* was published, the Roman Catholic church was in a very intolerant frame of mind, in contrast with the situation at the time of Copernicus, nearly a hundred years before. Kepler's treatise soon found itself on the *Index of Prohibited Books,* along with *De Revolutionibus.*

In 1627 Kepler published his last significant astronomical work, a table of planetary positions based on his laws of motion, which could be used to accurately predict the planetary motions. These tables, which he called the *Rudolphine Tables* in honor of a former benefactor, were used for the next several years. The *Rudolphine Tables* represented an improvement in ac-

FIGURE 2.7. *The Ellipse.* At left is an exaggerated ellipse representing a planetary orbit with the sun at one focus; at right is a similarly exaggerated ellipse, with lines drawn to illustrate Kepler's second law. If the numbers represent the planet's position at equal time intervals, then the areas of the triangular segments are equal.

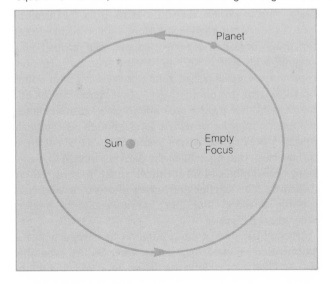
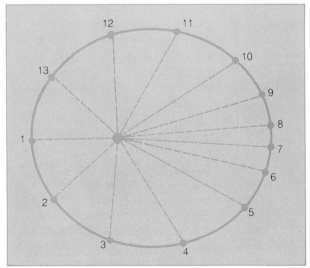

curacy over any previous tables by nearly a factor of 100, a resounding and remarkable confirmation of the validity of Kepler's laws. In a very real sense the *Rudolphine Tables* represented Kepler's life's work, since with their publication he completed the task set before him when he first went to work, for Tycho. Kepler died in 1630, at the age of fifty-nine.

Galileo, Experimental Physics, and the Telescope

Very strong contrasts can be drawn between Kepler and his great contemporary, Galileo Galilei (Fig. 2.8), born in Pisa, in northern Italy, in 1564. Whereas Kepler was fascinated with universal harmony and therefore with the underlying principles on which the universe operates, Galileo was primarily concerned with the nature of physical phenomena and was less devoted to finding fundamental causes. Galileo wanted to know *how* the laws of nature operated, whereas Kepler sought the *reason* for the existence of the laws.

Galileo's approach was level-headed and rational in

FIGURE 2.8. *Galileo Galilei.*

the extreme. He used simple experiment and deduction in advancing his perception of the universe, and has frequently been cited as the first truly modern scientist, although others of his time probably deserve a share of that recognition. Whereas a follower of Plato and Aristotle, whose works still dominated in Galileo's time, would proceed by rational thought from standard unproved assumptions, Galileo found it much more sensible to begin with experiment or observation, and to work from there toward a recognition of the underlying principles. In doing so, Galileo founded an entirely new basis for scientific inquiry, an achievement in many ways more profound than his contributions to astronomy, which were considerable.

Galileo's discoveries in physics, having to do with the motions of objects, were published in his later years, after his astronomical career had been forcibly ended by Church decreee. It was, in fact, an early interest in *mechanics,* the science of the laws of motion, that lured Galileo away from a career in medicine, the subject of his first studies. Galileo's contributions to physics will be briefly described later in this chapter.

Galileo's astronomical discoveries were quite sufficient to earn him a major place in history, and his flair for debate and his habit of ridiculing those whose arguments he disproved made him famous in his own time, although not universally loved.

In 1609 Galileo learned of the invention of the telescope, and devised one of his own, which he soon put to use by systematically viewing the heavens with it. Despite the poor quality of the instrument, Galileo made a number of important discoveries almost at once, and reported them in 1610 in a publication called the *Starry Messenger.* Here Galileo showed that the moon was not a perfect sphere, but instead was covered with craters and mountains (Fig. 2.9). He also reported the existence of many more stars than could be seen with the unaided eye. Most significant of all, he showed that Jupiter was attended by four satellites, whose motions he observed long enough to establish that they orbited the parent planet (Fig. 2.10). All these discoveries, but especially the latter, violated the ancient philosophies of an idealized universe centered on the earth. The satellites of Jupiter showed beyond reasonable question that there were centers of motion other than the earth.

Mostly because of the reputation he earned with the publication of the *Starry Messenger,* Galileo was able to negotiate successfully for the position of court math-

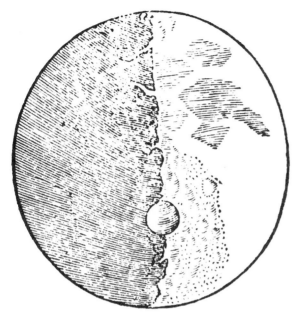

FIGURE 2.9. *Galileo's Sketch of the Moon.*

ematician to the Grand Duke of Tuscany, and he moved in 1610 to Florence, where he was to spend the remainder of his long career. Once established there, Galileo continued his observations, and soon added new discoveries to his list. The first of these was the varying aspect of Venus as its position changed with respect to the sun. Galileo quickly realized that this was analogous to the phases of the moon, as we see from the earth varying portions of the sunlit side of Venus (Fig. 2.11). This had two important implications: it showed that Venus shines by reflected sunlight, rather than by its own power; and it demonstrated that Venus is orbiting the sun instead of the earth, since the apparent size of Venus varies (from being smallest when Venus is fully sunlit to being largest when the dark side of Venus is facing the earth).

At about the same time, Galileo began to study the dark spots that occasionally appeared on the face of the sun. Although Galileo was not the discoverer of sunspots, as is sometimes claimed, he did establish, by geometrical arguments, that they really do lie on the

FIGURE 2.10. *Galileo's Sketches of the Moons of Jupiter.* This series of drawings by Galileo is often attributed to the *Starry Messenger,* but in fact was made some years later.

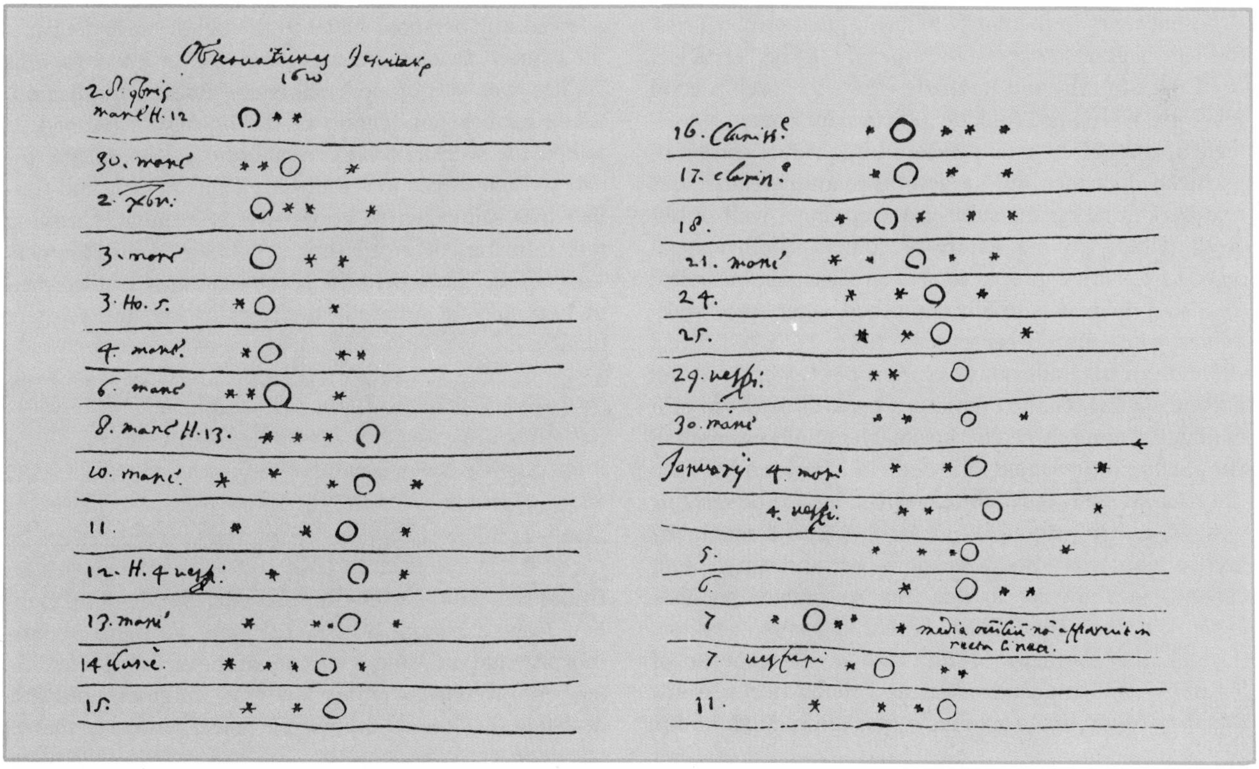

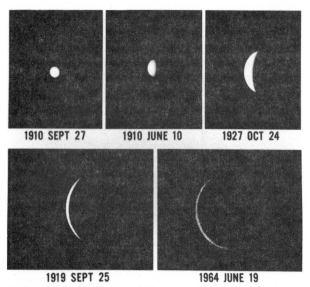

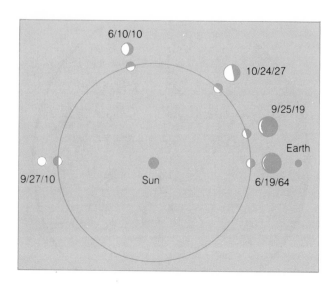

FIGURE 2.11. *The Phases of Venus.*
As Venus orbits the sun and its position changes relative to the sun-earth line, its phase varies as we
see differing portions of its sunlit side. In addition, its apparent size varies because of its varying distance from earth.

surface of the sun, and are not foreground objects such as small planets. That the sun should have blemishes on its surface, and rotate as well, was contrary to the established view.

In the years immediately following these discoveries, Galileo began to draw increasingly heavy criticism from the Church, and he made efforts to develop good relations with high-ranking officials in Rome. Nevertheless, in 1616 he was pressured into refuting the Copernican doctrine, and for several years thereafter was relatively quiet on the subject. Except for a well-publicized debate on the nature of comets, Galileo spent most of his time preparing his greatest astronomical treatise, which was finally published, after some difficulties with Church censors, in 1632. To avoid direct violation of his oath not to support the Copernican heliocentric view, Galileo wrote his book in the form of a dialogue among three characters, one of whom, named Simplicio, represented the official position of the Church; another, Salviati, that of Galileo (although this was, of course, not stated explicitly); and a third, Sagredo, who was always quick to see and agree with Salviati's arguments. In this treatise, called the *Dialogue on the Two Chief World Systems,* Galileo, through the character Salviati and at the expense of Simplicio, systematically destroyed many of the traditional astronomical teachings of the Church. The book

was published in Italian, rather than the scholarly Latin, and its contents were therefore accessible to the general populace.

Despite a lengthy preface in which Galileo disavowed any personal belief in the heliocentric doctrine, the Church reacted strongly, and within a few months Galileo was summoned before the Roman Inquisition, while further publication of the book was banned. It joined the works of Copernicus and Kepler on the *Index* (where it was to remain until the 1830s), and Galileo was sentenced to house arrest, eventually serving this punishment at his country home near Florence. This is where he spent the rest of his days. Galileo died in 1642, having suffered blindness the last four years of his life.

Isaac Newton and the Laws of Motion

In the year 1643, a few months after the death of Galileo, Isaac Newton (Fig. 2.12) was born, in Woolsthorpe, England, into a climate of inquiry and scientific ferment. Newton's childhood was unremarkable, except that he showed a growing interest in mathematics

Astronomical Insight 2.1

AN EXCERPT FROM *DIALOGUE ON THE TWO CHIEF WORLD SYSTEMS*

During the course of three days of discussion, the three characters in Galileo's *Dialogue* thoroughly air all the available evidence and arguments bearing on the understanding of mechanics and the structure and nature of the universe. About midway through, a discussion develops that is central to Galileo's principal point: the sun, not the earth, is at the center of the universe. In the following excerpt,* we see an example of Salviati's persuasive style, Simplicio's dogged reluctance to give up the ideas of Aristotle, and Sagredo's ready comprehension of Salviati's arguments.

Salviati: Now if it is true that the center of the world is the same about which the circles of the mundane bodies, that is to say, of the Planets, move, it is most certain that it is not the Earth but the Sun, rather, that is fixed in the center of the World. So that as to this first simple and general apprehension, the middle place belongs to the sun, and the Earth is as far remote from the center as it is from that same Sun.

Simplicio: But from whence do you argue that not the Earth but the Sun is in the center of planetary revolutions?

Salviati: I infer the same from the most evident and therefore necessarily conclusive observations, of which the most potent to exclude the Earth from the said center, and to place the Sun therein, are that we see all the planets sometimes nearer and sometimes farther off from the Earth, with so great differences, that, for example, Venus when it is at the farthest is six times more remote than when it is nearest, and Mars rises almost eight times as high at one time as at another. See therefore whether Aristotle was somewhat mistaken in thinking that it was at all times equidistant from us.

Simplicio: What in the next place are the tokens that their motions are about the sun?

Salviati: It is shown in the three superior planets, Mars, Jupiter, and Saturn, in that we find them always nearest to the Earth when they are in opposition to the Sun and farthest off when they are towards the conjunction; and this approximation and recession imports thus much, that Mars near at hand appears sixty times greater than when it is remote. As to Venus, in the next place, and to Mercury, we are certain that they revolve about the Sun in that they never move far from it, and in that we see them sometimes above and sometimes below it, as the mutations in the figure of Venus necessarily prove. Touching the Moon, it is certain that it cannot in any way separate itself from the Earth, for the reasons that shall be more distinctly alleged hereafter.

Sagredo: I expect that I shall hear more admirable things that depend upon this annual motion of the Earth than were those dependent upon the diurnal revolution.*

In this exchange, Galileo, in the guise of Salviati, advances several arguments based on observations, including the well-known phases of Venus, and the less widely quoted arguments having to do with the varying distances of the planets from the earth. In this and other passages, Galileo went out of his way to mock the followers of Aristotle, who, according to Salviati in an earlier paragraph, "would deny all the experiences and all the observations in the world, nay, would refuse to see them, that they might not be forced to acknowledge them, and would say that the world stands as Aristotle writes and not as Nature will have it."

*Reprinted from the translation of Galileo's *Dialogue* by G. D. Santillana, by permission of the University of Chicago Press. Copyright 1953 by the University of Chicago. All rights reserved.

FIGURE 2.12. *Isaac Newton.*

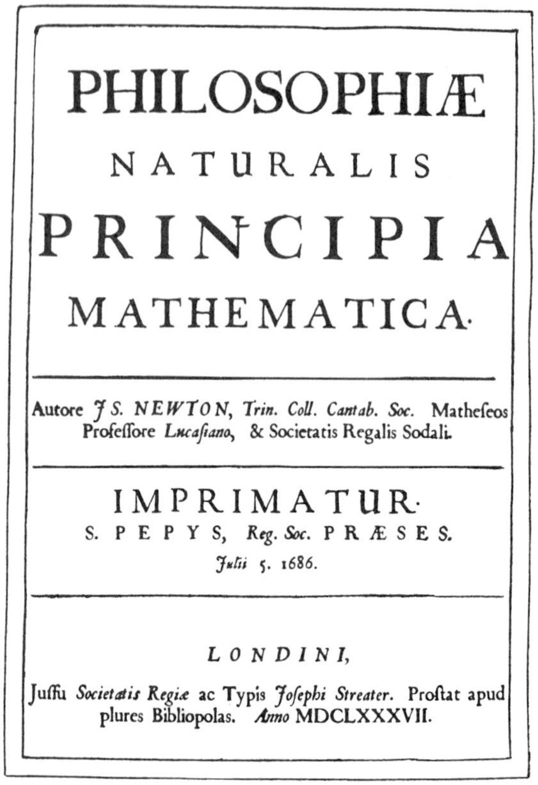

PHILOSOPHIÆ

N A T U R A L I S

PRINCIPIA

MATHEMATICA·

Autore *JS. NEWTON, Trin. Coll. Cantab. Soc.* Matheseos
Professore *Lucasiano,* & Societatis Regalis Sodali.

IMPRIMATUR·
S. P E P Y S, *Reg. Soc.* P R Æ S E S.
Julii 5. 1686.

L O N D I N I,

Jussu *Societatis Regiæ* ac Typis *Josephi* Streater. Prostat apud
plures Bibliopolas. *Anno* MDCLXXXVII.

FIGURE 2.13. *The Title Page from an Early Edition of the Principia.*

and science, and at age eighteen he entered college at Cambridge, receiving a bachelor's degree in early 1665. He then spent two years at his home in Woolsthorpe, largely because the Plague made city living rather dangerous, and it was during this time that he made a remarkable series of discoveries in the fields of physics, astronomy, optics, and mathematics, in what surely must have been one of the most intense and productive periods of individual intellectual effort in human history.

Newton had a tendency to exhaust a subject, get bored with it, and go on to new fields, so that nothing of his work was published for some time, during which some of his discoveries were repeated independently by others. Finally, after persistent urging by his friend (and fellow astronomer) Edmund Halley, in 1687 Newton published a massive work called *Philosophiae Naturalis Principia Mathematica* (Fig. 2.13), now usually referred to as the *Principia*. In this three-volume book, Newton established the science of mechanics and applied it to the motions of the moon and the planets, developing the law of gravitation as well. His work in optics was published separately in 1704, although it was probably written much earlier than that.

The *Principia* received great notice, particularly in England. As a result, Newton's later life was a public

one; he held various government positions which left him with less and less time for scientific discovery. With the help of younger associates, he did revise the *Principia* on two occasions, in 1713 and in 1726, making some improvements each time. Newton died in 1727, at the age of eighty-four. His impact lives today, in our modern understanding of physics and mathematics. Newton's conclusions on the nature of motions and gravity are still viewed as correct, although it is now realized that there are circumstances in which more complex theories (such as Einstein's relatively) must be used.

Newton put forth three laws of motion, principles he considered so self-evident that he relegated them to an introductory section of the *Principia*. The first of Newton's laws states the principle of **inertia,** a concept first recognized by Galileo, who realized that an object in motion tends to stay in motion unless something acts to slow or stop it. This was completely contrary to the teachings of Aristotle, who held that the natural tendency of any moving object was to stop, and that it would only continue moving if a force were applied. Aristotle was misled by his failure to recognize friction, a force that tends to stop motion in most everyday situations. Newton expanded the concept of inertia, recognizing it as just one in a series of physical principles that govern the motions of objects, and adding the all-important notion of **mass.**

The mass of an object reflects the amount of matter it contains, which in turn determines other properties such as weight and momentum. It is the other properties that are easily observed, not the mass, so mass can be a difficult concept. We can make useful illustrations by considering situations where we alter these other properties, but not the mass. For example, an object's weight varies according to the magnitude of the gravitational acceleration to which the object is subjected, but its mass does not change. An astronaut may be weightless, but in space he contains just as much mass as he does on the ground. Mass is usually measured in units of **grams** or **kilograms.** One gram is the mass of a cubic centimeter of water, and a kilogram, which weighs about 2.2 pounds at sea level, is the mass of 1,000 cubic centimeters (or one **liter**) of water.

Inertia is very closely related to mass. The more massive an object is, the more inertia it has, and the more difficult it is to start it moving, or to stop or alter its motion once it is moving (Fig. 2.14). Newton summarized the concept of inertia in what has become known as his first law of motion:

> A body at rest or in a state of uniform motion tends to stay at rest on in uniform motion unless an outside force acts on it.

Having stated that a force is required to change an object's state of rest or uniform motion, Newton went on to determine the relationship between force and the change in motion that it produces. To understand this, we must discuss the idea of acceleration.

Acceleration is a general work for *any* change in the motion of an object. It is acceleration when a moving object is speeded up or slowed down, or when its direction of motion is altered. It is also acceleration when an object at rest is put into motion. A planet orbiting the sun is undergoing constant acceleration; otherwise it would fly off in a straight line.

Another way of stating Newton's first law is that in order for an object to be accelerated, a force must be applied to it. Note that this force must be an unbalanced one; that is, there will be acceleration only when

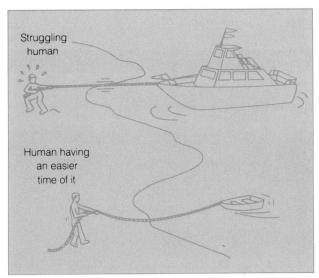

FIGURE 2.14. *Inertia.*
Here two people are attempting to pull to shore two boats of rather different masses. Both boats are at rest initially, and the more massive one resists being started in motion much more strongly than the less massive one.

a force is applied to an object with no other force to counteract it. Thus, a crate sitting on the floor is subject to a force caused by gravity, but this force is balanced by an upward force caused by the floor, and there is no acceleration. Newton's second law spells out the relationship among an unbalanced force, the resultant acceleration, and the mass of the object:

> The acceleration of an object is equal to the force applied to it divided by its mass.

This may be written mathematically as $a = F/m$, where a is the acceleration, F is the unbalanced force, and m is the mass. Usually it is written in the equivalent form $F = ma$.

We can visualize simple examples to help illustrate the second law. If one object has twice the mass of another, for example, and equal forces are applied to the two, the more massive one will be accelerated only half as much. Conversely, if unequal forces are applied to objects of equal mass, the one to which the greater force is applied will be accelerated to a greater speed.

Newton's third law of motion is probably more subtle that the first two, although in some circumstances it is quite obvious. It states:

> For every action there is an equal and opposite reaction.

In other words, when a force is applied to an object,

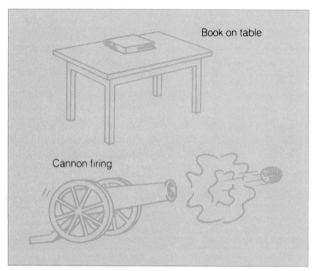

FIGURE 2.15. *Action and Reaction.*
Some manifestations of Newton's third law are rather subtle, whereas others are not. Here a book exerts a force on a table, and the table exerts an equal and opposite force on the book. This is a static situation; there is no motion. When a cannon is fired, however, the cannonball is accelerated one way and the cannon the other. The force applied to each is the same, but the cannon has more mass than the cannonball and is therefore accelerated less, as Newton's second law states.

it pushes back with an equal force (Fig. 2.15). This may sound confusing, because the acceleration is not necessarily equal, and because in most common situations, other forces such as friction complicate the picture. Furthermore, there are many static situations, in which forces are balanced and no motion occurs.

The third law can be most easily visualized by considering situations where friction is not important. For example, imagine standing in a small boat and throwing overboard a heavy object, such as an anchor. The boat will move in the opposite direction from the anchor, because the anchor exerts a force on you as you throw it. The "kick" of a gun when it is fired is another example of action and reaction. When a person jumps off the ground by pushing against the earth, both he and the earth are accelerated by the mutual force, but of course the immensely greater mass of the earth prevents it from being accelerated noticeably.

Newton's third law of motion explains how a rocket works. When a rocket is launched, hot, expanding gas is allowed to escape through a nozzle, creating a force on the rocket. The gas is accelerated in one direction, and the rocket is accelerated in the opposite direction

FIGURE 2.16. *Newton's Third Law Applied to the Launch of a Rocket.* Hot gases are forced out of a nozzle (or several, as in this case), and in return they exert a force on the rocket that accelerates it. This is the first launch of the Space Shuttle *Columbia.*

(Fig. 2.16). Anyone who has inflated a balloon and then let go of it, allowing it to zoom through the air, is familiar with the operating principle of a rocket.

The Law of Gravitation

Newton realized, as Galileo had before him, that inertia would cause the planets to fly off in straight lines if no force were acting on them. Newton was aware that the planets must be undergoing constant acceleration toward the center of their orbits; that is, toward the sun. He set out to understand the nature of the force that creates this acceleration.

If you feel confused about the direction of this force, remember what you have just learned about acceleration and inertia: a planet needs no force to keep it moving, but it does require a force to keep its path curving as it travels around the sun. What is needed is something to push the planets inward, toward the sun (Fig. 2.17). This force can be compared to the tension in a string tied to a rock that you whirl about your head; if you suddenly cut the string, the rock would fly off in whatever direction it happened to be going at the time.

What, then, is the string that keeps the planets whirling about the sun? Newton realized that the sun itself must be the source of this force, and he made use of Kepler's third law, as well as observations of the moon's orbit and falling objects at the earth's surface, to discover its properties. He was led to formulate his law of universal gravitation:

> Any two bodies in the universe are attracted to each other with a force that is proportional to the masses of the two bodies and inversely proportional to the square of the distance between them.

The law of gravitation is one of the fundamental rules by which the universe operates. As we shall see in later chapters, it describes the motions of stars about each other or about the center of the galaxy as well as it describes the motions of the moon and the planets, and it is applicable to the motions of galaxies about one another. Gravity, in fact, is the dominant factor that will determine the ultimate fate of the universe.

When we consider a planet's surface gravity, we must take into account the planet's size and mass. The moon, for example, has 0.273 times the radius of the earth, and 0.0123 times the mass. Thus, the acceleration of gravity at the moon's surface is $0.0123/0.273^2 = 0.165$ times the acceleration at the earth's surface. Therefore an astronaut on the moon weighs approximately one-sixth as much as he does on the earth.

Galileo had shown experimentally that the acceleration of gravity at the surface of the earth does not depend on the mass of the object that is falling. This was shown mathematically by Newton, using the law of gravitation and the second law of motion.

Energy, Angular Momentum, and Orbits; Kepler's Laws Revisited

Even though Newton made use of Kepler's third law in deriving the law of gravitation, the latter is in fact more fundamental. It was soon possible for Newton to show that all three of Kepler's laws follow directly from New-

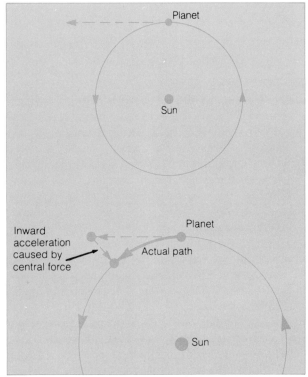

FIGURE 2.17. *Planetary Orbits and Central Force.*
A planet would fly off in a straight line if no force attracted it toward the center of its orbit. From diagrams like the one shown here, Newton was able to determine the amount of acceleration required by an orbiting body to keep it in its orbit.

ton's laws of motion and gravitation. Kepler's studies of planetary motions revealed the *result* of the laws of motion and gravitation, whereas Newton found the *cause* of the motions. To appreciate how this was done, we must further discuss some basic physical ideas.

An important concept in understanding not only orbital motions but also many other aspects of the universe is **energy.** In an intuitive sense, we can define energy as the ability to do work. Energy can exist in many possible forms, such as electrical energy, chemical energy, and heat, to name a few. All forms of energy can be classified as either **kinetic energy,** the energy of motion, or **potential energy,** which is stored energy that must be released (converted to kinetic energy) if it is to do work. A speeding car has kinetic energy because of its motion, whereas a tank of gasoline has potential energy in the form of its chemical reactivity, a tendency to release large amounts of kinetic energy if ignited. Thus a car operates when this potential energy is converted to kinetic energy in its cylinders.

Astronomical Insight 2.2

THE FORCES OF NATURE

One of the most difficult concepts to grasp is that of a force. We think of a force as a push or a pull, and we can visualize it quite easily in certain circumstances, such as when a gardener pushes a wheelbarrow and it moves. A force is perhaps less obvious when no movement results, and even less easy to visualize when no tangible agent exerts it. In the case of gravity, the force is exerted invisibly and over great distances.

In general terms, a force exerted by a concrete object is referred to as a **mechanical force,** whereas one exerted without any such agent is a **field force.** Gravity is the most familiar force created by a field.

There are four kinds of field forces in nature, and they form the basis for *all* forces, mechanical or field. Gravity was the first to be discovered, although in "discovering" gravity and in mathematically describing its behavior, Newton did not develop a real fundamental understanding of how it works, or why. The reason gravity was discovered first is simple: no special conditions are required for gravitational forces to be exerted (*all* masses attract each other), and it operates over very great distances. After all, Newton deduced its properties by noting how it controls the motions of planets that are separated from the sun by as much as a billion kilometers.

The second most obvious type of field force, and the second to be discovered and described mathematically, is the **electromagnetic force.** There is close relationship between electric and magnetic fields. The interaction of these fields with charged particles, first described mathematically by the Scottish physicist James Clerk Maxwell in the late 1800s, creates forces. The electromagnetic force is actually much stronger than gravity; when binding an electron to a proton in the nucleus of an atom, this force is 10^{39} times stronger than the gravitational force between the two particles. The electromagnetic force, like gravity, is inversely proportional to the square of the distance between charged particles. Therefore we might expect this force to always dominate over gravity, as it does on a subatomic scale, but it does not. The reason is that most objects in the universe, composed of vast numbers of atoms containing both electrons and protons, have little or no electrical charge, whereas they always have mass and are therefore subject to gravitational forces. If the planets and the sun had electrical charges in the same proportion to their masses as the electrons and protons do, then electromagnetic forces, rather than gravity, would control their motions. Besides being immensely stronger, the electromagnetic force also differs from gravity in that it can be either repulsive or attractive, depending on whether the electrical charges are the same or opposite.

The other two forces involve interactions at the subatomic level, and it was not until the science of quantum mechanics developed in the 1930s that they were discovered. One is the **strong nuclear force,** which is responsible for holding together the protons and neutrons in the nucleus of an atom, and the other is the **weak nuclear force,** some 10^{-5} times weaker than the strong nuclear force. The strong nuclear force in turn is about a hundred times stronger than the electromagnetic force, making it the strongest of all, but it operates only over very small distances. Within an atomic nucleus, where protons with their like electrical charges are held together despite their electromagnetic repulsion for each other, the strong nuclear force acts as the glue that keeps the nucleus from flying apart. The weak nuclear force plays a more subtle role, showing its effects primarily in certain modifications of atomic nuclei during radioactive decay.

From the smallest to the largest scales, these four forces appear to be responsible for all interactions of matter. It is ironic that we do not yet understand the mechanism that makes the forces work; to do so is one of the principal goals of modern physics. One hope is to develop a mathematical framework that encompasses all the forces, a framework first sought more than sixty years ago by Albert Einstein, and referred to as a **unified field theory.** Progress has been made since Einstein's time, particularly toward unifying the electromagnetic and nuclear forces, but the ultimate goal still eludes us.

The units used for measuring energy can be expressed in terms of the kinetic energy of specified masses moving at specified speeds. The kinetic energy of a moving object is $\frac{1}{2}mv^2$, where m is the mass of the object and v its velocity. In astronomy the most commonly used unit of energy is the *erg*, a small amount of energy that is equivalent to the kinetic energy of a mass of two grams moving at a speed of one centimeter per second.

We often speak in terms of **power,** which is simply energy expended per second. Astronomers tend to use units of ergs per second, which have no special name as a unit (the familiar watt is equal to 10^7 ergs per second). Thus, we will speak of the power (or, equivalently, the **luminosity**) of a star in terms of its energy output in ergs per second.

Using our understanding of energy, we can now discuss orbital motions in a much more general way than we have previously done. Two objects subject to each other's gravitational attraction have kinetic energy, caused by their motions, and potential energy, owing to the fact that they each feel a gravitational force. Just as a book on the table has potential energy that can be converted to kinetic energy if the book is allowed to fall, an orbiting body also has potential energy by virtue of the gravitational force acting on it.

Newton's laws and the concepts of kinetic and potential energy can be used to show that there are many types of possible orbits when two bodies interact gravitationally (Fig. 2.18). Not all are ellipses, because if one of the objects has too much kinetic energy (exceeding the potential energy caused by the gravity of the other), it will not stay in a closed orbit, but will instead follow an arcing path known as a *hyperbola,* and will escape after one brief encounter. Some comets have so much kinetic energy that after one trip close to the sun, they escape forever into space, following hyperbolic paths.

If the kinetic energy is less than the potential energy, as it is for all the planets, then the orbit is an ellipse, as Kepler found. It is technically correct to say that a planet and the sun orbit a common **center of mass** (Fig. 2.19), rather than to say that the planet orbits the sun. The center of mass is a point in space between the two bodies where their masses are essentially balanced; more specifically, it is the point where the product of mass times distance from this point is equal for the two objects. Since the sun is so much more massive than any of the planets, the center of mass for any sun–planet pair is always very close to the center of the sun, so the sun moves very little, and we do not easily see

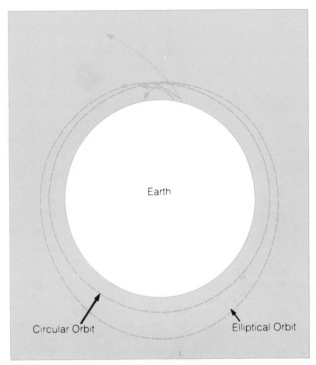

FIGURE 2.18. *Orbital and Escape Velocities.*
A rocket launched with insufficient speed for circular orbit would tend to orbit the earth's center in an ellipse, but would intersect the earth's surface. Given the correct velocity it will follow a circular orbit. A somewhat larger velocity will place the rocket in an elliptical orbit that does not intersect the earth. If given enough velocity so that its kinetic energy is greater than its gravitational potential energy, however, it will escape entirely.

its orbital motion. It is true, however, that the sun's position wiggles a little as it orbits the centers of mass established by its interaction with the planets, especially the most massive ones. In a double-star system, where the two masses are more nearly equal, it is easier to see that both stars orbit a point in space between them. Thus Kepler's first law as he stated it requires a slight modification: each planet has an elliptical orbit, with the center of mass between it and the sun at one focus.

The second law also can be restated in terms of Newton's mechanics. Any object that rotates or moves around some center has **angular momentum.** This is dependent on the object's mass, speed, and distance from the center of motion. In the simple case of an object in circular orbit, the angular momentum is the product mvr, where m is its mass, v its speed, and r its distance from the center of mass.

The total amount of angular momentum in a system is always constant. Because of this, a planet in an el-

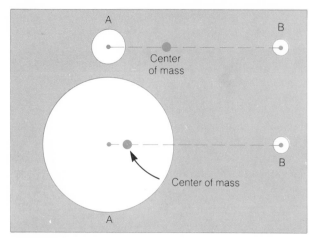

FIGURE 2.19. *Center of Mass.*
In the upper sketch, star A has twice the mass of star B, so the center of mass, about which the two stars orbit, is one-third of the way between the centers of the two stars. The lower sketch shows a case where star A has ten times the mass of star B, so the center of mass is very close to the center of star A. The sun is so much more massive than any of the planets that the center of mass for any sun-planet pair is very near the center of the sun, so the sun's orbital motion is very slight.

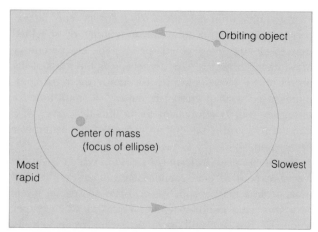

FIGURE 2.20. *Conservation of Angular Momentum.*
An object moves in an elliptical orbit with varying speed, and the product of the object's mass times its velocity times its distance from the center of mass (that is, its angular momentum) is constant. Kepler's second law of planetary motion is a rough statement of this fact.

liptical orbit must move faster when it is close to the center than when it is farther away, so that its velocity compensates for the changes in distance (Fig. 2.20). Thus a planet moves faster in its orbit near **perihelion** (its point of closest approach to the sun) than at **aphelion** (its point of greatest distance from the sun). Kepler's second law was really a statement that angular momentum is constant for a pair of orbiting objects.

Kepler's third law was also revised by Newton, and in this case the revision is especially important. Newton discovered that the relationship between period and semimajor axis depends on the masses of the two objects. Kepler had not realized this, primarily because the sun is much more massive than any of the planets, so that the differences among the masses of the planets have only a very small effect. Kepler's form of the third law can be written $P^2 = a^3$, where P is the planet's period in years and a is the semimajor axis in astronomical units. Newton revised this to:

$$(m_1 + m_2)P^2 = a^3,$$

where m_1 and m_2 represent the masses of the two bodies, for example the sun and one of the planets. The masses must be expressed in terms of the sun's mass

in this equation. If we use other units, such as grams for the masses, seconds for the period, and centimeters for the semimajor axis, then the equation is complicated by the addition of a numerical factor and is written $(m_1 + m_2)P^2 = 4\pi^2 a^3/G$, where G is the gravitational constant. This form of Kepler's third law can be solved for the sum of the masses, and we will see that this is a very important tool for astronomers in deducing the masses of distant objects.

The consideration of orbital motions in terms of kinetic and potential energy leads to the concept of an **escape velocity.** If an object in a gravitational field has greater kinetic than potential energy, it will escape the gravitational field entirely. To launch a rocket into space (that is, completely free of the earth) therefore requires giving it enough speed at launch so that its kinetic energy is greater than its potential energy caused by the earth's gravitational attraction (refer back to Fig. 2.18). It so happens that the speed required to accomplish this is the same for any mass of object. In equation form it is $v_e = \sqrt{GM/R}$, where v_e is the escape velocity, G is the gravitational constant, and M and R are the earth's mass and radius. For the earth, this velocity is 11.2 kilometers per second, or slightly more than 40,000 kilometers per hour. As we will see in later chapter, a planet's escape velocity plays a major role in determining the nature of its atmosphere.

Tidal Forces

We have seen that the gravitational force resulting from a distant body decreases with distance. This means that an object subjected to the gravitational pull of such a body feels a stronger pull on the side nearest that body, and a weaker pull elsewhere. For example, the part of the earth on the side facing the moon feels a stronger attraction toward the moon than do other points on the earth, and the point on the opposite side from the moon feels the weakest force. The earth is therefore subjected to a **differential gravitational force,** which tends to stretch it along the line toward the moon (Fig. 2.21). Of course, the sun exerts a similar stretching force on the earth, but it is too distant to have as strong an effect as the moon (but its *total* gravitational force on the earth is much greater than that of the moon). A **tidal force,** as differential gravitational forces are often called, depends on how close one body is to the other, because the key is how rapidly the gravitational force drops off over the diameter of the body subject to the tidal force, and it drops off most rapidly at close distances.

The earth is more or less a rigid body, so it does not stretch very much as a result of the differential gravitational force of the moon. Nevertheless, the tidal forces exerted on the earth by the moon create net forces that tend to make the liquid oceans flow in the direction of the points facing directly toward the moon and directly away from it. As the earth rotates, the water in the oceans tends to follow the tidal forces created by the moon, so that in effect the oceans have two huge ridges of water that flow around the earth as it rotates. Since there are two ridges of water on opposite sides, and the earth rotates in twenty-four hours, at any given point on its surface one of these ridges passes every twelve hours. Thus we have the oceans' tides, with high tides at any given location separated by about twelve hours.

It is interesting to consider what is happening to the moon at the same time. It is subjected to a more intense differential gravitational force than the earth, since the earth is more massive than the moon. Even though the moon is a solid body, its shape is deformed by this force, and it has tidal bulges. The moon is slightly elongated along the line toward the earth. As we shall see, this has had drastic effects on the moon's rotation.

There are many other examples of tidal forces, both within the solar system and outside it. The satellites of the massive outer planets are subjected to severe tidal forces, and there are double-star systems where the two stars are so close together that they are stretched into elongated shapes. We will discuss these in more detail later, along with star clusters and galaxies that are affected by tidal forces, sometimes even to the point of tearing each other apart.

FIGURE 2.21. *The Earth's Tides.*
The differential gravitational force caused by the moon tends to stretch the earth (large arrows). Seawater at any given point on the earth is subjected to a combination of vertical and horizontal forces, causing it to flow toward either the side of the earth facing the moon or the side opposite it (curved arrows).

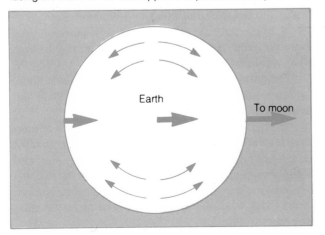

PERSPECTIVE

We have traced the development of astronomy from the introduction of the heliocentric concept to the point where the motions of the bodies in the solar system can be fully described in terms of a few simple laws of physics. The framework of these laws does more than that for us, however; it also provides a basis for understanding the motions of more distant astronomical objects. The law of universal gravitation will be invoked time and time again in our study of the solar system and the rest of the universe, because so many important phenomena are explained by it.

Newton's laws of motion and gravitation have stood the test of time rather well. Einstein's theory of general relativity, to be discussed later, may be viewed as a more complete description of gravity and its interaction with matter. For most situations on the earth's surface and in space, however, Newton's laws are perfectly adequate.

We are now ready to discuss other laws of physics, particularly those which govern the emission and absorption of light.

SUMMARY

1. Copernicus developed the heliocentric theory because it provided a unifying picture of the universe, not because it was more accurate or simpler than the earth-centered theory. With his new theory, Copernicus was able to calculate the relative distances of the planets from the sun, and he arrived at the correct explanation of the seasons and precession.

2. Tycho Brahe brought forth vast improvements in the quantity and quality of astronomical observations of stars and planets, by making more precise and complete measurements than his predecessors. Tycho was unable to detect stellar parallax, and therefore rejected the heliocentric hypothesis.

3. Kepler sought to discover the underlying harmony among the planets and its physical basis, and in the process discovered that each planet orbits the sun in an ellipse; that a line connecting a planet with the sun sweeps out equal areas in space in equal intervals of time; and that the square of the period of a planetary orbit is proportional to the cube of its semimajor axis.

4. Galileo used telescopic observations and deductive reason to argue for the heliocentric concept of the universe. Despite Church opposition, Galileo brought his ideas before the public in the form of a published fictional dialogue between characters of opposing points of view.

5. Isaac Newton, in the late 1600s, developed the laws of motion, the law of gravitation, and calculus, and made many important contributions to our knowledge of the nature of light and telescopes.

6. Newton's three laws of motion describe the concept of inertia; state that acceleration is proportional to the force exerted on a body and inversely proportional

to the body's mass; and state that for every action, there is an equal and opposite reaction.

7. The law of universal gravitation states that any two objects in the universe attract each other with a force that is proportional to the product of their masses and inversely proportional to the square of the distance between them.

8. Newton's laws, along with the concepts of kinetic and potential energy and angular momentum, can be used to explain orbital motions.

9. Kepler's third law was modified by Newton to show that the relationship between the period and semimajor axis of an orbit is dependent on the sum of the masses of the two objects; this is an important tool for measuring the masses of distant objects.

10. Every body such as a planet or a star has an escape speed, the speed at which a moving object has more kinetic energy than potential energy and will therefore escape into space.

11. Differential gravitational forces are responsible for tides on the earth and in the interiors of other planets and satellites.

REVIEW QUESTIONS

1. In what way did the heliocentric hypothesis of Copernicus provide a more unified overall concept of the universe than the earth-centered theory?

2. Kepler's outlook was similar to that of Plato and Aritotle in some ways, and rather different in others. Briefly discuss the similarities and differences.

3. Suppose there were a planet at a distance of 2 AU from the sun. According to Kepler's harmonic law, what would be its sidereal period?

4. How did Galileo's observational discoveries about the moon and the sun violate the teachings of Plato and Aristotle?

5. Explain, in your own words, why the law of inertia (Newton's first law) implies that the planets must experience a force attracting them toward the sun, and not in some other direction.

6. Assume that Saturn is ten times as far from the sun as the earth, and that it has a hundred times the mass

of the earth. Compare the gravitational force between Saturn and the sun with that between the earth and the sun.

7. If the diameter of the earth were suddenly reduced to one-fourth its present size (while the earth's mass remained constant), how would your weight be altered?

8. Imagine a double star in which star *A* has twice the mass of star *B* . If the two stars are 3 AU apart, how far from star *A* is the center of mass between the two stars?

9. When an ice skater in a spin pulls his arms in close to his body, his spin rate speeds up. Explain why this occurs, in terms of the conservation of angular momentum.

10. How many high tides per day would there be if the earth rotated once every twelve hours instead of once every twenty-four hours.

ADDITIONAL READINGS

The references listed here are primarily sources of biographical and historical data on the people discussed in this chapter; many of them contain additional lists of references. Readings on the principles of physics described in this chapter can most easily be found in elementary physics texts, which are available at all levels ranging from completely nonmathematical to any degree of mathematical sophistication desired

Beer, A., and Beer, P., eds. 1975. *Kepler*. Vistas in Astronomy, vol. 18. New York: Pergamon Press.

Beer, A., and Strand, K. A., eds. 1975. *Copernicus*. Vistas in Astronomy, vol 17. New York: Pergamon Press.

Cohen, I. B. 1960. Newton in light of recent scholarship. *Isis* 51:489.

———.1974. Newton. In *Dictionary of scientific biography*, vol.10, ed. C. G. Gillispie. New York: Scribner's.

Drake, S. 1980. Newton's apple and Galileo's *Dialogue*. *Scientific American* 243(2):150.

Galileo, G. 1632. *Dialogue on the two chief world systems*. Translated by G. de Santillana, 1953. Chicago: University of Chicago Press.

Gingerich, O. 1973. Kepler. In *Dictionary of scientific biography*, vol. 7, ed. C. G. Gillispie, p. 289. New York: Scribner's.

———.1973 Copernicus and Tycho. *Scientific American* 229(6):86.

———.1975. *The nature of scientific discovery*. Proc. of a symposium in honor of Copernicus. Washington, D.C., ed. Smithsonian Institution Press.

———.1983. The Galileo affair. *Scientific American* 247(2):132.

Draper, J. L. E. 1963. *Tycho Brahe: A picture of scientific inquiry in the sixteenth century*. New York: Dover.

Kuhn, T. 1957. *The Copernican revolution*. Cambridge: Harvard University Press.

Whiteside, D. T. 1962. The expanding world of Newtonian research. *History of Science* 1:15.

Learning Goals

Some of the tools for unlocking the secrets of the universe became available with the publication of Newton's *Principia* in the 1680s, but others had to wait two hundred years or more to be discovered. The laws of motion allow astronomers to understand how the heavenly bodies move, and were of fundamental importance in unraveling the clockwork mechanism of the solar system. To understand the essential nature of a distant object, however—to learn what it is made of and what its physical state is—requires an understanding of what light is and how it is emitted and absorbed, and requires tools with which to capture the light and analyze it. The only information we can obtain on the nature of a distant object is conveyed by the light that reaches us from it. Fortunately, an enormous amount of information is there, and astronomers have learned much about how to dig this information out.

The Electromagnetic Spectrum

One characteristic of light is that it acts as a wave (Fig. 3.1). It is possible to think of light as passing through space like ripples on a pond (although, as we will discuss shortly, the picture is actually somewhat more complicated than that). The distance from one wavecrest to the next, called the **wavelength,** distinguishes one color from another. Red light, for example, has a longer wavelength than blue light. It is possible to spread out the colors in order of wavelength, using a prism to obtain the traditional rainbow. Newton was the first to discover that sunlight contains all the colors, and he did so by carrying out experiments with a prism. Whenever light is spread out by wavelength, the result is called a **spectrum;** more technically, a spectrum is the arrangement of light from an object according to wavelength. The science of analyzing spectra is called **spectroscopy,** and will be discussed at some length later in this chapter.

The concept of **frequency** is often used as an alternative to wavelength in characterizing light waves. The frequency is the number of waves per second that pass a fixed point, and it is determined by the wavelength and the speed with which the waves move. The speed of light, usually designated c, is constant, and the frequency of light with wavelength λ is $\nu = c/\lambda$.

The standard unit for measuring frequency is the **hertz (Hz),** one hertz being equal to one wave per second. The frequency of visible light is typically about 10^{14} Hz.

We have discussed light as a wave, but there are situations where it acts more like a stream of particles. Newton developed a corpuscular theory of radiation in which he assumed that light consisted of particles; but at the same time other people, including most notably Huygens, carried out experiments showing light to have definite wave characteristics. There are good arguments for either point of view. For example, the way light waves seem to bend as they pass obstacles and the way they interfere with each other are wave characteristics. On the other hand, it has been found that light waves can carry only discrete, fixed quantities of energy, and it is known that they can travel in a complete vacuum rather than requiring a medium, and these are properties of particles.

Out of a variety of seemingly contradictory evidence has developed the concept of the **photon.** A photon is thought of as a particle of light that has a wavelength associated with it. The wavelength and the amount of energy contained in the photon are closely linked; in general terms, the longer the wavelength, the lower the energy. Thus a red photon carries less energy than a blue one. The energy can be expressed mathematically as $E = hc/\lambda$, where h is called the Planck constant, c is the speed of light, and λ is the wavelength. It is important to understand that a photon carries a precise amount of energy, not some arbitrary or random quantity, and that when light strikes a surface, this energy arrives in discrete bundles like bullets, rather than in a steady stream. When a photon is absorbed, this energy can be converted into other forms, such as heat.

FIGURE 3.1. *Properties of a Wave.*
Light can be envisioned as a wave moving through space at a constant speed, usually designated c. The distance from one wavecrest to the next is the wavelength, often denoted λ. The frequency ν is the number of wavecrests to pass a fixed point per second; it is related to the wavelength and the speed of light by ν = c/λ.

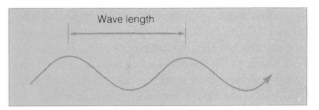

Astronomical Insight 3.1
ELECTROMAGNETIC WAVES AND POLARIZATION

Although we have spoken of the wave nature of light, we have not specified what it is that undulates or oscillates. It is easy to visualize waves in a fluid medium such as water, where we can see the motion of the water was the waves pass by. Electromagnetic waves are quite different, however; no medium is required. Light waves can travel through a total vacuum, unlike most waves with which we are familiar.

Rather than having physical motions, electromagnetic waves consist of electric or magnetic fields that oscillate back and forth at right angles to the direction of motion. The electric and magnetic fields lie in planes that are also perpendicular to each other, so that the electric field flips back and forth in one plane, with the magnetic field doing so in the plane that lies perpendicular to it, and both are transverse to the direction of travel.

A photon, therefore, can be viewed as a packet of electromagnetic energy consisting of alternating fields, which, by virtue of their wavelength and

frequency, provide the photon with its wave properties.

Light from most sources (such as stars) consists of vast numbers of photons, each with specific, constant orientation of its electric and magnetic planes. Ordinarily, the orientation of the various photons is random, but under some circumstances this is not the case. If light passes through or reflects from a medium that preferentially absorbs photons with a certain orientation, what is left is light in which all the photons are aligned; that is, the planes of the electric and magnetic fields of the photons are parallel. In this case, the light is said to be **polarized**. Many of the sunglasses sold today consist of polarizing filters, and they have the effect of screening out light in a given orientation. In astronomy, we will learn that polarization occurs naturally in many situations, sometimes because a natural filtering occurs, as in the interstellar medium (see chapter 18), and sometimes because the source of light emits it with a preferred orientation.

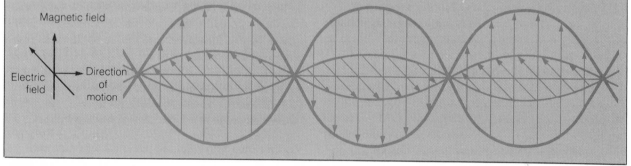

Let us consider for a moment what lies beyond red at one end of the spectrum or beyond violet at the other end. By the mid-1800s experiments had been carried out to demonstrate that there is invisible radiation from the sun at both ends of the spectrum. At long wavelengths, beyond red, is **infrared** radiation, and at short wavelengths is **ultraviolet** radiation. The spectrum continues in both directions, virtually without limit. Going toward long wavelengths, after infrared light, are microwave and radio waves, and toward

short wavelengths, after ultraviolet, come X rays and then γ-rays (gamma rays). All these kinds of radiation are just different forms of light, distinguished only by their wavelengths, and together they form the **electromagnetic spectrum** (Fig. 3.2). Electromagnetic radiation is a general term for all forms of light, whether it is visible, X rays, radio, or anything else.

The range of wavelengths from one end of the electromagnetic spectrum to the other is immense. Visible light has wavelengths ranging from 0.00004 to 0.00007

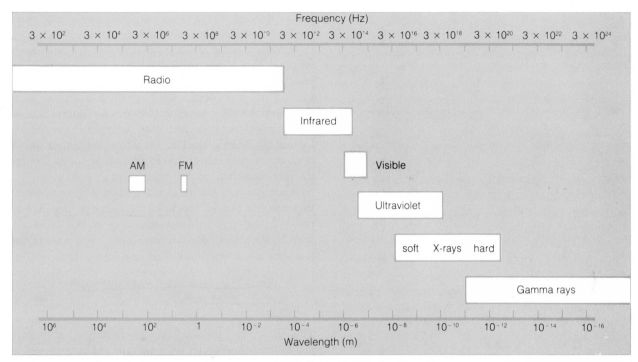

FIGURE 3.2. *The Electromagnetic Spectrum.* All the indicated forms of radiation are identical except for wavelength and frequency.

centimeters. A special unit called the **angstrom** (Å) is used to measure light, defined such that 1 cm = 100,000,000 Å, or 1Å = 0.00000001 cm = 10^{-8} cm. Thus, visible light lies between 4,000 Å and 7,000 Å in wavelength.

Infrared light has wavelengths between about 7,000 Å and a few million angstroms; that is, between 7×10^{-5} cm and $2 - 3 \times 10^{-2}$ cm. Microwave radiation (which includes radar wavelengths) lies roughly between 0.1 and 50 cm, with no well-defined boundary separating this region from radio, which simply includes all longer wavelengths, up to many meters or even kilometers. At the other end of the spectrum, ultraviolet light is usually considered to lie between 100 Å and 4,000 Å, whereas X rays are in the range of 1 to 100 Å, anything shorter than that being considered γ-rays.

Continuous Radiation

It was discovered by researchers who followed Newton that the sun's spectrum contains a number of dark lines, each one corresponding to a particular wavelength. These **spectral lines** provide a great deal of information about a source of light such as a star, and so does the **continuous radiation** (Figs. 3.3 and 3.4), the smooth distribution of light as a function of wavelength. Here we discuss the continuous radiation, and in the following sections the spectral lines.

Any object with a temperature above absolute zero emits radiation over a broad range of wavelengths, simply by virtue of the fact that it has a temperature. Thus, not only stars, but also such commonplace objects as the walls of a room or a human body, emit radiation. The continuous radiation produced because of an object's temperature is called **thermal radiation.** (In this chapter we will restrict ourselves to this type of continuous radiation; later in the book we will discuss nonthermal sources of radiation.)

For glowing objects such as stars, thermal radiation is emitted over a broad range of wavelengths (technically, in fact, at least some radiation is emitted at *all* wavelengths), with a peak in intensity at some particular wavelength. A simple relationship between the wavelength of maximum emission and the temperature of an object was discovered in 1893 by W. Wien, and is now referred to as Wien's law, which states:

The wavelength of maximum emission is inversely proportional to the absolute temperature.

$$\lambda_{max} = W\left(\frac{1}{T}\right)$$

$$W = .29 \text{ cm degree}$$

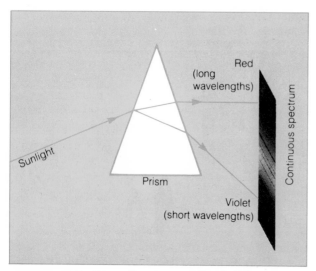

FIGURE 3.3. *Continuous Spectrum and Spectral Lines.*
When sunlight is dispersed by a prism, the light forms a smooth rainbow of continuous radiation, gradually merging from one color into the next, with a maximum intensity in the yellow portion of the spectrum. Superimposed on this continuous spectrum are numerous spectral lines, wavelengths where little or no light is emitted by the sun.

The hotter an object is, the shorter the wavelength of peak emission (Fig. 3.5). This explains the variety of stellar colors as being a result of the range in stellar temperatures. A hot star emits most of its radiation at relatively short wavelengths, and thus appears bluish in color, whereas a cool star emits most strongly at

longer wavelengths and appears red. The sun is intermediate in temperature and color.

When speaking of the colors of stars, we must keep in mind that a star emits light over a broad range of wavelengths, so we do not have pure red or pure blue stars. Our eyes receive light of all colors, and stars therefore are all essentially white. Our impression of color arises from the fact that there is a wavelength (given by Wien's law) at which a star emits more strongly than at other wavelengths. It is not as though the star emits *only* at that wavelength.

A second property of glowing objects, known as either Stefan's law or the Stefan-Boltzmann law, has to do with the total amount of energy emitted over all wavelengths, and how this total energy is related to the temperature of an object. This law says:

> The total energy radiated per square centimeter of surface area is proportional to the fourth power of the temperature.

This shows that the total energy emitted is very sensitive to the temperature; if we change the temperature a little, we change the energy a lot. If, for example, we double the temperature of an object (such as the electric burner on a stove), we increase the total energy it radiates by $2^4 = 2 \times 2 \times 2 \times 2 = 16$. If one star is three times hotter than another, it emits $3^4 = 3 \times 3 \times 3 \times 3 = 81$ times more total energy per square centimeter of surface area.

Notice that we have been careful to express this law in terms of the surface area. The total energy emitted

FIGURE 3.4. *An Intensity Plot of a Continuous Spectrum.*
This kind of diagram shows graphically how the brightness of a glowing object varies with wavelength. The curve roughly represents the sun, whose continuous radiation peaks near 5,500 Å, in the yellow-green portion of the spectrum.

FIGURE 3.5. *Continuous Spectra for Objects of Different Temperatures.*
This diagram illustrates Wien's law, which says that the wavelength of maximum emission is inversely proportional to the temperature (on the absolute scale).

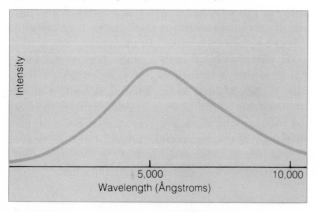

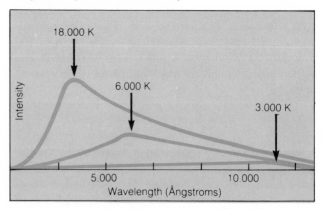

by an object depends also on how much surface area it has. If we talk of stars or other spherical objects, which have a surface area of $4\pi R^2$ (where R is the radius), we can say that the total energy emitted is proportional to the fourth power of the temperature and to the square of the radius. This can be illustrated by considering two stars, one that is twice as hot but has only half the radius of the other. The hotter star emits $2^4 = 16$ times more energy per square centimeter of surface, but has only $\frac{1}{2}^2 = \frac{1}{4}$ as much surface area, hence it is $16 \times \frac{1}{4} = 4$ times brighter overall. If, on the other hand, this star were twice as hot and three times as large in radius as the other, it would be $2^4 \times 3^2 = 144$ times brighter.

Both Wien's law and the Stefan-Boltzmann law were first derived experimentally, much in the same manner as Kepler's discovery of the laws of planetary motion. In the case of planetary motions, it remained for Newton to find the underlying reasons for the laws, and he was able to derive them strictly on a theoretical basis. Analogously, Max Planck, the great German physicist who was active early in the twentieth century, found a theoretical understanding of thermal emission, and was able to derive Wien's law and the Stefan-Boltzmann law from purely theoretical considerations. The basis of Planck's new understanding was the **quantum** nature of light: the fact, already discussed, that light has a particle nature and carries only discrete, fixed amounts of energy.

Finally, in addition to taking into account the temperature and size of a glowing object, we must consider the effect of its distance. So far we have discussed the energy as it is emitted at the surface, but not how bright it looks from afar. What we actually observe, of course, is affected by our distance from the object. For a spherical object that emits in all directions, the brightness decreases as the square of the distance, as specified by the **inverse square law.** (Fig. 3.6). Thus, if we double our distance from a source of radiation it will appear $\frac{1}{2}^2 = \frac{1}{4}$ as bright. If we approach, reducing the distance by a factor of two, it will appear $2^2 = 4$ times brighter. As we will see in later discussions of stellar properties, we must know something about the distances to stars before we can compare other properties having to do with their brightnesses.

The Atom and Spectral Lines

The first detailed cataloging of the sun's spectral lines was carried out in the early 1800s by Joseph Fraunhofer, and the lines are known to this day as **Fraunhofer lines.** In the late 1850s the German scientists R. Bunsen and G. Kirchhoff performed experiments and developed theories that made clear the importance of the Fraunhofer lines. Bunsen observed the spectra of flames created by burning various substances, and found that each chemical element produced light only at specific places in the spectrum (Fig. 3.7). The spectrum of such a flame in this case is dark everywhere except at these specific places, as though the flame were emitting light only at certain wavelengths. The bright lines seen in this situation are called **emission lines** for that reason.

FIGURE 3.6. *The Inverse-Square Law of Light Propagation.* This shows how the same total amount of radiant energy must illuminate an ever-increasing area with increasing distance from the light source. The area to be covered increases as the square of the distance; hence, the intensity of light per unit of area decreases as the square of the distance.

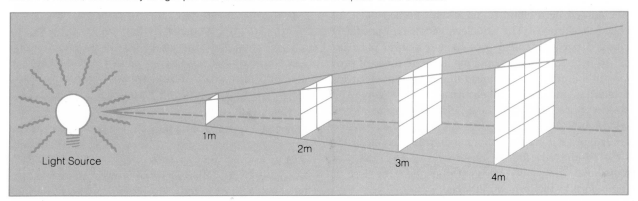

Light Source   1m   2m   3m   4m

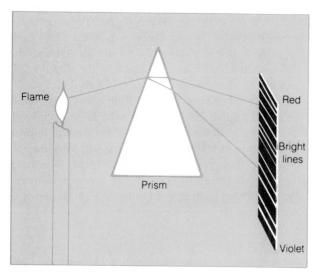

FIGURE 3.7. *Emission Lines.*
The spectrum of a flame is devoid of light except at specific wavelengths, where bright emission lines appear.

It was soon noticed that some of the dark lines seen in the sun's spectrum coincide exactly in position with some of the bright lines seen by Bunsen in his laboratory experiments. Kirchhoff studied this in detail, and was able to show that a number of common elements such as hydrogen, iron, sodium, and magnesium must be present in the sun because of the coincidence in wavelengths of the lines. This was the first hint that the chemical composition of a distant object could be determined.

Further studies of the spectral lines revealed various regularities in the arrangement of the lines for a given element. Although it was suspected that these regularities must reflect some aspect of the structure of atoms, it was not until 1913 that the true relationship was discovered by Niels Bohr. By that time it had been established that an atom consists of a small, dense nucleus surrounded by a cloud of negatively charged particles called electrons. The nucleus contains positively charged particles called protons and neutral particles called neutrons. In a normal atom, the number of protons in the nucleus is equal to the number of electrons orbiting it, and the overall electrical charge is zero. The electrons are held in orbit by electromagnetic forces, with laws of the science called **quantum mechanics** governing their motions. Bohr found that the electrons were responsible for the absorption and emission of light; that they gain energy (absorption) or lose it (emission) in the form of photons.

The key to the fixed pattern of spectral lines for each element lies in the fixed pattern of energy levels the electrons can be in (Fig. 3.8). We can visualize an atom as a miniature solar system, with electrons orbiting the nucleus. Each kind of atom (each element) has its own characteristic number of electrons, and in each case the electrons have a certain set of electron orbits. It may be helpful to visualize a ladder, with each rung representing an orbit, or energy level. The ladder for one element, say hydrogen, has different spacings between the rungs than the ladder for some other element. The energy associated with each level increases, the higher up the ladder, or the farther from the nucleus, the electron goes.

An electron can absorb a photon of light *only* if the photon carries the precise amount of energy needed to move the electron to some higher level than the one it is in. Imagine an atom sitting in space, with light

FIGURE 3.8. *The Formation of Spectral Lines.*
An electron must be in one of several possible orbits, each representing a different electron energy. If an electron absorbs exactly the amount of energy needed to jump to a higher (more outlying) orbit, it may do so. This is how absorption lines are formed, because the wavelength of photon absorbed corresponds to the energy difference between the two electron orbits, which is fixed for any given kind of atom. Conversely, when an electron drops from one orbit to a lower one, it emits a photon whose wavelength corresponds to the energy difference between orbits, and this is how emission lines form.

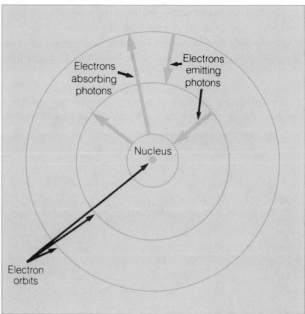

streaming past it: the atom will let pass every photon that comes by, until it finds one with just the right amount of energy to boost one of its electrons to a higher level. Since the energy of a photon is closely related to its wavelength, this means that the atom will only absorb photons that have specific wavelengths, corresponding to the spacings between its energy levels. Because each kind of atom has its own unique set of energy levels, each can absorb light only in its own unique set of wavelengths, forming **absorption lines**. Thus Bohr was able to explain the findings of Bunsen and Kirchhoff.

Emission is the reverse process of absorption. Emission will occur whenever an electron in an upper level drops down to a lower one, and the energy of the photon emitted corresponds, as before, to the energy difference between the two levels. Since the energy levels are fixed for a given element, the emission lines and the absorption lines occur at the same wavelengths. Whether an atom produces emission lines or absorption lines depends simply on whether the electrons are moving downward or upward.

We can now state three rules of spectroscopy, first discovered experimentally by Kirchoff (Fig 3.9) and later put in the context of atomic structure:

1. In a hot, dense gas or a hot solid, the atoms are crowded together so that their energy levels overlap

and their lines are all blended together,[1] and we see a continuous spectrum.

2. In a hot, rarefied gas, the electrons tend to be in high energy states, and create emission lines as they drop to lower levels.

3. In a relatively cool gas in front of a hot, continuous source of light, the electrons tend to be in low energy levels, and absorb radiation from the background continuous source.

Deriving Information from Spectra

A great wealth of information is stored in the spectrum of an object. By analyzing the spectral lines, we can learn such quantities as the temperature and density of a gas, and the velocity of the glowing object with respect to the earth.

One important property of a gas is its degree of **ionization** (Fig. 3.10). Ionization refers to the loss of one or more electrons from an atom, when the electrons receive so much energy that they jump entirely free of the atom. The gain in energy that frees an electron can come in either of two ways: it can be caused by the

[1]The production of a continuous spectrum is actually a bit more complex than stated here. When free electrons combine with ions, they emit at any wavelength, not just in spectral lines. See the discussion of ionization in the following section.

FIGURE 3.9. *Continuous, Emission-Line, and Absorption-Line Spectra.* The positions of the emission and absorption lines match because the same element emits or absorbs at the same wavelengths. Whether the element emits or absorbs depends on the physical conditions, as described in Kirchhoff's laws.

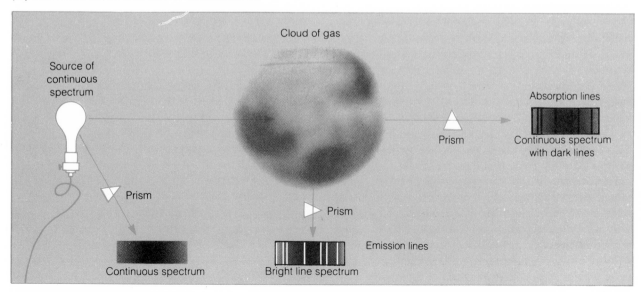

absorption of a photon with energy exceeding that of the highest electron energy level associated with the atom; or it can be caused by a collision between atoms. In either case, the remaining atom has a net positive electrical charge, having lost one or more negative charges, and is called an **ion.** The likelihood of collisional ionizations depends on the temperature of the gas, since the speed of collision depends on temperature. The hotter the gas, the more violent the collisions between atoms, and the greater the degree of ionization.

The spectrum of an atom changes drastically when it has been ionized, because the arrangement of energy levels is altered, and because different electrons are now available to do the absorbing and emitting of photons. The spectrum of atomic helium, for example, is quite different from that of ionized helium, so the astronomer not only can see that helium is present in the spectrum of a star, but also can tell whether it is ionized. This provides information on the temperature in the outer layers of the star, where the absorption lines are formed. By analyzing the degree of ionization of all the elements seen in the spectrum of a star, the astronomer can determine the gas temperature quite precisely.

When an electron moves to an energy state above the lowest possible one, it is said to be in an **excited**

state (Fig. 3.10). This can happen as the result of the absorption of a photon of light with appropriate energy, or as the result of a collision that is not energetic enough to cause ionization. The degree of excitation of a gas affects its spectrum, because the spectral lines created in the gas depend on which energy levels the electrons are in. A hydrogen atom with its electron in the first excited level can produce absorption lines corresponding to the energy separation between this excited level and higher levels, whereas a hydrogen atom with its electron in the lowest state can only produce absorption lines corresponding to transitions of the electron out of that state (Fig. 3.11). Therefore, analysis of the spectrum of a gas can tell us how highly excited the gas is, which in turn can provide information on other properties, such as density (the density affects the

FIGURE 3.11. *The Effect of Excitation on the Spectrum of Hydrogen.*
As shown here, the absorption lines that can be formed by an atom depend on the atom's degree of excitation. At left are the wavelengths of absorption lines originating in the lowest energy level of hydrogen, and at right are the wavelengths of absorption lines arising from an electron in the first excited level. Note that the Lyman lines are in ultraviolet wavelengths, whereas the Balmer lines are in the visible portion of the spectrum. This sketch shows only a few of the many energy levels of hydrogen.

FIGURE 3.10. *Excitation and Ionization.*
In the energy-level diagram at left, electrons are jumping up from the lowest level to higher ones. This is excitation, and it can be caused either by absorption of photons or by collisions between atoms. On the right, electrons are gaining sufficient energy to escape altogether, a process called ionization.

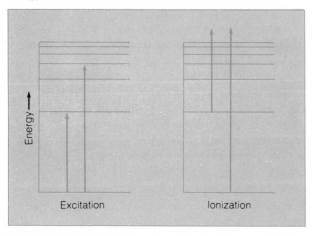

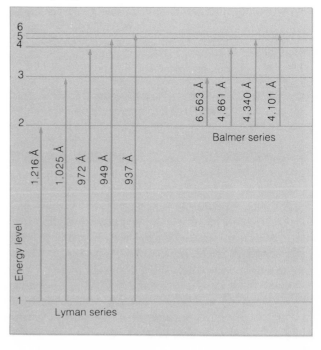

frequency of collisions between atoms, which in turn determines how many electrons are in excited levels).

From careful analysis of a star's spectrum, then, it is possible to determine both the temperature of the outer layers of the star (from the ionization) and the density (from the excitation). Although most of the examples used in this discussion have referred to the sun or the stars, the same considerations can be applied to the spectra of the planets. Because a planet glows only by reflected sunlight, however, its spectrum is the same as the sun's spectrum, except that some additional features are added by the gas in the planet's atmosphere. These extra spectral lines must be identified and analyzed if the properties of this gas are to be derived.

There is something else that spectral lines can tell us about a distant object, in addition to all the physical data we have been discussing. We can also learn how rapidly an object such as a star or a planet is moving toward or away from us.

When a source of light such as a star is approaching an observer, all the lines in its spectrum are shifted toward shorter wavelengths (than if the light source were at rest); and when the light source is moving away, all the lines are shifted toward longer wavelengths (Fig. 3.12). These two cases are called **blueshifts** (approach) and **redshifts** (recession), respectively, because the spectral lines are shifted toward either the blue or the red end of the spectrum.

A general term for any wavelength shift caused by relative motion between source and observer is **Doppler shift,** in honor of the German physicist who first explored the properties of such shifts. The effect does not apply just to light waves. Most of us, for example, have noticed the Doppler shift in sound waves when a source of sound passes by. The whistle on an approaching train will suddenly change to a lower pitch at the moment the train passes by us, because the wavelength we receive suddenly shifts to a longer one.

With the Doppler shift of light, we can determine the speed with which the source of light is approaching or receding, from the formula:

$$v = (\Delta\lambda/\lambda)\, c$$

where v is the relative velocity between source and observer, $\Delta\lambda$ is the shift in wavelength (the observed wavelength minus the rest, or laboratory, wavelength of the same line), λ is the laboratory wavelength of the line, and c is the speed of light. If the observed wave-

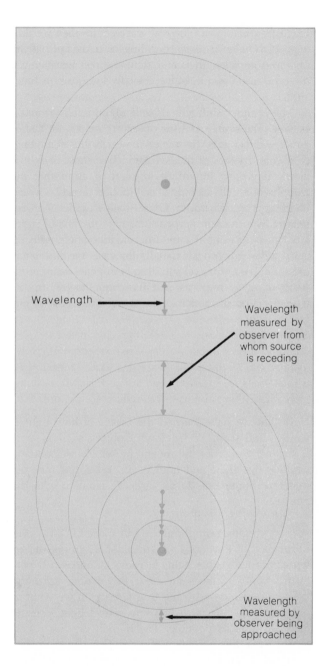

FIGURE 3.12. *The Doppler Effect.*
The light waves from a stationary source (upper) remain at constant separations (that is, constant wavelength) in all directions, whereas those from a moving source (lower) get "bunched up" in the forward direction and "stretched out" in the trailing direction. This causes a blueshift or a redshift for an observer who approaches or recedes from a source of light. Note that it does not matter which is moving, source or observer.

length is greater than the rest wavelength, then we find a positive velocity, corresponding to a redshift; if the observed wavelength is less than the rest wavelength, then the result is a negative velocity, indicating a blueshift.

The Doppler shift tells us only about relative motion between the source and the observer: we cannot distinguish whether it is the star or the earth that is moving, or a combination of the two (which is most likely the case). It is also important to keep in mind that the Doppler shift tells us only about motion directly toward or away from the earth. There is no Doppler shift resulting from motion perpendicular to our line of sight. If a star is moving at some intermediate angle with respect to the earth, as is usually the case, we can determine the part of its velocity that is directed straight toward or away from us, but we cannot determine its true direction of motion or its speed transverse to our line of sight.

The Principles of Telescopes

Now that we have learned the basics of how light is emitted and absorbed by glowing objects such as stars, it is time to discuss the means by which astronomers capture this light and analyze it. Telescopes are the material monuments to our great fascination with the heavens.

Although we will discuss a wide variety of telescopes, ranging from those used in space for X-ray and ultraviolet observations, to traditional ground-based telescopes for visible light, to radio antennae, all perform essentially the same task: they gather as much light as possible and bring it to a focus. If we want to observe faint objects, it helps to collect light from as large an area as possible. The human eye has a collecting area only a fraction of a centimeter in diameter, whereas the largest telescopes are several meters in diameter.

Another basic difference between the eye and the telescope is that the telescope can be equipped to record light over a long period of time, through the use of photographic film or a modern electronic detector, whereas the eye has no capability for storing light. A long-exposure photograph taken with a telescope allows us to obtain pictures of objects that we could not see even when looking through the same telescope.

Both their large size and the capability of making long exposures makes it possible for the largest telescopes to detect objects some 100,000,000 times fainter than what we can see with the unaided eye, even under the best conditions.

A third major advantage of building large telescopes is that they have superior **resolution,** the ability to discern fine detail. For visible-wavelength telescopes, as we will see, the earth's atmosphere creates practical limitations on how fine the resolution can be, but for radio telescopes and optical telescopes in space, the atmosphere is not a hindrance. We will discuss resolution in more detail shortly.

One problem must be overcome in making long exposures: the earth rotates, and if nothing were done to compensate for this, a star would quickly move out of the field of view. So that this problem is avoided, telescopes are mounted so that they can be moved by a motor in the direction opposite the earth's rotation, keeping a target object centered in the field of view. The telescope mounting usually allows the telescope to move independently in declination (north-south motion) and right ascension (east-west motion), so the drive mechanism that compensates for the earth's rotation has to operate only in the east-west direction. The need to do this, of course, is avoided for telescopes that are not attached to the earth, such as the various orbiting observatories (discussed later in the chapter).

Although it is possible to construct a telescope, called a **refractor,** that brings light to a focus using lenses (Fig. 3.13)—and many of the earliest telescopes were of this type—today nearly all telescopes are **reflectors,** using mirrors to focus light. Reflectors can be built much larger, and are not hampered by certain technical difficulties associated with lenses, such as the

FIGURE 3.13. *The Refracting Telescope.*
The path of light is bent when it passes through a surface such as that of glass. A properly shaped lens thus can bring parallel rays of light to a focus, where the image of a distant object can be examined. A second lens is used to magnify the image.

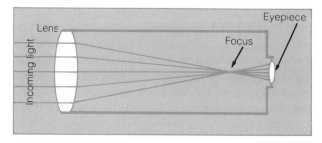

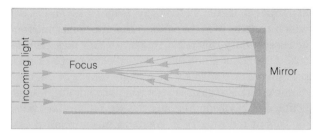

FIGURE 3.14. *The Reflecting Telescope.*
A properly shaped concave mirror can be used instead of a lens to bring light to a focus. Usually an additional mirror is used to reflect the image outside the telescope tube.

fact that different wavelengths of light are refracted at different angles by a lens.

In a reflector, light is focused by a concave mirror, called the **primary mirror** (Figs. 3.14 and 3.15). Because it is usually inconvenient to look at or record an image formed inside the telescope tube, there is most often a secondary mirror that deflects the image outside of the tube (Fig. 3.16). This can be a flat mirror, sending the image out the side (the **Newtonian focus**); or a convex mirror that reflects the image straight back out of a hole in the bottom of the telescope (the **cassegrain focus**); or a series of flat mirrors that send the image to a remote location where heavy or large equipment can be used to analyze the light (the **coudé**

FIGURE 3.15. *A Large Primary Mirror.*
This photo shows that the proper shape for a telescope mirror is not very highly concave. In this case only the distorted reflections reveal that the mirror is not flat. This is the 2.3-meter mirror of the University of Wyoming's infrared telescope.

focus). In very large telescopes, it is possible to observe at the **prime focus**, in the point where the light is focused inside the telescope structure by the primary mirror. This has the advantage of reducing the number of reflections, so that light loss is minimized, but the

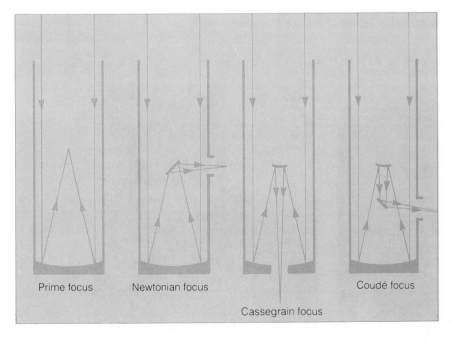

Prime focus   Newtonian focus   Coudé focus

Cassegrain focus

FIGURE 3.16. *Various Focal Arrangements for Reflecting Telescopes.*

FIGURE 3.17. *The Five-Meter Telescope.*
This is the largest telescope in the United States., the 200-inch instrument at Mt. Palomar, in southern California.

disadvantage that the observer must spend the night in a tiny, cramped cage mounted inside the telescope. All the focal arrangements have some structure, either a mirror or an observer's cage, mounted inside the telescope, blocking some of the incoming light, but the percentage of light that is lost this way is usually very small.

The biggest telescope for many years was the 200-inch (5-meter) reflector at Mt. Palomar, in southern California (Fig. 3.17), but recently a 6-meter telescope was completed in the Soviet Union (Fig. 3.18).

A very promising new kind of telescope was completed in 1979 at Mt. Hopkins, Arizona. This is the **Multiple-Mirror Telescope** (Figs. 3.19 and 3.20), which consists of six separate 72-inch primary mirrors, with secondary mirrors arranged so that they each focus at the same point. The total collecting area of these six mirrors is equivalent to a single mirror of 176-inch diameter. This system is very complicated, and the difficulties of perfectly aligning the six mirrors are enormous, but the savings in cost and in the relative ease with which the smaller mirrors can be constructed make the effort worthwhile.

In all modern telescopic observations, instruments other than the human eye are used to collect and analyze light. These instruments can be simple cameras, which make images of the sky; **photometers,** which measure intensities of light using photocells (devices that convert light into electrical currents); or **spectrographs,** which disperse light according to wavelength so that the spectrum can be recorded and analyzed. In all cases the light must be recorded. Traditionally, film has been (and still is) used for this, but today many kinds of electronic **detectors** are being developed. These have many advantages over film, both in their accuracy and in such practical matters as storing the information electronically, so that it can be transmitted directly into computers for analysis. The auxiliary instruments used in astronomy range from rather compact, lightweight devices that can easily be mounted directly on the telescope for use at the cassegrain focus, to bulky, large instruments (usually spectrographs) that must be used at the coudé focus.

FIGURE 3.18. *The Six-Meter Telescope.* This is the world's largest telescope, located in the Caucasus Mountains of the southern central Soviet Union.

FIGURE 3.19. *The Multiple-Mirror Telescope, Mt. Hopkins, Arizona.* The observatory at Mt. Hopkins is operated by the University of Arizona and the Smithsonian Institution.

Observatories on the Ground

Most major observatories are located in remote regions, for a number of reasons. It is best to minimize absorption of light by the atmosphere, so an observatory should be situated on a high mountain, above as much of the atmosphere as possible. Of course, a site with generally clear weather is imperative, for only radio telescopes can peer through clouds. It is also desirable to build observatories well away from large cities, because the diffuse light from a densely populated area can drown out faint stars. Another consideration is the latitude of the site, which should not be too far north or south, since that would exclude a large portion of the sky.

Other factors that affect the quality of an observing site are the cleanliness of the air, for pollution can reduce the transparency of the atmosphere; and the amount of turbulence in the air above the site, because turbulence creates the twinkling effect known to astronomers as "seeing." This twinkling, which tends to smear out an image, reduces the quality of photographs or spectra. On a good night, the diameter of a star image is 1 or 2 arcseconds, whereas when the seeing is bad, images may blow up to diameters of 5 arcseconds or more.

The largest observatories in the world generally are located along the western portions of North and South America, in Hawaii (Fig. 3.21), and in Australia, al-

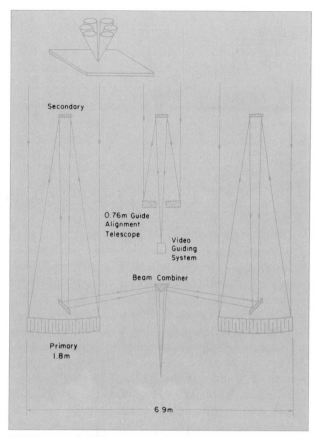

FIGURE 3.20. *A Schematic of the Multiple-Mirror Telescope Design.* This drawing shows how the images from two of the six mirrors are brought to a common focus.

FIGURE 3.21. *The Mauna Kea Observatory.*
Located atop Mauna Kea at an elevation of nearly 14,000 feet, this observatory has become one of the leading astronomical sites in the world. The total collecting area of all its telescopes is greater than that at any other site. Mauna Kea is considered a likely location for the next large instrument.

Astronomical Insight 3.2

A NIGHT AT THE OBSERVATORY

A typical astronomer's night at the observatory will vary quite a bit from one individual to another, and especially from one observatory to another. There is a big difference between sitting in a heated room viewing an array of electronics panels and standing in a cold dome peering into an eyepiece while recording data on a photographic plate, yet both extremes (and various possibilities in between) are part of modern observational astronomy.

Despite apparent differences in practice, however, the principles are very much the same, regardless of where the observations take place, or with what equipment. Here is a brief outline of a typical night at a large observatory, where a modern electronic detector is used to record spectra of stars.

By mid-afternoon the astronomer and the assistants (an engineer who knows the detector and its workings, and a graduate student) are in the dome, making sure the detector is working properly. To do this requires various tests of the electronics, and the recording of test spectra from special calibration lamps. Everything in order, they eat supper in the observatory dining hall a couple of hours before sunset, then return to the dome to complete preparations for the night's work. A night assistant (an observatory employee) arrives shortly before dark and is given a summary of the plans for the night. It is the night assistant's job to control the telescope and the dome, and he is responsible for the safety of the observatory equipment. At large observatories,

the telescope is considered so complex and valuable that a specially trained person is always on hand so that the astronomer (who may use a particular telescope only a few nights a year) does not handle the controls, except to make fine motions required to maintain accurate pointing during the observations.

Before it is completely dark, a number of calibration exposures are taken, using lamps and even the moon, if it is up. These measurements will help the astronomer analyze the data later, by providing information on the characteristics of the spectrograph and the detector.

Finally, everything is in order, and if things are going smoothly, the telescope will be pointed at the first target star just as the sky gets dark enough to begin work. Observing time on large telescopes is difficult to obtain and is very valuable; not a moment is to be wasted.

A pattern is quickly established and repeated throughout the night. First, after the night assistant has pointed the telescope at the requested coordinates, the astronomer looks at the field of view to confirm which star is the correct one (this is often done by consulting a chart prepared ahead of time). When the star is properly positioned so that its light enters the spectrograph, the observation begins. Depending on the brightness of the star and the efficiency of the spectrograph and the detector, the exposure time may be anywhere from seconds to hours. During this time, the astronomer or the grad-

though the biggest single telescope is in the Caucasus Mountains of the southern central Soviet Union. The U.S. government, through the National Science Foundation, supports large observatories in North America (Kitt Peak National Observatory, about fifty miles west of Tucson, Arizona) and South America (the Cerro Tololo Inter-American Observatory, on the crest of the Andes, some 200 miles north of Santiago, Chile).

Radio Telescopes

Radio emission from celestial sources was discovered as long ago as 1931 by the American scientist Karl Jansky (Fig. 3.22). However, it was not until after World War II that radio astronomy became established as a science. Today it is an integral part of astronomy, and a number of major radio observatories exist in many

uate-student assistant continually checks to see that the star is still properly positioned, making corrections as needed by pushing buttons on a remote-control device.

If the observations are made through the use of an instrument at the coudé focus, the astronomer and assistants spend the entire night in an interior room in the dome building, emerging occasionally only to view the skies, give instructions to the night assistant, or go to the dining hall for lunch (usually served around midnight).

As dawn approaches and the sky begins to brighten, a last-minute strategy is devised to get the most out of the remaining time. Finally, the last exposure is completed, and the night assistant is given the OK to close the dome and park the telescope in its rest position (usually pointed straight overhead,

to minimize stress on the gears and support system). The astronomer and assistants make final calibration exposures before shutting down their equipment and trudging off to get some sleep. The new day for them will begin at noon or shortly thereafter.

This scenario depicts a typical night at a major modern observatory, such as Kitt Peak, which has extensive resources and the largest telescopes. At smaller facilities, such as a university observatory, it is more common to find the astronomer working alone, most often using a cassegrain-focus instrument. In this case the observer must spend the night out in the dome with the telescope (where it can be very cold in the winter), and be content to eat a homemade lunch. The work pattern is similar, however, as is the desire not to waste any telescope time.

A typical night at a large, modern observatory.

parts of the world. The radio portion of the spectrum is being explored as fully as the visible wavelengths, and a great wealth of new information about the universe is being collected. Radio observatories have some practical advantages. For instance, the daytime sky is dark in radio wavelengths, so radio observatories need not shut down when the sun rises; in fact, most are on a 24-hour operating schedule. We will also see that radio telescopes can achieve much better **resolution**

(the ability to distinguish fine detail) than the current optical telescopes, so we will be able to obtain much more detailed pictures of celestial objects.

A radio telescope works much like a reflecting telescope for visible wavelengths, except that generally everything is on a larger scale (Fig. 3.23). The primary reflecting element need not be a smooth surface, but instead is usually a wire mesh. To radiation with such long wavelengths, a mesh acts like a shiny surface, as

FIGURE 3.22. *Karl Jansky With the Radio Telescope He Used to Discover the First Celestial Radio Source.*

FIGURE 3.23. *An Assortment of Radio Telescopes.* These antennae range in diameter from 26 meters to 91 meters, and are part of the U.S. National Radio Astronomy Observatory in Green Bank, West Virginia.

long as the wavelength observed is much longer than the spacing between wires in the mesh. At a point centered above the radio antenna, at what would be called the prime focus in a visible-wavelength reflector, is placed a **receiver,** a device that measures the intensity of the radiation. This receiver usually can be tuned to screen out unwanted wavelengths, serving in a sense as a filter. The intensity data can then be stored directly in a computer for later analysis.

A radio antenna must be very large in diameter to produce clear pictures of the sky. The resolution of a telescope is usually expressed in terms of how close together two objects can be and still be seen as separate objects, and is determined by the ratio of the wavelength being observed to the diameter of the telescope. Thus if long wavelengths, such as radio emission, are to be observed, the telescope must be very large in order to yield good resolution (sharp images with fine detail). A visible-wavelength telescope with a diameter of 10 centimeters (about 4 inches) achieves a resolution of about 1 arcsecond, but a radio telescope would need to have a diameter of several kilometers to achieve comparable resolution. Therefore even the largest radio telescopes (up to around 100 meters in diameter) have resolutions of several minutes of arc, and a radio picture of the sky tends to look fuzzy, with all the fine details smeared out.

A solution to this problem has been found and is being exploited at several radio observatories. By using two or more radio antennae simultaneously, astrono-

mers can simulate the effect of one very large antenna. The farther apart the antennae are, the better the resolution. Whereas a visible-wavelength telescope looking through the earth's atmosphere typically has a resolution of 1 to 2 seconds of arc (that is, two stars about 1 or 2 arcseconds apart can be distinguished as separate objects), radio telescopes used in combination have managed resolutions of a small fraction of an arcsecond. This technique, called **interferometry,** is obviously very powerful, but it is also very complex, requiring precise knowledge of when particular radio photons arrive at the separate antennae. Radio telescopes at various separations ranging from a few hundred feet to many thousands of miles have been employed. Imagine the complexity involved in simultaneously using radio telescopes on opposite sides of the Atlantic!

The largest radio observatory in the United States is the National Radio Astronomy Observatory, sponsored by the National Science Foundation, with a 300-foot antenna at Green Bank, West Virginia; a 30-foot dish at Kitt Peak (used mostly for observing interstellar molecules, whose emission lines occur at relatively short radio wavelengths; see chapter 18); and a new observatory, called the *Very Large Array* (VLA), near Socorro, New Mexico (Fig. 3.24). The VLA has 27 movable 82-foot antennae arranged in a Y-shaped pattern spanning some 15 miles of the New Mexico desert, and is designed for interferometry, ultimately achieving a resolution of better than 0.1 arcseconds. Another large U.S. radio observatory, at Arecibo, Puerto Rico, consists of a 1,000-foot dish built into a natural bowl in the mountains. This large diameter provides fairly high resolution and sensitivity, but since the antenna cannot be pointed (it is immovable), sky coverage is limited in the north-south direction. The earth's rotation allows full sky coverage in the east-west direction.

Telescopes for All Wavelengths

As we saw in our discussion of the emitting properties of glowing objects, the radiation from a star or planet extends beyond the limits detectable by the human eye. It also extends beyond the limits set by the earth's atmosphere, which absorbs all wavelengths shorter than about 3,100 Å, and absorbs in several wavelength regions in the infrared as well. Some portions of the infrared can penetrate to the ground, so infrared astronomy can be done at the major observatories (two of the four large telescopes on Mauna Kea, for example, were built specifically for infrared observations, as was the large telescope at the University of Wyoming). To cover

the entire infrared spectrum or any of the ultraviolet, we must, however, place telescopes in orbit above the earth's atmosphere. This is accomplished through the use of high-altitude planes or balloons (which do not quite get above all of the atmosphere), sounding rockets (which provide only five minutes or so of observing time), or satellites.

Simply putting an infrared telescope above the atmosphere does not solve all the problems. As we saw in our discussion of Wien's law, a telescope at room temperature glows at infrared wavelengths, and this can drown out the infrared light from a target object. We solve this problem, on the ground or in space, by cooling the telescope, or at least the associated instruments, so that the emission is shifted to much longer wavelengths and does not confuse the observations. The cooling is usually done by circulating liquid nitrogen (which has a temperature of 77 K) or liquid helium (4 K) through a series of tubes surrounding the instrument.

Until recently infrared astronomy was carried out from the ground whenever possible, and also from high-altitude balloons and planes. Only in the 1980s have there been satellite missions for this purpose. The *Infrared Astronomical Satellite* (usually simply called *IRAS;* Fig. 3.25), designed to map the sky in infrared

FIGURE 3.24. *The Very Large Array (VLA).*
This is an overall view of the VLA, showing the 27 antennae (each about 25 meters in diameter) in their Y-shaped arrangement. The longer arms of the Y are 21 kilometers long, and the antennae can be moved along rails to achieve optimum separation for interferometric observations.

FIGURE 3.25. *The Infrared Astronomy Satellite (IRAS).*
This satellite carries liquid helium to keep the instrument cool, so that it can observe the universe at long infrared wavelengths. A number of important discoveries on the structure of our galaxy and the nature of others have been made.

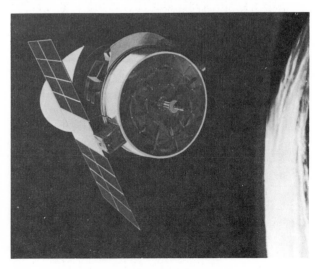

wavelengths, was launched in 1983, and has been a huge success, mapping thousands of very cool objects whose presence was unknown from visible-wavelength observations. Later in the 1980s the *Shuttle Infrared Telescope Facility* (*SIRTF*), a large infrared telescope with many auxilliary instruments, will be flown. Both *IRAS* and *SIRTF* have on-board liquid-cooling systems, a very difficult technological achievement.

Ultraviolet astronomy began in the late 1940s with sounding-rocket observations of the sun, and has progressed steadily ever since, with the development of a series of satellite observatories for ultraviolet spectroscopic observations. The principles on which an ultraviolet telescope operates are the same as for a visible-light telescope, except that special coatings must be used to preserve the reflectivity of the mirrors. Currently one ultraviolet spacecraft, the *International Ultraviolet Explorer* or *IUE* (Fig. 3.26) is in operation, and the *Space Telescope* (described briefly later in this chapter) will have a variety of ultraviolet capabilities.

Extremely hot gases, according to Wien's law, emit most strongly at very short wavelengths. Temperatures of a million degrees or more produce emission in the X-ray portion of the electromagnetic spectrum, between 1 Å and 100 Å. There are even spectral lines in this wavelength region, produced by electrons jumping between orbits with very large energy differences.

It is not simple to build an X-ray telescope, however, because most materials are not shiny enough for X rays; a mirror used in the usual way simply absorbs X rays rather than reflecting them. It is possible, though, to reflect X rays if the radiation strikes the reflecting surface at a very oblique angle, and X ray telescopes can be designed using this trick (Fig. 3.27). Sometimes several reflections are necessary to bring the X rays to a focus, because only a small deflection of the radiation can be accomplished at each grazing-incidence reflection. The process of dispersing X rays by wavelength is also more complicated than for visible light, again because most substances absorbs X rays.

Despite the difficulties, a number of rocket and satellite observatories have successfully probed the heavens in X-ray wavelengths, discovering a wide range of phenomena. Sites of violent explosions and other highly energetic regions are the main sources of X rays, so these space-borne observatories have opened a new universe for astronomers, one whose existence was only hinted at by more conventional observations.

Currently X-ray astronomy is in a lull between major instruments, but within a few years the *Advanced X-*

FIGURE 3.26. *The International Utraviolet Explorer (IUE).* This NASA-operated spacecraft has been in operation since 1978, obtaining ultraviolet spectroscopic data for hundreds of astronomers.

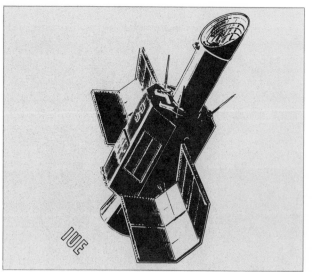

FIGURE 3.27. *A Typical X-Ray Telescope Design.* The radiation is brought to a focus by a series of grazing-incidence reflections; that is, reflections at oblique angles. Several mirrors may be "nested" concentrically, to increase the light-gathering power.

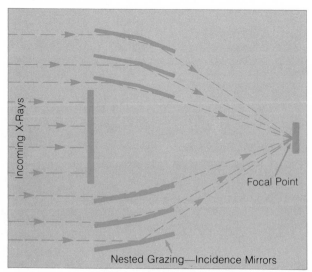

Incoming X-Rays

Focal Point

Nested Grazing—Incidence Mirrors

ray *Astrophysics Facility* (AXAF) will be launched, with a variety of X-ray detectors.

The shortest wavelengths of electromagnetic radiation, the γ-rays, also require special techniques for detection. Here even grazing-incidence reflections will not work, and as yet no instrument has been developed that can bring γ-rays to a precise focus. So far only rather primitive γ-ray telescopes have been flown, but plans are being made for a major satellite mission called the *Gamma Ray Observatory*, which promises to provide important new data on the extremely violent regions in space where nuclear reactions are taking place, such as inside stars and in supernova explosions.

The Next Generation

What kinds of telescopes will be developed? We can speculate about this; astronomers have a flair for dreaming up new and better ways to overcome technical difficulties, so in some cases plans for innovative instruments are already well developed.

There has been widespread discussion of a *Next Generation Telescope* (Fig. 3.28), a visible wavelength telescope with a diameter of 10 to 15 meters (remember, the largest telescope today is the Russian 6-meter one). It is unlikely that such a large mirror can be produced as a solid surface with the required precision, so two novel alternatives are being considered. One, the multiple-mirror idea, has already been put into successful operation on a smaller scale, as described earlier in this chapter. The other concept, not yet well tested, is to build a single mirror of a relatively thin, flexible metal, and place behind it a number of automatically activated pushrods that will bend the mirror into the proper shape, compensating for sag caused by gravity as the telescope is pointed in different directions. This *rubber mirror* concept obviously is complicated in practice, but it may be feasible. A merging of the two concepts, called a *segmented mirror*, may eventually prove best. In this design, a number of hexagonal mirrors are constructed and shaped separately, then mounted together in a large honeycomb-like structure, with each segment being adjustable in keeping all the light focused at a single point.

Many astronomers are looking forward to the *Space*

FIGURE 3.28. *The Next-Generation Telescope.* These models illustrate the two leading possibilities for design of the planned 15-meter telescope. One is a segmented-mirror design (left; the segments in the primary mirror cannot be seen in this view), and the other is a multiple-mirror concept (right; four 7.5-meter mirrors are shown here). At center is the existing 4-meter telescope at Kitt Peak National Observatory, to the same scale.

Telescope (Fig. 3.29), a 2.4-meter orbiting telescope that will have with it a variety of cameras, photometers, and spectrographs for visible and ultraviolet wavelengths. This facility, scheduled for launch on the Space Shuttle in the late 1980s, is intended to have a long lifetime, since it will be possible for astronauts to visit it occasionally to make repairs. It is even planned that the Space Telescope will be able to return to earth for repairs or refurbishing of the instruments. In addition to providing opportunities for ultraviolet observations, the Space Telescope will have greater resolution and be more sensitive in visible wavelengths than ground-based telescopes, because it will avoid problems introduced by the earth's atmosphere.

Even farther in the future, we might expect someday to see a very large visible- and ultraviolet-wavelength telescope in space, to capitalize further on the advantages of getting above the atmosphere. Speculative discussions are under way regarding an unmanned interstellar mission, in which a probe would be launched toward alpha Centauri (the nearest star) and would radio back information on interstellar gas and dust, as well as data on alpha Centauri itself, as seen from close up. Such a mission would take many years since alpha Centauri is four light-years away, so the scientists waiting on earth will have to be patient.

PERSPECTIVE

We have now learned how light is absorbed and emitted by natural processes, and how it is gathered and analyzed by various devices. We have seen how the continuous spectrum and overall intensity of light from a star depend on the star's temperature and size, and how the spectral lines provide information on temperature, density, motion, and chemical composition of a star.

Telescope technology allows astronomers to derive a remarkable amount of information about distant objects, something we will come to appreciate more and more in the chapters that follow. We are ready now to explore the universe, to see what has been learned with the techniques we have been discussing.

SUMMARY

1. Visible light is just one part of the electromagnetic spectrum, which extends from the γ-rays to radio wavelengths.

2. Light has properties associated with both waves and particles, and these aspects are combined in the concept of the photon.

3. The continuous radiation from a star provides information on the temperature, luminosity, and radius of a star, through the use of Wien's law, the Stefan-Boltzmann law, and the more general Planck's law.

4. The observed brightness of a glowing object is inversely proportional to the square of the distance between the object and the observer.

5. Spectral lines are produced by transitions of electrons between energy levels in atoms and ions. Absorption occurs when an electron gains energy, and emission occurs when it loses energy, the wavelength in both cases corresponding to the energy gained or lost as the electron changes levels.

FIGURE 3.29. *The Space Telescope.*
This is an artist's sketch of the Space Telescope in its operating configuration. The open end of the 2.4-meter telescope is pointed away from us in this view; the near end houses control systems, cameras, and other scientific instruments. The large "wings" are solar panels, which provide electrical power for the spacecraft. A Space Shuttle is seen in the distance.

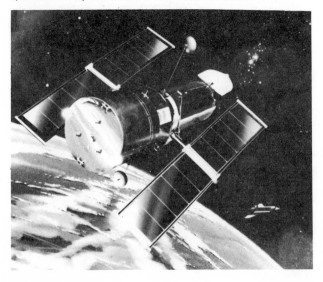

6. Each chemical element has its own distinct set of spectral-line wavelengths, and therefore the composition of a distant object can be determined from its spectral lines.

7. The ionization and excitation of a gas can be inferred from its spectrum, the former yielding information on the temperature of the gas, and the latter providing data on its density.

8. Any motion along the line of sight between an observer and a source of light produces shifts in the wavelengths of the observed spectral lines. Measurement of this Doppler effect, as it is called, can be used to determine the relative speed of source and observer.

9. Telescopes for visible light can use lenses or mirrors to bring light to a focus, but all large telescopes are reflectors.

10. Reflecting telescopes can bring light to a focus in a variety of ways for different purposes.

11. Auxiliary instruments are used to analyze the light once it is brought to a focus, by photographing the image, measuring its brightness, or dispersing the light for spectral analysis.

12. An observatory site must be chosen for clear weather, high atmospheric transmissivity, low turbulence in the air overhead, minimal pollution, remoteness from city lights, and the proper latitude for viewing the desired part of the sky.

13. Radio telescopes are similar in design to reflecting telescopes for visible light, but must be built very large in order to provide good resolution.

14. Infrared and ultraviolet observations can be made through the use of the same optical designs as visible-light telescopes, but must be carried out from above the earth's atmosphere, which blocks these wavelengths from reaching the ground.

15. Observations in X-ray and γ-ray wavelengths require novel telescope designs, and must be made from space.

REVIEW QUESTIONS

1. Suppose your favorite radio station has a frequency of 1,200 kHz (that is 1,200 kilohertz, where 1 kilohertz equals 1,000 hertz). What is the wavelength at which this station transmits?

2. Using Wien's law, calculate the wavelength of maximum emission for the following objects, and comment on whether they would appear to glow in visible light: (1) a star with a surface temperature of 25,000 K; (2) the walls of a room with a temperature of 300 K; (3) the surface of a planet whose temperature is 100 K; (4) gas being compressed as it falls into a black hole, so that its temperature is 10^6 K; and (5) liquid helium, with a temperature of 4 K.

3. Suppose a white dwarf star has a surface temperature of 10,000 K, and a red giant has a temperature of 2,000 K. Compare their surface brightnesses (the energy emitted per square centimeter). Now suppose that the white dwarf has a radius of 10^8 cm, and the red giant has a radius of 10^{13} cm. Compare the luminosities (the total energy emitted over the entire surface) of the two stars.

4. Normal stars have *absorption* lines in their spectra, because the relatively cool gas in the outer layers, lying above the hotter interior, absorbs light at specific wavelengths. What would you conclude about a star that has *emission* lines in this spectrum?

5. Suppose the elements iron and beryllium each have a spectral line at the same wavelength. If you found a line at this wavelength in the spectrum of a star, how could you decide which element was responsible?

6. Compare the light-gathering power of a 4-meter telescope with that of a 1-meter telescope.

7. Explain why the coudé-focus arrangement is usually not used for observing very faint objects.

8. Suppose you want to make observations of an interstellar dust cloud whose temperature is 500 K. Explain why this would be difficult to do with a telescope at room temperature. How would the situation be improved by cooling the telescope with liquid nitrogen, which has a temperature of 77 K?

9. Explain why ultraviolet telescopes are needed for observations of very hot stars.

10. Suppose the so-called Next Generation Telescope is to be built along the lines of the multiple-mirror concept. How many 4-meter mirrors would be required to build the equivalent of single 15-meter telescope?

11. The strong line of hydrogen, whose rest wavelength is 6,563 Å, is observed in a number of stars. Calculate the line-of-sight velocity of a star where the line is observed at (a) 6,561.8 Å; (b) 6,567 Å; and (c) 6,593 Å.

ADDITIONAL READINGS

The best sources for further explanation of the properties of light and radiation are elementary physics books, of the sort used as texts in introductory courses. In addition, there are a few nontechnical books on the nature of light, and a variety of books and articles on telescopes.

Bahcall, J. N., and Spitzer, L., Jr. 1982. The space telescope. *Scientific American* 247(1):40.

Bragg, W. 1959. *The universe of light.* New York: Dover.

Asimov, I. 1975. *Eyes on the universe.* Boston: Houghton Mifflin.

Harrington, S. 1982. Selecting your first telescope. *Mercury* 11(4):106.

King, H. C. 1979. *The history of the telescope.* New York: Dover.

Kirby-Smith, H. T. 1976. *U.S. observatories: a directory and travel guide.* New York: Van Nostrand.

Labeyrie, A. 1982. Stellar interferometry: a widening frontier. *Sky and Telescope* 63(4):334.

Meyer-Arendt, J. R. 1972. *Introduction to classical and modern optics.* Englewood Cliffs, N.J.: Prentice-Hall.

Miczaika, G. R. 1961. *Tools of the astronomer.* Cambridge: Harvard University Press.

Mims, S. S. 1980. Chasing rainbows: the early development of astronomical spectroscopy. *Griffith Observer* 44(8):2.

Physics Today, 35(11): (several review articles on telescopes for various wavelengths).

Rublowsky, J. 1964. *Light.* New York: Basic Books.

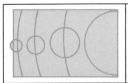

John N. Bahcall

Professor Bahcall, trained as a physicist, is a widely recognized leader in the application of atomic and nuclear physics to astrophysical problems. Since receiving his Ph. D. from Harvard in 1961, his work has always been at the very forefront of new science and technology. He has been especially prominent in the study of quasistellar objects (known more commonly as quasars; these are discussed in chapter 22 of this text), and has contributed in large measure to the great progress that has been made in understanding these mysterious objects. Professor Bahcall's many important analyses of atomic and subatomic processes in astrophysics have received perhaps less public attention than his work on quasars, but have been at least equally important to research in his field. In parallel with all of this largely theoretical work, he has also been prominently involved in the development of new telescope technology, and has played key roles in the Space Telescope program. After serving briefly on the faculty at the California Institute of Technology, Professor Bahcall has since 1968 been a member of the faculty at the Institute for Advanced Study in Princeton, where he leads the program in astrophysics. One of the problems of long-term interest to him has to do with nuclear reactions inside the sun, and a fantastic new kind of telescope that might enable us to "see" what is going on there.

Many of the great discoveries in the early history of science were made by people who did nothing more than use tools readily available and open their minds to new ideas. Galileo's experiments in mechanics required only simple equipment and an inquisitive mind. Other profound realizations became possible only when technology provided the means. Galileo's great astronomical discoveries were not conceivable before the telescope was invented, but followed quickly and inevitably once it was.

In modern observational and experimental science it is usually the case that sophisticated tools are required in order for new advances to be made. For the most part, we have already learned what our natural senses can tell us about the physical universe. Even very basic questions now may require immensely expensive or complex machinery in order to be answered.

Nowhere are the limits of technology pushed harder for the sake of pure research than in the development of new kinds of telescopes. In chapter 3 of this text are described some examples: the *Multiple Mirror Telescope,* the so-called *Next Generation Telescope,* and various orbiting observatories, most notably the *Space Telescope.* The latter is a multi-purpose instrument, designed and built at a cost approaching a billion dollars, which is expected to answer some very fundamental questions about the nature of the universe and its eventual fate.

Closer to home, one of the simplest yet most perplexing of astronomical mysteries is why the sun shines. To unambiguously answer this seemingly innocuous question may require one of the most unusual telescopes ever imagined.

The accepted explanation of the sun's glow is the "burning" of hydrogen nuclei in nuclear reactions (see chapter 11 of this book). The interior of the sun is believed to be a giant nuclear furnace in which hydrogen nuclei are fused together to form helium nuclei. Helium is a more complex element than hydrogen; when hydrogen particles are forced together under conditions of very high temperature and pressure, they can merge to form helium, releasing energy at the same time. In the process, elusive nuclear particles called neutrinos are produced. These strange particles interact only weakly with matter and hence can escape directly from the center of the sun where they are produced. On the other hand, light emitted at the sun's core is absorbed immediately by surrounding atomic nuclei, and does not make its way directly into space. Therefore, we cannot see what is happening in the solar interior, where the energy that makes the sun glow is generated, and the neutrino provides us our only

chance to verify our theoretical understanding of what goes on there.

Can we use neutrinos to look inside the sun? The answer is yes, but with difficulty. Neutrinos are tough to capture on earth for the same reason that they can escape readily from the interior of the sun: they interact very weakly with whatever is in their path. There is only about one chance in a hundred billion that a neutrino produced in a nuclear reaction in the sun will be stopped as it passes through the entire earth.

Nevertheless, neutrinos can be captured. Certain atomic elements have unusually high probabilities of interacting with neutrinos, and one such element is chlorine, which can be altered to a radioactive form of argon when struck by a neutrino. A huge tank of perchlorethylene (a hundred thousand gallons of cleaning fluid) has been installed in a gold mine—the famous Homestake Mine—in South Dakota. Scientists from Brookhaven National Laboratory, led by R. Davis, Jr., have developed sensitive chemical methods of extracting a few atoms of radioactive argon from this huge tank and counting them accurately.

What does this tell us about the sun? A lot, provided only that nothing happens to the neutrinos on their way to the earth from the sun. One can calculate the rate at which neutrinos ought to be captured in the tank of cleaning fluid if the conventional theory of how the sun shines is correct. The result is about twelve captures per month. To everyone's initial surprise, the observed rate is at least a factor of

three less than the predicted rate.

This discrepancy between theory and observation has persisted for more than fifteen years. As is usual in science when a fundamental experiment conflicts with the conventional wisdom, there have been three types of response. Most astronomers have ignored the conflict and continued to use the standard theory of stellar structure and evolution, having no alternative. Astronomers need to invoke some theory of how stars shine in order to interpret their observations. On the other hand, some have taken the discrepancy seriously and have published "cocktail hour" explanations, explanations which they would in normal circumstances only discuss under informal social conditions. One imaginative explanation in this category is the suggestion that the sun contains a massive black hole at its center that is gobbling up gas around it and emitting heat as the gas falls into the hole.

A few physicists and chemists have responded in the third possible manner: they have proposed new experiments to try to capture the elusive neutrinos with other materials. Since many tons of any material are needed in order to capture just a few hundred neutrinos per year, these propsed experiments are all difficult, costly, and lengthy. A team of scientists will require roughly a decade of work and many millions of dollars in order to design and carry out one of the proposed experiments.

Is a new neutrino experiment worth all the trouble? To some people, yes. It is a chance to solve

a fundamental problem, one that will continue to be discussed in textbooks like this one. Ten years is a long time—and millions of dollars is a lot of money. But how often can a scientist be sure that he or she is helping write a bit of scientific history?

What question would you try to answer with the next neutrino experiment? Scientists believe that the most important question is: do the neutrinos actually make it to earth unaffected by their trip from the sun? Or conversely, is our theory of how the sun shines wrong in some significant but unknown way? Another way of phrasing the question is this: who is at fault, the physicists (masters of the neutrinos), or astrophysicists (explainers of the sun)?

An experiment requiring tens of tons of an element called gallium is being worked on by scientists in several countries. This is not a common element, and the needed amount is a significant fraction of the available world's supply. To carry out this experiment would be rewarding, however, because it would detect the neutrinos emitted in one of the basic reaction steps thought to occur in the sun (in contrast with the neutrinos that can be captured by the cleaning-fluid experiment, which are produced in a minor reaction not responsible for much of the sun's energy). Every astronomer and physicist I know believes that these particular neutrinos are being produced in the sun at the rate predicted by the standard theory. Given that they are correct, the gallium experiment will discriminate between the

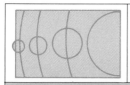
two general classes of possible explanations of the cleaning-fluid results: that the neutrinos can't make it to the earth from the sun on one hand, or that something is wrong with our theory of how the sun shines on the other. The first class of explanations will be indicated if too few neutrinos are detected in the gallium experiment. The second class of explanations will be indicated if the expected numbers of neutrinos are captured with gallium, but—as noted earlier—not with perchlorethylene.

Scientists in the United States, West Germany, and Israel have joined together to form one experimental team. A larger effort—but with a less efficient chemical process—is going on in the Soviet Union. There are frequent scientific exchanges and discussions between the two groups. The Soviets are ahead of their western colleagues in the amount of gallium they have managed to acquire, but the westerners are ahead in scientific experience and technical facilities.

The outcome of this friendly scientific race will probably be decided by financial considerations. If the Americans, West Germans, and Israelis get sufficient funding over the next several years to obtain the required gallium, they will probably be able to capitalize on their technical advantages. Otherwise, scientists in the Soviet Union—with a guaranteed supply of gallium—will have the privilege of being first to solve one of the current fundamental problems of science.

THE SOLAR SYSTEM

Introduction to Section II

In this portion of the book, we will examine the properties of our system of sun and planets, with a view toward understanding how this system came to be. We will also learn about the fascinating phenomena that have been revealed through its exploration.

The solar system is complex, and we will see that it is made up of a variety of individual parts. We will also find, however, that there are a number of underlying general processes at work, so that as we go through the system, examining individual objects, we will be able to make comparisons, to see how these general processes have influenced different situations.

Among the events that we will encounter time and again are tidal forces and other gravitational effects which are responsible for an astonishing assortment of phenomena, such as the spin rates of many of the planets and satellites, the locations of their orbits, and the failure of the asteroids and the ring particles of the outer planets to form large bodies. We will also allude often to the process by which the solar system formed (namely, from the collapse of a rotating interstellar cloud), because on our tour of the universe we will encounter many recognizable artifacts of the process. Another phenomenon that will be mentioned often is the solar wind, a steady stream of subatomic particles from the sun that pervades the solar system, creating important effects in the outer atmospheres of many of the planets.

As we study the planets, we will make a number of comparisons among them. To establish some of the standards for these comparisons, we will begin with the best-studied and best-understood of all the planets, the earth. We will then tour the moon and the other terrestrial planets, the innermost four (including the earth), which have similar overall properties. After this, we will turn our attention to the outer, giant planets, starting with Jupiter, the biggest of them all. Again, the first example will serve as the standard for comparison with the others. Finally, we will discuss other bodies in the solar system, which are important leftovers from its formation, and then the sun itself, before wrapping up the section with a description of the formation and evolution of the solar system.

THE EARTH AND ITS COMPANION

4

Learning Goals

Of all the bodies in the heavens, the earth, of course, has been the best studied, although it is equally true that many mysteries remain. To understand the planet we live on has practical, as well as philosophical, importance. Furthermore, working at such close quarters with the object of study has numerous advantages, including the ability to observe in great detail and over long periods of time, and the possibility of making direct experiments by probing and sampling the surface of the earth.

The earth's satellite, the moon, has been more thoroughly probed than any other remote body, and it, too, has much to tell us about the past and future of our planet. As we will learn in this chapter, the earth and moon probably formed together, but the moon has been relatively little affected by the forces that have since shaped and altered the earth. Therefore the moon is a useful laboratory for developing and testing theories of the history and formation of the earth-moon system.

We will begin by discussing the earth and the moon as separate entities, but we will draw parallels and contrasts between them, and at the end of the chapter we will treat their formation jointly.

The Earth and Its Atmosphere

The earth is a brilliant sight as seen from space (Fig. 4.1). It has an overall blue color, created by the scattering of light in its atmosphere and by its oceans. Much of it is white, where there is cloud cover, and here and there brown land masses can be seen. The earth is so bright because it reflects a large fraction of the incident sunlight. The **albedo** of the earth, the fraction of incoming light that is reflected back into space, is 0.39. This is much higher than what is typically for bodies without atmospheres. The moon, for example, has an albedo of only 0.07, and is far outshone by the earth when the two are viewed together from space.

We begin by examining the thin layer of gas that surrounds the solid earth. The earth's atmosphere is composed of a variety of gases, as well as a distribution of suspended particles called **aerosols.** The gases, which are evenly mixed together up to an altitude of about 80 kilometers, are primarily nitrogen and oxygen, in the form of the molecules N_2 and O_2. Nearly 80 percent of the gas is nitrogen, about 20 percent is oxygen, and all the other species exist only as traces, although in many cases they are important traces. In addition to this standard mixture of gases, a number of others appear in the atmosphere, but are highly variable in quantity. Among these, water vapor (H_2O) is important, because of its direct role in supporting life, and others, such as ozone and carbon dioxide, are also essential, but for less direct reasons. Ozone (designated O_3) is a form of oxygen, in which three oxygen atoms are bound together instead of two. Unlike ordinary oxygen (O_2), ozone in the upper atmosphere acts as a shield against ultraviolet light from the sun, which could be harmful to life forms if it penetrated to the ground.

The principal atmospheric constituents, nitrogen and oxygen, are both released into the air by life forms on the surface. This process has been important in the evolution of the earth's atmosphere, and leads to the expectation that other planets in the solar system do not have the same composition (although nitrogen is produced also by processes that do not involve life forms, so we might find it elsewhere). Nitrogen is released into the earth's atmosphere by the decay of biological material and by emission from volcanic eruptions. Oxygen comes almost exclusively from the photosynthesis process in plants, a process in which carbon dioxide is converted into oxygen with the assistance of radiant energy from the sun.

TABLE 4.1 Earth

Orbital semimajor axis : 1.000 AU (149,600,000 km)
  Perihelion distance: 0.983 AU
  Apehelion distance: 1.017 AU
Orbital period: 365.256 days (1.000 years)
Orbital inclination: 0°0′0″

Rotation period: $23^h 56^m 4.1^s$
Tilt of axis: 23°27′

Diameter: 12,756 km (1.000 $D_{\oplus}$)
Mass: 5.976×10^{27} grams (1.000 $M_{\oplus}$)
Density: 5.518 grams/cm^3
Surface gravity: 980 cm/sec^2 (1.000 earth gravity)
Escape velocity: 11.2 km/sec

Surface temperature: 200–300 K
Albedo: 0.39 (average)

Satellites: 1

FIGURE 4.1. *The Earth as Seen From Space.*

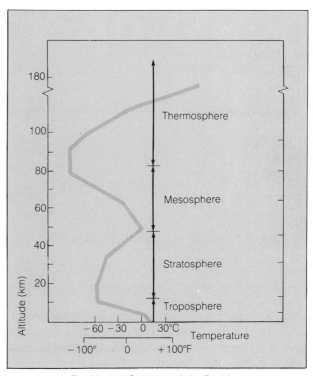

FIGURE 4.2. *The Vertical Structure of the Earth's Atmosphere.*

Although there are obvious fluctuations in the temperature of the atmosphere, particularly near the surface, there is a definite, stable temperature structure as a function of height (Fig. 4.2). This structure includes several distinct layers in the atmosphere: the **troposphere**, from the surface to about 10 km altitude, where the temperature decreases slowly with height and where the phenomenon that we call **weather** occurs; the **stratosphere**, between 10 and 50 km, where the temperature increases with height because of the absorption of sunlight by ozone; the **mesosphere**, extending from 50 to 80 km, where the temperature again decreases; and the **thermosphere**, above 80 km, where the temperature gradually rises with height to a constant value above 200 km. In layers where the temperature decreases with height, there can be substantial vertical motions of the air, whereas in layers where it gets warmer with height, the air is stable, with only lateral motions.

The primary influences on the global motions in the earth's atmosphere are heating from the sun, which is most effective at middle latitudes, and the rotation of the earth. The heating creates regions where the air rises. The air must later cool and fall somewhere else, and a pattern of overturning motions called **convection** is created (Fig. 4.3). Generally, air rises in the tropics and descends at more northerly latitudes, al-

though the situation is more complicated than that (Fig. 4.4). For example, the continents tend to be warmer than the oceans, so that high-pressure regions (characterized by descending air) preferentially lie over water, whereas low-pressure regions (where air rises and cools) tend to be over land. Air that is rising or

FIGURE 4.3. *Convection in the Earth's Atmosphere.*
This is a simplified illustration of the principle of convection, in which warm air rises, cools, and then descends.

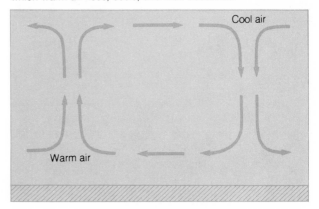

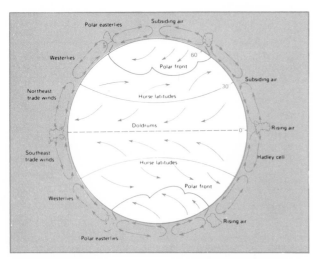

FIGURE 4.4. *The General Circulation of the Earth's Atmosphere.*

falling is forced into a rotary pattern by the earth's rotation (Fig. 4.5), creating circular flows called **cyclones** (where air is rising) and **anticyclones** (where it descends). Seasonal shifts in the distribution of the sun's energy input cause changes in the flow patterns, creating our well-known seasonal weather variations.

FIGURE 4.5. *A Storm System on Earth.*
This is in the southern hemisphere, so the circulation about this low-pressure region is clockwise.

The Earth's Interior and Magnetic Field

The primary probes of the earth's interior are **seismic waves** created in the earth as a result of major shocks, most commonly earthquakes. These waves take three possible forms, called the P, S, and L waves. Studies of wave transmission in solids and liquids have shown that the P waves, which are the first to arrive at a site remote from the earthquake location, are **compressional** waves, which means that the oscillating motions occur parallel to the direction of wave motion, creating alternating regions of high and low density without any sideways motions (Fig. 4.6). Sound waves are examples of compressional waves. The S waves are **transverse,** or **shear,** waves, the vibrations occurring at right angles to the direction of motion. These waves require that the material they pass through have some rigidity and, unlike the P waves, they cannot be

FIGURE 4.6. S *and* P *Waves.*
An *S*, or shear, wave consists of alternating motions transverse (perpendicular) to the direction of propagation. Waves in a tight string or on the surface of water are *S* waves. Compressional, or *P*, waves have no transverse motion, but instead consist of alternating dense and rarefied regions created by motions along the direction of propagation. The arrows shown at the top of the *P* waves represent the relative speeds in and between the dense zones. Sound waves are *P* waves.

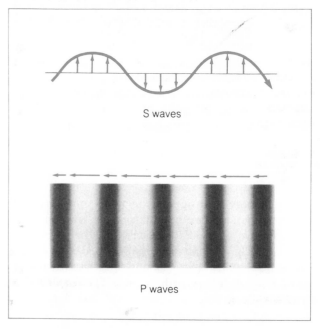

S waves

P waves

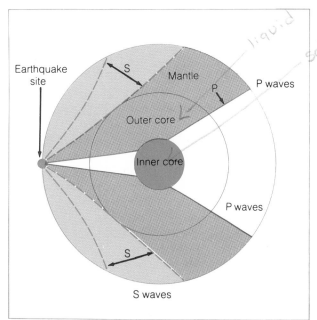

liquid   *solid*

FIGURE 4.7. *Seismic Waves in the Earth.*
This simplified sketch shows that *P* waves can pass through the core regions (although their paths may be bent), whereas *S* waves cannot. This led to the deduction that the earth's outer core is liquid, because *S* waves, which require an elastic or solid medium, cannot penetrate liquids.

transmitted through a liquid. The L waves travel only along the surface of the earth, and thus do not provide much information on the deep interior.

surface waves

By measuring both the timing and the intensity of these seismic waves at various locations away from the site of an earthquake, scientists can determine what the earth's interior is like (Fig. 4.7). The speed of the P waves depends on the density of the material they pass through, and the distribution of the P and S waves reaching remote sites provides data on the location of liquid zones in the interior.

The general picture that has developed from these studies is that of a layered earth, something like an onion (Fig. 4.8). At the surface is a crust whose thickness varies from a few kilometers beneath the oceans to perhaps 60 kilometers under the continents. There is a sharp break between the crust and the underlying material, which is called the **mantle.** The mantle transmits S waves, so it must be solid, but on the other hand, it undergoes slow, steady flowing motions in its uppermost regions. Perhaps it is best viewed as a plastic material, one that has some rigidity but which can be deformed, given sufficient time.

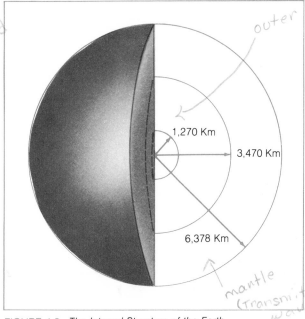

outer

1,270 Km

3,470 Km

6,378 Km

mantle (transmits waves)

solid material

FIGURE 4.8. *The Internal Structure of the Earth.*

The uppermost part of the mantle and the crust together form a rigid zone called the **lithosphere** (Fig. 4.9). The part of the mantle itself where the fluid motions occur, just below the lithospere, is called the **asthenosphere.** Below the asthenosphere, there is a more rigid portion of the mantle that extends nearly

FIGURE 4.9. *The Structure of the Earth's Outer Layers.*
This illustrates the relative thickness of the crust in continental areas as compared to the seafloor.

mantle

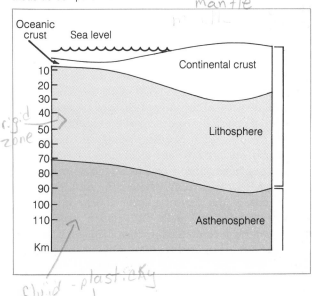

Oceanic crust   Sea level

10
20
30
40
50
60
70
80
90
100
110
Km

Continental crust

Lithosphere

Asthenosphere

rigid zone

fluid - plasticity material

halfway to the center of the earth. The lower mantle is called the **mesosphere** (not to be confused with the level in the earth's atmosphere bearing the same name).

Beneath the mantle is the **core,** consisting of an outer core   (between 2,900 and 5,100 km in depth), and an inner core. The S waves not are transmitted through the outer core, which is therefore thought to be liquid. As a result, the inner core can be probed only with the P waves. Since these waves travel more rapidly through the inner core, its density is thought to be greater than that of the outer core, and the inner-most region is thought to be solid. It is known that the core has great density, and its composition is thought to be primarily iron and nickel.

The high density and probable metallic composition of the core is quite significant, for they show that the earth has undergone **differentiation,** a sorting out of elements according to their weight. This can only happen in a planet when it is in a molten state, so it shows that the earth once was largely liquid, probably early in its history, soon after it formed. The earth's molten state resulted from heating caused by its formation, and radioactivity. Several naturally occurring elements, such as uranium, thorium, and some forms of potassium, are radioactive, meaning that their nuclei spontaneously emit subatomic particles and, over a long period of time, produce substantial quantities of heat. We shall see that differentiation has not occurred in all the bodies of the solar system, and this tells us something significant about their histories.

The fluid interior portions of the earth give rise to its magnetic field. It is convenient to visualize the structure of the field by imagining magnetic lines of force connecting the two poles: these correspond to the lines along which iron filings lie when placed near a small bar magnet. In cross-sectional view, the earth's magnetic field is reminiscent of a cut apple, but one that is lopsided, because of the flow of charged particles from the sun that constantly sweeps past the earth (Fig. 4.10; also see the discussion of **solar wind** in chapter 11). The region enclosed by the field lines is called the **magnetosphere,** and it acts as a shield, preventing the charged particles from reaching the surface.

The axis of the earth's magnetic field is aligned closely (within 11½°) with the rotation axis of the planet, but this has not always been the case. The past alignment of the magnetic field can be ascertained from studies of certain rocks that contain iron-bearing minerals whose crystalline structure is aligned with the

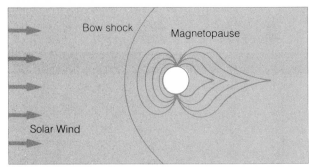

FIGURE 4.10. *The Earth's Magnetic-Field Structure.*
This cross-section shows the earth's field and how its shape is affected by the stream of charged particles from the sun known as the solar wind.

direction of the magnetic field at the time the rocks solidify from a molten state. Traces of the earth's ancient magnetic activity, called **paleomagnetism,** thus can be deduced from the analysis of magnetic alignments of rocks. Such studies reveal that the magnetic poles have moved about during the earth's history, and that the north and south magnetic poles have completely and rather suddenly reversed from time to time, so that the north pole has moved to the south and vice versa. This flip-flop of the magnetic poles seems to have happened at irregular intervals, typically thousands or hundreds of thousands of years apart.

The source of the earth's magnetic field is a mystery, although it is nearly certain that it has to do with the nickel-iron core. A magnetic field is produced by flowing electrical charges, as in the current flowing through a wire wound around a metal rod, which is the basis of an electromagnet. Convection and the earth's rotation may combine to create systematic flows in the liquid outer core, giving rise to the magnetic field, if the core material carries an electrical charge. This general type of mechanism is called a **magnetic dynamo**. The cause of the reversals of the poles is not well understood, but presumably could be related to occasional changes in the direction of the flow of material in the core.

We have referred to the magnetosphere and its ability to control the motions of charged particles. When the first U.S. satellite was launched in 1958, zones high above the earth's surface were discovered to contain intense concentrations of charged particles, primarily protons and electrons. There are several distinct zones containing these particles, and they are now called the **Van Allen belts,** after the physicist who in the late

1950s first recognized their existence and deduced their properties.

The charged particles, or ions, in the Van Allen belts are captured from space (primarily from the solar wind). and are forced by the magnetic field to spiral around the lines of force. This is also true in the uppermost portion of the outer atmosphere, called the **ionosphere,** which extends upward from a height of about 60 kilometers. While a reversal of the earth's magnetic poles is taking place, there is, for a short period of time, a much-weakened field and consequently a major disruption of the Van Allen belts and the ionosphere. The magnetosphere is greatly diminished, and charged particles from space are more likely to penetrate to the ground. These particles, particularly the very rapidly moving ones called **cosmic rays,** can cause important effects on life forms, including genetic mutations. The sporadic reversals of the earth's magnetic field may have played a major role in shaping the evolution of life on the surface of our planet.

The ionosphere has important effects for us on the surface, including enhanced radio communications and the beautiful light displays known as **aurorae borealis** (northern lights) and **aurorae australis** (southern lights). The aurorae are caused by charged particles entering the atmosphere, and occur most commonly near the poles, where the magnetic field lines allow particles to penetrate closest to the ground. The ionosphere's role in radio communications is that it reflects signals in the short-wave band, so that they can travel around the earth. When there are fluctuations in the solar wind, particularly after solar flares, enhanced fluxes of charged particles entering the ionosphere from space can disrupt radio communications.

A Crust in Action

About three-fourths of the earth's surface is covered by water, the rest taking the form of several major continents. The basic substance of the earth's crust is rock, and rocks are classified in three groups according to their origin; they are either **igneous, sedimentary,** or **metamorphic.** Igneous rocks, formed from volcanic activity, consist of cooled and solidified **magma,** the molten material that flows to the surface during volcanic eruptions. Sedimentary rocks are formed from deposits of gravel and soil that have hardened, usually

in layers where old seabeds or coastlines lay. Metamorphic rocks are those which have been altered in structure by heat and pressure created by movements in the earth's crust. All three forms can be changed from one to another in a continual recycling process. Rocks are also classified according to their chemical compositions, as **minerals** of various types. The most common of all minerals are the silicates, which account for some 90 percent of all rock on the earth's surface.

Because of various evolutionary processes, the surface of the earth is continually being renewed. Whereas the age of the earth itself is thought to be some 4.5 billion years, the ages of most surface rocks can be measured in the millions or hundreds of millions of years. The distinction between the age of a planet and the age of its surface is an important one, and it will reappear as we discuss the surfaces of other bodies.

The crust of the earth is not static, but rather is in constant motion. The continents themselves move about, and the world map is variable on geological time scales (Fig. 4.11). The evidence favoring the idea that continental drift occurs takes on several forms, ranging from the obvious fit between land masses on

FIGURE 4.11. *Continental Drift.*
These maps show the distribution of the continents today and as it was some 200 million years ago, before continental drift rearranged things.

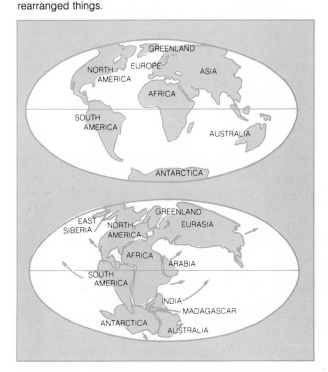

THE AGES OF ROCK

We have spoken of the ages of rocks and of the earth itself, but we have said little about how these ages are measured. The most direct technique, **radio-active dating,** involves the measurement of relative abundances of closely related atomic elements present in rocks.

Recall from our earlier discussions of atomic structure (chapter 3) that the nucleus of an atom consists of protons and neutrons, and that the number of protons (called the **atomic number**) determines the identity of the element. Different **isotopes** of an element have the same number of protons, but differing numbers of neutrons. In most elements the number of protons is not very different from the number of neutrons, but there are exceptions. If the imbalance between protons and neutrons is large enough, the element is unstable, which means that it has a natural tendency to correct the imbalance by undergoing a spontaneous nuclear reaction. The reactions that occur in this way usually involve the emission of a subatomic particle, and the element is said to be **radioactive.** The emitted particle may be an **alpha particle** (that is, a helium nucleus, consisting of two protons and two neutrons), or, as is more often the case, an electron or a **positron,** a tiny particle with the mass of an electron but with a positive electrical charge. When a positron or an electron is emitted, the reaction is called a **beta decay,** and at the same time a proton in the nucleus is converted into a neutron (if a positron is emitted) or a neutron is changed into a proton (if an electron is released). If the number of protons in the nucleus is altered, the identity of the element is changed. A typical example is the conversion of potassium 40 (^{40}K, with 19 protons and 21 neutrons in its nucleus) into argon 40 (^{40}Ar, with 18 protons and 22 neutrons).

Several elements are known to be radioactive, naturally changing their identity by emitting particles. The rate of change, expressed in terms of the **half-life,** is known in most cases. The half-life is the time it takes for half of the original element to be converted, and it may be as short as fractions of a second or as long as billions of years. The very slow reactions are the ones that are useful in measuring the ages of rocks.

If we know the relative abundances of the elements that existed in a rock when it formed, then measurement of the ratio of the elements in that rock at the present time can tell us how long the radioactive decay has been at work; that is, the age of the rock. In a very old rock, for example, there might be almost no ^{40}K, but a lot of ^{40}Ar. The decay of ^{40}K to ^{40}Ar has a half-life of 1.3 billion years, so from the exact ratio of these two species in a rock, we can infer how many periods of 1.3 billion years have passed since the rock formed.

Other decay processes that are useful in dating rocks include the decay of rubidium 87 (^{87}Rb) into strontium 87 (^{87}Sr), which has a half-life of 47 billion years; and the decays of two different isotopes of uranium to isotopes of lead, ^{235}U to ^{207}Pb and ^{238}U to ^{206}Pb, with half-lives of 700 million years and 4.5 billion years, respectively (these decays each involve several intermediate steps; the half-lives given represent the total time for all the steps).

Radioactive dating techniques have shown that the oldest rocks on the earth's surface are about 3.5 billion years old, and that ages of a few hundred million years are more common. The age of the solar system, and therefore of the earth itself, is estimated (from isotope ratios in meteorites) to be about 4.5 billion years.

---

opposite sides of the Atlantic, to similarities of mineral types and fossils, to the alignment of vestigial magnetic fields in rocks that once were together in the same place. In modern times, more direct evidence, such as the detection of seafloor spreading away from undersea ridges and the actual measurement (through the use of sophisticated laser-ranging techniques) of continental motions, has removed any trace of doubt that pieces of the earth's crust are in motion.

From all this evidence has arisen a theory of **plate**

tectonics, which postulates that the earth's crust (that is, the lithosphere) is made of a few large, thin, pieces that float on top of the asthenosphere (Fig. 4.12). The flowing motions in the asthenosphere cause these plates to constantly move about and occasionally to crash into each other. The rate of motion is only a few centimeters per year at the most, and the major rearrangements of the continents have taken many millions of years to occur (the Americas and the European-African system became separated from each other 150 to 200 million years ago).

The driving force in the shifting of the plates is not well understood. The most widely suspected cause is convection currents in the asthenosphere (Fig. 4.13). This is the same process that occurs in the earth's atmosphere: temperature differences between levels cause an overturning motion. If a fluid is hot at the bottom and much cooler at the top, warm material will rise, cool, and descend again, creating a constant churning. The speed of the overturning motions depends largely on the **viscosity** of the fluid; that is, the degree to which it resists flowing freely. The earth's mantle, as we have already seen, is sufficiently rigid to transmit S waves, and must therefore have a high viscosity as well. Hence if convection is occurring in the mantle, it is reasonable for the motions to be very slow.

Whether convection is the cause or not, the effects of tectonic activity are becoming well known. When plates collide, one may submerge below the other in a process called **subduction,** creating an undersea trench; or, particularly if both plates carry continents, the collision may force the uplifting of mountain ranges. The lofty Himalayas are thought to have been created when the Indian plate collided with the Eurasion plate some fifty million years ago. The boundaries where plates either collide or separate are marked by a wide variety of geological activity. Earthquakes are attributed to the sporadic shifting of adjacent plates along fault lines, and volcanic activity is common where material from the mantle can reach the surface, most often along plate boundaries. It has been long recognized, for example, that a great deal of earthquake and volcanic activity is concentrated around the shores of the Pacific Ocean, defining the so-called Ring of Fire. Now it is understood that this zone represents the boundaries of the Pacific plate.

FIGURE 4.12. *The Earth's Crustal Plates. This map shows the plate boundaries and the directions in which the plates are moving.*

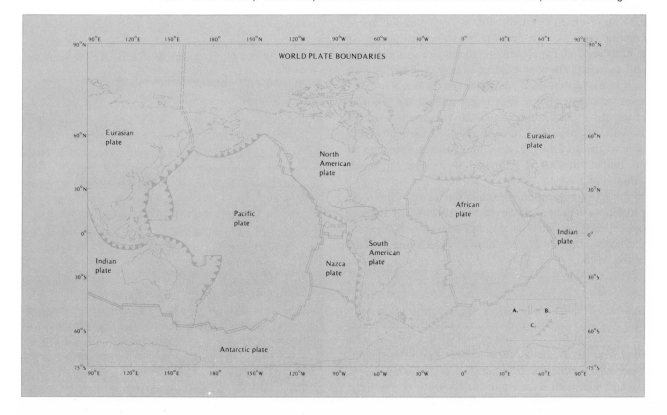

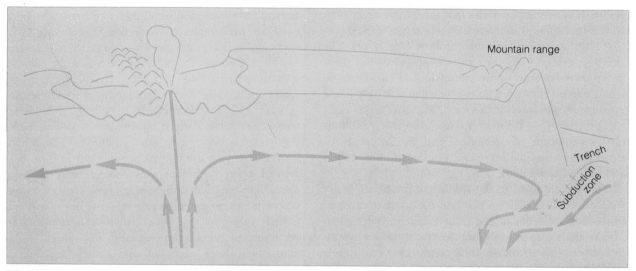

FIGURE 4.13. *Schematic of the Mechanisms of Continental Drift.* This sketch shows how continental drift is responsible for uplifted mountain ranges and parallel undersea trenches, where one crustal plate sinks below another (subduction zone); and how a mid-ocean ridge is built up where two plates move away from each other.

Chains of volcanoes, such as the Hawaiian Islands, are also attributed to the action of plate tectonics. In several locations around the world, volcanic hot spots that lie deep and are fixed in place bring molten rock to the surface. As the crust passes over one of these hot spots, volcanoes are formed and then carried away, creating a chain of mountains as new material keeps coming to the surface over the hot spot. This process is still active, and the Hawaiian Islands, for example, are still growing. There is volcanic activity on the southeast part of the largest (and youngest) island, Hawaii. Futhermore, a new island, already given the name Loihi, is rising from the seafloor about 20 kilometers south of Hawaii. Its peak has already risen 80 percent of the way to the surface and only has about 1 kilometer to go, but it will take an estimated 50,000 years to break through.

Plate tectonics, then, accounts for many of the most prominent features of the earth's surface. Of course, other processes, such as running water, wind erosion, and glaciation, are also important in modifying the face of the earth.

Exploring the Moon

Seen from afar, the moon is a very impressive sight, especially when full (Fig. 4.14). Its ½° diameter is pockmarked with a variety of surface features, the most prominent of which are the light and dark areas. The latter, thought by Renaissance astronomers to be bodies of water, are called **maria** (singular: *mare*), after the Latin word for seas. Closer examination of the moon with primitive telescopes revealed numerous circular

FIGURE 4.14. *The Full Moon.*
This is a high-quality photograph taken through a telescope on the earth.

TABLE 4.2 Moon

Mean distance from earth: 384,401 km (60.4 R$_{\oplus}$)
   Closest approach: 363,297 km
   Greatest distance: 405,505 km
Orbital sidereal period: 27^d 7^h 43^m 12^s
Synodic period (lunar month): 29^d 12^h 44^m 3^s
Orbital inclination: 5°8′43″

Rotation period: 27.32 days
Tilt of axis: 6°41′ with respect to orbital plane)

Diameter: 3,476 km (0.273 D$_{\oplus}$)
Mass: 7.35 × 10^{25} grams (0.0123 M$_{\oplus}$)
Density: 3.34 grams/cm^3
Surface gravity: 0.165 earth gravity
Escape velocity: 2.4 km/sec

Surface temperature: 400 K(day side) 100 K(dark side)
Albedo: 0.07

FIGURE 4.15. *The Moon as Seen From Space.*
Here is a view obtained by one of the *Apollo* missions, showing portions of the near (left) and far (right) sides of the moon. On the left horizon is Mare Crisium, and in left center are Mare Marginis (upper) and Mare Smythii (lower). No maria are seen in the right half of this image; there are almost none on the entire lunar far side.

features called **craters,** named after similar looking structures on volcanoes. The lunar craters are not volcanic, however, but were formed by the impacts of bodies that crashed onto the surface from space.

The pattern of surface markings on the face of the moon is quite distinctive, and because it never changes, it was recognized long ago that the moon always keeps one face toward the earth. The far side cannot be observed from the earth, but it has been thoroughly mapped by spacecraft (Fig. 4.15).

The lunar surface in general consists of lowlands, primarily the maria, and highlands, which are vast mountainous regions, not as well organized into chains or ranges as mountains on the earth. With no trace of an atmosphere (because the escape velocity is so low that all the volatile gases were able to escape long ago), the moon does not experience erosion caused by weathering, so large-scale features such as craters and mountain ranges retain a jagged appearance when seen from afar. Seen close-up, the mountains and crater walls appear relatively smooth, because the surface consists everywhere of loosely piled rocks ejected and redistributed by impacts. Other types of features seen on the moon include **rays,** light-colored streaks emanating from some of the large craters, and **rilles,** winding valleys that resemble earthly canyons.

The moon's mass is only about 1.2 percent of the earth's mass, and its radius is a little more than one-fourth the earth's radius. The moon's density, 3.34 grams/cm^3, is below that of the earth as a whole; it

more closely resembles the density of ordinary surface rock. This indicates that the moon probably does not have a dense core, and that it therefore most likely did not undergo strong differentiation. Infrared measurements showed, even before the space program allowed direct measurements, that the lunar surface is subject to hostile temperatures, ranging from 100 K (−274° F) during the two-week night to 400 K (273° F) during the lunar day.

Of course, all the long-range studies of the moon were almost instantly antiquated by the development of the space program, which has featured extensive exploration of the moon, first by unmanned robots, then by men (Figs. 4.16 and 4.17). The historic first manned landing occurred on July 20, 1969. This was the *Apollo 11* mission, and was followed by five more manned landings, the last being *Apollo 17*, which took place in late 1972. Each mission incorporated a number of scientific experiments; some involved observations of the sun and other celestial bodies from the airless moon, but most were devoted to the study of the moon itself.

Astronomical Insight 4.2.
LUNAR GEOGRAPHY*

All the major topographic features on the moon, and many minor ones, have names. These names are used in maps of the moon and are recognized by the international community of astronomers. It is interesting to examine the history of how they were assigned.

Galileo was the first to have a chance to name lunar features, because he was the first to look at the moon through a telescope that enabled him to see them. He is responsible for the terms **maria** (the dark areas that he thought were seas) and **terrae** (the highlands). Following Galileo's lead, as telescope technology improved and more and more people examined the moon, a number of early lunar cartographers chose names for prominent craters, mountain ranges, and other regions. Some of the names, such as those assigned by the court astronomer to the King of Spain (who named features after Spanish nobility), were doomed to be forgotten. Others, such as those based on the suggestion by a German astronomer that lunar mountain ranges be named after terrestrial features, have persisted to the present time.

The most influential of the lunar cartographers of medieval times was the Italian priest Joannes Riccioli, who, with his pupil Francesco Grimaldi, named maria after human moods and experiences, and craters after famous scientists. Thus we have maria named Mare Imbrium (Sea of Showers), Mare Tranquilitatis (Sea of Tranquility, where the first manned moon landing occurred), and Mare Serenitatis (Sea of Serenity); and craters named Tycho, Hipparchus, and Archimedes. There were political overtones to Riccioli and Grimaldi's nomenclature; they published their map in 1651, at a time when the heliocentric hypothesis was still not something that people publicly embraced. With this in mind, Riccioli and Grimaldi gave geocentrists like Ptolemy and Tycho very large, prominent craters, but assigned Galileo only a very small one, and gave the name of Copernicus to a crater in the Sea of Storms (Mare Nubium, which can also be translated as Sea of Clouds).

Other names were added during the eighteenth and nineteenth centuries, but until 1921, no formal international agreement was made to honor a specific, uniform naming system. In that year, the newly formed International Astronomical Union designated a committee to oversee and standardize lunar nomenclature. Things were uneventful there-

FIGURE 4.16. *Man on the Moon.*
The *Apollo* missions, six of which included successful manned landings on the moon, represent humankind's only attempt so far to visit another world.

A Battle-worn Surface and a Dormant Interior

Viewed on any size scale, from the largest to the smallest, the moon's surface is irregular, marked throughout by a variety of features. We have already mentioned the maria, the large, relatively smooth, dark areas (Fig. 4.18). The maria appear darker than their surroundings because they have a relatively low albedo (that is, they reflect less sunlight). Despite their smooth appear-

after, until 1959, when the Soviet Union obtained the first photographs of the far side of the moon, and promptly named numerous features after prominent Russians. Many of these names (but not all) were later approved by the International Astronomical Union. Perhaps the most controversial was the Sea of Moscow, which reportedly was finally approved (amid laughter) when the Soviet representative declared, in response to criticism that it was inconsistent with tradition to name a mare after a city, that Moscow is a state of mind, just as tranquility and serenity are.

More recently, numerous additional features have been mapped and named, as the sophistication of lunar probes has improved, and especially since manned exploration of the moon has occured. The *Apollo* astronauts assigned rather colloquial names, and because of all the attendant publicity, these names became well known before there was time for the International Astronomical Union to even consider them for approval. Most of them were based on characteristics of the features themselves, such as a terraced crater that was called Bench, and one with a bright rim that was dubbed Halo. Groups of craters were named for their pattern, such as a pair called Doublet and a group named Snowman, which included individual craters with names such as Head. Some of the features of the terrain mapped by the Apollo astronauts were given names of prominent scientists, in keeping with the tradition established by Riccioli and Grimaldi more than three hundred years earlier. Many of the names invented by the Apollo astronauts were eventually approved by the International Astronomical Union, as were a dozen craters named for American and Soviet astronauts and cosmonauts.

Just about the time that lunar cartography was getting resolved (after the years of confusion brought on by the lunar exploration program), unmanned probes began to obtain images of Mercury, Venus, and Mars, and the whole problem of naming extraterrestrial landmarks arose again. Except for Mars, whose large-scale light and dark regions had been named on the basis of telescopic observations made from earth, no tradition existed for naming features on other planets. After considerable discussion, the International Astronomical Union adopted rules for doing so. Newly discovered features on Mars will be named after towns and villages on earth; Venus will bear the names of prominent women and radio and radar scientists; Mercury's features will be named after people famous for pursuits other than science; and features of the outer planets (or, as we shall see, the satellites thereof) will have names assigned from mythology. Perhaps it will thus be possible to map the rest of the solar system with a minimum of controversy.

*Based on information from: El-Baz, F. 1979. Naming moon's features created "Oceans of Storms." Smithsonian 9(10):96.

ance relative to the more chaotic terrain seen elsewhere on the moon, the maria are marked here and there by craters.

Much of the rest of the lunar surface is covered by rough, mountainous terrain. Even though the maria dominate the near side of the moon, there are none on the far side, and the highland regions actually cover most of the lunar surface.

Craters are seen literally everywhere (Figs. 4.19 and 4.20). They range in diameter from hundreds of kilometers to microscopic pits that can be seen only under intense magnification. In some regions the craters are so densely packed together that they overlap. The lack of erosion on the moon means that craters can survive for billions of years, and there is plenty of time for younger craters to form within the older ones. The fact that relatively few craters are seen in the maria indicates that the surface in these regions has been transformed in recent times, after most of the cratering had already occurred.

All the craters on the moon are **impact craters,** formed by collisions of interplanetary rocks and debris with the lunar surface, rather than by volcanic eruptions. The particular shapes of the craters, the central peaks in some of them, and the trails left by **ejecta,** or cast-off material created by the impacts (Fig. 4.21), all suggest that the craters were formed in this manner. The rays seen stretching away from some craters are strings of smaller craters formed by the ejecta from the larger, central crater.

FIGURE 4.17. *The Lunar Rover*.
The later *Apollo* missions used these vehicles to travel over the moon's surface, allowing the astronauts to explore widely in the vicinity of the landing sites.

FIGURE 4.18. *The Lunar "Seas."*
Here is a broad vista encompassing portions of three maria: Mare Crisium (foreground); Mare Tranquilitatis (beyond Mare Crisium); and Mare Serenitatis (on the horizon at upper right). These relatively smooth areas are younger than most of the lunar surface, having been formed by lava flows after much of the cratering had already occurred.

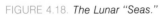

FIGURE 4.19. *Lunar Craters*.
This is a view of the far side of the moon, where there is little to interrupt the nearly total coverage by impact craters. The prominent crater in the upper center is Kohlschutter, named by Soviet scientists who first charted the far side of the moon.

FIGURE 4.20. *The Crater Eratosthenes*.
This is a prominent feature on the near side of the moon. Here the raised rim and central peak of the crater are clearly visible.

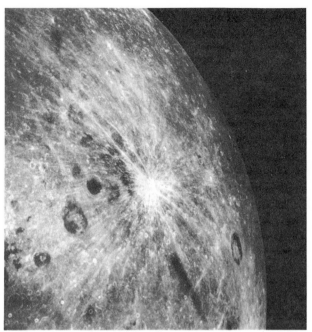

FIGURE 4.21. *A Crater with Ejecta.*
This crater on the lunar far side is a good example of a case where material ejected by the impact has created rays of light-colored ejecta. Close examination of such features often reveals secondary craters, formed by the impacts of debris blasted out of the lunar surface by the primary impact.

FIGURE 4.22. *A Rille.*
The sinuous feature meandering through this image is Hadley Rille, one of many such features seen in the maria. One of the *Apollo* missions landed at the edge of this rille, so that it could be examined at close range. The adjacent mountains are the Apennines.

Some of the lunar mountain ranges reach heights greater than any on earth. They are more jagged than earthly mountains, again because there is no erosion, and they lack the prominent drainage features usually found in terrestrial ranges.

The *rilles* are rather interesting features, resembling dry riverbeds (Fig. 4.22). Apparently they were formed by flowing liquid, but rather than water, lava was responsible. In some cases there are even lava tubes that have partially collapsed, leaving trails of sinkholes. The rilles and, as we shall see, the maria as well, indicate that the moon has undergone stages when large portions of its surface were molten.

The *Apollo* astronauts found a surface strewn in many areas with loose rock, ranging in size from pebbles to boulders as big as a house (Figs. 4.23 and 4.24). The rocks generally show sharp edges (owing to the lack of erosion), and occasional cracks and fractures. In most cases the large boulders appear to have been ejected from nearby craters, and are therefore thought to represent material originally from beneath the surface.

More than 4,500 pounds of smaller lunar rocks were returned to earth by the *Apollo* astronauts. Thus scientists were given the opportunity to study the lunar surface characteristics in detail comparable to that possible for earth rocks and soils (Fig. 4.25). Extensive chemical analysis, as well as close-up observation of rocks in place on the moon, were possible. At present, the rock samples from the moon are housed in numerous scientific laboratories around the world, where analysis continues. Specimens have also found their way into museums, and in at least one case (the Smithsonian Institution's Air and Space Museum in Washington, D.C.), a lunar rock can be touched by visitors.

The lunar soil, called the **regolith,** consists of loosely packed rock fragments and small glassy minerals probably created by the heat of meteor impacts. In addition to the loose soil, a few distinct types of surface specimens were recognized on the basis of morphol-

FIGURE 4.23. *A Field of Boulders.*
This jumbled region is inside a relatively young crater. The rocks lying around were disrupted by the impact.

FIGURE 4.24. *A Large Boulder.*
Rocks on the lunar surface range in size from tiny pebbles to massive objects such as this.

ogy. The most common of these are the **breccias** (Fig. 4.25), which consist of small rock fragments cemented together and resembling chunks of concrete. There are similar kinds of rock on earth, except that those on earth are formed in stream beds where water plays a role in shaping them. The lunar breccias contain jagged, sharp rock fragments, and they are probably fused together by pressure created in meteor impacts.

All lunar rocks are pitted, on the side that is exposed to space, with tiny craters called **micrometeorite craters.** These are formed by the impact of tiny bits of interplanetary material no bigger than grains of dust.

Radioactive dating techniques reveal the ages of lunar rocks to be very old by earthly standards; as old as 3.5 to 4 billion years. Rocks from the maria are not quite so old, but still date back several billion years.

Some of the experiments carried out on the moon by the *Apollo* astronauts were aimed at revealing the interior conditions by monitoring seismic waves caused by earthquakes on the moon. Therefore the astronauts carried devices for sensing vibrations in the lunar crust and, because it was not known whether natural moonquakes occurred frequently, they also brought along devices for thumping the surface to make it vibrate. It turned out that natural moonquakes do occur, although not with great violence. The seismic measurements continued after the *Apollo* landings, with data

radioed to earth by instruments left in place on the lunar surface.

The measurements showed that the regolith, or surface soil, is typically about 10 meters thick, and is supported underneath by a thicker layer of loose rubble.

FIGURE 4.25. *A Moon Rock.*
This is one of thousands of lunar samples brought back to earth by the *Apollo* astronauts. This is an example of a breccia.

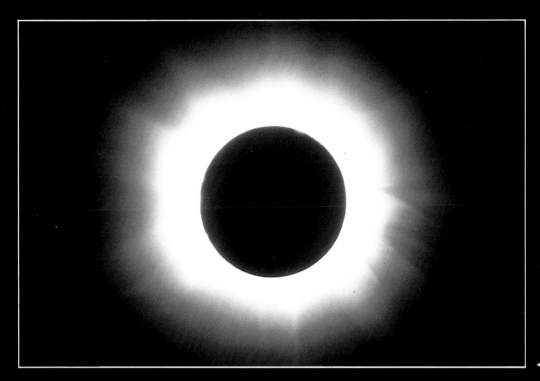

◀ A total solar eclipse

The 1982 total ▶
lunar eclipse

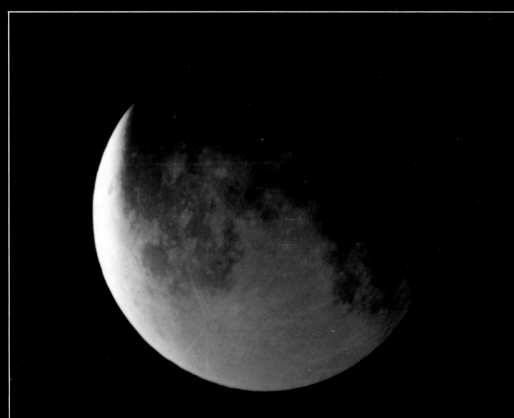

VISIBLE LIGHT SPECTRUM

| X-rays and gamma rays | Ultraviolet rays (beyond violet) | | | | | | | Infrared rays (less than re |

INVISIBLE SHORT WAVES　　　　　　　　　　　　　　　　　　**INVISIBL**

When a white light is directed through a prism, the visible light spectrum results.

The electromagnetic spectrum ▲

The 120-in telescope at
Lick Observatory ▼

The Cerro Tololo
Inter-American Observatory ▼

The Multiple-Mirror Telescope ▼

Color Plate 4

The ultraviolet glow from hydrogen in the earth's upper atmosphere ▼

The earth ▲

A storm system on the earth ▼

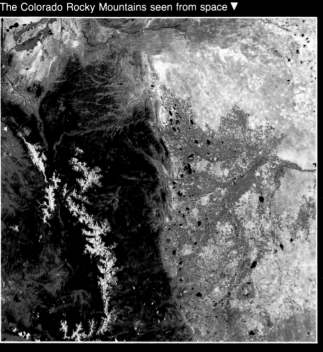

The first launch of the space shuttle *Columbia* ▲

Earthrise ▼

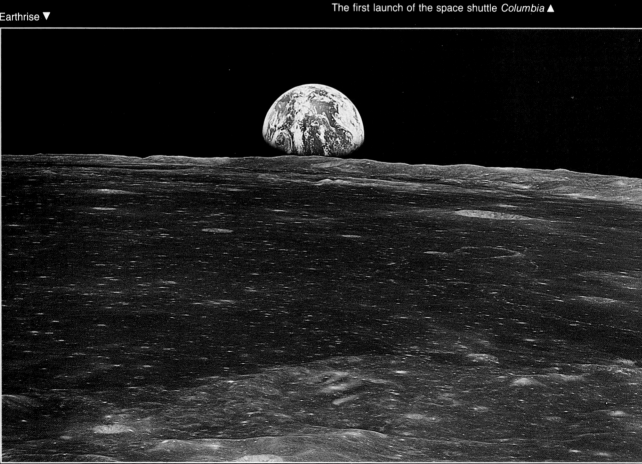

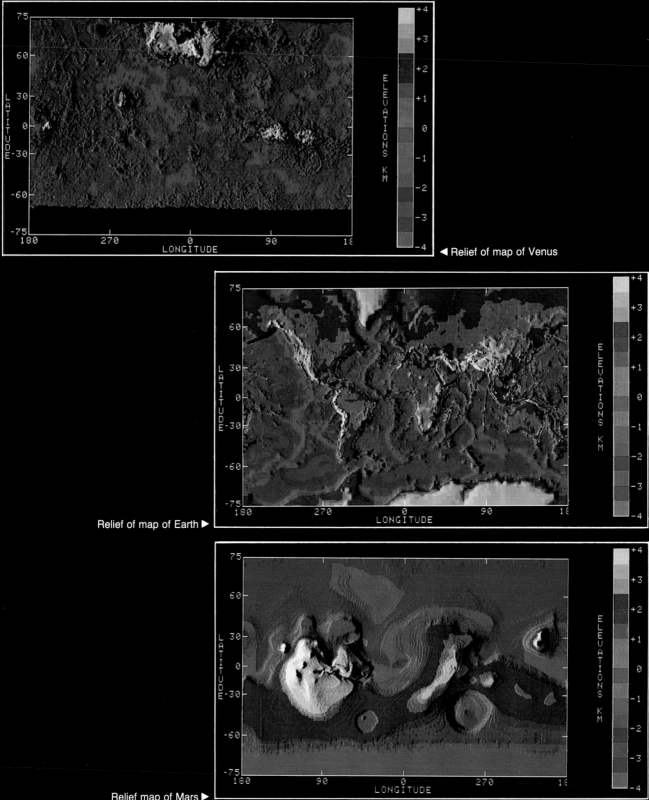

◀ Relief of map of Venus

Relief of map of Earth ▶

Relief map of Mars ▶

◄ The surface of
Venus as photographed
by *Veneras 13* and *14*

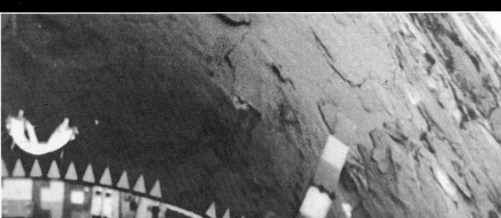

The full disk of Venus ▼

Nightglow on Venus ▼

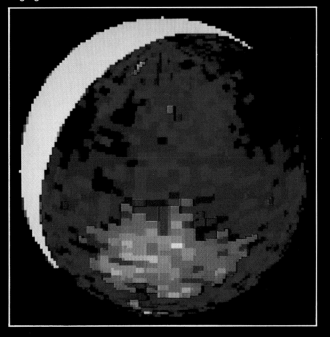

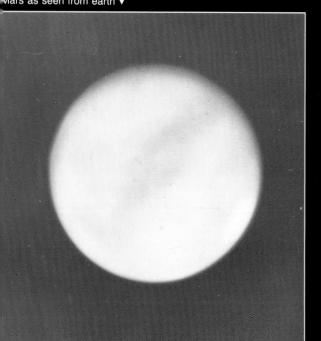

Mars as seen from earth ▼

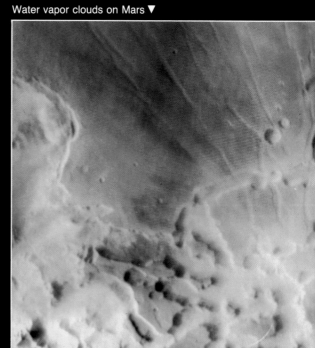

Water vapor clouds on Mars ▼

The crust is 50 to 100 kilometers thick at the *Apollo* sites, and may be somewhat thicker on the far side, where there are no maria (Fig. 4.26). Beneath the crust is a mantle, consisting of a well-defined lithosphere, which is rigid, and beneath it an asthenosphere, which is semiliquid. The innermost 500 kilometers consist of a relatively dense core, but not as dense as that of the earth.

The seismic measurements indicate no truly molten zones in the moon at the present time, although temperature sensors on the surface discovered a substantial heat flow from the interior. This is probably caused by radioactive minerals below the surface. There is no detectable overall magnetic field, further evidence against a molten core. The chief distinction among the internal zones in the lunar interior is density, with the densest material being closest to the center. Thus moderate differentiation has occurred in the moon which implies that it was once molten.

There is no trace of present-day lunar tectonic activity, perhaps the greatest single departure from the geology of the earth. The force that has had the greatest influence in shaping the face of the earth has no role today on the moon. Instead, the lunar surface is entirely the result of the way the moon formed (which may have involved some tectonic activity in the early stages) and the manner in which it has been altered by

FIGURE 4.26. *The Internal Structure of the Moon.* This cross-section illustrates the interior zones inferred from seismological studies. The existence of a dense core is not certain. Note that the maria lie almost exclusively on the side facing the earth, where the lunar crust is relatively thin.

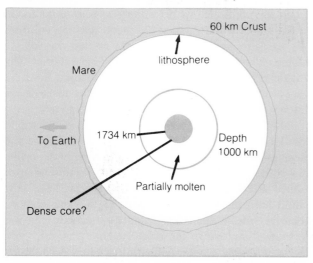

lava flows and by the incessant bombardment of rocks and debris from space.

The History of the Earth-Moon System

Whereas the formation of the earth (and of planets in general) is thought to be reasonably well understood, the moon's beginnings are fairly obscure to us. We will discuss the earth's history first, then delve into the complexities of the moon's origin.

The age of the earth-moon system, estimated from geological evidence, is about 4.5 billion years. The earth is thought to have formed from the coalescence of a number of **planetesimals,** small bodies that were the first to condense in the earliest days of the solar system. Most or all of the earth was molten at some point during the first billion years. The earth became highly differentiated during its molten period, as the heavy elements tended to sink toward the planetary core.

At the same time, volatile gases were emitted by the newly formed rocks at the surface. These gases, including hydrogen (H_2), ammonia (NH_3), methane (CH_4), and water vapor (H_2O), formed the earliest atmosphere of the earth. The water vapor was able to collect as surface water, and from its earliest days, our planet had seas. Another gas that probably was abundant for a time was carbon dioxide (CO_2), but this species was eventually removed from the atmosphere by being absorbed into rocks. This absorption process, of critical importance to the further evolution of the earth's atmosphere, depended on the presence of liquid water; if the early earth had not had oceans, the carbon dioxide might never have left the atmosphere. This would have had fateful consequences for the earth, which will become clear in our discussion of the evolution of Venus in the next chapter.

Nearly all the hydrogen escaped into space by the time the earth was about a billion years old (this is about when the first simple life forms apparently arose). The individual molecules in a gas whiz around randomly at speeds that depend on both the temperature of the gas and the mass of the particles. At a given temperature, the lightweight molecules move fastest and are therefore most likely to escape the planet's gravitational attraction. This is what happened to the hydrogen in the early atmosphere of the earth (whereas the heavier gases that now dominate the atmosphere are too massive to escape). Another gas that escaped early in the earth's history is helium, the second most abundant element in the universe, but one so rare on earth that it was not discovered until spectroscopic measurements revealed its presence in the sun.

The critical reactions that led to the development of life must have taken place before all the hydrogen escaped, because the types of reactions that were probably responsible involve this element. The earliest fossil evidence for primitive life dates back at least three billion years, when some hydrogen was still left.

Essentially no free oxygen was present in the atmosphere until life forms had developed that release this element as a by-product of their metabolic activities. Most plants release oxygen into the atmosphere, and as a supply of this element built up, the opportunity arose for complex animal forms to evolve. Besides providing the oxygen necessary for the metabolisms of living animals, the buildup of oxygen created a reservoir of ozone (O_3) in the upper atmosphere, which in turn began to screen out the harmful ultraviolet rays from the sun. Once life forms gained a toehold on the continental land masses, the process of converting the atmosphere to its present state began to accelerate. Soon nitrogen from the decomposition of organic matter began to be released into the air in large quantities, and by the time the earth was perhaps two billion years old, the atmosphere had reached approximately the composition it has today.

By this time also the mantle had solidified and the crust had hardened (the oldest known surface rocks are nearly 3.5 billion years old). The interior has remained warmer than the surface, because heat can only escape slowly through the crust, and because radioactive heating of the interior took place over a long period of time and probably is still effective today.

The moon's beginnings are much more obscure to us, despite the close-up examination that has been afforded by the *Apollo* program. Several theories regarding the origin of the moon have been proposed; (1) that the moon and the earth formed together out of the same material, (2) that the moon consists of material that split off from the earth some time after its formation, and (3) that the moon formed elsewhere in the solar system and was later captured by the earth's gravitational field. For a long time, the simultaneous-

formation theory was most widely accepted because of difficulties with the other two, but careful analysis of lunar samples has recently swung the opinions of many researchers in other directions.

The moon has major chemical similarities with the earth, but also some important differences. Certain forms of oxygen and other elements, for example, are very similar in the two bodies, which offers support for the theory of simultaneous formation. On the other hand, the moon has a lower overall iron content and a higher abundance in general of the so-called **refractory** elements, those which are not easily vaporized. This, the low abundance of easily vaporized volatile elements on the moon, and evidence found in the radioactive isotopes and apparent melting history of lunar surface rocks, indicate that the entire surface layer of the moon, down to a depth of some 300 kilometers, was molten for awhile during the first few million years after the moon's formation. These facts argue that the moon somehow was subjected to much more heating early in its history than was the earth.

There is little doubt that the moon formed from a coalescence of planetesimals in a manner similar to the earth's formation. The major uncertainties have to do with *where* the moon formed, and how it came to be orbiting the earth. Until recently, the most widely accepted idea was that the moon and the earth formed together; that debris which surrounded the infant earth in a disk merged together (Fig. 4.27). This simple theory has problems, however. Because of chemical contrasts between the moon and earth, some scientists have argued that the lunar material was actually once part of the young earth's outer layers. This could explain the relatively low abundance of iron in the moon if the material that formed the moon was ejected from the earth following differentiation, so that the volatile elements were eliminated from the lunar material. The major problem with this idea is explaining *how* the moon was then ejected from the earth; there is no known mechanism for doing this.

The capture hypothesis has long been subjected to similar criticism, because if two bodies simply come near each other as they travel their separate orbits about the sun, they cannot be trapped into mutual orbit unless excess kinetic energy is somehow lost. This could happen in a collision, and the modern capture theories invoke such an event. If the moon, having formed elsewhere in the solar system (probably near the center, where high temperatures could account for

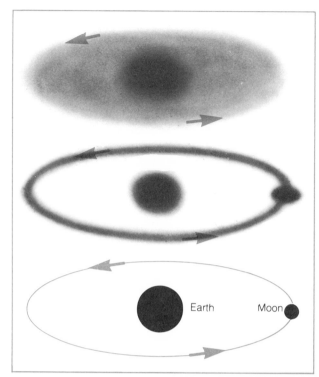

FIGURE 4.27. *Coeval Formation of the Earth and Moon.* In this theory, the moon condensed from a disk of debris that orbited the newly formed earth.

its chemical differences with the earth), encountered the earth and collided with debris surrounding it, then it could have been slowed enough to be trapped into orbit. Although such a collision may seem unlikely, it cannot be ruled out, and this scenario has the advantage of explaining most of the chemical differences between the earth and the moon. It is perhaps noteworthy that only the earth among the inner planets has any significant satellite, so it may be that such planets ordinarily form without moons. [Mars has two tiny moonlets that are almost certainly captured asteroids.)

It is ironic that so little is known for certain about the formation of our nearest neighbor in space. There are many contradictory bits of data, and it is still an open question whether the moon formed by capture or by one of the other mechanisms mentioned here. We do know something about the moon's history since its formation, however.

After the moon's formation, its crust was molten for a time because of heating caused by the heavy bombardment of the surface by objects from space. There undoubtedly was a great deal of outgassing during this period, but the moon's low escape velocity allowed all gases to dissipate into space, and thus the moon has no atmosphere. There was little differentiation, because the moon's low mass provided little gravitational force to bring about significant separation of the heavy elements. After a time, the crust cooled and hardened, but gradually the interior became molten owing to radioactive heating.

The next major stage in the moon's life involved extensive volcanic activity. Probably triggered by large impacts from space that cracked the young moon's crust, molten material from the interior flowed over large areas of the surface, particularly in the lowlands. Thus the maria were formed. It appears possible that the extensive maria on the near side of the moon (the only significant maria the moon has) were created as the result of a single tremendous impact. This also caused the near side of the moon to have a somewhat higher surface density than the far side, and this helped the earth's tidal forces alter the moon's spin. The moon's original rotation rate was apparently much faster than it is today, but the earth's tidal force has acted to slow the rotation, and now the heavy side of the moon is pointed permanently toward the earth. The internal friction that had been acting to slow the moon's rotation has been eliminated. As we will learn in later chapters, synchronous rotation such as this is the rule for all satellites that are reasonably close to their parent planets.

It is known that the earth and moon were once much closer together than they are now, and that they gradually separated as the moon's rotation was slowed by tidal forces. Because the total amount of angular momentum in a system must remain constant, the loss of angular momentum as the moon's spin slowed was compensated by the moon's growing distance from the earth, which increased the angular momentum of its orbital motion.

Throughout the moon's early history, it was continually bombarded by the rocky material still floating around the young solar system, as was the earth. On our planet, erosion and tectonic processes have erased most evidence of cratering, but on the moon, once the crust cooled, craters that were formed by the impacts of material hitting the surface were able to survive. The heavily cratered regions of the moon we see today were shaped at the time, within the first billion years after the moon formed. When the lava flows that created the maria occurred, they obliterated craters in the lowlands, so today there are few craters in the maria. Those which are seen there were formed by meteorites that hit the moon after the lava flows cooled.

After the maria were created, some three billion years ago, little else happened to modify the moon's structure. The rate of cratering decreased as the interplanetary debris either escaped the solar system or got swept up by the planets and other large bodies, so only a moderate amount of cratering has occurred since the maria formed. The moon's interior has gradually cooled, eventually reaching its present non-molten state.

PERSPECTIVE

We have itemized the overall properties of the earth and the moon, the two best-studied objects in the solar system. In doing so we have discussed many principles of planetary structure and evolution that will be applied to the other planets in the coming chapters. We have seen that the earth and moon have diverged radically in their evolutions: the earth has remained a vital, dynamic body; whereas the moon has been nearly dormant for billions of years. In the next chapter we discuss the two inferior planets, Mercury and Venus.

SUMMARY

1. The earth's atmosphere is 80 percent nitrogen and about 20 percent oxygen, with only traces of H_2O and CO_2.
2. The atmosphere is divided vertically into four temperature zones: the troposphere, stratosphere, mesosphere, and thermosphere.
3. The sun's heating and the earth's rotation create the global wind patterns.
4. The earth's interior, explored with seismic waves, consists of a solid inner core, a liquid outer core, a mantle, and a crust.
5. The earth has a magnetic field, probably created by currents in the molten core, which traps charged

particles in zones called radiation belts above the atmosphere.
6. The crust is broken into tectonic plates that shift around, which accounts for continental drift and most of the major surface features of the earth.
7. The lunar surface consists of relatively smooth areas called maria, as well as highland, mountainous regions, and is marked everywhere by impact craters.
8. The lunar soil is called the regolith, and all the rocks on the surface are igneous (most are silicates), with low abundances of volatile gases.
9. Seismic data show that the moon has a crust 50 to 100 kilometers thick, a mantle, and a core extending about 500 kilometers from the center.
10. The moon has no present-day tectonic activity and no magnetic field, indicating that there is probably no liquid core.
11. The earth's evolution from a largely molten planet with a hydrogen-dominated atmosphere to its present state was caused by the presence of liquid water on its surface, the loss of lightweight gases into space, and the development of life forms on its surface, which helped convert the atmospheric composition to nitrogen and oxygen.
12. There are significant chemical differences between the earth and the moon, this fact argues against the theory that the two share a common origin. The true mechanism for the moon's formation is unknown.
13. The moon's evolution consisted of a molten state, followed by hardening of the crust and subsequent large-scale lava flows, which created the maria. Since that time (about one billion years after the moon's formation), the moon has been geologically quiet.

REVIEW QUESTIONS

1. In the first photographs from deep space that showed the earth and moon together, the earth was very bright but the moon was dark and difficult to see. Both receive about the same intensity of sunlight. Explain why the moon looked so much darker.
2. Explain why the sun heats the earth's surface more effectively near the equator than near the poles.
3. Suppose the earth's interior had a liquid layer just below the crust but above the mantle, so that no S

waves could penetrate into the deeper zones. Would it be possible in that case to learn anything at all about the interior from observations of seismic waves?

4. If exploration of a planet revealed no sedimentary or metamorphic rocks, what would you conclude about the geological history of that planet?

5. Would you expect a planet with no molten core to have radiation belts?

6. Suppose you examine a portion of the moon where craters are so closely packed together that they actually overlap. How could you determine which were formed most recently?

7. The moon has no sedimentary or metamorphic rocks on its surface. What does this tell you about its geological history?

8. Why is the moon's surface constantly being hit by charged particles from space, whereas the earth's surface is not?

9. Summarize the differences in the way the surfaces of the moon and the earth have been shaped.

10. Explain in your own words, using information from chapter 2, how the orbit and spin of the moon have evolved since the formation of the earth-moon system.

ADDITIONAL READINGS

Anderson, D. L. 1974. The interior of the moon. *Physics Today* 27(3):44.

Battan, L. J. 1979. *Fundamentals of meteorology.* Englewood Cliffs, N.J.: Prentice-Hall.

Ben-Avraham, Z. 1981. The movement of continents. *American Scientist* 69:291.

Cadogan, P. 1983. The moon's origin, *Mercury* 12(2):34.

Carrigan, C. R., and Gubbins, D. 1979. The source of the earth's magnetic field. *Scientific American* 240(1):118.

Goldreich, P. 1972. Tides and the earth-moon system. *Scientific American* 226(4):42.

Leet, L. D., Judson, S., and Kauffman, M. E. 1978. *Physical geology.* 5th ed. Englewood cliffs, N.J.: Prentice-Hall.

Siever, R. 1975. The earth. *Scientific American* 233(3):82.

Wilson, J. T., ed. 1972. *Continents adrift* San Francisco: W. H. Freeman. (A collection of readings from *Scientific American*)

Wood, J. A. 1975. The moon. *Scientific American* 233(3):92.

THE INFERIOR PLANETS: VENUS AND MERCURY

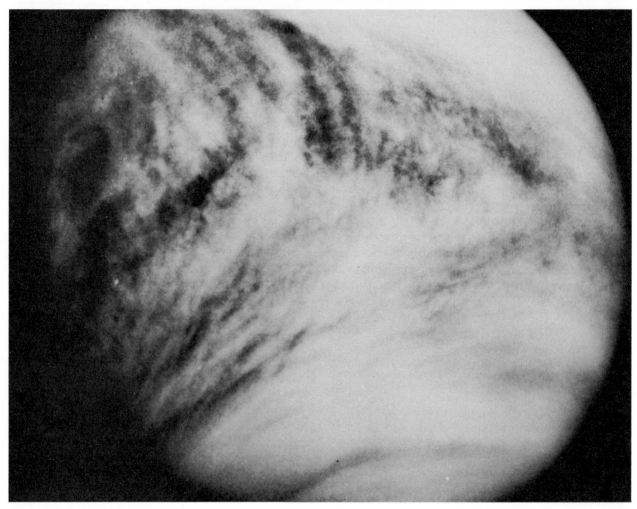

Learning Goals

5

The two planets that orbit closer than the earth to the sun belong to the class known as **terrestrial planets,** because both Venus and Mercury are similar to the earth in many general properties. Their average densities are similar to the earth's, and so are their compositions. Their formations and geological histories also bear some resemblance to that of the earth, but there are important contrasts as well.

Both Venus and Mercury can be observed only near sunrise and sunset, with greatest elongations of 47° and 28°, respectively. Each was once thought to be two separate planets, because the orbital motions could not be followed throughout a full cycle.

Venus is the brightest of all the planets, as seen from earth, and it played an important role in the development of astronomy, particularly when Galileo cited its phases as strong evidence for the heliocentric theory (Fig. 5.1). Mercury is also quite bright, but is difficult to see because of its proximity to the sun. In modern times both planets have been observed at close range by unmanned probes, and robot vehicles have landed on Venus. Thus we will be able to discuss Venus and Mercury in some detail, and we can make useful comparisons with the earth.

Probing Venus: Exploration and General Properties

Venus is very similar to the earth, more so than any other planet. Its diameter and average density are nearly identical to those of the earth. However, its surface conditions and atmosphere are vastly different from the earth's. One of the things we need to learn from the study of Venus is how these conditions came about, so that we can be sure not to turn the earth onto a similar path.

Venus is so brilliant because it is shrouded in clouds that efficiently reflect sunlight. Its albedo is 0.76. The same clouds that create the brilliance also hide the surface from telescopic view, and for this reason special techniques have been required to penetrate the clouds. A great deal was learned from remote observations, particularly at radio wavelengths, and in the last two decades spacecraft have probed Venus from the top of its atmosphere down to the surface. Both the U.S. and

FIGURE 5.1. *The Crescent Venus.*
This photograph, taken from earth, shows Venus when it is near inferior conjunction. Thus we see only a sliver of its sunlit side.

Soviet space programs have focused on Venus (**Fig. 5.2**), with each nation launching one or two missions to the clouded planet at nearly every opportunity. The crowning achievements so far have been the Soviet *Venera* landings on the surface, which have resulted in fine photographs of the landscape and useful data from other experiments; and the American *Pioneer Venus* mission, which involved several probes dropped into the atmosphere, and an orbiter that is still in operation.

Although it was once popular to imagine that Venus is a pleasantly tropical planet, with lots of water and, no doubt, plant and animal life, what was discovered from early radio observations was very different. Radio waves penetrated the cloud cover, and the message they brought to earth is that the surface of Venus is very hot, about 750 K (477°C, or about 900°F). This discovery made it obvious that, despite outward appearances of similarity, Venus and the earth were very different. No longer could Venus be viewed as a likely abode for life.

The reason for the high temperature was soon ascertained (Fig. 5.3). Spectroscopic measurements dating back to the 1930s showed that carbon dioxide (CO_2) is a major constituent of the atmosphere. This molecule has many spectral lines in the infrared part of the spectrum; thus, a CO_2 atmosphere is effectively opaque to infrared light.

FIGURE 5.2. *A Close-Up Portrait of Venus.*
This is the first full-disk image of Venus obtained by the *Pioneer Venus* spacecraft; it shows structure in the clouds.

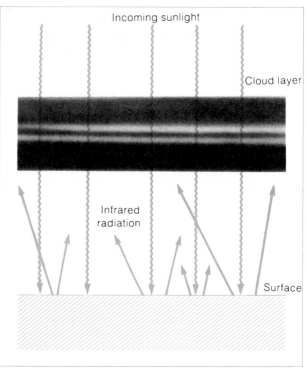

FIGURE 5.3. *The Greenhouse Effect.*
Visible light from the sun reaches the surface of Venus and heats it, causing the ground to emit infrared radiation. The CO_2 in the atmosphere of Venus efficiently absorbs infrared radiation, however, so heat is trapped near the surface.

Visible light from the sun can penetrate the clouds of Venus from above, and is absorbed at the surface, heating it. When the ground responds by emitting infrared radiation, however, this radiant energy cannot rapidly escape because of absorption by the CO_2, and heat is trapped near the surface. This heating mechanism is called the **greenhouse effect,** because it is similar to what happens in a greenhouse when light enters through the glass and heats the interior.

A second surprising discovery was made from earth, again through the use of radio (specifically radar) wavelengths. The first radar measurements of the surface were made in 1961, and the Doppler effect was used to determine the rate of rotation. Contrary to expectations, the planet was found to rotate very slowly, and even more astonishingly, to do so in a backward direction, the direction opposite that of the other orbital and rotational motions in the solar system. This is called **retrograde rotation** .

The spin of Venus takes 243 earth days, so slow that its day is actually longer than its year (the orbital period is 225 days). Here we must distinguish between the sidereal day, the 243 days just mentioned, and the solar day, the apparent time it takes for the sun to go through one daily cycle, as seen from a point on the surface of Venus. The length of the solar day on Venus is 116.8 days. The reason that the rotation of Venus is so very different from the spins of the other planets is not understood, but it seems likely that something unusual happened during the formation of Venus or at some later time.

Other data on the atmosphere of Venus were obtained from earth-based observations. Ultraviolet photographs, which reveal features in the cloud structure, showed that the cloud tops are moving rapidly (Fig. 5.4), circling the planet every four days at speeds of about 100 meters per second (360 kilometers per hour). This atmospheric circulation is also in the retrograde direction. The temperature of the cloud tops was derived from infrared measurements, which led to an estimate of 240 K ($-27°F$).

Spacecraft sent to Venus revealed a great deal more about the planet, especially about its atmospheric properties and geological history. The U.S. *Mariner* probes obtained close-up photography and other measurements that uncovered many more details about atmo-

TABLE 5.1. Venus

Orbital semimajor axis: 0.723 AU (108,200,000 km)
  Perihelion distance: 0.718 AU
  Aphelion distance: 0.728 AU
Orbital period: 224.7 days (0.615 years)
Orbital inclination: 3°23′40″

Rotation period: 243 days (retrograde)
Tilt of axis: 3°

Diameter: 12,102 km (0.949 $D_{\oplus}$)
Mass: 4.87 x 10^{27} grams (0.815 $M_{\oplus}$)
Density: 5.25 grams/cm^3
Surface gravity: 0.90 earth gravity
Escape velocity: 10.3 km/sec

Surface temperature: 750 K
Albedo: 0.76

Satellites: none

spheric circulation; the *Pioneer Venus* orbiter obtained radar maps of the surface that tell of past tectonic activity; and the *Pioneer Venus* and *Venera* landers discovered the secrets of the cloud layers and the nature of the surface rocks.

The Atmosphere of Venus

The most important ingredient in the atmosphere of Venus, already identified from earth- based measurements, is carbon dioxide. The *Venera* and *Pioneer Venus* probes showed that CO_2 dominates throughout the atmosphere all the way to the ground, and makes up about 96 percent of the atmosphere. It is interesting that the overall quantity of CO_2 on the earth is about the same as that on Venus, except that on earth the CO_2 is mostly in carbonate rocks or dissolved in seawater, and not in the atmosphere.

The next most abundant gas in the atmosphere of Venus is nitrogen (in the form of N_2), which accounts for most of the remaining 4 percent or so. In addition, traces of water vapor and other species such as oxygen, argon, neon, and sulfur (in the form of compounds such as sulfur dioxide and sulfuric acid) exist there. The sulfuric acid (H_2SO_4) apparently condenses out of the lower clouds, forming droplets like rain.

Data from the *Pioneer Venus* orbiter have recently shown that the sulfur dioxide (SO_2) abundance is

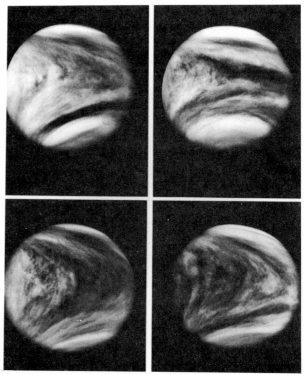

FIGURE 5.4. *Circulation of the Atmosphere of Venus.*
These four ultraviolet views cover two rotations of the planetary cloud cover. The upper two are separated by one day, and the lower two, obtained about a week after the first pair, are also separated by a day. The atmospheric motion is from right to left in these images.

highly variable, implying that this gas is periodically injected into the atmosphere. It has been suggested that active volcanoes on the surface may be responsible, since SO_2 is a major component of volcanic gases.

The atmospheric conditions on Venus vary quite a lot from the top of the clouds to the surface. The temperature increases from 240 K at cloud tops to 750 K at the ground, and the pressure at the bottom of the atmosphere reaches the rather incredible value of 90 earth atmospheres, or more than 1,300 pounds per square inch, equivalent to the pressure about 3,000 feet below the surface of the ocean. The temperature decreases smoothly from the surface to the top of the clouds.

The clouds in the atmosphere of Venus are composed of small droplets of sulfuric acid and some other, as yet unidentified, kinds of particles. The H_2SO_4 droplets are about 2 millionths of a meter (about 0.0002 centimeters) in diameter.

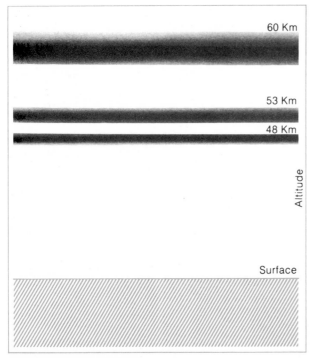

FIGURE 5.5. *The Clouds of Venus.*
The sulfuric-acid clouds are separated into three distinct layers, as shown here.

There are three distinct zones of clouds, at separate altitudes (Fig. 5.5). The uppermost layer, the one visible from afar, lies some 60 kilometers above the surface of Venus and averages about 10 km in thickness. The middle cloud layer, about 6 km thick, is suspended at an altitude of roughly 53 km, and the lowest layer lies at 48 km and is relatively thin, extending only 2 km or so vertically. Below the lower cloud is clear atmosphere, all the way to the ground. Above the uppermost clouds is a haze layer, consisting of particles much smaller than the droplets that form the clouds; their composition is unknown.

The clouds also contain quantities of sulfur dioxide which is an efficient absorber of ultraviolet light. We see from ultraviolet photographs that there is considerable structure in the clouds, because the SO_2 abundance varies with depth. Therefore, in places where the wind stirs up regions with quantities of SO_2, or where the cloud structure allows us to see to greater depths, the clouds look dark. Thus, ultraviolet photographs show some contrast in the clouds, and are useful for mapping the motions of the upper regions (see Figs. 5.2 and 5.4).

Clouds form in the atmosphere wherever the com-

bination of temperature, pressure, and relative abundance of H_2SO_4 corresponds to the conditions required for this gas to condense. This is exactly analogous to the formation of clouds in the earth's atmosphere, except that on the earth, the gas that condenses is water vapor rather than sulfuric acid. It happens that the proper conditions for condensation occur at three different levels in the atmosphere of Venus.

The atmosphere seems to be less changeable than that of the earth, where cloud structures come and go with changes in the weather. The relative stability of Venus is probably caused by a combination of factors, including the greater density and pressure of the atmosphere, and the slow rotation of the planet, which minimizes circulation patterns driven by rotation.

Another important distinction between the earth and Venus is that there is no magnetic field around Venus, a discovery made by the early *Mariner* missions. There is no magnetosphere, and no protective shield of magnetic-field lines to prevent charged particles from entering the atmosphere. This creates a complex electrical environment at the top of the atmosphere, including lightning discharges, which probably affects the chemical processes there as well.

We have seen that the upper atmosphere of Venus has a general circulation pattern at the cloud tops, moving in the same direction as the planet's rotation. The velocity is high at the cloud tops, but diminishes to nearly zero at the surface.

There are complex vertical motions in addition to the horizontal ones. Because of solar heating near the equator, there is a rising current there that spreads out in all directions, descending when it reaches cooler regions near the poles and on the dark side of the planet (Fig. 5.6). This pattern is similar to that of the circulation cells in the earth's atmosphere, except that on Venus there is insufficient rotation of the planet to break up the flow into numerous small cells. On Venus there is more of a global circulation pattern, with warm gas rising in the location called the **subsolar point,** where the sun is directly overhead, and descending in a very broad region away from that point. Besides minimizing rotational forces that might otherwise break up this simple flow pattern into smaller cells, the slow spin of Venus also means that the subsolar point moves very slowly as the planet rotates. Thus the heating from the sun lingers, enhancing the flow pattern.

Some of the gas that rises at the subsolar point flows all the way around to the dark side of venus before cooling and descending. As it does so, free atoms com-

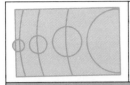

VELIKOVSKY AND VENUS

The planet Venus played a major role in a scientific controversy that began in the 1950s. The controversy was not so much about the proper interpretation of scientific data, although that played a role, but centered instead on questions of the proper role of science in responding to pseudoscientific theories, which occasionally catch the public imagination. Science must be open to new ideas, but at the same time it must be on watch against those which are not supported by evidence or rational analysis. In the case of Immanuel Velikovsky and the origin of Venus, scientists perhaps stepped beyond the boundaries of proper response.

In 1950, after years of studying ancient records, Velikovsky, trained as a medical doctor and psychologist, published a book called *Worlds in Collision*. In this book, extrapolating from records of catastrophic upheavals that apparently affected many different ancient cultures at about the same time (between the fifteenth and eighth centuries B.C.), Velikovsky suggested that these catastrophes had extraterrestrial origins. He described a sequence of events involving a "comet" released from the atmosphere of Jupiter, which twice nearly collided with the earth, causing its rotation to be stopped temporarily, and which then lost its tail in a collision with Mars before entering a circular orbit about the sun and becoming the planet Venus. Velikovsky hypothesized that the interactions between the sun and planets were dominated by electromagnetic forces, that the motions of both Mars and the earth were altered to their present state by the near-collisions with the comet, and that these near-collisions (there was also supposed to have been a collision between the earth and Mars in 687 B.C.) accounted for the upheavals in human society whose occurrence had originally inspired Velikovsky.

Despite its absolute and total disregard for the laws of physics and for astronomical observation, *Worlds in Collision* attracted a great deal of public attention and interest. Fearing that this attention would be followed by general acceptance, the sci-

entific community reacted quickly. Numerous well-known astronomers, historians, and physicists made public statements criticizing the book, its publisher was effectively boycotted, and in general the scientists gave the appearance of feeling very threatened. Unfortunately, critics of Velikovsky often used questionable tactics. Also, in many cases they were not as well-versed in his writings as they should have been in order to refute him effectively, and they frequently left themselves open for countercriticisms. Matters became more heated when new evidence, in the form of data from planetary space probes launched in the early 1960s, supported some of Velikovsky's predictions. These included the fact that Venus has a high surface temperature, that the earth's magnetic field extends very far into space, and that Jupiter is an emitter of radio waves.

The Velikovsky affair has by now faded from public view, but it has not been forgotten. There are probably people who still believe the ideas expressed in *Worlds in Collision* and in Velikovsky's second book, *Earth in Upheaval*. The continued development of the space program and the successful probes of other planets (including, of course, Venus) have helped a great deal, for other predictions of Velikovsky, such as his notion that electromagnetic forces control planetary motions, have been refuted by direct measurement. There is much less room now for Velikovsky's supporters to avoid conflict with observation.

Responding to works such as *Worlds in Collision* and similar unfounded pseudoscientific treatises (popular books on ancient astronauts and astrology, for example) is a difficult problem for scientists. There is clearly an obligation to keep the public as well-informed as possible, so that people can see for themselves the folly of unsupportable and unprovable ideas, but this must be done in a dignified manner, without creating the impression of having something to hide. Thus, Velikovsky could not be ignored, but he and his ideas probably should have been treated with less emotion and more logic.

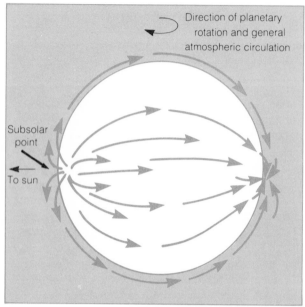

FIGURE 5.6. *High-Altitude Circulation.*
Well above the clouds (which are rotating from right to left in this sketch), gases heated at the subsolar point circulate around the planet toward the dark side, where they cool and descend.

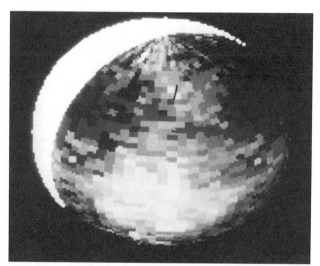

FIGURE 5.7. *Airglow on the Night Side of Venus.*
Atoms produced by the disruption of molecules on the sunlit side circulate around Venus to the dark side, where they cool and combine once again into molecules. The newly formed molecules have excess energy which is released in the form of light, creating a glow in the upper atmosphere on the night side.

bine to reform molecules, which then emit light as they cool further. This creates aurorae on the dark side of Venus (Fig. 5.7), a phenomenon that was suspected from earth-based observations, but that was not confirmed until the flight of *Pioneer Venus.* (These aurorae are quite different from those in the earth's upper atmosphere, which are created by charged particles from space.)

In and between clouds, additional vertical circulation patterns exist, depending on where the sun's heat energy is deposited. This in turn depends on the cloud composition; we have already seen that SO_2 is particularly effective as an absorber of solar energy, so there is extra heating in levels where SO_2 is abundant. There appear to be at least four flow patterns at different levels involving circulation between the equatorial and polar regions.

The Surface and Interior of Venus

Radar experiments have provided most of the information available concerning the surface of Venus. Observations made from earth had insufficient resolution to reveal much in the way of surface features, although there were some indications of highlands and lowlands. The radar instrument on the *Pioneer Venus* orbiter, however, is capable of making fairly detailed measurements, and can detect features as small as 30 kilometers. As a result, global relief maps of Venus have been constructed, showing many interesting features (Fig. 5.8).

The surface has three general types of terrain: (1) rolling plains, covering about 65 percent of the planet; (2) highlands, covering about 8 percent; and (3) lowlands, occupying the remaining 27 percent.

The rolling plains are characterized by numerous craters and circular basins, many of them apparently lava-filled, resembling small-scale lunar maria. There is uncertainty about whether all the craters are caused by meteor impacts, or whether some are volcanic in origin.

The highlands are concentrated in three major regions, two of which, called Ishtar Terra and Aphrodite Terra, are comparable in size to the continents of Africa and Australia; the third, Beta Regio, is much smaller. The maximum elevation above the mean surface level, which is found at the summit of a mountain called Maxwell Montes (Fig. 5.9), is greater than that of Mt. Everest above sea level, but the numerous can-

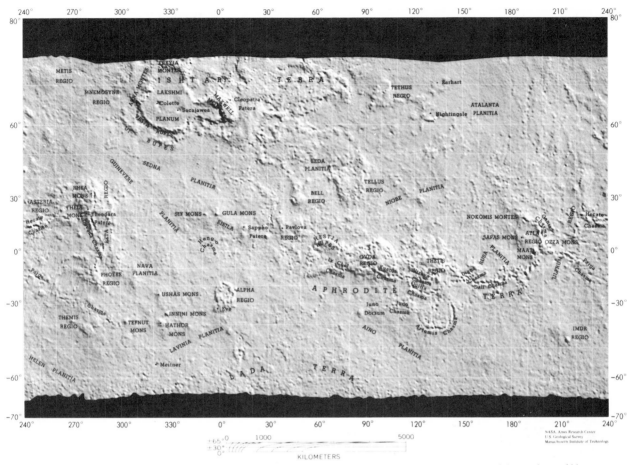

FIGURE 5.8. *A Relief Map of Venus*. This image, constructed on the basis of *Pioneer Venus* radar maps of the surface of Venus, shows the three major highland areas; Ishtar Terra at the top, Aphrodite Terra at right center, and Beta Regio at left center.

yons are much shallower than the deepest undersea trenches on earth. There are many adjacent parallel valley and trench systems in the highlands of Venus, forming systems up to 9,000 kilometers in length.

Given the similarities in size and average density of Venus and the earth, it is interesting to consider whether the internal structure and processes are also similar. One point of comparison is tectonic activity. Venus apparently does not have such active continental drift as the earth; there is not a planetwide system of plate boundaries, ridges, and trenches as there is on the earth. The reasons for this difference between the earth and Venus are not clear.

There is evidence, however, in the form of the valley and trench systems in the highlands, that tectonic activity has played some role in the formation of these regions. The highlands in part are composed of large vol-

canic mountains that built up as a result of prolonged lava-flow episodes. The smallest of the three highland areas, Beta Regio, appears to consist entirely of a few large volcanoes. The combination of the valley and ridge systems and the large volcanoes indicates that there are (or were in the past) convection currents in the upper mantle, and that there are (or were) volcanic hot spots where lava flowed upward for long periods. Thus, the internal activity and structure of Venus may be much like that of the earth, but the crust on Venus has not broken up into plates that move about, in contrast with the earth. We noted earlier that apparent variations in the atmospheric sulfur dioxide content imply that there may be active volcanoes on Venus. This would show that some tectonic activity is occuring at the present time, although it does not involve crustal movements.

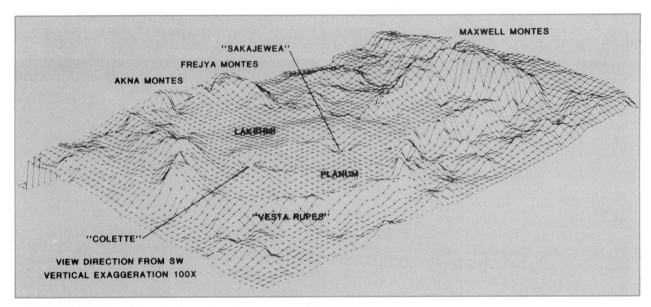

FIGURE 5.9. *A Portion of Ishtar Terra.* This computer reconstruction illustrates (on an exaggerated vertical scale) the vertical relief of this highland region, and includes the largest mountain on Venus, Maxwell Montes, more than 11 kilometers above the mean surface level.

The reason for the lack of continental drift on Venus is not known, although there are some interesting speculations. One suggestion is that because Venus has such a high surface temperature, its crust is warmer and therefore more buoyant than that of the earth. On both planets, the highlands (or continents) consist of old material that essentially floats, in equilibrium with the denser material of the lowlands (or seafloors). On earth, the seafloor material is as dense as the underlying asthenosphere, and therefore is easily forced below continental margins in the subduction zones that produce the great undersea trenches. On Venus, the lowland material is less dense than the underlying mantle, and therefore cannot easily be forced downward. The result is that convection currents inside Venus have not succeeded in creating crustal plates that can move about by slipping over one another, because subduction cannot take place.

Some information about surface rocks was derived from the photographs (Figs. 5.10 and 5.11) and other measurements obtained by the *Venera* landers. Somewhat to the surprise of scientists studying the photographs, the rocks have sharp edges, indicating a lack of erosion. This was understood when it was realized that winds at the surface of Venus are almost nonexistent.

Measurements of surface rocks reveal a basaltic composition, similar to the predominant crustal rocks on the earth and the moon. Basalt is an igneous silicate rock with a moderate density. This, combined with the fact that the average density of Venus may be comparable to that of the earth (5.3 grams/cm^3 compared to 5.5 for the earth), implies that Venus is differentiated, with a large, dense core.

Other than this very indirect evidence, very little is known about the interior of Venus. The fact that the internal structure of Venus is similar to that of the earth might lead us to suspect that Venus has a magnetic field, but as we have seen, it does not. Perhaps the rotation of the planet is too slow to allow a magnetic dynamo to be created in its core.

The Earth and Venus: So Near and Yet So Different

We are now equipped with enough information to discuss the history of Venus, and to develop some understanding of what might be responsible for the extreme differences between conditions there and on the earth.

FIGURE 5.10. *Surface Rocks on Venus*. These views are from the first Soviet *Venera* landers. They show that sunlight penetrates strongly to the surface, and that the surface rocks have surprisingly sharp edges, indicating a lack of erosion.

FIGURE 5.11. Venera 13 *and* 14 *Photos of Surface Rocks*. These two missions, which took place in the spring of 1982, obtained more extensive and better-quality images of surface rocks than had their predecessors, *Venera 9* and *10*.

This is important, for we need to know how precarious our situation is, so that we know how to avoid turning the earth onto the path that Venus has followed.

Venus is thought to have formed from rocky debris orbiting the infant sun in a great disk, just as the earth formed. The early evolution also must have been similar, with Venus undergoing a molten period when its dense elements sank to the center, leaving a lighter crust composed largely of carbonates and silicates. Before the crust cooled, volatile gases escaped from the surface, forming a primitive atmosphere of hydrogen compounds and carbon dioxide much like the earliest atmosphere on earth.

At this point a major contrast developed between the earth and Venus. As oxygen escaped from the rocks on earth and combined with hydrogen to form H_2O, much of it could persist in the liquid state, and the earth had oceans from that time onward. On Venus, however, which is closer to the sun, liquid water could not exist, but instead stayed in the atmosphere as water vapor. Venus is 0.72 AU from the sun, so the intensity of sunlight at its surface is $1/.72^2 = 1/.52 = 1.92$ times greater that at the surface of the earth (recall that the intensity of light from a source like the sun drops off as the square of the distance). This is not a very great difference, but it was extremely significant.

Once the atmosphere of Venus became contaminated with water vapor, the greenhouse effect went into operation, further heating the surface. More important, the lack of liquid water meant that there was no mechanism by which the atmosphere could lose its carbon dioxide, which was continuing to build up because of outgassing. Recall that on the earth, the carbon dioxide in the primitive atmosphere was dissolved in the oceans and deposited in rocks, where it still resides. The development of life on earth also had a major effect that was absent on Venus, gradually altering the earth's atmospheric composition to one dominated by oxygen and nitrogen, which do not create a strong greenhouse effect.

The increasing CO_2 content in the atmosphere of Venus enhanced the heating caused by the greenhouse effect, leading to the extreme surface conditions known today. Eventually the water vapor in the atmosphere was dissociated, releasing the hydrogen atoms so that they escaped into space.

Meanwhile the surface of Venus started to follow an evolution much like that of the earth. Convection apparently developed soon after the interior differentiated,

giving rise to the tectonic activity that has shaped the highland masses. On Venus, these continental areas did not begin to move about as they did on earth, for reasons discussed in the previous section. Volcanic activity has been a continuing feature of the environment, however, just as it has been on earth. In a strictly geological sense, Venus may almost be a twin of the earth.

The Mysteries of Mercury

Many people never see Mercury, despite the fact that it is rather bright (a little brighter than Sirius, the brightest star in the sky). The difficulty of observing it has made Mercury the most enigmatic of the terrestrial planets; little of its nature has been revealed by earth-based telescopes. The best photographs (Fig. 5.12), taken in the daylight so that Mercury can be viewed well above the distortion caused by the earth's atmosphere near the horizon, are degraded by the high degree of atmospheric turbulence or poor "seeing" that prevails during the day, and show little except a fuzzy disk with a hint of surface markings possibly resembling the lunar maria.

Mercury has been visited by only one space probe, the U.S. *Mariner 10* mission, but this spacecraft was able to observe the planet on no less than three separate occasions (Fig. 5.13). The *Mariner 10* trajectory first took it past Venus, which it observed at close

FIGURE 5.12. *Mercury as Seen From the Earth.*
The planet is never far from the sun, and is therefore difficult to observe.

TABLE 5.2. Mercury

Orbital semimajor axis: 0.387 AU (57,900,000 km)
 Perihelion distance: 0.307 AU
 Aphelion distance: 0.467 AU
Orbital period: 87.97 days (0.241 years)
Orbital inclination: 7°0'15"

Rotation period: 58.65 days
Tilt of axis: ? (less than 28°)

Diameter: 4,878 km (0.382 $D_{\oplus}$)
Mass: 3.302 x 10^{26} grams (0.0553 $M_{\oplus}$)
Density: 5.43 grams/cm^3
Surface gravity: 0.38 earth gravity
Escape velocity: 4.3 km/sec

Surface temperature: 700 K (day side); 100 K (dark side)
Albedo: 0.06

Satellites: none

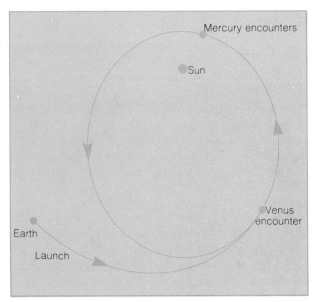

FIGURE 5.13. *The Trajectory of* Mariner 10.
The spacecraft flew by Venus, and then encountered Mercury repeatedly, as its orbital period was adjusted to coincide with twice the orbital period of Mercury.

range, and then on to Mercury, which it encountered in March 1974. It proved possible for the flight controllers to alter its course after the Mercury encounter so that it went into a solar orbit whose period was exactly twice that of Mercury itself. This meant that every other time Mercury came around the sun, there would be *Mariner 10,* flying past the same location. Thus scientists were provided with ample opportunities to make measurements of the planetary properties, although there was one shortcoming: for reasons having to do with Mercury's complex spin-orbit coupling (see next section), *Mariner 10* always faced the same side of the planet on its flybys; the other half of Mercury has yet to be observed at close range.

The orbit of Mercury is unusually eccentric; its elliptical shape is more highly elongated than that of all the other planets except distant Pluto. At its greatest distance from the sun, Mercury is almost 50 percent farther away than it is at its closest approach to the sun. As we will see, this variation in distance from the sun has had some important consequences. The orbit is also tilted by the unusually large angle of 7 degrees with respect to the ecliptic. At an average distance from the sun of about 0.4 AU, Mercury has an orbital period of 88 days, the shortest of all the planets.

Mercury is small, as planets go, with a diameter only slightly larger than the moon. Two satellites in the solar system, Ganymede of Jupiter and Titan of Saturn, are bigger. The mass of Mercury has been determined from its gravitational influence on nearby objects such

as comets, and especially the asteroid Icarus, which passed nearby in 1968. More accurate measurements were later possible, when *Mariner 10* flew by Mercury. Mercury's mass is about 5.5 percent of the earth's mass, and its density is 5.4 grams/cm^3, very similar to that of the earth, implying that Mercury probably has a large, dense core. No trace of an atmosphere was detected from earth-based observations.

Orbital Eccentricity and Rotational Resonance of Mercury

As we have seen, observations of Mercury are difficult to make, and no surface features on the planet can be seen distinctly. Careful observations of the fuzzy markings that do show up led to the conclusion that Mercury was in synchronous rotation, always keeping the same side facing the sun. As close to the sun as Mercury is, this was not a surprising idea, for the tidal forces acting on it are immense.

Radio observations of Mercury carried out in the early 1960s allowed its surface temperature to be measured, though the use of Wien's law (see chapter 3).

In chapter 2 there was some discussion about the mechanics of placing a satellite in the earth's orbit, but nothing about how to aim a spacecraft that is designed to visit another planet. The principles involved are particularly simple when an inferior planet is to be the target.

The concept of energy was discussed in chapter 2, where we pointed out that an orbiting body has both kinetic energy owing to its motion, and potential energy owing to the gravitational field of the object that it orbits. For a planet orbiting the sun, the total energy (the sum of kinetic plus potential) is greater, the larger the orbit. Thus, for an object to go from the orbit of the earth to that of an inner planet, it must lose energy.

A spacecraft sitting on the launchpad is moving with the earth in its orbit, at a speed of 29 km/sec. To make the rocket fall into a path that intercepts the orbit of an inner planet, we must diminish this speed. Therefore, we launch the rocket backward with respect to the earth's orbital motion, thereby lowering its velocity with respect to the sun, and decreasing its orbital energy. If the launch speed is properly chosen, the spacecraft will fall into an elliptical orbit that just meets the orbit of the target planet. In order to reach Mercury from the earth, the rocket must be launched backward with a speed of 7.3 km/sec relative to earth.

There are, of course, a few additional considerations. For one thing, the rocket has to be launched with enough speed to escape the earth's gravity. The escape velocity for the earth is 11.2 km/sec, so in fact we have to launch our Mercury probe with a speed in excess of this value, which is more than the speed needed to attain our trajectory to Mercury. The launch speed must therefore be calculated to take the earth's gravity into account, so that the rocket escapes the earth, but in the process is slowed just the right amount to give it the proper course. The gravitational pull of the target planet must also be taken into account, for this speeds up the spacecraft as it approaches.

Another important consideration is timing; the target planet must be in the right spot in its orbit at just the moment when the spacecraft arrives. It is for this reason that the term *launch window* is used. The earth and the target planet must be in specific relative positions at the time of the launch. The interval between launch windows for a given planet is simply its synodic period. For a probe to Mercury, the travel time is about 106 days, so Mercury actually makes a little more than one complete orbit while the probe is on its way.

The gravitational pull of the target planet can be used to good advantage in modifying the orbit of a spacecraft that flies by. If the probe is aimed properly, its trajectory may be altered in just the right way to send it on to some other target. This technique was used with *Mariner 10,* first to send it from Venus to Mercury, and then to modify its orbit again so that it returned to Mercury repeatedly thereafter.

Sending spacecraft to the outer planets is a bit more difficult, because the spacecraft has to have more energy than what it gets from the earth's motion. The launch is therefore made in the forward direction, so that the rocket has the combined speed of the earth in its orbit plus its own launch speed with respect to the earth.

Ironically, the seemingly simple task of launching a probe to the sun is one of the most difficult. The most straightforward solution would be to launch the spacecraft backward from the earth with a speed of 29 km/sec, entirely canceling out the earth's orbital speed, so that the probe would then fall straight in toward the sun. This is a prohibitively high launch speed, however, so in practice we will probe the sun by first sending the spacecraft out around Jupiter. The spacecraft will fly by the massive planet in such a way (backward with respect to Jupiter's orbital motion) that its own orbital energy is reduced; the craft can then fall into the center of the solar system. A mission to explore the outer layers of the sun in this manner is in the planning stages.

The daylit side was hot, as expected, with a temperature of about 700 K. Because observers thought that the other side never saw the light of day, they expected to find very low temperatures there. It was a surprise, therefore, when a temperature of about 100 K was deduced for the dark side. This is certainly cold, but not as cold as it should have been if this side of the planet never faced the sun.

Radar observations soon provided the solution to this mystery. The Doppler shift of radar signals reflected off of Mercury showed that the planet's rotation period is shorter than its 88-day orbital period. Mercury is not in synchronous rotation, but instead rotates once every 59 days, so that all portions of its surface are exposed to sunlight some of the time.

But why 59 days? This was a bit of a puzzle in itself, because most planets spin much more rapidly. Again, the answer was forthcoming, when it was noted that 59, or more precisely, 58.65 days, is exactly two-thirds of the orbital period of 88 (actually 87.97) days. Thus Mercury rotates exactly three times for every two trips around the sun.

This is no mere coincidence; clearly the gravitational tidal forces exerted by the sun are at work. The key is what happens at the point where Mercury is closest to the sun, called **perihelion,** for it is here that the tidal forces are the strongest. If Mercury has a heavy side, like our moon, then the tidal forces exerted by the sun will act to ensure that this side is aligned with the sun's direction when Mercury is at perihelion. One way to accomplish this, of course, would be synchronous rotation, so that this side would always point toward the sun, but because Mercury's orbit is so elongated, the urge to keep its heavy side facing the sun is most significant at perihelion.

The planet spins one-and-a-half times during each orbit (Fig. 5.14); thus, the heavy side either faces the sun or exactly the opposite direction at each perihelion passage. In either case, the tidal forces on Mercury are balanced, so that there is no tendency to alter the rotation further. Apparently the planet once rotated much faster, but whenever it passed through perihelion with the heavy side pointed in some random direction, tidal forces exerted a tug that tended to change its spin. This went on until the rotation slowed to the present rate. If the orbit had not been so elongated, the tidal forces could have been more uniform throughout, and Mercury would no doubt be in synchronous rotation.

The three-to-two relationship between Mercury's orbital and spin periods is an example to **spin-orbit**

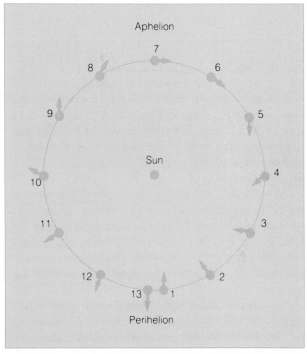

FIGURE 5.14. *Spin-Orbit Coupling.*
This sketch shows how Mercury spins 1½ times while completing one orbit around the sun. At perihelion, it always has the same axis aligned with the sun.

coupling, a general term applied to any situation where the spin of a body has been modified by gravitational forces so that a special relationship between the orbital and spin periods is maintained. Synchronous rotation is the most common example, and we will find several cases of it as we explore the rest of the solar system.

One remaining question regarding Mercury and its spin is why it should have a heavy side, rather than being a symmetric sphere. We will find a hint of a possible answer in a later section.

The combination of orbital and rotation speeds on Mercury produces a very unusual occurrence for anyone who might visit the planet. At closest approach to the sun, when Mercury is moving most rapidly in its orbit, the orbital speed is actually greater than the rotation speed at its surface. For a short time (lasting a few days) the sun would appear to turn around and move backward across the sky, from west to east. Imagine the difficult time our ancestors would have had explaining this retrograde motion of the sun if a similar phenomenon had occurred on the earth!

Mercury: A Moonlike Exterior and a Dense Interior

The relatively high density of Mercury, known before the *Mariner 10* encounters, indicated the likelihood of a dense core inside this planet, similar to the earth and Venus. This might have led us to believe Mercury had a magnetic field, except for the fact that the slow rotation was thought insufficient to produce the internal electrical currents neeeded to activate a magnetic dynamo.

Thus it was a bit of a surprise when *Mariner 10* detected a magnetic field around Mercury. Its strength is not great, about 1 percent as strong as that of the earth, but nevertheless it is there. Mercury has a magnetosphere, just as the earth does, whose shape is modified by the charged particles flowing past from the sun.

Scientists concluded that Mercury's core must be relatively large (Fig. 5.15), in order to produce a magnetic field despite the slow rotation of the planet. The core apparently extends about three-fourths of the way out from the center to the surface, whereas the earth's core is contained within the inner half of its radius.

The implication of such a large core is not only that Mercury is differentiated, but also that it contains a relatively high overall abundance of heavy elements to begin with. This says something about its formation, namely, that a greater portion of its volatile, lightweight gases escaped than was the case for the other terrestrial planets. Thus, the temperature in the vicinity of Mercury must have been relatively high in the early history of the solar system, higher that at the places where Venus, the earth, and Mars formed. Like the other planets where well-developed cores exist, Mercury must have been fully molten, staying in that state long enough for differentiation to occur.

Not much is known about the mantle of Mercury, which occupies the outer 25 percent or so of its radius. The density there is lower than in the core; its other properties, however, are unknown. The photographs taken by *Mariner 10* show no obvious evidence of tectonic or volcanic activity, although we must keep in mind that half of the surface has yet to be examined.

The surface of Mercury, at first glance, strongly resembles the lunar landscape (Fig. 5.16). It even looks quite similar on second glance; the differences are revealed only through rather detailed scrutiny.

Perhaps the most obvious distinction is found in the

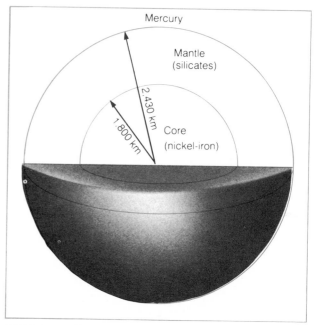

FIGURE 5.15. *Mercury's Internal Structure.*
The presence of a magnetic field, along with the planet's relatively high density, implies that Mercury has a large core.

prominent and extensive cliffs, referred to as scarps, which exist in many locations and form a network that girdles the planet (Fig. 5.17). These are quite unlike anything seen on the other terrestrial planets or the moon. Their appearance suggests that Mercury shrank

FIGURE 5.16. *Mercury as Seen From Space.*
This *Mariner 10* view is a mosaic of several images. The planet bears a strong resemblance to the moon, although there are significant contrasts, particularly in internal structure.

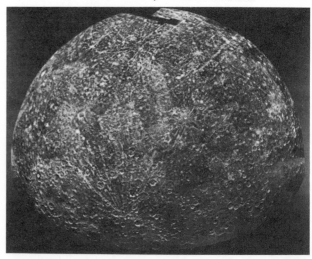

FIGURE 5.17. *A Scarp System on Mercury.*
This lengthy series of cliffs (seen here extending from upper left to lower right) is about 300 km long, and is one of many on the planet's surface.

FIGURE 5.18. *Craters on Mercury.*
Because Mercury has a greater surface gravity than the moon, impact craters have lower rims and are shallower, and ejecta do not travel as far. A scarp is also visible in this photo, in the upper left.

a little after its crust hardened, causing it to shrivel and crack.

Another less obvious difference between Mercury and the moon is that on Mercury the craters tend to be more widely separated, with more smooth space between them (Fig. 5.18). Apparently when impacts occur on Mercury and form craters, the ejecta do not travel as far as on the moon, so there are less extensive rays or other ejected debris around the large craters. The reason for this is clear: Mercury has a higher surface gravity than the moon, by about a factor of three, so that the rubble blasted out of the surface by an impact does not travel as far before falling back down (Fig. 5.19). The craters on Mercury also tend to have a flatter appearance than those on the moon, for the same reason. There are large smooth areas on Mercury, much like the lunar maria but not so extensive, which were formed by lava flows.

One very large impact crater was seen by *Mariner 10*, just at the dividing line between daylight and darkness. The basin, formed by the tremendous crash of a massive body hitting the surface, resembles a giant bull's-eye, with concentric rings around it, and it extends over a diameter of some 1,400 kilometers (Fig. 5.20). The gigantic crater, called Caloris Planitia, happens to lie on the side of Mercury that is facing the sun at perihelion every other orbit (the name Caloris was given to the crater for this reason, because this position is the hottest place on Mercury every other time it

passes close to the sun). We noted earlier that Mercury is apparently lopsided, with an asymmetric distribution of mass; the position of this huge impact basin suggests that perhaps the object that crashed into Mercury was responsible for making the planet heavier on one side.

Directly opposite Caloris Planitia, on the far side of Mercury, is a curiously wavy region, so strange and unprecedented in appearance that it is called the *weird terrain*. Its rippled appearance was probably created by seismic waves that raced around the planet when

FIGURE 5.19. *Crater Formation on the Moon and on Mercury.*
This shows how moon's lower surface gravity allows higher rim walls to remain, and allows ejecta to travel farther from the point of impact than on Mercury.

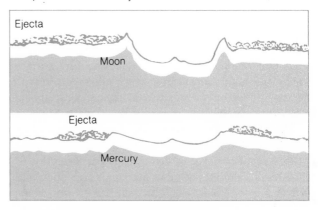

FIGURE 5.20. *Caloris Planitia.*
This photo shows half of the immense impact basin known as Caloris Planitia. This region is directly facing the sun at perihelion on every other orbit.

the impact that formed Caloris Planitia occurred. When the waves met on the far side of the planet, they distorted the surface there, resulting in the odd terrain seen today. Similar but less prominent features are seen on the moon, at positions opposite some of the largest impact craters.

Infrared measurements made by *Mariner 10* indicate that there is a fine layer of dust on Mercury, much like that on the moon. The surface rocks themselves are probably also similar to earth and moon rocks, except perhaps for a lower abundance of volatile elements.

Despite the combination of relatively low surface gravity and high temperature, which allowed nearly all the gases present on Mercury to escape long ago, a trace of an atmosphere does appear to exist. *Mariner 10* detected very small quantities of a number of gases that are thought to originate in the solar wind, the steady stream of particles that sweeps past Mercury.

All these data indicate that Mercury's formation and evolution must have been quite similar to that of the moon. Mercury and the moon went through comparable stages of accretion and melting at the beginning, followed by hardening of the crust along with major lava flows (perhaps triggered by large meteor impacts), and succeeded finally by a long period of cooling and decreased cratering.

We are not sure just when the tremendous impact that created Caloris Planitia took place, nor how long ago tidal forces locked the planet's rotation into its 3:2 coupling with the orbital period. We can surmise that there must have been substantial internal heating before then, as tidal forces exerted by the sun created internal friction. Perhaps there is still some excess internal heating caused by tidal forces, helping to maintain the large molten core needed to create the planet's magnetic field.

PERSPECTIVE

The study of Venus and Mercury has been very instructive, not only for what it tells us of the planets themselves, but also for the principles that can be applied to other planets. Venus is almost identical to the earth, yet has a very different atmosphere and different surface conditions, and the reason for these differences helps teach us how precarious the environment here on earth could be. Mercury illustrates for us the principles of spin-orbit coupling, a phenomenon that we will see many more times in our exploration of the solar system.

We turn our attention now to the last of the terrestrial planets, Mars.

SUMMARY

1. Earth-based observations of Venus, particularly radio data, revealed that the planet is extremely hot at the surface, and rotates slowly in the retrograde direction.

2. The atmosphere of Venus consists mostly of carbon dioxide, with a bit of nitrogen and traces of sulfur dioxide and sulfuric acid.

3. The atmospheric temperature of Venus increases

from 240 K at the cloud tops to 750 K at the surface, where the pressure is ninety times that at sea level on the earth. The heating is caused by the greenhouse effect.

4. The clouds of Venus, composed of droplets of sulfuric acid, form three distinct layers between 48 and 60 km altitude. The atmosphere is clear below the clouds.

5. At Venus's cloud tops, high winds circle the planet in four days, while it is calm at the surface. Above the clouds there are global convection currents.

6. The surface of Venus, consisting of rolling plains, highlands, and lowlands, shows evidence of limited tectonic activity but no continental drift.

7. Venus evolved in a manner very different from the earth, primarily because it is so close to the sun that liquid water could not exist on its surface. The lack of water allowed CO_2 to stay in the atmosphere and discouraged the development of life.

8. The *Mariner 10* space probe made three close flybys of Mercury, and provided the most detailed information regarding the planet's properties.

9. Radar measurements showed that Mercury rotates exactly one-and-a-half times for each orbit around the sun, in a form of spin-orbit coupling that was created by a combination of the sun's tidal force on Mercury and the highly elliptical orbit of the planet.

10. Mercury has a magnetic field and a high density, indicating that it has a large, dense core, probably partially molten.

11. The surface of Mercury is nearly identical to that of the moon, except for a planetwide system of scarps, and minor differences in the heights of impact craters and the distances traveled by ejecta.

12. The great impact basin Caloris Planitia lies at the subsolar point at perihelion on alternate orbits of the sun, and may be the site of the mass imbalance that caused Mercury's spin to fall into resonance with its orbital period.

REVIEW QUESTIONS

1. Why could the mass of Venus not be deduced precisely from earth-based observations in the same manner as the masses of most of the other planets?

2. On the earth a solar day is a little longer than a sidereal day, but on Venus it is shorter. Explain why.

3. Compare the role of SO_2 in the atmosphere of Venus to that of ozone in the atmosphere of earth.

4. Why does the fact that Venus has a large, dense core lead us to suspect that the surface rocks have a low iron content?

5. Why do conditions on Venus cause scientists to be concerned about the increasing levels of CO_2 in the earth's atmosphere?

6. How much more intense is sunlight on the surface of Mercury at perihelion than at aphelion? Assume that the planet is 50 percent farther from the sun at aphelion.

7. Given what you know of the surface albedos and the diameters of the moon and Mercury, would the moon be visible from earth if it orbited Mercury?

8. What do you think the rotation period of Mercury would be if the planet had a perfectly circular orbit?

9. Compare Mercury and the moon in terms of surface terrain and probable internal structure.

10. Mercury has about the same average density as the earth, yet its core occupies a greater fraction of its volume than does that of the earth. What does this tell you about the density of Mercury's core, compared with the density of the earth's core?

ADDITIONAL READINGS

Beatty, J. K., O'Leary, B., and Chaikin, A., eds. 1981. *The new solar system.* Cambridge, England: Cambridge University Press.

Hartmann, W. K. 1976. The significance of the planet Mercury. *Sky and Telescope* 51(5):307.

Head, J. W., Wood, C. A., and Mutch, T. A. 1977. Geologic evolution of the terrestrial planets. *American Scientist* 65:21.

Murray, B. C. 1975. Mercury. *Scientific American* 233(3):58.

Pettingill, G. H., Campbell, D. B., and Masursky, H. 1980. The surface of Venus. *Scientific American* 243(2):54.

Schubert, G., and Covey, C. 1981. The atmosphere of Venus. *Scientific American* 245(1):66.

Weaver, K. F. 1975. *Mariner* unveils Venus and Mercury. *National Geographic* 147:848.

Young, A., and Young, L. 1975. Venus. *Scientific American* 233(3):70.

MARS AND THE SEARCH FOR LIFE

6

Learning Goals

The planet Mars, although not as brilliant as Venus, is nevertheless a prominent object in the heavens (Fig. 6.1). The fourth planet from the sun, and our next nearest neighbor after Venus, Mars has played an important role in the development of astronomy, first lending itself to mythological interpretations, and later providing the basis for Kepler's discoveries on the motions of the planets. In rather recent times, Mars again became a center of considerable speculation, as people mused over the possibility that life might exist there. The most recent chapter in the story, the exploration of Mars by unmanned spacecraft, has dispelled such notions but has not diminished our fascination.

Historical Notions about Mars

From antiquity, Mars has attracted special attention because of its distinct and unusual reddish color, apparent to even the unaided eye. The Greeks named this planet Ares, after their god of war, and the Roman name Mars has the same connotations.

The first telescopic observations of the red planet were made by Galileo, who noted that its disk varies in angular diameter as its position with respect to the sun varies. About fifty years later, Huygens became the first to sketch surface features as he saw them on the planet's disk, something that later led to one of the more fascinating scientific debates since medieval times. Successive generations of astronomers attempted to study the surface markings—a very subjective business, especially since there was no photographic film with which to record the image. Photographs are not necessarily superior to the human eye in this case anyway, because variations in the turbulence of the earth's atmosphere may allow momentary clear glimpses, but these few instances of high clarity will be smeared out in a photograph obtained over a long exposure time.

Huygens noted a dark patch on the surface, which was named Syrtis Major. Additional details were seen later, and by the 1870s observers had named a number of features, using Latin words related to bodies of water, which they thought the features to be. At about this time, Father Secchi, the Roman Catholic priest who later was to play a pioneering role in the development of stellar spectrum analysis (see chapter 13), saw what

TABLE 6.1. MARS

Orbital semimajor axis: 1.524 AU (277,900,000 km)
  Perihelion distance: 1.382 AU
  Aphelion distance: 1.666 AU
Orbital period: 1.881 years (687.0 days)
Orbital inclination: 1°51′0″

Rotation period: 24^h 37^m 22.^{s}6
Tilt of axis: 23°59′

Diameter: 6,786 km (0.532 D$_\oplus$)
Mass: 6.42 × 10^{26} grams (0.107 M$_\oplus$)
Density: 3.95 grams/cm^3
Surface gravity: 0.38 earth gravity
Escape velocity: 5.0 km/sec

Surface temperature: 130–290 K
Albedo: 0.15 (average value)

Satellites: 2

he believed to be linear features on the surface of Mars, and he referred to them as *canali*, an Italian word referring generally to natural channels of water. Giovanni Schiaparelli, who was then the director of an observatory in Milan, imagined that he saw a network of these features on the surface of Mars, and in 1877, when Mars was at opposition, he produced a detailed sketch mapping them (Fig. 6.2). Soon other observers around the world were seeing the canali, and in due course

FIGURE 6.1. *The Earth-Based View of Mars.*
A red filter was used to make this photograph, in order to enhance the clarity of surface markings.

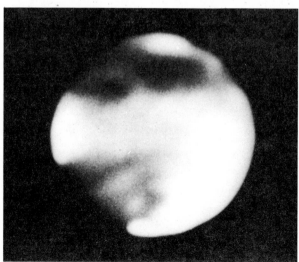

the name began to be interpreted as meaning canals, artificially constructed channels for transporting water.

This notion was particularly fascinating to a wealthy Boston aristocrat named Percival Lowell, who fantasized an extensive and well-developed Martian civilization. Lowell published a book full of these speculations in 1895, which created a groundswell of public interest in the idea. Despite skepticism on the part of some astronomers who were unable to see the canals or who realized how unreasonable it was that Martian canals should be visible from earth even if they really existed, the idea of a Martian civilization became fixed in the public mind. To this day a wide variety of science-fiction novels and movies (starting with H. G. Welles's *The War of the Worlds*) have been inspired by these beliefs.

When close-up photographs of Mars were eventually obtained by space probes that flew by the red planet, nothing was seen resembling the linear features that gave rise to the idea of canals. Apparently these were entirely nonexistent; it was the tendency of the human eye to connect separated dark areas that led to the early depictions of Mars as covered with dark lines.

Observations and General Properties

As long ago as 1666, the rotation period of Mars (slightly more than twenty-four hours) was determined from observations of motions of the surface features. In the late 1700s the British astronomer Sir William Her-

schel deduced the angle of inclination of the Martian axis of rotation, finding it very similar to the tilt of the earth's axis; this fact explained the seasonal variations already seen in the polar ice caps. In 1877 the two moons of Mars, Phobos and Deimos, were discovered, which allowed, through the use of Kepler's third law, accurate determinations of the mass of the planet to be made.

Spectral analysis of the light reflected to earth from Mars started in the first decades of the twentieth century. Photographs taken through red and blue filters revealed a haze in the Martian atmosphere that scatters blue light, and the first spectroscopic measurements in the 1930s showed an absence of water vapor and oxygen, of particular interest in view of the prevailing idea that life might exist on Mars. More recently, absorption lines of carbon dioxide were identified in the spectrum of Mars.

Seasonal changes in the coloration of certain regions on Mars (Fig. 6.3), well away from the polar caps, were once interpreted as being caused by foliation of green plants, which were thought to undergo the same sort of annual cycle as earthly plants. By the 1960s photographs had revealed evidence of large-scale dust storms in the Martian atmosphere, and this inspired the American astronomer Carl Sagan to explain the seasonal color variations in terms of seasonal shifts in the Martian wind patterns, which cover and uncover some areas of the planetary surface. This explanation was later verified by the space probes sent to observe Mars at close range.

Infrared observations of Mars yielded data on its surface temperature, which was found to vary between 130 K (about −225°F) and 290 K (63°F). At high noon

FIGURE 6.2. *The Martian "Canals."* This sketch, made by Schiaparelli, shows the network of linear features that were for a time interpreted as artificial water channels on the surface of Mars.

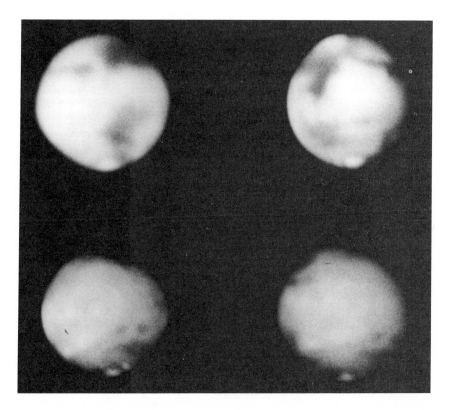

FIGURE 6.3. *Variations in the Martian Surface Markings*. This series of photos shows the changes in the polar ice caps and surface markings that occur with the seasons on Mars.

in the summer near the equator of Mars, the temperature is comfortable by earthly standards.

Given our profound interest in Mars and the history of speculation over the possibility that life exists there, it is not surprising that the red planet became an early target for unmanned interplanetary probes. The first successful mission to Mars was *Mariner 4,* launched in 1964, and this was followed by other *Mariner* missions, most notably *Mariner 9,* which orbited the red planet and surveyed it photographically. This task was stymied for several months by a global dust storm, which completely obscured the surface of Mars (Fig. 6.4). The storm finally abated, however, allowing *Mariner 9* to make many profound discoveries, including several gigantic volcanoes and a valley that dwarfs anything on earth.

The crowning achievement of the exploration program came some four years later, with the launching of two *Viking* spacecraft in the late summer of 1975. Each consisted of an orbiter, destined to circle Mars as *Mariner 9* had done; and a lander, designed to descend to the surface and land intact (Fig. 6.5). The orbiters and landers each carried a complex assortment of instruments designed for a variety of measurements. The orbiters improved on the quality of the older *Mariner 9* photographic survey of the entire planet, and the landers were able to carry out many sophisticated experiments, including most notably panoramic photographs of the Martian landscape and sensitive tests for the presence of microscopic life forms in the soil. The results are discussed later in this chapter.

The Martian Atmosphere and Seasonal Variations

Mars has a very thin atmosphere compared to the earth, with a surface pressure of 0.006 times the sea-level pressure on earth. The Martian air is about 95 percent carbon dioxide (CO_2), and is similar to Venus in this regard. Besides CO_2, other species include nitrogen, which accounts for 2.7 percent; argon, which constitutes another 1.6 percent; and other trace gases, which together total less than 1 percent. Water vapor

Astronomical Insight 6.1

PLANETARY SPECTROSCOPY AND MARTIAN WATER.

We have made numerous references, both in this chapter and in our discussion of Venus, to analyzing the composition of a planet's atmosphere by measuring its spectrum. This is in fact a fairly complex procedure, primarily because a planet shines by reflected sunlight, and does not emit light of its own. This means that its spectrum is basically a solar spectrum, which is complex in its own right, with spectral lines caused by the planet's atmosphere mixed in. A second complication is that when observing a planet from a telescope on the earth's surface, we are looking through the earth's atmosphere, which may contain some of the same gases that are in the atmosphere of the planet. Therefore, when absorption lines caused by a given gas are seen, it is difficult to ascertain whether they form in the earth's atmosphere or that of the distant planet or both.

The first problem, that of distinguishing between solar lines and those caused by a planetary atmosphere, is solved in two ways. First, taking into account the high temperature of the sun compared to any planet's atmosphere, we know that the sun's spectrum should not show lines of molecules (which cannot exist at high temperatures), whereas most of the gas in a planetary atmosphere is in molecular form. Second, confusion can be reduced by comparing the sun's spectrum, taken directly, with that of the planet, which contains the sun's spectral features plus those caused by the planetary atmosphere. Features that are identical in both spectra may be attributed to the sun (or to the earth's atmosphere), whereas those which appear only in the spectrum of the planet must form in the planet's atmosphere.

This comparison technique is also a partial solution to the second problem, that of distinguishing between planetary spectral lines and those formed by the earth's atmosphere. The latter (called **telluric lines**) would show up in both the solar and the planetary spectra, and would therefore be recognizable as not being caused by the planet alone.

---

is present in variable quantities, reaching levels of a few percent in the Martian summer. This is occasionally sufficient to form clouds.

The Martian polar caps have been somewhat controversial, because of our difficulty in determining whether they are made of water ice or frozen CO_2. Temperature measurements have shown, however, that dry ice (frozen CO_2) accounts for most of the cap material, but that some water ice is also present. Apparently the seasonal variations in the polar caps are caused by the accumulation of CO_2 ice during the Martian winter, followed by evaporation in the summer. The residual caps that remain throughout the year are probably water ice.

The general question of water on Mars is complex, and for a long time was the basis of a serious scientific mystery, quite separate from the earlier speculations about canals. Enough water should have been released from surface rocks in the early history of the planet to cover its entire surface to a depth of several hundred meters; thus, the lack of water on the planet is difficult to explain. There probably was never an energy source on Mars to dissociate H_2O molecules, as happened on Venus, but on the other hand, the atmospheric pressure is too low to allow liquid water to exist on the surface. The atmosphere, which gets rather cold at night, cannot support much water in the form of vapor, and there is only a small amount in the ice caps. The answer had to be sought elsewhere. We will discuss current ideas about the fate of the Martian water supply in the next section.

The existence of clouds in the atmosphere of Mars was recognized long ago from earth-based observations. Some are dust clouds, many are white clouds formed of ice crystals (primarily H_2O), and some are CO_2 clouds. The clouds are found in lowlands in the early Martian morning, and in regions where winds push the air up mountain slopes, forcing the H_2O va-

There still can be confusion, however, when the same gas is present in the atmospheres of both the earth and the planet. The best technique for resolving this is to take advantage of the fact that a planet, under most circumstances (except at the instant of opposition or conjunction), will have a small velocity toward or away from the earth. The Doppler effect will therefore shift the positions of the planet's spectral lines with respect to those caused by the earth's atmosphere. The shift is always small, so that the planetary and telluric lines will be close together in the spectrum. Thus it is necessary to use high-resolution spectroscopy, to spread out the wavelengths as widely as possible.

All the techniques described here were used in the attempt to measure the quantity of water vapor in the atmosphere of Mars. It was clear from the start that lines caused by water vapor were present in the spectrum of Mars, but they also appear in the spectrum of the sun, showing that the earth's atmosphere contains water vapor (which was no surprise, of course). Very early results, obtained just after the turn of the century, seemed to indicate that these lines were much stronger in the Martian spectrum than in the solar spectrum, showing that there

must be considerable quantities of water vapor on Mars. This conclusion, however, was negated by later studies showing little or no difference between the solar and Martian spectra. In the 1930s, when high-resolution techniques began to be used, no absorption lines caused by water vapor were found at the Doppler-shifted position where they were expected (at that time the earth was receding from Mars at a speed of 17 km/sec). It was finally concluded that Mars had little water vapor in its atmosphere.

Despite the results of the spectroscopic observations, other data indicated that there must be some water vapor on Mars, even though it may be in miniscule amounts. This suspicion was based on the presence of what appeared to be water-vapor clouds or mist, and the notion that the polar ice caps contained water ice. And long before the launching of the first space probes to Mars, it was suggested that there might be large amounts of water below the surface, in the form of permafrost. All these tentative ideas have been confirmed or at least strongly supported by close-up inspection of the red planet.

por to cool and condense. In late summer in each hemisphere, extensive cloud systems form over the polar caps.

Winds on Mars are always present, often with substantial velocities (although the thinness of the atmosphere reduces the winds' impact on surface features). The global flow patterns are created by the daily variations in solar heating at the surface, and are strongly modified by the terrain, such as the large mountains. The prevailing winds tend to move from west to east, but with a veer either toward or away from the poles, depending on the season. Typical wind velocities are 35 to 50 kilometers per hour.

The fact that Mars has seasons was revealed a long time ago, when variations in the sizes of the polar ice caps were noticed. The cause of the seasons is a bit more complex than on earth, however, because of the elliptical shape of the orbit of Mars. Earth's orbit is so nearly circular that the intensity of sunlight hitting its

surface is almost constant year-round. (In fact, the earth is closest to the sun in early January, and the slight extra solar input surely does not moderate the winter weather in the northern hemisphere.) The orbit of Mars is noticeably elongated, as we have already learned, with the result that at closest approach to the sun Mars is about 17 percent nearer than when it is farthest away, on the other side of its orbit (Fig. 6.6). Therefore the sun's intensity on the Martian surface is $1/.83^2 = 1.45$ times greater at closest approach, which is enough to have a pronounced effect on the climate. Mars approaches the sun most closely during winter in the northern hemisphere, and is farthest away during the northern summer; hence the shape of the orbit tends to moderate both seasons in the north. By the same token, the southern seasons are enhanced by the orbital shape, so that this hemisphere has much more extreme variations from summer to winter. The southern polar cap grows larger than the northern one in

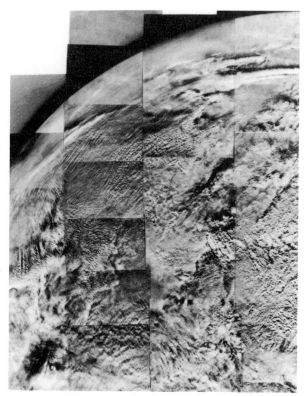

FIGURE 6.4. *A Global Dust Storm.*
This is the view of Mars that confronted the *Mariner 9* spacecraft for the first several weeks after it went into orbit about the planet.

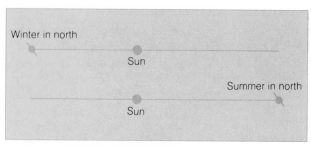

FIGURE 6.6. *The Martian Seasons.*
This exaggerated view shows that in the northern hemisphere, both summer and winter are moderated by the varying distance of the planet from the sun, whereas in the southern hemisphere, both seasons are enhanced. The extreme temperature fluctuations in the southern hemisphere give rise to the winds that cause the seasonal dust storms.

winter, but diminishes to a smaller remnant in summer.

The most striking seasonal variations on Mars are the widespread dust storms, which for several decades have been observed to occur when Mars is closest to the sun, during the southern summer. The extra solar heating in the south causes very rapid evaporation of the polar cap, and this in turn creates strong winds between the cap and the surrounding territory, winds driven by the extreme temperature difference's of the two adjacent regions. Before long, the winds are rapid enough to raise fine particles off the ground, and

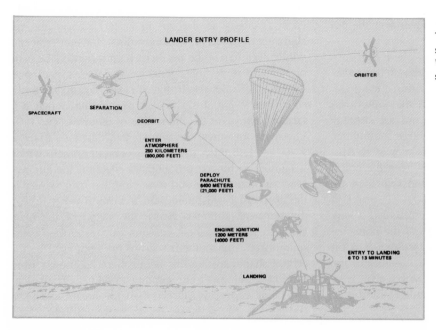

FIGURE 6.5. *The* Viking *Landers.*
This schematic illustration shows the sequence of events as one of the *Viking* spacecraft descended toward its soft landing on the Martian surface.

quickly the air is filled with dust, which reaches high altitudes. The storm grows, usually covering the southern hemisphere; it often expands to smother the entire planet and lasts for weeks. As we noted earlier, one result of these annual storms is seasonal variations in the dust cover on the ground in certain regions, creating the well-known seasonal color variations seen from earth.

Water, Tectonics, and the Martian Surface

The surface of Mars (Fig. 6.7) can be classified into two types of areas: plains, which dominate much of the northern hemisphere; and rough, cratered terrain, which covers much of the south. The plains are low-elevation regions covered with lava flows, and altogether they spread over some 40 percent of the planet. The rest of Mars is covered by the rougher cratered areas, which mostly consist of highlands, and which are older than the plains. One of the highland regions, called Tharsis, encompasses about a quarter of the entire Martian surface, and averages some 6 kilometers' elevation above the mean surface level. This massive plateau, which is very ancient, may be a Martian analogue to earthly continents, although there are some differences. The formation of this gigantic bulge on Mars is not well understood, but most theories suggest it is a result of tectonic activity driven by convection in the mantle that occurred very early in the planet's history.

A major valley system (Fig. 6.8), extending some 4,500 kilometers in length, was named Valles Marineris in honor of *Mariner 9,* whose photographs revealed its gargantuan proportions. This valley would

FIGURE 6.7. *A Barren Surface.*
The first close-up photos of Mars revealed it to be arid and cratered, more closely resembling the moon than the site ot teeming civilization it was once imagined to be. This photo was taken by one of the *Viking* orbiters.

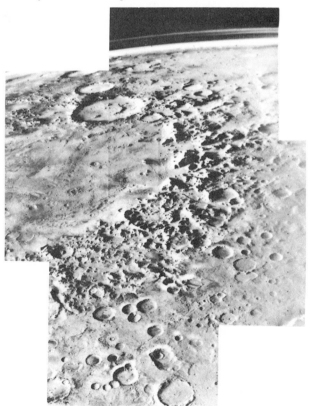

FIGURE 6.8. *Valles Marineris.*
This tremendous valley, as long as the breadth of North America, is put into perspective in this global mosaic of *Viking* photos. The three dark spots at left are the giant volcanoes Ascraeus Mons, Pavonis Mons, and Arsia Mons (top to bottom).

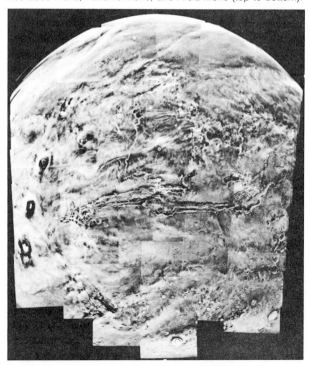

dwarf the Grand Canyon, which is five to ten times smaller in every dimension. Valles Marineris is 700 kilometers across at its widest, and 7 kilometers deep at its deepest. It probably was formed when the Tharsis highland was uplifted, creating extensive cracks in the crust.

Another major feature of the Martian surface is the astounding volcanoes, which exist primarily in three locations. These are the most conspicuous features on the planet, except for the polar caps. The grand champion is Olympus Mons (Fig. 6.9), a monster with a height of 27 kilometers (nearly 90,000 feet) above the mean surface level, and a base some 700 kilometers across. Olympus Mons is in the northern hemisphere, not far from the Tharsis plateau. The other great volcanic structures are formed by chains of peaks stretching across vast areas of the plains.

The giant volcanoes apparently formed because of a lack of continental drift on Mars. If they had formed on moving plates in the crust, they should have moved away from the subsurface hot spots that created them, before they got so big. This and other evidence indicate that Mars probably has a thicker and more rigid crust than the earth, so that there is not a planetwide system of moving tectonic plates. The Tharsis highland region was uplifted some 4 billion years ago and shows some effects of tectonic activity, with rift valleys and ridges. However, when the great volcanoes formed, between 2.5 and 3.5 billion years ago, there certainly was no widespread crustal motion. The present-day lack of tectonic activity is also consistent with the absence of any detectable magnetic field. Most theories of planetary magnetism invoke a fluid core, something else that Mars may not have.

Close examination of the surface of Mars reveals some interesting details. There are craters in most regions, created by impacts, as on the moon. Some regions appear completely disorganized, resembling loosely piled rocks and rubble, as though landslides had occurred there. In many locations, often adjacent to this **chaotic terrain,** there are winding valleys and flood plains that appear to have been formed by flowing water (Figs. 6.10, 6.11, and 6.12). These features have caused a great deal of excitement because, as we learned earlier, the lack of water on Mars has been a major puzzle.

It now appears that a large amount of water has been there all along, but hidden underground. The exact form it may take is still uncertain, but the prevailing suggestion is that Mars has a thick layer of **permafrost** beneath the surface, comparable to the permafrost found in the frozen tundra of the far north on the

FIGURE 6.9. *Olympus Mons.*
This is the largest mountain known to exist in the entire solar system. Its base is comparable in size to the state of Colorado, and its height is about three times that of Mt. Everest.

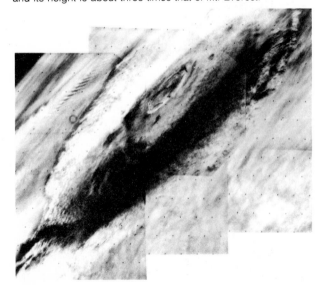

FIGURE 6.10. *Chaotic Terrain.*
This *Viking* orbiter photo of Mars shows a region that apparently subsided when permafrost below the surface suddenly melted and flowed away (toward the left in this photo).

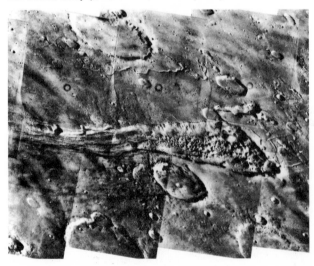

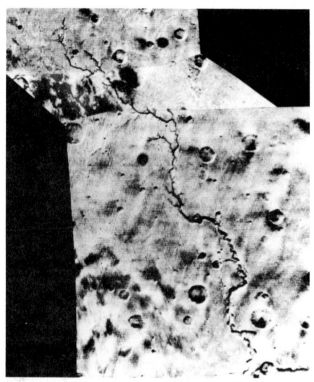

FIGURE 6.11. *An Ancient River Channel on Mars.*
This is one of the most striking examples of evidence for flowing water in the Martian past.

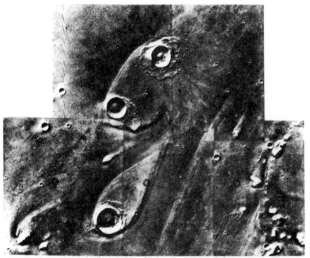

FIGURE 6.12. *A Flood Plain.*
Here is seen clear evidence that water once inundated this region of the Martian surface. The surface patterns indicate that the prominent craters already existed when the water flowed.

earth. The presence of underground water on Mars, whether it is in the form of permafrost or not, was deduced from the close association of the dry river valleys and the chaotic terrain.

The idea is that a large reservoir of underground water is suddenly released in liquid form, most likely as a result of volcanic heating. It quickly runs off, carving a channel in the surface and leaving behind empty caverns or porous soil in the subsurface volume that it evacuated. The soil there collapses, creating the kind of jumbled, disorganized appearance associated with the chaotic terrain. This theory explains the fact that the dry riverbeds most often are found leading away from the borders of chaotic regions.

Another class of surface features on Mars are those created by winds. We have already learned about the seasonal dust storms and the variations in the surface coloration that they cause. But there are other features created by the winds, including light and dark streaks and the **laminated terrain** (Fig. 6.13), curious areas where the surface has apparently been deposited in

layers and then eroded at the edges. The stair-step structures that result are especially prominent around the edges of the polar caps, and may be formed of thin layers of dust and ice that are precipitated by the annual storms.

There are also sand-covered regions, where the sur-

FIGURE 6.13. *Laminated Terrain.*
This photo shows some of the curious layered surface features that exist near the edges of the polar ice caps on Mars.

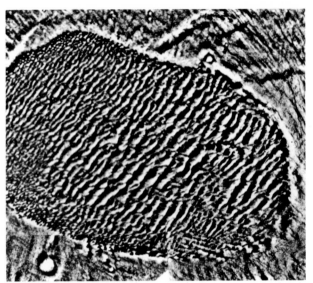

FIGURE 6.14. *Martian Sand Dunes.*
These dunes are testimonials to the incessant action of winds on Mars.

FIGURE 6.15. *A Ground-Level Panorama.*
This is the first photo of Mars made by the *Viking 1* lander. It shows a rock-strewn plain extending in all directions.

face has been shaped into dunes just like those seen on earthly deserts and beaches (Fig. 6.14). The sand in the dune regions is distinct from the much finer dust that is suspended in the air during the major storms. A vast field of dunes girdles the planet just outside the north polar cap, creating both a dark belt visible from afar and a mystery for scientists, who can offer no plausible explanation of its origin.

Surface Rocks and the Interior of Mars

The *Viking* landers were able to carry out some close-up analyses of Martian soil and surface rocks. Both landing sites were in the plains regions of the northern hemisphere, and at both places the landscape consisted of dusty soil, with low ridges, a dense scattering of rocks and boulders, and craters here and there (Figs. 6.15 and 6.16). The rocks lying around were probably ejected by the impacts that formed the nearby craters.

The chemical composition of the surface minerals was measured, and striking differences were found between Mars and the earth. The soil of Mars appears to have formed from the breakdown of igneous rocks such as basalt, but it contains unusually high concentrations of silicon and iron. The iron, in the form of iron oxide (rust) is responsible for the reddish color.

The high abundance of iron is quite significant, for it indicates that Mars is not highly differentiated. Apparently the crust was not molten long enough to allow the heavy elements to sink to the center of the planet, and Mars does not have a dense nickel-iron core, in contrast with the earth and Venus. The lack of a large, dense core is also indicated by the relatively low aver-

FIGURE 6.16. *Drifting Dust on Mars.*
The large boulder in this view is about two meters across. The fine dust shows evidence of nearly continuous change, as winds rearrange the drifts.

age density of Mars (3.9 grams/cm^3, whereas the earth's density is 5.5 grams/cm^3 and that of Venus is 5.3 grams/cm^3), and the absence of a detectable magnetic field.

These data lead to a picture of an early Mars which, because of its low mass or its distance from the sun, was not fully molten for long, and which formed a very thick crust early in its history. Some tectonic activity occurred, but there were no extensive motions of the crust. Thus massive volcanoes formed, building up to tremendous heights because there was no continental drift to carry them away from the hot spots that created them.

Prospecting for Life: The *Viking* Experiments

The most widely publicized aspect of the dual *Viking* missions was the attempt made by the robot landers to detect evidence of life forms on Mars. This was done in several ways, the first and most straightforward being simply to look with the television cameras for any

large plants or animals. None were seen, and more sophisticated tests were tried.

The tests for life were carried out by three different experiments on board the landers, each of which used soil scooped up by a mechanical arm (Fig. 6.17) and deposited in containers for the analysis. The basic idea of all three was similar: to look for signs of metabolic activity in the sample. All living organisms on earth, even microscopic ones, alter their environment in some way just by existing. Usually the effects involve chemical changes as the organism derives sustenance from its surroundings and ejects waste material.

In the **labeled release experiment,** a sample of Martian soil was sealed in a container to which a nutrient solution was added. The nutrient contained a trace of radioactive carbon (the isotope ^{14}C, consisting of 6 protons and 8 neutrons in the nucleus instead of the usual 6 of each). After the sample was incubated at a warm temperature for several days, detectors were activated to look for traces of ^{14}C in the gas in the container, on the theory that if life forms were present, they would have ingested some of the nutrient and excreted waste products containing ^{14}C. As a control on the experiment, separate samples were sterilized by high temperatures and then treated with the nutrient and

FIGURE 6.17. *Sampling the Martian Soil.* Here the scoop on *Viking 2* digs up a sample of Martian soil for one of the life-detection experiments.

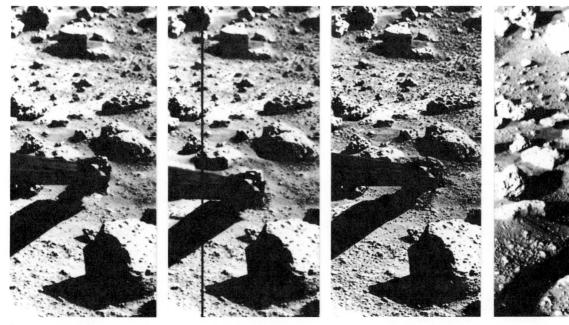

checked for ^{14}C. This was done to be sure that the soil did not undergo nonorganic chemical reactions that mimicked life forms by releasing ^{14}C.

Much to everyone's surprise, large quantities of ^{14}C were released in the sample that was not sterilized, and very little ^{14}C was released in the sample that was sterilized. This is just what had been expected from life forms, except that the amount of ^{14}C that was released was far more than had been thought possible if microscopic organisms in the soil had been responsible. There were other inconsistencies as well, particularly in the fact that a given sample would only produce the released ^{14}C once, even if nutrient were added a second time. It began to look as though some kind of rapid chemical reaction was responsible after all, one that was hindered by the sterilization process. If there were highly oxidized materials in the Martian soil, they might have reacted by violently bubbling and fizzing when the liquid nutrient was first added, and in the process have released large quantities of gas. If the reaction used up all the oxidized material the first time, it would not occur again when more nutrient was added to the sample.

The labeled release experiment might have been viewed more optimistically if the other two experiments had also given positive signs of life, but they did not. The **pyrolytic release experiment** was designed to look for a reverse of the reaction sought in the labeled release experiment. In this case a soil sample was placed in a container filled with gases (mostly carbon monoxide and carbon dioxide that contained ^{14}C), with the hope that any organisms in the soil would ingest these gases, just as earthly plants do. A lamp was used to illuminate the sample, on the theory that it might stimulate photosynthesis. After several days of exposure to the soil, the tracer gases were flushed out of the container so that there would be no ^{14}C left inside it, unless some had been absorbed or ingested into the soil sample. The sample was then heated, so that any ^{14}C in it would be baked out and detected. A little ^{14}C was absorbed into the soil, but in quantities too small to likely be caused by living organisms. Furthermore, the addition of some water to the samples did not increase the amount of ^{14}C that was absorbed, whereas it should have had some effect if there had been organisms in the soil.

The third attempt to detect life was carried out by the **gas exchange experiment,** which looked for signs of respiration. The soil sample was placed in a container of inert gases, which do not easily combine with others to form molecules and compounds. A nutrient was then added, and detectors in the container searched for any new gases that might appear as a result of biological activity. Some carbon dioxide, nitrogen, and oxygen were released from the sample, but only in quantities expected from nonorganic chemical reactions. Again, no traces of life forms were found.

Another test for evidence of life on Mars was carried out by a device called a **mass spectrometer,** which is capable of analyzing a sample to determine the types of molecules it contains. The mass spectrometers aboard the *Viking* landers found no evidence of **organic molecules** (those molecules which contain certain combinations of carbon atoms always found in plant or animal matter) in Martian soil samples.

The lack of evidence for life on Mars is not the same as evidence for lack of life there. After all, these experiments only tested samples at two localities on the planet, and furthermore, the tests were predicated on the assumption that Martian life forms, if they exist, would in some way be similar to those on earth. Now that a great deal is known about the chemical properties of the Martian soil, new experiments probably could be devised that would be relatively free of the confusion created by the nonbiological activity detected in the *Viking* experiments. Perhaps one day soon further attempts to find living organisms on Mars will be made.

The Martian Moons

The two satellites of Mars, discovered telescopically in 1877, are rather insignificant compared with the earth's moon and the major satellites of the outer planets. Both are very near to the parent planet and therefore have relatively short periods: Phobos (Fig. 6.18), the inner moon, circles Mars in just 7 hours, 39 minutes; and Deimos (Fig. 6.19) has a period of 30 hours, 18 minutes. Because Phobos orbits the planet in less than a Martian day, to an observer on the planet's surface, this satellite would cross the sky twice daily, in the west-to-east direction.

Close-up photographs of Phobos and Deimos reveal that each is a small, irregularly shaped chunk of rock, pitted with craters; and that Phobos is covered by linear grooves. Both satellites are somewhat elongated: Phobos has a length of some 28 km and a width of 20

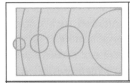

Astronomical Insight 6.2

RE-CREATING MARS IN EARTH'S IMAGE?

Although speculations that an advanced civilization might exist on Mars have been laid to rest, the possibility that one might prosper there in the distant future is very much alive. We speak here not of a native culture that might arise to populate the planet, but of a future time when the human race might settle there in large numbers.

Mars has a relatively hospitable climate, by extraterrestrial standards, but it is hostile enough that human settlers there would have to rely on sealed buildings, pressure suits (or at least helmets), and imported food and other supplies. Although this may prove feasible, another idea is developing that might someday make it possible for humans to live on Mars without depending on artificial means.

The idea, familiar to science-fiction readers, is that it might be possible to deliberately alter the atmosphere and climate of Mars to simulate that of the earth. A new word, *terraforming*, has been added to the English language to describe this process. In science-fiction novels this is usually represented as a nearly instantaneous transformation achieved with powerful machinery and unlimited supplies of energy, but in the more serious schemes being considered by astronomers and engineers, natural forces would be used to remodel Mars in earth's image.

Recall that the earth's atmosphere was modified by the presence of life forms that released oxygen and nitrogen into the air. It has been hypothesized that the introduction of life on Mars might have the same effect there, given enough time. To that end, experiments were performed in which earth plants and microorganisms were sealed in containers that simulated the atmospheric conditions of Mars, and the life forms grew successfully.

It is possible to simulate the low pressure, the cold temperatures, and the less intense sunlight of Mars; the only essential ingredient missing is genuine Martian soil. This may not be a serious problem, however, in view of the vigor with which plants have grown in samples of lunar soil. Perhaps if widespread crops of plants could be made to grow on Mars, eventually enough oxygen would be produced to make the Martian air breathable without artificial aid. Calculations show that this is possible, despite the lower gravity (and consequently low escape velocity) of Mars. The increased atmospheric pressure, along with the addition of water vapor and other gases to the Martian atmosphere, would produce a greenhouse effect that could warm up the surface to comfortable temperatures.

Other necessities for life could be derived from the ground. Water, as we have seen, is probably present in plentiful quantities beneath the surface, in the form of permafrost. Food could be produced in sufficient supply through farming, perhaps initially indoors but eventually out in the open.

The scenario just described is, of course, very speculative, and would take centuries to run its course. For the time being, there are less ambitious steps that could be taken, starting with the first manned missions to Mars. Although it would take a very long time to terraform the entire planet, and doing so would require of the human race great patience and vision, it is perhaps not unrealistic to start by producing localized habitable climates within sealed buildings. To do so would greatly enhance the practicality of maintaining at least an outpost on Mars, even if we did not yet colonize it wholesale.

km, and Deimos measures roughly 16 by 10 km. Both are in synchronous rotation, keeping one end permanently pointed toward Mars.

A striking and unusual characteristic of the surface of Phobos is that most of it is covered with parallel grooves some 100 to 200 meters wide and 10 to 20 meters deep. These features are most prominent in the vicinity of a very large crater on Phobos, and least evident on the opposite side. There are several proposed explanations of their origin, but the one that appears to

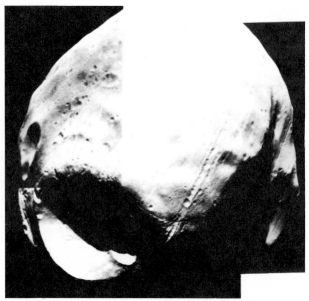

FIGURE 6.18. *The Martian Moon Phobos.*
This *Viking* mosaic shows two interesting features: the parallel grooves that mark the surface of Phobos, and the major impact crater whose formation may have caused the stresses that created the grooves.

fit the data best is that they were created by the same impact that carved out the major crater. It has been suggested that this impact fractured Phobos along planes where its material was weakest, and that heat from the impact caused some melting of subsurface rock, which then partially filled the fractures. The rea-

FIGURE 6.19. *Deimos.*
The more distant of the two moons from Mars, Deimos has a relatively featureless surface.

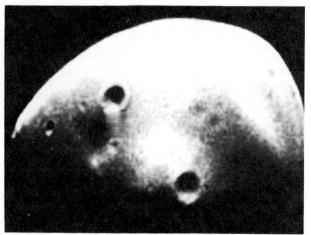

son Deimos does not have similar grooves (according to this view), is that it has never suffered such a major impact, as shown by its lack of large craters.

Clearly Phobos and Deimos were not formed in the same way as the earth's moon, which has a much larger mass relative to its parent planet, and which evidently went through some geological evolution. Phobos and Deimos are tiny and show no signs of having been acted on by the kinds of forces, such as volcanism, that have forged the surface of the earth's moon. The origin of the two Martian moons is not well understood, but many people have pointed out the general similarities between them and typical asteriods.

PERSPECTIVE

The mysterious red planet is still mysterious, but in new ways. Most of the age-old questions related to its changing appearance, the possible civilizations thriving on its surface, and its distinctive color have been answered through close-up examination. What we have found is a planet like the earth, except that it did not develop as fully, in the geological sense. Tectonic activity was arrested in the planet's youth, and the low temperatures and low escape velocity combined to leave Mars with only a thin atmosphere of CO_2, its water hidden away underground.

We are done with our tour of the terrestrial planets, and we are ready to move outward, to discover the fantastic giant planets in the outer solar system.

SUMMARY

1. Telescopic observations of Mars showed a thin carbon-dioxide atmosphere, polar caps that vary in size with the Martian seasons, and a very low abundance of water vapor.

2. The atmosphere on Mars has almost the same composition as that of Venus, but has only 0.6 percent of earth's sea-level pressure.

3. Seasonal variations on Mars are strongly influenced by the elliptical shape of its orbit.

4. The Martian surface consists of plains and rough highlands, with evidence of ancient tectonic activity but no continental drift, similar to Venus.

5. Water on Mars is probably stored beneath the surface, in the form of permafrost.

6. Surface rocks on Mars have a high iron content, and Mars has only a moderate average density, both facts indicating that the planet has undergone little differentiation.

7. The *Viking* experiments found no evidence of life forms in the Martian soil.

8. Mars has two tiny satellites, resembling asteroids in size and shape.

REVIEW QUESTIONS

1. Before the *Mariner* and *Viking* missions brought back close-up photographs of the Martian surface, what were the best arguments against the idea that the dark markings seen on Mars were canals filled with water?

2. Summarize the similarities and contrasts between the atmospheres of Mars and Venus.

3. To show why the earth's seasons are not strongly affected by the shape of its orbit, calculate how much more intense sunlight is at the earth's surface when the earth is closest to the sun than when it is farthest away. At closest approach, the earth is 0.983 AU from the sun, and at its farthest is 1.017 AU from it. How does this compare with the situation for Mars?

4. Summarize the similarities and differences between the winds on Mars and on the earth.

5. Compare Mars, Venus, and the earth in terms of tectonic activity.

6. Mars has no detectable magnetic field. What effect would this have on any life forms that might exist there, and on humans who might settle there?

7. Summarize the evidence supporting the suggestion that the Martian water supply exists just beneath the surface, in the form of permafrost.

8. The reason given for Mars' lacking a magnetic field was different from the explanation as to why Venus lacks one. What are the two explanations?

9. In testing for the possible presence of life on Mars, the designers of the *Viking* experiments made certain assumptions about the nature of life forms that might exist there. Summarize these assumptions.

10. Both Maritan moons are somewhat elongated. Explain why they each keep one end pointed toward Mars at all times, rather than being in synchronous rotation with some other orientation.

ADDITIONAL READINGS

Arvidson, R. E., Binder, A. B., and Jones, K. L. 1978. The surface of Mars. *Scientific American* 238(3):76.

Beatty, J. K., O'Leary, B., and Chaikin, A., eds. 1981. *The new solar system.* Cambridge, England: Cambridge University Press.

Carr, M. S. 1976. The volcanoes of Mars. *Scientific American* 234(1):32.

————— 1983. The surface of Mars: a post-*Viking* view. *Mercury* 12(1):2.

Horowitz, N. H. 1977. The search for life on Mars. *Scientific American* 237(5):52.

Leovy, C. B. 1977. The atmosphere of Mars. *Scientific American* 237(1):34.

Pollack, J. B. 1975. Mars. *Scientific American* 233(3):106.

Veverka, J. 1977. Phobos and Deimos. *Scientific American* 236(2):30.

Young, R. S. 1976. *Viking* on Mars: a preliminary survey. *American Scientist* 64:620.

Learning Goals

The largest planet in our solar system, Jupiter is completely unlike any of the inner four. In discussing the terrestrial planets we could draw comparisons with the earth, but our modest home is not in the same league with Jupiter, and we will find that almost no aspect of this gaseous giant can be compared with the earth. Entirely new standards must be adopted here, and Jupiter, the closest of the giants to earth and the best studied, will serve as the model for our later discussions of the others.

The Jovian system is very complex, with its many fascinating satellites, its ring, the extensive and intense radiation belts, and the fantastic appearance of the planet itself. Like Venus and Mars, Jupiter has played a role in the historical development of astronomy, particularly with Galileo's discovery of the planet's four large moons, whose existence he cited as an argument against the earth being the center of all heavenly motions.

Jupiter truly is a giant planet, containing more mass than all the other planets together. It orbits at an average distance of 5.2 AU from the sun, so it is never closer to the earth than 4.2 AU. Its mass (318 times that of the earth) is spread over such a large volume that the average density is only 1.3 grams/cm³, barely greater than that of water.

Observation and Exploration

Even a small telescope reveals colorful structure on the Jovian surface; reddish-brown **belts** and light-colored **zones** encircle the planet (Fig. 7.1). The Great Red Spot, a giant reddish oval in the southern hemisphere that dwarfs the earth in size, has apparently been a regular feature for at least three hundred years. The thick atmosphere of Jupiter makes it impossible for us to see a solid surface underneath; as we will learn, modern theories hold that there is no real surface anyway. Spectroscopy of Jupiter's atmosphere reveals compounds of the general type that are thought to have been present in the primitive atmospheres of the terrestrial planets: hydrogen-bearing molecules such as methane (CH_4), ammonia (NH_3), and hydrogen itself (H_2), as well as helium. This composition is rather significant, for it means that the evolution of the Jovian atmosphere was arrested at a very early stage. Jupiter

TABLE 7.1. Jupiter

Orbital semimajor axis: 5.203 AU (778,300,000 km)
  Perihelion distance: 4.951 AU
  Aphelion distance: 5.455 AU
Orbital period: 11.86 years (4,333 days)
Orbital inclination: 1°18′17″

Rotation period: 9^{h}50^m
Tilt of axis: 3°5′

Mean diameter: 138,500 km (10.79 D$_\oplus$)
  Polar diameter: 134,100 km
  Equatorial diameter: 142,800 km
Mass: 1.90 × 10³⁰ grams (318.1 M$_\oplus$)
Density: 1.33 grams/cm³
Surface gravity: 2.64 earth gravities
Escape velocity: 59.5 km/sec

Surface temperature: 130 K (cloud tops)
Albedo: 0.51

Satellites: 16

is a cold planet by earthly standards, with a temperature at the cloud tops of about 130 K.

Jupiter's rotation is very rapid, despite its large size. The rotation period is difficult to pinpoint because the planet spins differentially, meaning that it goes around more quickly at the equator than near the poles. This can happen only in a nonrigid object. The Jovian day

FIGURE 7.1 *The View of Jupiter From Earth.*
This is a good photograph from an earth-based telescope.

is just under ten hours long, which means that the cloud tops near the equator must travel at a speed close to 45,000 kilometers per hour. This rapid spin has flattened the planet a bit, giving it an oblate shape, so that the diameter through the equator is slightly larger than the diameter through the poles (Fig. 7.2).

Quite accidentally, Jupiter was discovered to be a source of radio emission; further observations have shown a variety of activity in the radio portion of the spectrum, some of it understood, and some of it not. A

FIGURE 7.2. *The Effect of Rapid Rotation.*
An elastic or fluid body is distorted out of a spherical shape by rotation. Jupiter is "flattened" by its rotation so much that its equatorial diameter is 6 percent larger than its polar diameter.

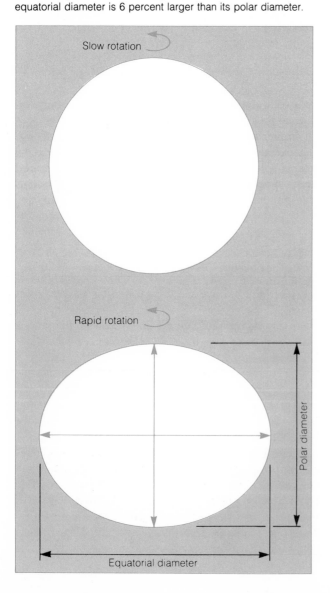

general radio glow from the planet, at a wavelength of a few centimeters, is the thermal radiation expected according to Wien's law (although the peak emission occurs at shorter wavelengths, in the infrared).

In addition to the thermal radiation, Jupiter emits bursts of radio static much like the emission associated with lightning in the earth's atmosphere, and it is now known that there is lightning on Jupiter. A form of radio emission that is caused by charged particles moving through a magnetic field was also detected. This **synchrotron radiation** showed that Jupiter has both a strong magnetic field and intense radiation belts surrounding the planet, analogous to the Van Allen belts of earth.

The intensity and spectrum of the infrared radiation from Jupiter imply that the giant planet is producing some excess energy in its interior. The total amount of energy being emitted by the planet is about two-and-a-half times greater than the amount it receives from the sun. The probable explanation for this is discussed later.

Observations from earth revealed that Io, the innermost of the four Galilean satellites, has strange properties. This satellite modulates the Jovian radio emission as it passes in front of the planet. Also, Io is accompanied by a cloud of atomic and ionized gas that streams along behind it in its orbital path around Jupiter.

Like the terrestrial planets, Jupiter has been a target of the U.S. space program. Sending probes to the outer planets requires some modifications to the technology that has been so successful in reaching the inner planets. The spacecraft must endure a passage through the asteroid belt, have their own internal power source (small nuclear reactors have been used) because they travel too far from the sun to use solar panels, and be able to communicate with the earth over great distances.

These challenges were successfully met by the *Pioneer 10* and *11* missions (Fig. 7.3), and more recently by the *Voyager 1* and *2* spacecraft (Fig. 7.4). *Pioneer 10* and *11* reached Jupiter in late 1973 and late 1974, and carried out an assortment of studies of the planet's atmosphere, magnetic field, radiation belts, and satellites. The *Voyager* spacecraft, which flew past Jupiter in March and July of 1979, carried out similar types of research and sent back hundreds of sensational images of Jupiter and its moons, obtained with their fine high-resolution cameras. Perhaps more than any other facet of the planetary exploration program, these pictures of

FIGURE 7.3. *A Close-Up View.*
This image of Jupiter was obtained by the *Pioneer 11* spacecraft from a distance of slightly more than a million kilometers. Although far superior to any earth-based photos, this image was soon to be surpassed by those from the *Voyager* missions.

FIGURE 7.4. *A Voyager Image.*
This photo was obtained by *Voyager 1* when it was some 35 million kilometers from Jupiter. A vast amount of detail was already evident in the atmospheric structure.

Jupiter (and later of Saturn) have caught the public imagination, and today copies are seen everywhere.

Pioneer 11, as well as both *Voyager* spacecraft, went on to visit Saturn, using the gravitational pull of Jupiter to correct their courses and provide a boost along the way. Throughout much of the 1980s, all the outer planets will be on the same side of the sun, which has led to the idea of a "grand tour." It was realized that a space probe could visit all the outer planets, using each along the way to provide a gravitational boost toward the next. Although the full tour is not being taken, most of it is being traveled by *Voyager 2.* This spacecraft, having flown by Saturn in August of 1981, will reach Uranus in 1986, and, if all goes well, will visit Neptune in 1989.

Atmosphere in Motion

Let us now return to Jupiter itself, and consider the structure and especially the complex motions in its atmosphere. The upper portion of the atmosphere is made of hydrogen and hydrogen compounds, as already mentioned. There are also large quantities of he-

lium. The presence of these gases represents a major contrast with the terrestrial planets, which long ago lost to space all of their free hydrogen and helium. The overall chemical composition of Jupiter is essentially identical to that of the sun, except that Jupiter is cold enough for most of the material to be in molecular form, whereas the sun is so hot that few molecules can exist, and most atoms are ionized. It is noteworthy that the principal species in the Jovian atmosphere, CH_4, NH_3, H_2, and helium, are the ones that are thought to have been present in the earth's atmosphere when life forms first appeared.

Heavy elements, such as metals, have not been observed in the Jovian atmosphere, but are no doubt present, having sunk to the core of the planet. Jupiter, being fluid nearly all the way to the center, is highly differentiated.

The bright and dark colors in the upper atmosphere are probably caused by slight differences in molecular composition, which in turn are created by subtle differences in temperature and pressure. The dark belts are regions of low pressure at the cloud tops, where the atmospheric gas is descending, and the light-colored zones are high-pressure regions where the gas is rising. The banded appearance of the planet (Fig. 7.5) is cre-

When we discussed weather patterns and atmospheric motions on the earth, in chapter 4, we put everything in the perspective of someone living at the bottom of the atmosphere, looking up. When we observe a distant planet, we see things from the top, looking down.

There are some problems with terminology that arise because of these different perspectives. The most potentially confusing has to do with high-and low-pressure regions. When we spoke of the earth, we described a high-pressure area as one where cool air descends, creating an anticyclonic flow. When discussing Jupiter, on the other hand, we said that cool air descends in *low*-pressure regions, which seems to contradict what we said about the

earth. The difference is that in the case of Jupiter, we are considering the top of the atmosphere, whereas in the case of earth, we are talking about conditions at the bottom. Everything becomes clear when we realize that when a column of cool air descends, it creates high pressure at the bottom and low pressure at the top. Conversely, when warm air rises, low pressure is created at the bottom and high pressure is created at the top, and we have an anticyclonic-flow pattern.

The terms *clockwise* and *counterclockwise* are used in the same way for both the earth and Jupiter, referring to the direction of flow as seen from a distance, as though the hypothetical clock lies face-up on the planetary surface.

ated by the rapid rotation, which stretches what would otherwise be circular flow patterns into elongated strips that circle the planet. Just as on the earth, air that is descending forms a cyclonic flow, and it rotates in a clockwise direction in the southern hemisphere, and counterclockwise in the north (Figs. 7.6 and 7.7). When cool gas is ascending, the anticyclonic flow, as it is called, occurs in the opposite direction.

Besides the regular pattern of belts and zones in Jupiter's atmosphere, there is a wide variety of spots and other features, the Great Red Spot being the largest and

most prominent (Fig. 7.8). There are also a number of smaller spots, both dark and light in color. Just like the belts and zones, these spots are cyclonic (if dark-colored) or anticyclonic (if light). Thus the Great Red Spot may be characterized as a storm system, although it is very long-lived, having been present for at least three hundred years. The Great Red Spot is not well understood; it has properties that are unusual even for Jupiter. One surprising fact is that the central portion of the spot consists of gas that is rising, and this gas reaches much greater heights than the gases in the belts and

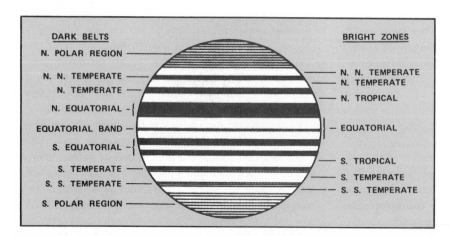

FIGURE 7.5. *Belts and Zones On Jupiter.* This schematic drawing names the major features that give Jupiter its banded appearance.

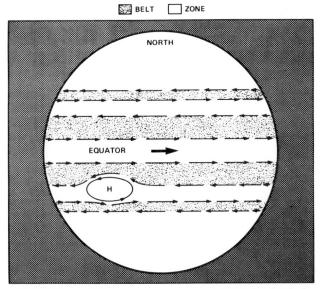

FIGURE 7.6. *Circulation of the Jovian Atmosphere.*
This sketch illustrates the direction of flow in Jupiter's belts and zones.

zones, yet it is dark-colored, in contrast with the rising gas in the light-colored zones.

Very detailed photographs of Jupiter's cloud tops reveal a fantastic picture of turbulent, chaotic flows where the belts, zones, and spots interact. Movie sequences reconstructed from images obtained from the *Voyager* spacecraft as they approached Jupiter show an

FIGURE 7.7. *Vertical Motions in the Jovian Atmosphere.*
This cross-section indicates how the horizontal circulation pattern is related to vertical convective motions in the outer layers of Jupiter's atmosphere. This is analogous to circulation in the earth's atmosphere, except that the rapid rotation of Jupiter stretches rotary patterns into belts and zones.

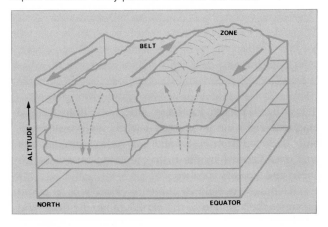

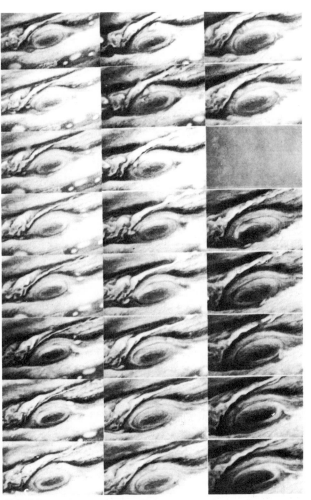

FIGURE 7.8. *Rotation of the Great Red Spot.*
The sequence starts at the upper left and goes down each column. Each frame is separated from the ones above and below it by about twenty hours. Note the two white spots in the upper left, and trace how they enter the spot, move around it counterclockwise, and then move out of view at the lower left.

incredible complexity of motions all happening at once, as in some gigantic machinery of gears and wheels.

Internal Structure and Excess Radiation

No direct probes of the interior of Jupiter have been made, so what we know has largely been surmised from theoretical calculations. This may seem like an exceedingly uncertain basis for any detailed theories about interior conditions, and of course there is room for error, but in fact the laws of physics dictate rather explicitly what the internal structure must be like. The general procedure for calculating a theoretical model of Jupiter is very much like that described in chapter 14 for determining the internal properties of a star. A set of equations describing the relationships among various properties such as pressure and temperature, as well as processes such as heat flow, is solved, starting with the known size, mass, composition, and surface characteristics of the planet.

Although different assumptions made by people working in this field lead to somewhat different results, the major features turn out to be the same (Fig. 7.9). Below a relatively thin layer of clouds (perhaps 1,000 kilometers thick) is a zone consisting of hydrogen in liquid form, extending more than one-third of the way toward the center. The pressure in this region is as great as three million times sea-level pressure on earth, and the temperature is as high as 10,000 K or more. An even thicker zone below the liquid hydrogen has such extreme pressure that hydrogen is forced into a state called **liquid metallic hydrogen,** which has a somewhat rigid structure like certain metals. Liquid metallic hydrogen cannot be produced on earth, be-

cause there is no known way to create the necessary pressure, but the simplicity of hydrogen lends credence to the theoretical calculations that predict its behavior under these extreme conditions. The central core of Jupiter, below the liquid metallic hydrogen, is probably solid rocky material of very high density, containing all the heavy elements of the entire planet, which sank to the center as Jupiter differentiated. In the core the temperature is expected to be as great as 30,000 K. The mass of the core is probably in the range of ten to twenty times the mass of the earth, indicating that as much as 6 percent of the planet's mass is contained in only 0.5 percent of its volume. Except possibly for the boundary of this rocky central zone, Jupiter has no solid surface. Rather, there is (inward from the top) gas, then liquid, and finally, metallic liquid.

As noted earlier, the planetary radio emission shows the presence of an internal heat source, and this was included in the theoretical calculations just described. The source of this internal heat is difficult to understand, however, because Jupiter is not sufficiently massive to have nuclear reactions taking place in its core, as in a star. For a while it was thought that Jupiter might be gradually shrinking and in the process converting gravitational energy into heat, but this idea now appears to be ruled out. The most likely explanation, based on calculations of the heat capacity and compressibility of the interior, is that the interior is still hot from the time of the planet's formation, some 4.5 billion years ago. Such a prolonged retention of heat is made possible by the characteristics of the outer layers, which do not efficiently transmit heat from the interior to be lost to space. The atmospheric circulation, already described, is caused in part by heating of the lower atmosphere from below.

The Magnetic Field and Radiation Belts

It has been known for some time that Jupiter has a magnetic field, as mentioned earlier. This was originally deduced from observations of radio emission, and was later verified by the instruments on board the *Pioneer* and *Voyager* probes.

The magnetic field at Jupiter's cloud tops is about ten times the strength of the earth's magnetic field, and

FIGURE 7.9. *Internal Structure of Jupiter.*
This cutaway illustrates the rough model of the Jovian interior described in the text.

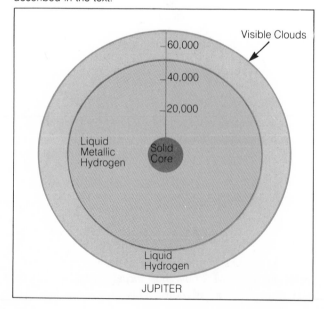

is oriented in the opposite direction, with the north magnetic pole pointing south, and vice versa. This is not too surprising, since we know that the earth's magnetic poles have reversed themselves from time to time. Perhaps the Jovian field does the same, and has its poles aligned with those of the earth some of the time.

The magnetic axis of Jupiter does not exactly parallel the rotation axis, but is instead tilted by 11 degrees. This situation is also similar to the earth's, but on Jupiter it has some important effects on the distribution of charged particles in the radiation belts.

The origin of the magnetic field is thought to be roughly similar to that of the earth's field; that is, circular motions of electrical currents inside the planet create a magnetic dynamo. The structure of the field indicates that the dynamo operates not far below the surface of the planet, probably in the liquid hydrogen zone.

Within the magnetosphere of Jupiter are very intense radiation belts (Fig. 7.10), with much higher densities of charged particles than in the Van Allen belts of earth. The passage of the *Pioneer* and *Voyager* spacecraft through these zones was rather perilous, since electronic circuits can be damaged by an environment full of speeding electrons and ions. Some data transmissions were garbled or lost, but on the whole the craft survived intact, and were able to continue functioning normally.

One mystery concerning the Jovian radiation belts was the origin of the great quantities of charged particles. A number of possibilities were considered, but most were not feasible, and gradually it was realized that the answer had to do with the inner Jovian satellites. The five closest to the planet all orbit within the magnetosphere, where they sweep up charged particles as they move along in their paths. The discovery from earth-based observations that Io is accompanied by a cloud of glowing gas led to the speculation that some of the satellites also emit particles that enrich the radiation belts. As we shall see in the next section, the *Voyager* missions were soon to provide spectacular confirmation that just such a process is occurring.

The rapid rotation of Jupiter forces the radiation belts to be confined to a thin layer lying in the equatorial plane of the planet. This layer is called the **current sheet,** because it is sheetlike and the electrons and ions move about within it, creating electrical currents.

The Satellites of Jupiter

Jupiter has sixteen known satellites, ranging from Metis (also designated J–16), the innermost, to faraway Sinope, the outermost. Most of the satellites are small, rocky objects, with diameters of less than 100 kilometers in many cases. The outermost four have retrograde

FIGURE 7.10. *The Jovian Magnetosphere and Radiation Belts.* The shape of the magnetosphere is influenced by the solar wind and by the rapid rotation of Jupiter. The result is a sheet of ionized gas that is closely confined to the equatorial plane, but which wobbles as the planet's off-axis magnetic field rotates.

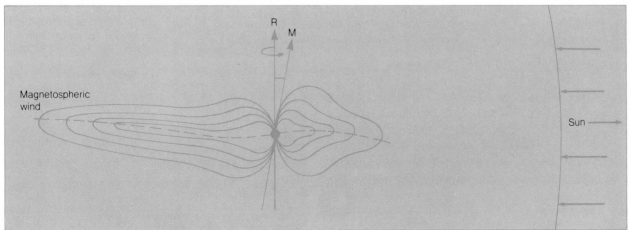

TABLE 7.2. Satellites of Jupiter

NO.	NAME	DISTANCE	PERIOD	DIAMETER	MASS	DENSITY
16	Metis	1.7922 R_J*	0^d.295	40 km		
14	Adrastea	1.8064	0.298	25 × 20 × 15		
5	Amalthea	2.55	0.498	270 × 170 × 150		
15	Thebe	3.11	0.675	110 × ? × 90		
1	Io	5.95	1.769	3,630	8.92 × 10^{22}gm	3.53 gm/cm^3
2	Europa	9.47	3.551	3,138	4.87 × 10^{22}	3.03
3	Ganymede	15.1	7.155	5,262	1.49 × 10^{23}	1.93
4	Callisto	26.6	16.689	4,800	1.08 × 10^{23}	1.70
13	Leda	156	240	10:		
6	Himalia	161	251	180		
10	Lysithia	164	260	20:		
7	Elara	165	260	80		
12	Ananke	291	617	20:		
11	Carme	314	692	30:		
8	Pasiphae	327	735	40:		
9	Sinope	333	758	30:		

* The symbol R_J stands for the equatorial radius of Jupiter, which is 71,398 km.

orbits that are substantially elongated in shape, and inclined by several degrees to the orbital plane of the more regular inner satellites. The fact that the outer four vary from the norm may suggest their origin is different from the others, which probably formed from interplanetary debris swirling around the infant Jupiter at the time the solar system coalesced.

Four of the Jovian satellites, the fifth through the eighth in order from Jupiter, are much larger than the rest, all being comparable to or larger than the earth's moon (Fig. 7.11). These are the four satellites Galileo spotted with his primitive telescope, and they are bright enough to be seen with the unaided eye, except that the brilliance of nearby Jupiter drowns them out.

The *Pioneer* and *Voyager* spacecraft were sent along trajectories aimed at providing close-up views of these Galilean satellites. Of particular interest was Io, the innermost of the four, because of its mysterious influence on radio emissions from Jupiter and because of the glowing cloud of gas that surrounds it.

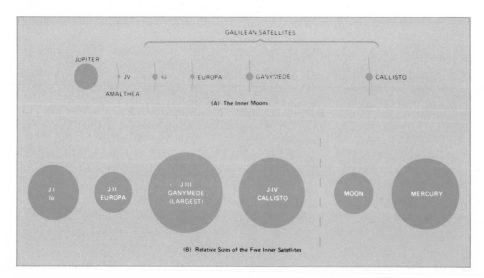

FIGURE 7.11. *The Major Jovian Satellites.* These profiles indicate the relative distances of the satellites from Jupiter, as well as their relative sizes. The earth's moon and the planet Mercury are included for comparison.

The sizes of the Galilean satellites, as well as their masses, have been most accurately determined by the *Pioneer* and *Voyager* probes. Although each of the four has its own distinctive character, the Galilean satellites fit a pattern, with systematic trends from one to the next. Their densities decrease steadily with distance from Jupiter, Io having the greatest density, and Callisto, the outermost of the four, the least. Their albedos also decrease in the same manner. These two facts, along with measurements of the surface temperatures, indicate that the outer satellites contain large quantities of water ice, which can be very dark when seen from afar, especially if the ice contains some gravel and grit. The inner Galilean satellites contain less ice and greater fractions of rock. The surfaces of these bodies also show marked contrasts with one another. The color photographs taken by the *Voyager* probes showed tremendous complexity, revealing the Galilean satellites to be wondrous worlds, each with its own special effects.

Callisto, the farthest from Jupiter, is covered entirely with craters, and has the appearance of a celestial target range that for ages has been bombarded with rocks and missiles from space (Fig. 7.12). The impact points look white, in contrast to the overall dark coloration of the surface. This contrast occurs because the impacts have removed the weathered, old surface layer of the

ice, and because the fractured ice has created multifaceted cracks and fissures, which reflect and scatter sunlight. At one point on the surface of Callisto is a tremendous circular feature resembling a giant bull's-eye, where a very large chunk of material from space left its mark.

Ganymede, next in from Callisto, is the largest satellite in the solar system. Until recently Titan, the largest of the satellites orbiting Saturn, was thought to be bigger, but *Voyager* measurements showed that the solid body hidden within the dense atmosphere of Titan is actually smaller than Ganymede. Triton, a satellite of Neptune, was thought to be the largest of all, but recent measurements have greatly reduced its estimated size. The surface of Ganymede is dark, but is marked with linear, light-colored streaks that show evidence of tectonic activity in the satellite's past, with discontinuities and fractures where the crustal plates have shifted (Figs. 7.13 and 7.14). Numerous impact craters are seen, though they are not as dense as on Callisto, indicating that some geological process has eradicated the oldest craters on Ganymede.

The next satellite in sequence in Europa, whose sur-

FIGURE 7.13. *Ganymede.*
The surface of this moon shows evidence for less geological activity than the inner two Galilean satellites, but it does have banded segments indicative of surface motions and stresses.

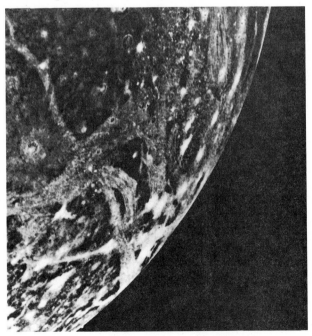

FIGURE 7.12. *Callisto.*
The outermost of the Galilean satellites, Callisto has the oldest surface. Here numerous white impact craters are seen, as is most of a huge impact basin (at right).

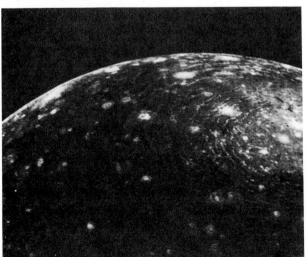

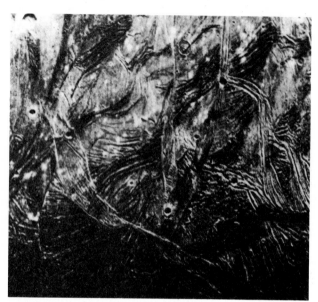

FIGURE 7.14. *Features on Ganymede.*
This close-up of Ganymede's surface shows that a complex pattern of grooves and ridges probably was created by deformations of an icy crust. Some impact craters are seen.

FIGURE 7.15. *Europa.*
This is the second of the Galilean satellites (in progression outward from Jupiter). The linear features and relative lack of craters indicate that this is a tectonically active moon with a young surface.

face is reminiscent of the old sketches of canals on Mars (Fig. 7.15). Europa is covered by random linear features, and resembles nothing so much as a cloudy cystal ball that has been cracked throughout. The dark lines appear to be rifts in the rocky crust where water has oozed to the surface and frozen, leaving a smooth surface, but one with strong color contrasts. Recent theoretical studies show that Europa may have liquid water underneath the surface ice. Thus, this moon of Jupiter may be the one place in the solar system other than the earth where oceans exist. Like Ganymede, Europa shows evidence of tectonic activity, particularly in the pattern of linear fractures. The surface of Europa has very few craters, which implies that some process has removed the traces of most of the impacts that occurred in the past.

We finally reach Io, an object of great mystery. Seen close-up, Io is a brilliantly colorful object with a wonderfully variegated red-orange-yellow surface, marked here and there by black smudges and circular features resembling volcanic craters (Fig. 7.16). White frosty patches are seen in places. No impact craters have been found, indicating that Io has a very young surface, constantly being reprocessed.

The explanation for all these fantastic features was discovered by *Voyager 1,* in a photograph taken as the spacecraft left Io behind and looked back toward it,

with the sunlight passing the edge of the satellite. There, illuminated by the sun's rays, was a plume of gas streaming out of Io's interior, in a volcanic eruption (Fig. 7.17). This was the first detection of active volcanism on any body other than the earth, and soon nearly ten active regions were found on Io's surface (Fig. 7.18).

Apparently Io is undergoing incessant volcanic activity, and the surface is continually being coated with deposits ejected from its interior. The dominant colors are caused by sulfur compounds, which can range from black to red to yellow, depending largely on temperature. Perhaps even more so than Venus, Io resembles the classical picture of hell, with its tortured landscape and sulfurous fumes.

Tidal Forces and the Volcanoes on Io

We have seen that the four Galilean satellites show some clear-cut trends in properties as we progress outward from one to the next: densities and surface albedos decrease, as the ice contents and the ages of the surfaces increase steadily. When all of this was discovered by the *Voyager* probes, no one had to wait long

FIGURE 7.16. *Io.*
The innermost of the four Galilean satellites, Io has long been a source of mystery. Here we see a number of volcanic vents (dark markings) amid a surface of sulfur compounds. No impact craters are visible, indicating that this is a very young surface.

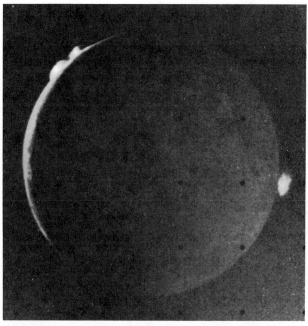

FIGURE 7.18. *Multiple Eruptions.*
Here three volcanoes are seen simultaneously erupting on Io; *Voyager* scientists found as many as ten to be active. Individual vents appear to stay active continuously over periods of at least several months.

for an explanation, for it had already been published. In considering how the tremendous gravitational force of Jupiter affects Io, scientists had been able to predict the presence of volcanoes there. As close to the massive planet as Io is, it is subject to very strong tidal forces.

FIGURE 7.17. *An Eruption on Io.*
This image dramatically illustrates Io's present state of dynamic activity, as it shows an eruption from a volcanic vent, with a plume of ejected gas above.

One result is that Io, as well as the other Galilean satellites and Amalthea, (Fig. 7.19) are in synchronous rotation, always keeping the same side facing the parent planet, just as the earth's moon does. In addition to this, because of the force exerted by Jupiter and the smaller tugs of the other satellites, Io is being con-

FIGURE 7.19. *Amalthea.*
Orbiting very close to Jupiter, this tiny, potato-shaped moon always keeps the same end pointed toward Jupiter.

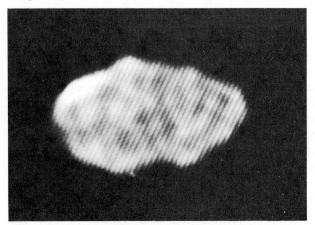

Astronomical Insight 7.2

LIFE FORMS IN THE JOVIAN SYSTEM?

Any hope we might have had of finding living organisms on the terrestrial planets has been dashed by close-up examination. Venus is far too hot to support life on its surface, and Mars, apparently not so hostile, has revealed not a trace of biological activity. At first thought, it might seem exceedingly unlikely that faraway Jupiter, so much colder at its cloud tops than any of the terrestrial planets, could be an abode for life. On second thought, however, there is some basis for considering the possibility, both for Jupiter itself and for its satellite Europa.

The clouds on Jupiter contain the same chemical compounds thought to have fostered the creation of life on earth; these are mainly hydrogen-bearing molecules such as methane and ammonia. Lightning discharges are common in the Jovian clouds, so the energy necessary to spark the chemical reactions could also be readily available. Temperatures within the clouds may be hospitable because, as we have seen, the atmosphere gets increasingly warmer with depth.

Of course, any life forms that might exist on Jupiter would be radically different from those we are familiar with on earth. Their chemical makeup and respiration would be different, and they probably would have physical shapes adapted for floating about in the currents of the Jovian atmosphere. Perhaps they would subsist on molecular species abundant in the atmosphere, or on smaller life forms.

Maybe the Jovian atmosphere hosts a biological zoo as complex as that on earth.

One ingredient not present in the clouds of Jupiter, but possibly abundant in the moon Europa, is a liquid medium. It has been argued that such a medium is necessary in order to support the complex chemical reactions that lead to the development of life. Not only is Europa suspected of having liquid in its interior, but it is thought to be the same liquid that nurtures life on earth: water.

The *Voyager* missions revealed that on the surface of Europa there are extensive fissures filled with water ice. It is also known that Europa undergoes severe tidal stresses, caused by its frequent alignments with Io and the gravitational forces of Jupiter. Thus it is hypothesized that the temperature in parts of Europa's interior is warm enough to maintain liquid water. If so, it is interesting to speculate about whether this water has provided the basis for the development of an organic chemistry.

Perhaps some future probes to the Jovian system will be equipped with sensors designed to seek out life forms. At present, however, the case for doing so is too weak to justify such an expense, and the upcoming *Galileo* mission, for example, will not be equipped with any instruments for this purpose. For now, the possibility of life in the Jovian system will remain in the realm of the science-fiction writer.

stantly squeezed and stretched in different directions, and this has created internal friction and heating. Europa, which has a period almost exactly twice that of Io and therefore is aligned with it frequently, is particularly effective in helping keep Io in a vise of tidal forces. As a result, the interior of Io is probably fully molten, accounting for all the volcanic activity. The heating and continual eruptions also explain why Io has such a high density compared with the other Galilean satellites, since any water or other volatiles that

initially might have been present were vaporized and expelled from the satellite long ago.

Similar processes have affected the other three large satelites, but to a decreasing degree for each one in sequence going outward from Jupiter. Thus Europa, Ganymede, and Callisto (in order) show fewer signs of volcanic and tectonic activity; have older, less processed surfaces; and contain higher abundances of ice and volatiles.

The eruptions on Io help explain another puzzle

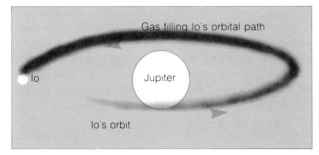

FIGURE 7.20. *The Io Torus.*
This sketch illustrates the geometry of the ring of gas that fills
the orbit of Io.

that we have mentioned. Io is attended by a cloud of
gas that fills its orbital path in a gigantic doughnut-
shape called a *torus* (Fig. 7.20). The discovery of volcanoes on Io offered an immediate explanation for the
source of this cloud. Recall that Jupiter rotates very rapidly, taking only about ten hours to spin once. This rotation period is much shorter than the orbital period of
Io, so as Jupiter rotates, its magnetic field sweeps past
Io at high speed. Charged particles from the volcanoes
are captured by the magnetic field and carried along
with it, quickly being spread out along the full length
of Io's orbital path. The particles become more highly
ionized in the acceleration process, and eventually become part of the Jovian radiation belts. Since Io lies in
Jupiter's equatorial plane, the torus is thickened by the
wobbling of the magnetic equator of Jupiter, in turn
caused by the 11-degree misalignment of the magnetic
and rotation axes. The surface and the immediate surroundings of Io are filled with charged particles, accounting for the electrical properties that modify radio
emissions from Jupiter as Io plows through the radiation belts.

The *Voyager* missions were not yet finished making
major discoveries about the complex and fascinating
Jovian system. A bright streak seen in some of the photographs turned out to be a ring around the planet; it
is very close in, well within the orbit of Io and between
the orbits at Metis and Adrastea (Figs. 7.21 and 7.22).
The ring is very thin and is not bright enough to be
seen from earth (although its presence was suspected
from *Pioneer* observations). There are indications of
some structure, consisting of a relatively bright outer
rim and a diffuse disk extending inward toward the
planet's atmosphere. Its discovery made Jupiter the
third of the four giant planets found to have at least

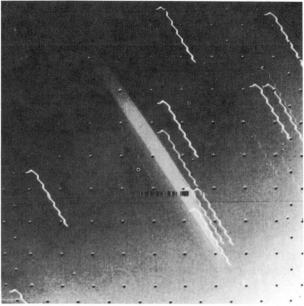

FIGURE 7.21. *The Jovian Ring.*
This *Voyager 1* image revealed the first evidence of a ring
girdling Jupiter. The ring is the broad light-colored band
extending from upper left to lower right. The bright, wavy lines
are star images, distorted by spacecraft motion during the
exposure.

one ring. Whatever the origin of planetary rings, they
could no longer be viewed as special, singular features,
but rather as a natural consequence of the formation of
the giant planets.

FIGURE 7.22. *The Geometry of the Jovian Ring.*
Here the ring has been sketched in, showing its size relative to
Jupiter. The ring radius is about 1.8 times the radius of the
planet, and it is estimated to be no more than 30 kilometers
thick.

PERSPECTIVE

The planet Jupiter, with its satellites, ring, and radiation belts, is the most complex and varied member of the solar system. It is like a universe all its own, with a distinctive and unique family of celestial bodies. Although it is modified by the sun, which provides some heat energy as well as solar wind particles to deform and feed the Jovian magnetosphere, Jupiter would probably exist in very much the same condition with no help from the sun. It is interesting to speculate about whether the universe contains lots of objects like Jupiter, just short of being stars in their own right.

Our journey through the realm of the giant planets now continues, with our sights set next on Saturn, another fascinating world that has recently been studied by space probes.

SUMMARY

1. Jupiter is gigantic, with 318 times the earth's mass, but so large that its average density is only one-fourth that of the earth.

2. Jupiter is totally unlike the terrestrial planets, having no solid surface, a thick atmosphere of hydrogen compounds, and rapid rotation.

3. Jupiter's atmosphere has a pattern of belts and zones, which are planet-girdling wind systems created by a combination of convection and rapid rotation of the planet.

4. Internally, Jupiter has a small, solid core formed by differentiation, a thick zone of liquid metallic hydrogen outside the core, an even thicker layer of liquid hydrogen above that, and a thin layer of clouds at the top.

5. Jupiter's strong magnetic field and its rapid rotation combine to create and maintain an intense radiation belt that is confined to the planet's equatorial plane.

6. Jupiter has sixteen satellites; of these, the four Galilean moons are by far the largest.

7. The Galilean satellites show a range of properties

indicating a decreasing degree of tectonic and geological activity from the innermost (Io) to the outermost (Callisto).

8. Io, the innermost Galilean moon, has volcanic activity and is accompanied in its orbit by a cloud of gas that forms a torus around Jupiter. Io's size and orientation is controlled by Jupiter's magnetic field.

9. Jupiter has a thin, dim ring.

REVIEW QUESTIONS

1. Summarize the differences in the compositions of Jupiter and Mercury, and comment on the cause of the differences.

2. Using Wien's law, calculate the wavelength of maximum emission for the Jovian cloud tops, whose temperature is 130 K. Why would it be difficult to observe this radiation from the earth?

3. Since Jupiter and the sun have the same composition, and the light from Jupiter is reflected sunlight, you might expect a great deal of confusion in attempting to distinguish lines formed in the Jovian atmosphere from those formed in the sun. Explain why this is not a problem.

4. Summarize the differences in the atmospheric circulation patterns of the earth and Jupiter, and give the principal reasons for these differences.

5. Suppose a spot seen on Jupiter is found to be an anticyclone. If it is in the southern hemisphere of the planet, which way does it rotate, and is the material within the spot rising or falling? How does the behavior of the Great Red Spot depart from the behavior expected of an anticyclone?

6. If Jupiter's mass were twenty times that of the earth and its radius 20,000 kilometers, what would the average density of its core be?

7. What would you conclude about the formation of the solar system if Jupiter had the same composition as the earth?

8. Why are the particle belts of Jupiter confined largely to the equatorial plane of the planet, whereas those of earth are distributed all around it?

9. Why does the density of surface craters decrease steadily from Callisto, the outermost Galilean satellite, to Io, the innermost?

10. Contrast the properties and probable geological histories of the earth's moon and Europa, which are very nearly the same size.

ADDITIONAL READINGS

Allen, D. A. 1983. Infrared views of the giant planets. *Sky and Telescope* 65(2):110.

Beatty, J. K. 1979. The far-out worlds of Voyager 1. *Sky and Telescope* 57:423 (pt. 1); 57:516 (pt. 2).

Beatty, J. K., O'Leary, B., and Chaikin, A., eds. 1981. *The new solar system.* Cambridge England: Cambridge University Press.

Cruikshank, D. P., and Morrison, D. 1976. The Galilean satellites of Jupiter. *Scientific American* 234(5):108.

Ingersoll, A. P. 1981. The meteorology of Jupiter. *Scientific American* 245(6):90.

Morrison, D., and Samz, J. 1980. *Voyage to Jupiter.* NASA Special Publication SP–439. Washington, D.C.: NASA.

Soderblom, L. A. 1980. The Galilean moons of Jupiter. *Scientific American* 242(1):68.

Squyres, S. W. 1983. Ganymede and Callisto. *American Scientist* 71(1):56.

Wolfe, R. E. 1975. Jupiter. *Scientific American* 233(3):118.

SATURN AND ITS ATTENDANTS

Learning Goals

8

Perhaps the most awe-inspiring sight in the solar system, one that can be appreciated with even a modest telescope, is the planet Saturn (Fig. 8.1). With its banded atmosphere and delicate ring system (Fig. 8.2), this planet more than any other celestial object has come to represent the realm of outer space in our imagination. Saturn is unusual in almost every respect. Besides the fantastic ring structure, it has a less dense constitution than any other planet, it has an asortment of mysterious moons, and it has among these satellites one called Titan, which is one of only two nonplanetary objects in the solar system known to have an atmosphere.

For a long time Saturn, at its distance of nearly 10 AU from the sun, was the outermost planet known. Its long orbital period, close to thirty earth years, means that its synodic period is not much longer than a year, and we get annual opportunities to view it in our nighttime skies.

In this chapter we will discover Saturn's nature, both in the distant view and as it is seen from nearby, for space probes have visited it, too. Many comparisons will be made with the other giant planet we have studied, and some similarities with Jupiter will be noted, along with numerous differences.

General Properties

Galileo made the first important discoveries about Saturn, using his telescope to observe it in 1610. He saw a brilliant yellowish object that appeared greatly elongated. He later referred to Saturn as the planet with "ears." Some fifty years after Galileo's observations, Huygens, with better data, was able to deduce the ring-like structure of the extensions on the planet, and thus explained the shape noted by Galileo. Extended observations showed the rings to be so thin that when viewed edge-on, they are nearly impossible to see. Saturn's rotation axis is tilted to its orbital plane by 27 degrees and the rings, which lie in its equatorial plane, are therefore tilted alternately 27 degrees above and 27 degrees below the orbital plane as Saturn goes around the sun.

Huygens also discovered Titan, the largest of Saturn's satellites, and shortly thereafter the French astronomer G. D. Cassini found four more satellites and

FIGURE 8.1. *Saturn as Seen From Earth.*
This photo was taken under good conditions, and shows some structure in the ring system.

a prominent gap in the rings (to this day called the Cassini division). From the satellite orbits, observers using Kepler's third law deduced the mass of Saturn, and this led to the realization that the planet's density is very low, only about 0.7 grams/cm^3. This is about half the density of the sun and a little more than half that of Jupiter. The density of Saturn is actually less than that of water.

Careful study of photographs revealed that the rings, as seen from the earth, appear to be three in number, with two inner ones (one of which is brighter than the other), and a third that lies outside of Cassini's division. From observations made when the rings are

FIGURE 8.2. *A Sketch of Saturn as Seen From Earth.*
This drawing, done by E. E. Barnard in 1898, shows the rings tilted to the greatest possible extent.

TABLE 8.1. Saturn

Orbital semimajor axis: 9.555 AU (1,427,000,000 km)
  Perihelion distance: 9.020 AU
  Aphelion distance: 10.090 AU
Orbital period: 29.46 years (10,759 days)
Orbital inclination: 2°29′33″

Rotation period: $10^h13^m59^s$
Tilt of axis: 26°44′

Mean diameter: 113,450 Km (8.91 $D_\oplus$)
  Polar diameter: 106,900 km
  Equatorial diameter: 120,660 km
Mass: 5.69 × 10^{29} grams (95.2 $M_\oplus$)
Density: 0.69 grams/cm^3
Surface gravity: 1.13 earth gravities
Escape velocity: 35.6 km/sec

Surface temperature: 95 K (at cloud tops)
Albedo: 0.50

Satellites: 17 (three additional satellites are suspected)

viewed edge-on, it was determined that the rings can be no more than about 10 kilometers thick. It was also discovered that when they are seen nearly face-on, background stars and the disk of Saturn itself are visible through them, as though the rings were made of sheer fabric. This supported a suggestion made in the late 1800s that the rings actually consist of innumerable small chunks of debris, each orbiting Saturn in accordance with Kepler's laws. Spectroscopic measurements (and later radar observations), in which the Doppler effect was used to determine the speeds of the particles, confirmed this idea by showing that the particles do travel at orbital speeds consistent with Kepler's laws. The radar data also revealed that at least some of the particles are chunks of material some centimeters or meters in size.

The planet itself rotates very rapidly, with a period of about ten hours, similar to that of Jupiter, and resulting in a similarly oblate shape. The rotation is also differential, with a shorter period at the equator than at the poles.

When spectroscopic measurements of the atmosphere were first carried out, familiar chemical compounds such as methane (CH_4), molecular hydrogen (H_2), and ammonia (NH_3) were found. The lower temperature of Saturn compared to Jupiter causes some differences, however, in that much of the ammonia on Saturn has apparently crystallized and precipitated out

of the atmosphere in the form of snow, and a few more complex molecular species are present on Saturn. No doubt the overall composition of Saturn is the same as that of Jupiter, since both formed out of the same material originally, and neither has lost any significant fraction of its gases to space. With their large masses and low temperatures (about 130 K at the Jovian cloud tops, and 95 K in the upper atmosphere of Saturn), the combination of high escape velocity and low average particle speed has prevented any large quantity of particles from escaping.

Like Jupiter, Saturn is a source of radio emission, with a combination of different origins. Synchrotron emission signifies the presence of radiation belts, where the planetary magnetic field has captured an extensive cloud of electrons and ions, and thermal radiation shows that Saturn also emits excess energy from its interior.

Another similarity of Saturn and Jupiter is the attention each has received from the U.S. space program. Three of the four probes sent to Jupiter so far have also visited Saturn, using the gravitational pull of Jupiter to assist and guide them on their way. Only *Pioneer 10* did not do this, having taken a path through the Jovian system that was unsuited for such a maneuver.

Pioneer 11, with its charged-particle monitors, radio antennae, and camera, uncovered a number of new details about Saturn and its attendant rings and satellites. At least one new moon was discovered and an additional ring was found, lying outside of those previously known or suspected. The magnetic field of Saturn was directly measured for the first time, and was found to be somewhat stronger than that of the earth, with the north and south magnetic poles upside down compared to earth (as in the case of Jupiter), and the magnetic axis well aligned with the rotation axis, unlike both Jupiter and the earth. The radiation belts are largely confined to the equatorial plane by the rapid rotation of Saturn, again mimicking Jupiter.

Hardly more than a year after *Pioneer 11* encountered Saturn in September 1979, *Voyager 1* arrived, and a great wealth of additional imformation quickly followed *Pioneer's* discoveries. The high-resolution color cameras on *Voyager* provided new revelations as sweeping in impact as their previous portraits of Jupiter. *Voyager 2* reached Saturn in August 1981, following a different trajectory and viewing it from a different perspective (Fig. 8.3).

As in the case of Jupiter, the *Voyager* missions have

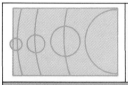

Astronomical Insight 8.1

THE ELUSIVE RINGS OF SATURN*

In the text we said little about the discovery of the ring system encircling Saturn, except that Galileo was perplexed by the appearance of the planet, and that some years after his death the problem was solved by Christiaan Huygens. The story is actually much more complex and interesting. The mystery of Saturn's appearance and the extensive efforts devoted to solving it were the leading concerns in astronomy from the 1630s to the end of the seventeenth century.

Galileo first observed Saturn through a telescope in 1610, and he immediately saw that its appearance was strange. He thought he was observing a spherical planet attended on each side by a smaller spherical body, and he concluded that these were two satellites of Saturn. He made repeated observations thereafter, but to his surprise he saw no changes in the positions of the suspected satellites, at least not for awhile. Frustrated by the lack of change, Galileo stopped observing Saturn regularly, and was therefore shocked when, in 1612, he found it to be a simple sphere, with no trace of the peculiar extensions or satellites seen earlier (the rings were edge-on to the earth at that time). Galileo resumed frequent observations and saw the shape of Saturn change steadily. In 1616, he observed an elliptical overall shape with dark spots near each end, and this is what gave him the impression of ears (the dark spots were the small gaps between the rings and the planet, where the blackness of space can be seen in the background).

Galileo exchanged correspondence with other observers, and soon Saturn was being watched by a number of astronomers, all of whom were puzzled by its unorthodox behavior. When the planet regained a simple spherical appearance in 1642, interest was further heightened, and by this time Saturn was the central problem in astronomical research and discussion. After 1642, numerous sketches of the planet were published by a number of observers, and soon a variety of suggestions were made as to the cause of Saturn's variable and strange appearance.

It is interesting that the resolution of most of the telescopes used was actually good enough to provide some detail, and the explanation of the phenomenon in terms of a ring circling Saturn quite possibly could have been developed years before it actually was put forth by Huygens. The problem was that the idea of a ring was completely alien in terms of the usual astronomical phenomena, and nobody made the mental leap to this concept. Meanwhile, the data on the question continued to accumulate.

In early 1655, soon before the ring was to disappear again as it presented itself edge-on to the earth-based observer, Huygens had his first look at Saturn. His telescope, contrary to popular belief, was really little better than those of many of his contemporaries, and he was able to see no more detail than they. Furthermore, he made his first observation at a time when the rings were so nearly edge-on that he had no chance of seeing the true nature of Saturn's peculiar shape. Nevertheless, on the basis of his own limited data and especially his knowledge of Saturn's appearance as seen by others, Huygens deduced that all the observations could be explained by a flat ring that circled the planet in its equatorial plane (the inclination of the equator of Saturn had been deduced from the orbital motions of the satellite, Titan, which had been discovered by Huygens during his observations in 1655).

A number of alternative views were proposed at about the same time, and it took years for Huygens's solution to become generally accepted. It is interesting that it was essentially a theoretical solution, rather than something that he saw simply by looking through a good telescope, as is commonly assumed.

*Based on information from Van Helden, A. 1974. Saturn and his anses. *Journal Hist. Astronomy* 5(2):105; and Annulo Cingitur: The solution to the problem of Saturn. *Journal Hist. Astronomy* 5(3):155.

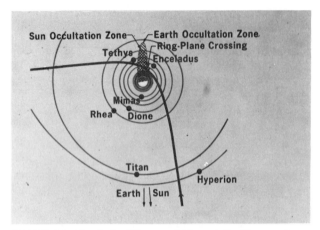

FIGURE 8.3. *The Path of Voyager 2 Through the Rings and Satellites of Saturn.*

FIGURE 8.4. *A Voyager Portrait.*
This image of Saturn was obtained at a distance of 5.3 million kilometers. It shows much more detail than can be seen in the best earth-based photos.

added breadth, color and a certain amount of fantasy to our picture of a newly explored planetary system (Fig. 8.4). Now we can discuss in some detail the weather patterns on Saturn, the composition and structure of its atmosphere and interior, and the nature of its dazzling array of rings and satellites.

Atmosphere and Interior: Jupiter's Little Brother?

To the faraway observer, the atmospheres of Jupiter and Saturn look similar, in terms of their overall color and banded appearance. There are also many contrasts, however.

As noted earlier, the chemical compositions of the two atmospheres are much alike, except that the colder temperature on Saturn has caused much of the ammonia to precipitate out and has allowed the formation of somewhat more complex molecules. An example that was detected from earth-based spectroscopic data is ethane (C_2H_6), and other derivatives of methane are probably present.

Whereas Jupiter has a colorful and distinct contrast between belts and zones, and turbulent, swirling winds that show up with great clarity, Saturn has a muted overall appearance. The belts and zones are not so distinct, and the dark and light spots that can be seen do not contrast strongly with their surroundings.

Observations show that Saturn's cloud layer is much thicker than Jupiter's (Fig. 8.5), although it is lacking

the high haze layer that characterizes the larger planet. The lower temperature of Saturn allows the cloud layer to persist to much greater heights, with the result that the structure of the lower atmosphere is much more heavily obscured than on Jupiter. Thus Saturn does not

FIGURE 8.5. *The Contrasting Cloud Thicknesses on Jupiter and Saturn.* The deeper cloud zone on Saturn is largely responsible for the relative lack of contrast in its visible surface features.

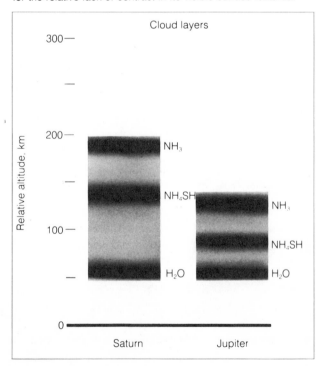

have the striking banded appearance of Jupiter, and the atmospheric flow patterns are correspondingly more difficult to discern.

Enough can be seen through the haze on Saturn for us to determine that the atmospheric flow patterns, although similar in some ways to those on Jupiter, also display significant differences. As on Jupiter, rising gas forms cyclonic flows, and descending cool gas creates anticyclones, in each case stretched by the rapid planetary rotation into elongated strips circling the planet.

FIGURE 8.6. *Circulation in the Atmosphere of Saturn.* Even though the contrast is not as vivid on Saturn as it is on Jupiter, the ringed planet does have complex atmospheric motions, driven by a combination of convection and rapid planetary rotation. Bright spots in the upper left of the upper frame are seen here to move in a counterclockwise (anticyclonic) direction.

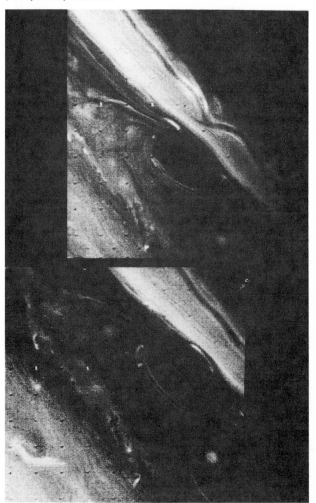

Like Jupiter, Saturn has many spots and oval flow patterns (Fig. 8.6).

The speed of the flow patterns on Saturn is much greater than on Jupiter. At the equator, for example, Jupiter has wind velocities of about 100 meters/sec (360 km/hr), whereas Saturn has winds moving as rapidly as 400 to 500 meters/sec (up to 1,800 km/hr). The belt and zone structure on Saturn extends closer to the poles than on Jupiter (Fig. 8.7), and the light and dark strips are broader in the north and narrower in the south. These contrasts between Saturn and Jupiter may be caused by the fact that Saturn, with its axial tilt of nearly 27 degrees, has seasons and therefore variations in the sun's intensity, whereas Jupiter, whose axis is nearly perpendicular to the ecliptic (its tilt is only a little more than 1 degree), does not. To directly observe the effects of Saturn's seasons will take a long time, in view of its nearly thirty year orbital period.

There are differences in the internal structures of the two planets as well. First and foremost, the density inside Saturn is much lower, because its much smaller mass is spread out over nearly the same volume as Jupiter. The solid, rocky core inside Saturn is probably not as dense as the core of Jupiter, and may contain some ice (Fig. 8.8). Outside of this core, the internal pressure, although not as great as inside Jupiter, is suf-

FIGURE 8.7. *Bands in the Atmosphere of Saturn.* This view of the planet's northern polar region shows the muted appearance of the belts and zones.

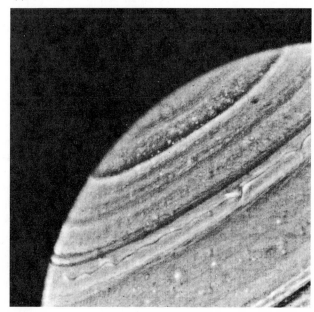

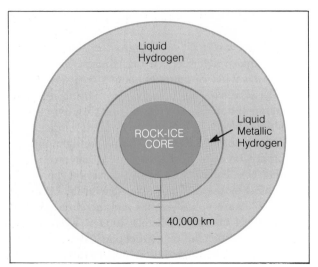

FIGURE 8.8. *The Internal Structure of Saturn.*
This cross-sectional view of Saturn's internal structure should be
compared with the model of Jupiter's interior shown in Fig. 7.9.

ficient to form liquid metallic hydrogen, which is
thought to inhabit a zone extending about halfway out
to the surface. Above this is a thick layer of liquid hydrogen and helium, topped off by the layer of clouds
that is visible from afar.

Like Jupiter, Saturn also radiates substantially more
energy than it receives from the sun, by a factor of 2.2
in this case. The explanation offered for Jupiter does
not work as well here; Saturn, with its lower mass,
could not have retained enough primordial heat from
its formation to still be as warm internally as the excess
radio emission implies. Some additional source of heat
must be present, and its nature is not clear. One suggestion is that differentiation is still going on in the metallic and liquid hydrogen zones, as the heavier helium
atoms gradually sink toward the center of the planet,
releasing gravitational energy as they do so. The likelihood that this process is really at work inside Saturn is
enhanced by the observation that Saturn has relatively
less helium in its outer layers than does Jupiter, consistent with the idea that Saturn's helium has sunk,
whereas Jupiter's has not. The probable reason that
this has not occurred inside Jupiter is that the interior
of the larger planet is warmer and more turbulent, preventing the helium from quietly sinking toward the
center.

The magnetic field of Saturn is, no doubt, formed by
a dynamo in the interior. The structure of Saturn's
magnetic field suggests that the depth at which the field

forms is greater than in the case of Jupiter. The zone
of trapped ions and electrons around Saturn is still
quite large, however; and most of the satellites and all
of the rings are within this zone, sweeping up charged
particles as they move along in their orbits. Because of
variations in the flow of ions from the sun, the magnetosphere of Saturn fluctuates in size, with the result that
Titan, the largest satellite, is sometimes within the
magnetosphere and sometimes outside it.

Filling the orbit of Titan, and extending quite a bit
farther inward and outward, is a cloud of hydrogen
atoms resembling in shape the Io torus that girdles Jupiter. In this case, however, the gas is not ionized, so
its structure is not controlled by the planetary magnetic
field.

The Satellites of Saturn

Before the *Pioneer* and *Voyager* encounters, Saturn was
thought to have at least ten satellites, some of them
small and rather difficult to see from earth. The first of
Saturn's moons to be discovered was Titan, noted by
Huygens in 1655. Four more were found by Cassini in
the latter part of the seventeenth century, and in the
next three hundred years, an additional five were seen.
The innermost was thought for a time to be the satellite
Janus, first reported in 1966, but a suspected eleventh
satellite, called S–11, appeared to be even closer in, orbiting very near the outer portion of the rings visible
from earth.

The *Pioneer 11* and *Voyager 1* and *2* encounters
have clarified the situation quite a bit. *Pioneer 11* confirmed the presence of S–11 (now called Epimetheus;
Fig. 8.9). In addition, four new satellites were discovered by the *Voyager* missions and two from earth-
based observations, so the current total is seventeen,
with three more suspected on the basis of *Voyager*
data. The confounding former eleventh satellite turned
out to be sharing the same orbit with another satellite
(Janus), something previously never seen. Both of these
moons are elongated in shape, and orbit with their
long axes pointed toward Saturn, in synchronous rotation. Two of the new satellites, also very small, were
found orbiting in the same path as Dione, one of the
larger satellites. These moons, designated 1980S6 and
No Name II in Table 8.2 are held in position about 60
degrees ahead of Dione by the combined gravitational

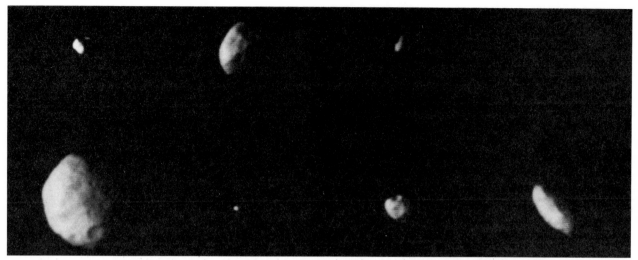

FIGURE 8.9. *Some of the Small Satellites of Saturn*. Taken at various ranges, these images do not reflect the true relative sizes. The upper middle and lower left moons are the co-orbital satellites (Epimetheus and Janus), and the two at lower right (1980S27 and 1980S26) are the pair that "shepherd" the F ring (discussed later in this chapter). The upper left satellite (1980S56) is the one that shares the orbit of Dione, and at upper right and lower left center are the pair (Telesto and Calypso) that share the orbit of Tethys.

forces of Saturn and Dione. Two small satellites, Telesto and Calypso share an orbit with Tethys, another of the relatively large moons.

Another pair of small satellites, designated 1980S27 and 1980S26, were found to be orbiting at nearly the same distance from Saturn, with a thin ring, itself discovered by *Pioneer 11*, between them. As we shall see in the next section, these satellites probably formed and now maintain this ring by their gravitational forces.

The innermost of the new satellites discovered by the

TABLE 8.2 Satellites of Saturn

NO.	NAME	DISTANCE	PERIOD	DIAMETER	MASS	DENSITY	ALBEDO
17	Atlas	2.282 R_S*	0^d.602	38 × ? × 26 km			0.4
16	1980S27	2.310	0.613	140 × 100 × 74			0.6
15	1980S26	2.349	0.629	110 × 84 × 66			0.6
10	Janus	2.51	0.694	220 × 190 × 160			0.5
11	Epimetheus	2.51	0.695	140 × 114 × 100			0.5
1	Mimas	3.075	0.942	392	3.75×10^{22} gm	1.19 gm/cm^3	0.7
2	Enceladus	3.946	1.370	500	8.4×10^{22}	1.13	1.0
13	Telesto	4.884	1.888	30 × 24 × 16			0.6
14	Calypso	4.884	1.888	? × 24 × 22			0.8
3	Tethys	4.884	1.888	1,060	7.55×10^{23}	1.20	0.8
	No Name I	5.47:		15–20			
4	Dione	6.256	2.737	1,120	1.05×10^{24}	1.43	0.6
12	1980S6	6.267	2.739	36 × ? × 30			0.5
	No Name II	6.27:	?	15–20			
	No Name III	7.79:	?	15–20			
5	Rhea	8.737	4.518	1,530	2.49×10^{24}	1.33	0.6
6	Titan	20.25	15.945	5,150	1.35×10^{26}	1.88	0.2
7	Hyperion	24.55	21.277	350 × 234 × 200			0.3
8	Iapetus	59.02	79.331	1,460	1.88×10^{24}	1.16	0.5/0.04
9	Phoebe	214.7	550.4	220			0.06

*The symbol R_S stands for the equatorial radius of Saturn, taken here to be equal to 60,330 km.

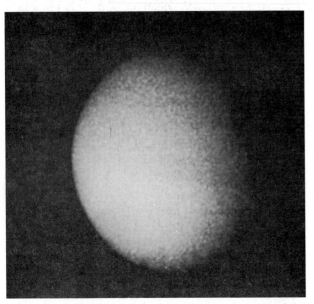

FIGURE 8.10. *Titan.*
The largest of Saturn's moons, Titan is almost unique among all the satellites in the solar system in that it has an atmosphere.

Voyager probes, named Atlas actually orbits well inside the three rings that lie farthest out, although it is still just outside the broad, bright ring that is the outermost one visible from earth. A small moon (No Name I) is suspected to exist between the larger satellites Tethys and Dione, and there may be two more (No Name II and No Name III) between Dione and Rhea.

The major satellites, those easily observed from earth, are as varied in their individual qualities as the Galilean satellites of Jupiter. The biggest of them all, Titan (Fig. 8.10), was especially important for *Voyager 1* to examine closely, since it was already know to have an atmosphere, making it nearly unique in that respect among all the satellites of the solar system. Recall that until *Voyager 1* revealed the great depth of Titan's atmosphere, this satellite was thought to be larger than Ganymede, whereas in fact Ganymede has the honor of being the largest satellite in the solar system.

It was hoped that the vertical structure of Titan's atmosphere (Fig. 8.11) could be probed and the surface below could be examined, but the atmosphere is so

FIGURE 8.11. *Titan's Atmosphere.* This diagram compares conditions in the atmospheres of Titan and the earth. Note that the vertical scales are not the same; Titan's atmosphere is much deeper.

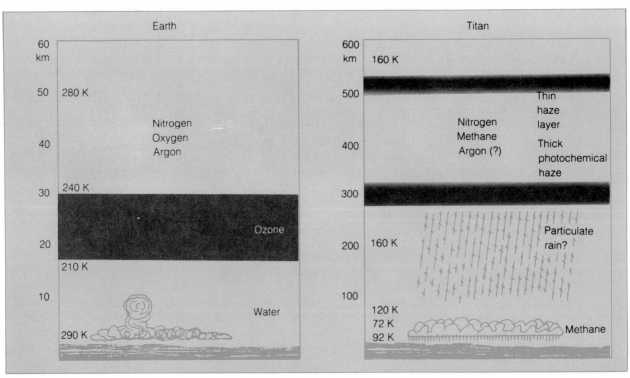

thick that it was impossible for the *Voyager* instruments to see to any great depth. In addition, there is an obscuring haze layer at the top of the atmosphere, which makes direct observations more difficult. Nevertheless, radio and infrared sensors were able to penetrate deeply, providing data which, along with spectroscopy of the upper layers, were sufficient to allow some rather detailed estimates of the surface conditions to be made.

First, the density of the solid body of Titan itself was ascertained to be just under 2 grams/cm^3, corresponding to an internal compostion of about half rock and half ice. This is very similar to Ganymede.

We can only speculate as to what the surface of Titan might be like. There is methane in the atmosphere, and radio data showed that the surface temperature is about 90 K, very near the value at which methane can exist in any of three forms: liquid, solid, or vapor. Thus there may be liquid methane on the surfce, along with methane ice formations such as glaciers, as well as gaseous methane in the atmosphere. In short, methane on Titan may play very much the same role as water on the earth. If it does, there is probably substantial erosion, so that old impact craters and tectonic formations, if they exist, are continually being smoothed over.

There is also the possibility of liquid nitrogen lakes on Titan, since it was discovered that nitrogen is by far the most abundant gas in the atmosphere, and that the surface temperature is not far above the point where nitrogen condenses into a liquid form. Besides the earth, only Mars, Venus, and now Titan are known to have any significant abundance of nitrogen in their atmospheres. The origin of the nitrogen on Titan may be volcanic, since it is a volatile gas that can be emitted from certain kinds of rocks when they are heated. Therefore, there may also be volcanic features on the surface of this satellite.

The atmospheric pressure is about 50 percent greater at the surface of Titan than at sea level on earth, and Titan's atmosphere extends to five times the height of earth's atmosphere above the surface. It is not well understood why Titan has been able to retain such a thick atmosphere, when no other satellite (except Triton) has any atmosphere at all. Certainly the low temperature and large mass have helped, but in these regards Titan does not appear to have any great advantage over Ganymede, for example. Perhaps the atmosphere of Titan is being continually replenished by some ongoing process, such as volcanic activity.

The satellites of Saturn are generally classified into four groups: the tiny inner moons that orbit near to and between the rings; Titan, which is in a class by itself; distant Phoebe, a weird misfit of a satellite that also is without peer; and the remaining satellites, which are intermediate in size and similar in many ways. We have already discussed the first two of these categories, and we turn our attention now to the intermediate-sized moons.

These seven, named Mimas, Enceladus, Tethys, Dione, Rhea, Iapetus, and Hyperion, are actually quite varied in some respects. They have diameters ranging from about 400 to 1,500 km (Titan's diameter is 5,150 km, by comparison). Masses for most of them were determined from their gravitational influences on the trajectory of the *Pioneer* and *Voyager* spacecraft, and from these measurements the densities were estimated. In all cases the densities are rather low, indicating a high ice content, and a few of these moons may be nearly pure ice, with densities near 1 gram/cm^3. In this regard they are quite unlike any other bodies in the solar system.

Among the intermediate-sized satellites, a variety of surface features were seen in the *Voyager* photographs, including heavy cratering in many cases, as well as some peculiar structures. On Dione (Fig. 8.12) were found sinuous, branching valley systems and white, wispy patterns that look deceptively like the high-altitude cirrus clouds of earth. These may be frozen gases that crystallized immediately upon being emitted from surface rocks or volcanoes, falling to the ground in cloudlike arrangements. A few of the moons, including Dione, Tethys (Fig. 8.13), and Rhea (Fig. 8.14), have linear trenches, possibly the result of fractures caused by massive impacts. The innermost of this group of satellites, Mimas (Fig. 8.15), is marred by a tremendous impact crater that is fully one-fourth as large as the diameter of the satellite itself.

Mysteries abound. Iapetus (Fig. 8.16), one of the larger members of this group, has a bright and a dark side, the one some ten times more reflective than the other. *Voyager* 2 images indicate that the dark side of Iapetus may be covered by some sort of deposit, perhaps resembling soot. Enceladus (Fig. 8.17) is the shiniest object in the solar system, with an albedo near 1, meaning that it reflects just about all the light that hits it. The orbit of this satellite is coupled with that of Dione, with Enceladus orbiting Saturn once for every two orbits of Dione; thus the two are frequently lined up. This may create sufficient tidal friction inside En-

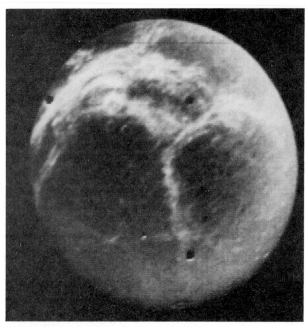

FIGURE 8.12. *Dione*.
This intermediate-sized moon of Saturn has wispy, light-colored features on its surface that may be icy deposits of material that escaped from the interior and crystallized. This side of Dione, which is in synchronous rotation, always trails as the satellite orbits Saturn; the leading side is more heavily cratered.

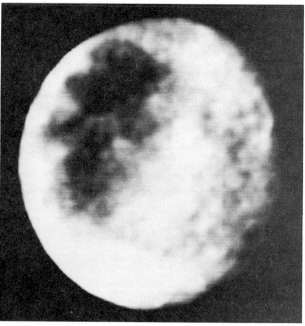

FIGURE 8.14. *Rhea*.
One of the seven intermediate-sized moons, Rhea shows evidence of cratering, as well as light-colored areas that probably represent ice.

FIGURE 8.15. *Mimas*.
This satellite shows many impact craters, including one very large one that has about one-fourth the diameter of the satellite itself. This moon is probably composed chiefly of ice, as are several of the other satellites of Saturn.

FIGURE 8.13. *Tethys*.
Another of the intermediate-sized moons, Tethys is characterized by large linear trenches, an example of which is seen here.

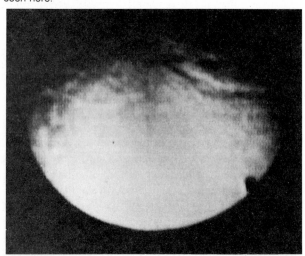

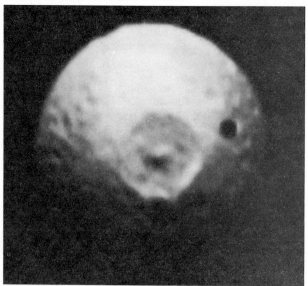

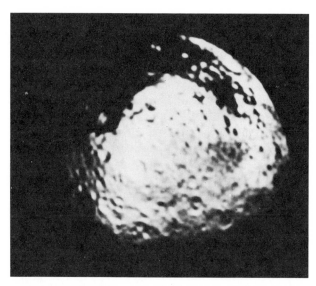

FIGURE 8.16. *Iapetus.*
This satellite has very unusual bright and dark areas. It appears that the dark material may be some sort of deposit.

FIGURE 8.17. *Enceladus.*
This is the shiniest object in the solar system, with an albedo of approximately 1.0, just like a mirror. The surface may have been coated with deposits resulting from internal melting and outgassing.

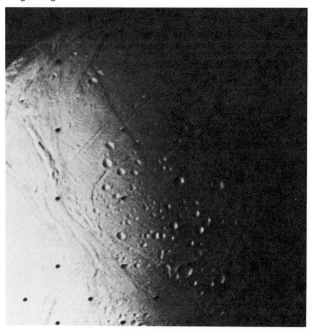

celadus to make it volcanically active, much like Io, so its surface may be continually recoated by volcanic ejecta. Another hypothesis is that Enceladus is being continuously coated by ice particles from the tenuous E ring, within which its orbit lies. The satellite Hyperion (Fig. 8.18) was found to be strangely asymmetric in

FIGURE 8.18. *Three Views of Hyperion.*
This satellite has a very unusual shape, and is marked by several impact craters.

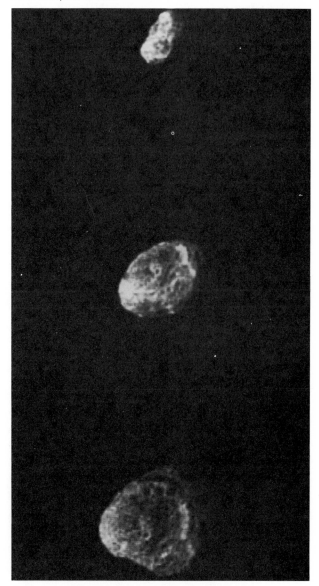

shape despite its moderately large size, indicating that it possibly was never molten.

Perhaps the strangest of all the satellites of Saturn is Pheobe, nearly four times farther out than all the rest, and orbiting in the retrograde direction, with the plane of its orbit tilted 30° with respect to the others. This oddball among oddballs, which has a dark surface, may be a stray asteroid that was captured by Saturn's gravitational field and forced into orbit. It has also been suggested that Pheobe is a cometary nucleus that was captured.

The Rings

We turn our attention now to the stunning ring system of Saturn, probably the most spectacular sight of all (Fig. 8.19). The *Voyager* data proved these structures

FIGURE 8.19. *The Rings of Saturn.*
This *Voyager* image reveals some of the fantastically complex structure of the ring system. A small satellite (1980S27) is visible just inside the F ring at upper left.

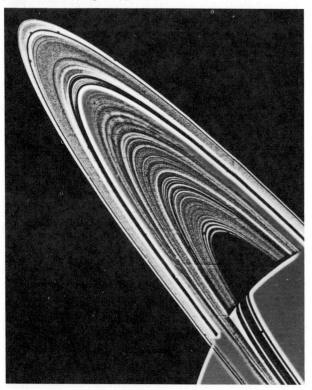

to be even more complex and fascinating than had been imagined.

The rings consist of countless individual particles, ranging in diameter from a fraction of a centimeter in some portions of the system to a few meters in others. Each particle has its own orbit, with its period determined by its distance from Saturn, in accordance with Kepler's third law. Thus the rings do not rotate like rigid hoops around the planet, in which case all the particles would have the same period (meaning that the outer ones would have to travel fastest). Instead, they are fluid structures, moving slowest in the outermost portions.

The three distinct rings visible from earth were labeled the A, B, and C rings (in order, from the outer one inward). The possibility of a faint ring inside the C ring was raised some time ago, and this D ring was actually found, as were E, F, and G rings outside the A ring (these letters are assigned in the order of discovery).

Even before the full complexity of the ring system was known, some idea of how the rings formed had been developed. The rings are all very close to Saturn, and this led to the suspicion that tidal forces have had something to do with their formation. There is a certain distance from any massive body, called the **Roche limit,** inside of which the tidal forces pulling apart an object of a given size are greater than the gravitational force holding it together. The exact location of the Roche limit depends on the mass and size of the orbiting object. We have discussed the fact that our moon

TABLE 8.3. The Rings of Saturn

FEATURE	DISTANCE
D ring, inner edge	1.11 R_s*
C ring, inner edge	1.233
B ring, inner edge	1.524
B ring, outer edge	1.946
A ring, inner edge	2.021
A ring, gap center	2.212
A ring, outer edge	2.265
F ring, center	2.326
G ring, center	2.8
E ring, inner edge	3:
E ring, outer edge	8:

*The symbol R_s stands for the equatorial radius of Saturn, taken here to be equal to 60,330 km.

has a tidal bulge caused by the earth's gravitational pull; to envision what the Roche limit means, imagine that the bulge becomes so severe that the moon is actually pulled apart and broken into pieces.

All the rings of Saturn are within the Roche limit for a satellite of any significant size. Therefore, there once might have been a satellite that wandered too close to Saturn and was broken apart, leaving behind the debris that now forms the rings. More likely, the rings are made of particles that have been orbiting Saturn since its formation, but these particles have stayed so close to the planet that tidal forces have prevented their ever merging together to form a satellite.

Gravitational forces are also responsible for the major gap in the ring system of Saturn. The Cassini division, between the A and B rings, lies at just the right distance from Saturn that any particle orbiting there would be in orbital resonance with Mimas, the innermost of the larger satellites. By this we mean that the orbital period of such a particle would be precisely half the period of Mimas, so that for every time the particle made two trips around, the particle and Mimas would line up on the same side of Saturn. On each of these occasions, the particle would feel a little extra tug exerted by Mimas. Over a long period of time, the cumulative effect of these tugs would be sufficient to change the shape of the particle's orbit such that it would stray out of its former circular path and eventually collide with other particles and be deflected out of that part of the ring system entirely. In this way Mimas has maintained a clear gap at the position of Cassini's division. Other satellites, with other fractions of the orbital period of Mimas (such as the location where an orbiting particle would have exactly one-third the period of Mimas), could create other gaps in the rings, and indeed such gaps have been found. Such discoveries led scientists to suspect that the rings have complex structure, even before *Pioneer* and *Voyager* data were available.

Despite this expectation, when the first *Voyager* images of the rings were obtained, scientists were amazed by what they saw. None of the rings had a smooth appearance when viewed from nearby; the prominent ones visible to us from earth are each composed of a number of thin rings that came to be called "ringlets" (Fig. 8.19). The photopolarimeter experiment on *Voyager 2* revealed much more detail in the rings, showing structures as small as 100 meters. There are thousands of ringlets throughout the entire system, some appearing dark and others light. Even Cassini's division is not completely empty, but rather is inhabited by several dark ringlets. There apparently is also a narrow gap that truly is empty, probably created by Mimas in the manner just described.

Whether a ring appears dark or light depends on how the particles in it reflect sunlight, and the angle from which the ring is viewed. Some of the rings tend to reflect light straight back, whereas others allow it to go through in a forward direction. A ring that reflects sunlight back will appear bright when viewed from the earth, but will look dark when viewed from the far side of Saturn, looking toward the sun. Conversely, a ring that allows light to pass through will appear dark from the earth (as do the narrow ringlets in Cassini's division) and bright if viewed from behind. Whether a given ring will scatter light forward or backward depends on the size, shape, and composition of the particles it is made of.

By considering the light-scattering properties of the particles, observers have been able to deduce the average sizes of the particles in the various rings of Saturn. The outermost rings tend to be made of very tiny particles, a few ten-thousandths of a centimeter in diameter, whereas the inner rings, in contrast, seem to consist mostly of chunks of material a few meters across. All the particles are probably composed mostly of ice, and their shapes are irregular. The total mass of all the ring material is very small, only about a millionth of the mass of our moon, so if the ring particles could have formed into a satellite of Saturn, it would have been a rather small one.

The complex pattern of rings and ringlets seen by the *Voyager* cameras required a more sophisticated explanation than the orbital-resonance hypothesis described earlier, although orbital resonance undoubtedly occurs in the manner we discussed. Apparently there are at least three other mechanisms responsible for shaping the rings of Saturn.

The first of the mechanisms we will examine is also caused by gravitational effects of Saturn's moons. The discovery by *Voyager 1* of the two small satellites that orbit just inside and just outside the F ring (Fig. 8.20) led scientists to understand how this mechanism works. The gravitational interaction between the ring particles and these satellites keeps the particles confined to orbits between them. The moon just outside the ring moves a little more slowly than the ring particles, in accordance with Kepler's laws. Whenever a

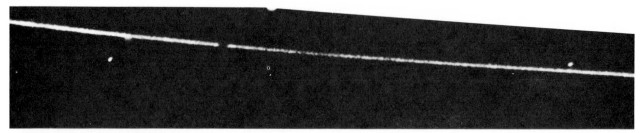

FIGURE 8.20. *The Shepherd Satellites and the F ring.* The two small satellites (1980S26 and 1980S27) that maintain the F ring through gravitational interaction are both visible in this image.

ring particle overtakes this satellite at a close distance, the particle loses energy because of the gravitational tug of the satellite, and drops to a slightly lower orbit. On the other hand, the satellite on the inside of the ring slowly overtakes the ring particles, and when it passes close by one, the particle gains energy and moves out to a higher orbit. In this way the two satellites ensure that the ring particles stay in orbit between them, because any particle that strays too close to either one will have its orbit altered in just such a way that it moves back into the ring. For this reason the two moons are called "shepherd" satellites, because they behave like sheep dogs, keeping their flock in line by nipping at their heels.

Voyager 2 data showed that another mechanism (which was suspected after the *Voyager 1* flyby) was also influencing the structure of the rings. Mathematical calculations show that in any flattened disk system such as Saturn's rings, small gravitational forces can create a spiral wave pattern, something like the grooves on a phonograph record. The waves take the form of alternating dense and rarefied regions, and the entire spiral pattern rotates about the center of the system. Similar spiral density waves have long been thought responsible for the spiral-arm structure of galaxies like our own (see chapter 19); the *Voyager* data have shown that the rings of Saturn are a small-scale model of a spiral galaxy.

It was predicted from calculations that the spacing between spiral density waves increases steadily with distance from the center. The photopolarimeter data from *Voyager 2* (Fig. 8.21) revealed that the expected increase in spacing between ringlets does occur in several portions of the ring system. This confirms that at least some of the ring structure is created by spiral density waves. The ringlets that are formed by this mechanism are not separate, circular structures, but instead are part of a continuous, tightly-wound spiral pattern

that slowly rotates. Individual ring particles, by contrast, follow elliptical orbits, and alternately pass through the ringlets and the spaces between. (It is important here to distinguish between the motion of the wave and the motions of the particles, which do not move with it; an analogy is a cork bobbing on water waves, but not moving along with them.) The reason there are more particles in the ringlets than between them at any given moment is that they move more

FIGURE 8.21. *Very Fine Structure in the Rings.* This is a synthetic image of a small section of Saturn's ring system, reconstructed from stellar occultation data obtained by the photopolarimeter experiment on *Voyager 2*. The finest details visible here are only about 500 meters in size, whereas the best *Voyager* camera images have a resolution of about 10 kilometers. Data like these have established that spiral density waves play a role in shaping Saturn's rings.

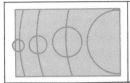

The *Voyager* missions to Saturn have produced a wealth of new information on the rings, primarily because of the high resolution of the images they have sent back to earth. Recall from chapter 3 that resolution is the ability to make out details. We emphasized that resolution can be enhanced by using large telescopes, placing observatories above the earth's turbulent atmosphere, or using special interferometry techniques. There is another way to achieve high resolution, however: get close to the subject of the observations. This is what the *Voyager* spacecraft did.

The cameras aboard *Voyager 1* and *2* yielded images of Saturn's rings with a maximum resolution of about 10 km; that is, details as small as 10 km in size could be seen. This was quite an improvement over the best photographs taken from earth, which have a resolution of about 4,000 km. The Space Telescope, when it is placed in the earth's orbit in the late 1980s, will resolve details in the ring system about ten times smaller than this, roughly 400 km, still far inferior to *Voyager* images.

There was an instrument on the *Voyager* spacecraft that provided much better resolution than even the cameras. This was the photopolarimeter, a device for measuring the intensity of light as well as its polarization. The photopolarimeter on *Voyager 1* failed before the spacecraft reached Saturn, but the one on *Voyager 2* was able to carry out an observation that has provided by far the most detailed information yet on the ring system.

The observation was rather simple: as *Voyager 2* approached Saturn, the photopolarimeter measured the brightness of a star (Delta Scorpii) that happened to lie behind the rings, from the point of view of the spacecraft. As *Voyager 2* swept by the planet with the photopolarimeter trained on the star, the spacecraft motion caused the ring system to move in front of the star. As it did so, the ringlets and gaps caused the brightness of the star to fluctuate. When a dense ringlet passed in front of the star, the star's brightness temporarily dimmed; when a gap passed in front, the star appeared brighter. This kind of observation is known as a *stellar occultation*.

The photopolarimeter was able to record the changes in the star's brightness very rapidly. This, combined with the rate at which the rings appeared to move in front of the star, determined how fine the details were that could be detected in the ring structure. As it turned out, the resolution of this experiment was about 100 meters, a factor of 100 better than the best images obtained by the *Voyager* cameras! This meant that ringlets or gaps as small as the length of a football field could be detected.

Of course, the photopolarimeter could only measure the ring structure along a single cross-section; it could not provide a full picture of the entire system. Its image of the rings was one-dimensional, but it nevertheless led to profound new understanding, particularly of the spiral density waves described in the text.

The data produced by the photopolarimeter were originally in the form of plots showing the variations in the star's intensity as a function of time. Dips in the intensity corresponded to the passage of ringlets in front of the star. This information was then translated into a plot of ring density as a function of distance from Saturn. The data were processed a step further, so that things were easier to visualize; synthetic images of the rings, simulating those taken by a camera, were produced (it was assumed that the rings were circular, and different colors were used to represent varying degrees of ring density). In this way, false-color images of the rings were obtained, somewhat similar to the actual photographs taken by the cameras but showing details a factor of 100 smaller.

slowly as they pass through the density waves, and tend to congregate there, like cars in a traffic jam that builds up at a point where the flow is slowed down.

The gravitational disturbances that create the density waves in portions of Saturn's ring system are caused by the satellites. The strongest disturbances occur at resonance points where the particles have orbital periods that are a simple fraction of the periods of satellites orbiting at greater distances. In some cases, these are also the locations of gaps in the rings (as explained earlier), so we sometimes find that a gap lies just at the inner edge of a series of ringlets that are created and maintained by a spiral density wave.

A fourth type of force affecting the shape of the rings may not be gravitational at all. Some of the rings have nonsymmetric shapes (Figs. 8.22 and 8.23) that cannot be formed by any simple gravitational forces exerted by Saturn or its moons. For example, one of the inner ringlets is not circular, but is instead flattened into a slightly oval shape (although it does not seem to be part of a spiral density wave); and the F ring, the one that is held in place by the pair of shepherd satellites, has a very peculiar braided structure, as though three rings had been interwoven. No reasonable explanation has been found for either kind of behavior. Comparison of *Voyager 1* and *Voyager 2* photographs showed that the irregular structure of the F ring varies with time.

One possiblity has been raised, however, but not yet worked out in detail, which takes note of the fact that all the ring particles orbit Saturn within the radiation belts. Electrons from the belts colliding with ring particles can cause the buildup of negative charges on the particles. Once these particles have electrical charges, they are subject to electromagnetic forces from the planet's magnetic field. If a charged ring partical is sufficiently small, its motion can actually be controlled by eletromagnetic forces, rather than gravitational ones. It is therefore possible that Saturn's magnetic field exerts some influence over the smaller particles in the ring system, distorting the shapes of some of the rings. The magnetic field rotates with Saturn much faster than the particles move in their orbits, and the magnetosphere itself varies in size and intensity with fluctuations in the stream of charged particles coming from the sun. Thus it is possible to imagine nonsymmetric forces acting to shape the rings.

It is almost certain that electromagnetic forces are responsible for the distribution of particles in another phenomenon associated with the rings. A haze layer of very fine particles is suspended above and below the plane of the rings, apparently held there by the mag-

FIGURE 8.23. *The F Ring.*
Here we see the unusual braided appearance of this thin, outlying ring. The cause of the asymmetric shape is not well understood.

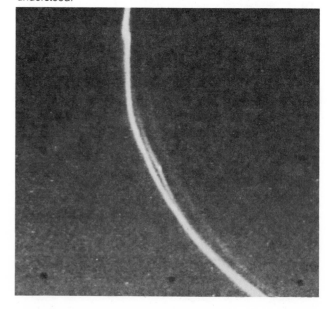

FIGURE 8.22. *Asymmetries in the Rings.*
This composite shows that the rings of Saturn are not perfectly circular. We see here that the thin ring within the dark gap is thinner on one side of the planet than on the other, and slightly displaced.

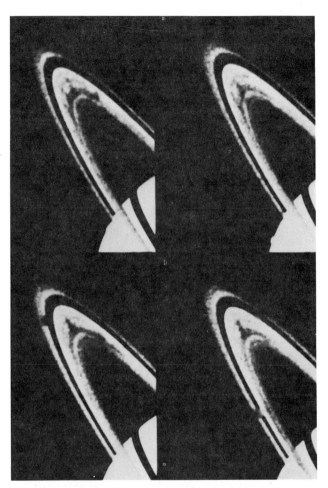

FIGURE 8.24. *The Dark Spokes.*
This sequence shows the dark, hazy, radial features referred to as "spokes." These features, probably a result of very fine particles suspended above and below the ring plane by electromagnetic forces, are seen here to vary in appearance with time.

netic field. These suspended particles were first noticed because they create spokelike structures that point away from Saturn and move around it (Fig. 8.24). The origin of these dark, asymmetric features was quite a mystery until it was realized that the particles there are so tiny that they easily could acquire enough electrical charge to have their motions governed by the magnetic field.

A great deal of work lies ahead before astronomers fully understand all the influences acting on the myriad particles and rings orbiting Saturn. Meanwhile we can be content to marvel at their wondrous appearance, and to speculate about new mysteries yet to be revealed.

PERSPECTIVE

Our visit to the ringed planet has taught us much about how complex nature can be. Every planet, even every satellite, that we have discussed has been revealed as a distinct individual, and Saturn is perhaps the most unique of the lot.

Although Saturn and Jupiter are similar in almost every respect—internal structure, atmospheric motions, satellites, and rings—they also are different in many ways. Saturn may be a cousin of Jupiter, but it is certainly not an indentical twin.

We have almost completed our tour of the planets, for the remaining three are not yet well understood. We can only discuss them as seen from afar, which we do in the next chapter.

SUMMARY

1. Saturn's atmosphere and internal structure are different from Jupiter's: Saturn lacks contrast in its global wind patterns, and it has seasonal variations, greater wind velocities, and a different source of excess internal heat.

2. Saturn has at least seventeen satellites, including the giant Titan, seven intermediate-sized moons, and a large number of tiny, irregular satellites.

3. Titan's thick atmosphere is dominated by nitrogen, but it also includes methane, which may exist in gaseous, liquid, and frozen states on the surface.

4. The intermediate-sized moons all have very low densities, indicating that they have icy compositions; and some show evidence of having been resurfaced, either by frozen internal gases that were released or by debris from space.

5. The complex structure of Saturn's ring system is

created by a combination of effects: orbital resonances with the satellites, spiral density waves, shepherding by pairs of satellites, and electromagnetic forces acting on charged particles.

REVIEW QUESTIONS

1. How often are Saturn's rings seen edge-on, as viewed from the earth?

2. As a check on the mass of Saturn, as given in table 8.1, calculate it by using satellite orbital data from table 8.2.

3. Rank, in order of decreasing orbital speed, the following: the satellite Mimas, a particle at the inner edge of the A ring, Tethys, a particle in the F ring, and Atlas. Explain how you decided on the ranking. Hint: you will need to use data from tables 8.2 and 8.3.

4. Compare the atmosphere of Saturn with that of the primitive earth.

5. Summarize the differences between the atmospheres of Jupiter and Saturn, in terms of structure and motions.

6. Compare the internal structures and heating mechanisms of Jupiter and Saturn.

7. Why do none of the satellites of Saturn have an ionized torus like that of Io?

8. Why does the large abundance of nitrogen in the atmosphere of Titan imply that there might be present-day geological activity occurring on this satellite?

9. Calculate the location of a gap in the ring system that might be created by the 3:1 orbital resonance with the satellite Mimas (that is, calculate the orbital radius where a ring particle would have exactly one-third the orbital period of Mimas).

10. Summarize the processes thought to be responsible for the complex structure in the rings of Saturn.

ADDITIONAL READINGS

Allen, D. A. 1983. Infrared views of the giant planets. *Sky and Telescope* 65(2):110.

Beatty, J. K., O'Leary, B., and Chaikin, A., eds. 1981. *The new solar system.* Cambridge, England: Cambridge University Press.

Hunten, D. M. 1975. Saturn. *Scientific American* 233(3):130.

Ingersoll, A. P. 1981. Jupiter and Saturn. *Scientific American* 145(6):90.

Morrison, D. 1981. The new Saturn system. *Mercury* 10(6):162.

———1982. *Voyages to Saturn.* NASA Special Publication SP–451. Washington, D.C.: NASA.

Pollack, J. B., and Cuzzi, J. N. 1981. Rings in the solar system. *Scientific American* 145(5):104.

Soderblom, L. A., and Johnson, T. V. 1982. The moons of Saturn. *Scientific American* 246(1):100.

THE OUTER PLANETS

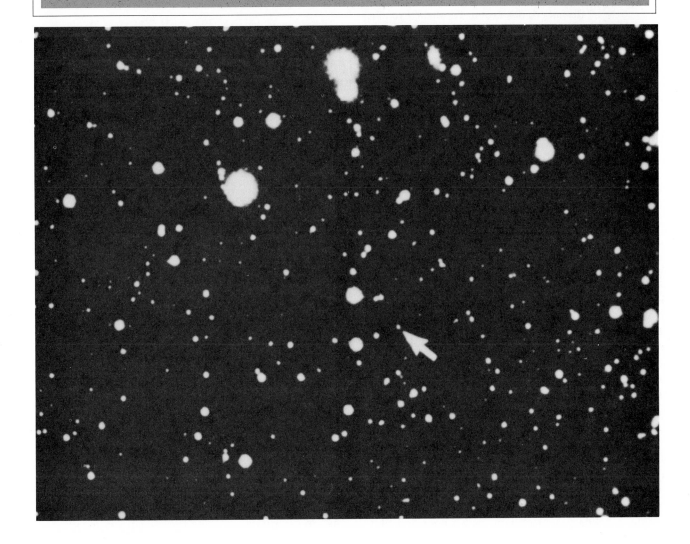

Learning Goals

The three planets that we have not yet discussed, orbiting in the distant reaches of the solar system far from the warmth of the sun, are not well understood. Uranus, Neptune, and Pluto all lie well beyond the orbit of Saturn, and were not known to the ancient astronomers. They are grouped together in this chapter, not because of intrinsic similarities (although, as we will see, two of the three are virtually identical in many respects), but because of our similarly limited knowledge of them. In discussing the outer planets, we will gain an appreciation of the precision and power of Newton's laws of motion on the one hand, and an understanding of the perversity of human intuition on the other.

Uranus

The seventh planet from the sun, Uranus is marginally bright enough to be seen with the naked eye; it was even included in a number of star charts compiled from telescopic observations in the seventeenth and eighteenth centuries. At its distance of nearly 20 AU from the sun, its period is so long (about eighty-four years) and its motion so slow that for quite awhile observers did not notice that its position constantly changes with respect to the background stars.

The planet was finally discovered when, in 1781, the English astronomer William Herschel found during a routine sky survey an object that seemed unusual for a star. It had a disklike appearance, rather than being a single, twinkling point of light, and it appeared bluegreen in color. Herschel at first suspected he had found a comet, but after observing the object for an extended period of time and then calculating its orbit, he found it to be moving about the sun in a nearly circular path, about twice as far out as Saturn. This established that Herschel had found a planet, for no comet is visible at such a great distance from the sun, nor would a comet move in a circular orbit. Herschel initially named the new planet after King George III of England, but this was never widely accepted. The recognized name became Uranus, after the Greek god representing the heavens.

The size of Uranus was not easily estimated, because its image is very indistinct in even the best telescopic views. Photographs taken from a high-altitude balloon in 1972 (Fig. 9.1) allowed a fairly accurate measurement to be made of the planet's angular diameter, which turned out to be about three seconds of arc, corresponding to a true diameter of 51,800 kilometers. A more accurate size determination was made in 1977, when Uranus moved in front of a background star; its diameter could then be deduced directly from knowledge of its speed and how long it took to pass in front of the star, occulting the star's light. This led to an estimated diameter of 52,300 kilometers, only about one third the diameter of Jupiter.

Uranus has five satellites, and these allowed scientists to determine, using Kepler's third law, the mass of the planet. It was found that Uranus is about fifteen times more massive than the earth; it has an estimated density of 1.2 grams/cm^3, very similar to that of Jupiter.

In 1986, the *Voyager 2* spacecraft will reach Uranus on its long trek through the solar system. If all goes well, and the spacecraft systems are still operating, we will get close-up portraits of yet another member of the sun's family.

Anomalous Inclination

The rotation periods of Uranus and the other outer planets are not easy to measure. There are substantial uncertainties in values based on data using the Doppler

TABLE 9.1 Uranus

Orbital semimajor axis: 19.218 AU (2,869,000,000 km)
Perihelion distance: 18.31 AU
Aphelion distance: 20.12 AU
Orbital period: 84.01 years (30,685 days)
Orbital inclination: 0°46′23″

Rotation period: 15^h $36.^m$
Tilt of axis: 97°55′

Mean diameter: 51,200 km (4.01 D$_\oplus$)
Mass: 8.67×10^{28} grams (14.5 M$_\oplus$)
Density: 1.2 grams/cm^3
Surface gravity: 1.07 earth gravities
Escape velocity: 21.2 km/sec

Surface temperature: 95 K
Albedo: 0.66

Satellites: 5

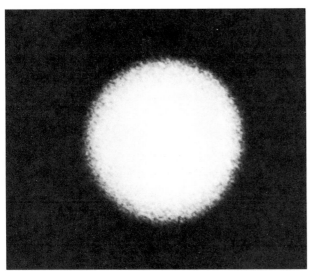

FIGURE 9.1. *Uranus.*
This exceptionally fine photo of Uranus was obtained from a high-altitude balloon, which was able to avoid much of the blurring effect of the earth's atmosphere. No surface markings are evident, except for some darkening at the edges of the disk.

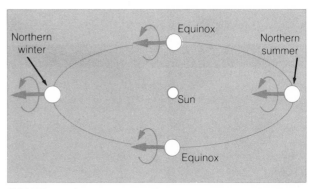

FIGURE 9.2. *Seasons on Uranus.*
The unusual tilt of Uranus creates bizarre seasonal effects. The solid arrow in each case indicates the direction of the planet's north pole, and the curved arrow, the direction of its rotation.

shift, largely because the disks of these planets appear very small when observed from earth. Currently, the best value for the rotation period of Uranus is 15.6 hours. This is similar to the other giant planets, which was no surprise. However, the data revealed something else about the planet's rotation that is very peculiar indeed: Uranus is tipped over so far on its axis that its north pole actually points a little below the plane of the ecliptic. The inclination of its rotation axis is 98 degrees, compared with 30 degrees for most of the other planets. The north pole of Uranus points almost directly at the sun at one point in its 84-year orbit, and directly away 42 years later. In between, the sun is overhead at middle latitudes on the planet. Needless to say, this must create rather strange seasonal variations (Fig. 9.2).

The extreme tilt of Uranus stands as an anomaly in the solar system, and is not easily explained by theories of planetary formation (see chapter 12). The fact that all of the known satellites, as well as the recently discovered rings of Uranus, orbit in the equatorial plane of the planet (that is, the plane of the satellite and ring orbits is nearly perpendicular to the ecliptic) suggests that whatever made the planet deviate from the usual orientation must have happened before the satellites

formed, perhaps when they were still part of a disk of orbiting debris.

Atmospheric Properties

Even the best photographs of Uranus show a featureless disk, with no obvious bands or other markings of the sort that appear on Jupiter and Saturn. Because Uranus does have a similarly rapid rotation, however, we might expect to find some similarities in the atmospheric flow patterns, and indeed some observers claim to have seen banded surface features during moments of unusual clarity. Infrared observations have shown variations that can be attributed to motions in the atmosphere (Fig. 9.3).

Infrared observations (based on Wien's law) have also shown the temperature to be rather low, as we might expect, considering the planet's great distance from the sun. The best estimate is 95 K at the top of the atmosphere, which means Uranus is comparable in temperature to Saturn. Spectroscopic observations show that hydrogen (H_2) and methane (CH_4) are present, but not ammonia (NH_3). This molecule probably is there, but as in the case of Saturn, has precipitated out of the atmosphere in the form of crystalline snow.

Although little is known about the atmosphere of Uranus, even less is known about the internal structure. It probably resembles that of Jupiter and Saturn, with the possible exception that inside Uranus the pressure may not be sufficient to form liquid metallic hydrogen. There is probably a rocky core, since differen-

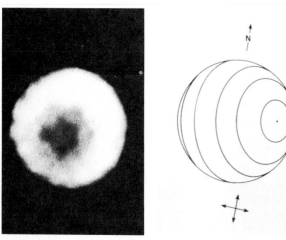

FIGURE 9.3. *Surface Markings on Uranus.*
This photo, obtained at the infrared wavelength of an absorption band of gaseous methane, shows some contrast between different portions of the planetary surface. The sketch at right shows the orientation of the planet at the time.

tiation has most likely occurred, allowing the heavy elements to sink to the center. Outside of this there may be a zone of ice, and above that liquid and then gaseous hydrogen, mixed throughout with other light elements such as helium.

Recent observations made from earth orbit provide circumstantial evidence that Uranus has a magnetic field. Enhanced ultraviolet emission from the planet is apparently caused by the impact of charged particles on the planet's upper atmosphere. The most likely explanation for the presence of such charged particles is that they are confined in the vicinity of Uranus by a magnetic field. It should be possible for *Voyager 2* to verify the discovery directly when it flies by Uranus in early 1986.

Satellites and Rings

Uranus has five known satellites (Fig. 9.4). The two largest, named Oberon and Titania, were found by Herschel in 1787, and the most recently discovered, Miranda, was spotted in 1948. The diameters of the five satellites range from about 400 to 1,600 kilometers. If Uranus, like Saturn and Jupiter, has a number of other, smaller moons, perhaps *Voyager 2* will reveal their presence.

Little is known about the satellites except their distances from Uranus and their approximate sizes. No mass or density measurements will be possible until their gravitational effects on a passing object can be determined, and this will not be feasible until *Voyager 2* arrives in 1986. It is not known whether the satellites are in synchronous rotation, but in view of their distances from Uranus and the large mass of the planet, it seems very likely that at least the innermost satellites do keep one face pointing perpetually inward.

Uranus was the second planet discovered to have a ring system (Fig. 9.5). In 1977, observers measured the diameter of the planet by timing how long it took Uranus to pass in front of a background star (the stellar occultation mentioned earlier in this chapter) and found, much to their surprise, that the star blinked off and on five times during the minutes before the planet was to pass in front of it, and five times afterward. The only reasonable explanation was that Uranus is encircled by five rings, with gaps between them. After this accidental discovery was made, careful observations revealed the five rings as well as four very faint additional ones. The ring system is difficult to see because the particles do not reflect light as efficiently as the ring particles around Saturn, and because the rings of Uranus are much thinner, perhaps more closely resembling Saturn's ringlets than its main rings.

The rings of Uranus lie well inside the orbit of Miranda, the innermost of the known satellites. They probably originated in the same way that the rings of Saturn did (see chapter 8), where it appears that tidal forces prevented particles inside the Roche limit from coalescing into satellites. However, repeated observations of the rings, based on the stellar occultation technique, have revealed some contrasts with the rings of Saturn. In general, the rings of Uranus are thin with wide gaps between them, as opposed to Saturn's rings, which have only very narrow gaps. Also, some of the rings of Uranus are elliptical rather than circular. These noncircular rings tend to be tilted slightly out of the equatorial plane of the planet, and at least one of them varies in breadth, being thinnest where it is closest to Uranus, and thickest at its greatest distance from the planet. Finally, and perhaps most significant, the rings of Uranus are very dark, and are therefore probably not composed of ice particles as are the rings of Saturn. Candidate materials for the rings of Uranus include silicates and iron-bearing compounds.

We can expect further information on the ring structure of Uranus when *Voyager 2* reaches the planet in early 1986. Perhaps light-scattering measurements and

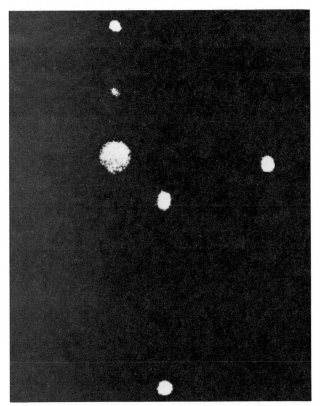

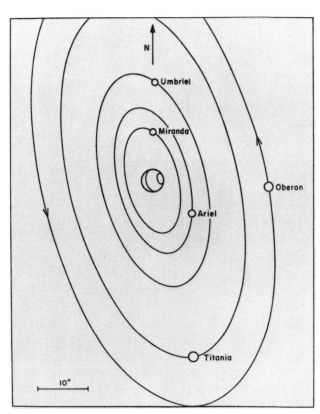

FIGURE 9.4. *Uranus and its Satellites*. All five moons are visible in the photo (left), and are identified in the sketch (right). The direction indicated as north is in the earth's reference frame; Uranus and its entire satellite (and ring) system are tilted nearly perpendicular.

FIGURE 9.5. *The Rings of Uranus*.
This artist's sketch illustrates the diameters but exaggerates the thicknesses of the rings of Uranus. Also shown is the *Voyager 2* spacecraft as it will approach the planet in 1986.

direct images will provide information on the sizes of the ring particles (something not known at all) and details on the ring structure, including the possibility that small satellites are playing roles similar to those of the small moons of Saturn.

Neptune

Perhaps more than any other two planets, Uranus and Neptune can be regarded as twins. Neptune, the eighth planet, orbits at a distance of some 30 AU from the sun, taking 165 years to make one circuit. Its size is very similar to that of Uranus, but its mass is somewhat greater, giving it a density of 1.7 grams/cm^3, the greatest of the gaseous giants.

Since Neptune is even farther from the sun than Uranus, less is known about it. To find out more about

TABLE 9.2 Neptune

Orbital semimajor axis: 30.110 AU (4,498,000,000 km)
Perihelion distance: 29.84 AU
Aphelion distance: 30.38 AU
Orbital period: 164.8 years (60,188 days)
Orbital inclination: 1°46′22″

Rotation period: $18^h 12^{:m}$
Tilt of axis: 28°48′

Mean diameter: 50,460 km (3.96 $D_{\oplus}$)
Mass: 1.03×10^{29} grams (17.2 $M_{\oplus}$)
Density: 1.66 grams/cm^3
Surface gravity: 1.08 earth gravities
Escape velocity: 23.6 km/sec

Surface temperature: 50 K
Albedo: 0.62

Satellites: 2 (a third is suspected)

this planet we will have to wait until 1989, when the well-traveled *Voyager 2* spacecraft is to fly by Neptune.

The Discovery: A Victory for Newton

Neptune is a rather dim object, too faint to be seen with the unaided eye, and even too inconspicuous to have been included on star charts made before the mid-1800s. Rather than being found by accident, as Uranus was, Neptune's existence was first deduced indirectly.

When Uranus was found by Herschel in 1781, and its orbit subsequently computed, astronomers naturally were eager to see whether its path conformed perfectly to the predictions of Newton's laws of motion and gravitation. Since Uranus had been included on star charts dating back some ninety years before Herschel's discovery, it was possible to compare its observed and computed positions over a long period. Much to the dismay of the scientists who made the comparisons, a discrepancy was found. The mismatch between the expected and observed positions of the planet was not very large, amounting to two minutes of arc by 1840, but it was considered very serious, for this was too large an error to be allowed by the laws of motion. Something was amiss, and people began to question the validity of Newton's work.

It was soon suggested, however, that if there were another body exerting a gravitational influence on the motion of Uranus, the discrepancy could be accounted for. Calculations carried out independently in England and France (see astronomical insight 9.1) showed that an eighth planet could be responsible, if it were located at a particular position in the constellation Aries. In 1846 a planet was found at the predicted position, and what had begun as a major problem for Newton's laws turned into a resounding victory. This experience showed that the laws of motion and gravitation were so accurate and universal that they could actually be used to deduce the presence of unseen objects. The new planet was given the name Neptune, after the Greek god of the seas.

Properties of Neptune

This planet is difficult to examine in any great detail, even more so than Uranus. The best photographs show little more than a spot of light, and only rather large-scale features can be detected. The fine resolution of the *Space Telescope* will allow smaller details to be seen on Neptune, if they exist. The angular diameter of Neptune, as seen from the earth at opposition, is 2.3 seconds of arc; the *Space Telescope* will be capable of discerning features as small as about one-tenth of an arcsecond in size. Of course, this instrument will also provide better views of Uranus than any now available, but presumably *Voyager 2* will be sending back close-ups of the planet at about the time the *Space Telescope* is launched.

For now, we surmise that Neptune closely resembles Uranus, since the two are so similar in all respects that can be measured from earth. The diameter of Neptune, like that of Uranus, was measured from a stellar occultation, and Neptune's mass was deduced from the application of Kepler's third law to the orbits of its satellites. It has proved difficult to measure the rotation period of Neptune, but recent observations indicate that it is 18.2 hours and that Neptune's rotation axis is tilted by 29 degrees, similar to the tilts of several other planets.

The atmosphere may be even colder than that of Uranus, with a temperature of perhaps 50 K. Spectroscopy reveals the presence of hydrogen and methane, and again it is suspected that ammonia has precipitated out. Neptune's internal structure is probably similar to that of Uranus, but Neptune most likely has a larger core, since its overall density is greater.

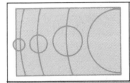

Astronomical Insight 9.1

SCIENCE, POLITICS, AND THE DISCOVERY OF NEPTUNE

There have been many cases throughout the history of science where a great discovery was just waiting to be made, as soon as the necessary technology or preliminary information had been developed. If Galileo had not turned his telescope to the heavens, surely someone else would have done so very soon and found the wonders he discovered. If Newton had not developed the laws of mechanics, some other student of the motions of the moon and planets would have.

The discovery of Neptune was such a case. The irregularities in the orbit of Uranus were well known by the mid-1800s, and the mathematical techniques that were needed to calculate the position of the eighth planet were also standard. It was only a matter of someone recognizing the possibility and carrying out the computation. As it happened, two people did so, quite unbeknownst to each other, which led to one of the more fascinating disputes in the history of science.

In England, a young college graduate named John C. Adams made the prediction in 1845 that an eighth planet was at a certain location in the constellation Aries. Being young and unknown, Adams had difficulty convincing the established astronomers of England that an effort should be made to find the predicted planet. A contributing factor was that Adams developed a tiff with the Astronomer Royal, with the result that the latter simply ignored the affair.

Within a few months, a French scientist named Urbain Leverrier carried out calculations that led him to the same conclusion, although he was totally unaware of Adams's work, which was not published. Leverrier's work did appear in scientific journals, however, and this prompted the Astronomer Royal in England to take the idea of an eighth planet a bit more seriously. The search began.

The English astronomers were hindered by the lack of good star charts for the appropriate region of the sky, and substantial preliminary work was necessary before an earnest search could begin. Meanwhile, Leverrier contacted an acquaintance at the observatory in Berlin, who had excellent charts of Aries, and discovered Neptune on the very first night he looked for it. This was in 1846.

England and France were international rivals, and years of contentious arguments followed as to who should be credited with the prediction of Neptune's existence. There never was an amicable settlement of this question; it just gradually faded from the forefront of French-English affairs. Today, the heat of the debate having cooled with the passage of time, Adams and Leverrier generally are credited equally with the discovery.

Surprising as it may seem, we have some data regarding atmospheric motions on Neptune. With the use of special filters that match the wavelengths of methane-gas absorption lines, observers can infer, from brightness variations measured through these filters, the flow pattern of methane on Neptune (Fig. 9.6). The information is not very detailed, but there are indications of a pattern of zonal winds like those on Jupiter and Saturn, with velocity differences of about 110 kilometers per second between zones. This tends to confirm our natural expectation that there are strong similarities among all the giant, gaseous planets.

Satellites in Chaos

Neptune has two known satellites, and a third is suspected. Recent spectroscopic observations have shown that Triton, the largest, has a gaseous atmosphere containing methane; thus it is the second satellite (along with Titan) known to have an atmosphere. The other

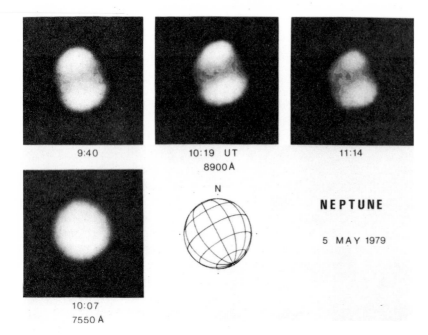

FIGURE 9.6. *Neptune.*
This series of images, like the photo of Uranus in Fig. 9.3, was obtained in infrared light. The three upper images were made at a wavelength where methane absorbs light, so the dark areas are regions of methane concentration. The bright areas north and south of the equator are thought to have ice crystals and haze above the methane.

9:40   10:19 UT   11:14
            8900 A

N

NEPTUNE

5 MAY 1979

10:07
7550 A

confirmed satellite of Neptune, Nereid, is very small, with a diameter of roughly 400 kilometers (compared with Triton's diameter of 3,500 kilometers). It was not detected until 1949.

Neptune has a fairly normal rotation, but the motions of its two satellites are extremely unusual (Fig. 9.7). Triton orbits the planet backward, taking about six days to complete each retrograde orbit. Furthermore, the orbit is tilted by an unusually large angle,

about 20 degrees with respect to the planet's equatorial plane.

The orbit of Nereid is perhaps even more bizarre. It goes around in the normal direction (opposite of the motion of Triton), and its orbital plane is inclined 28 degrees to Neptune's equatorial plane. Nereid is very distant from Neptune, and has a very long period for a satellite, about one earth year. The strangest aspect of its orbit, however, is its shape, which is very highly elongated. Its distance from Neptune varies by a factor of five between its closest approach and the farthest reaches of its orbit. Newton's laws of motion allow for the possibility of such highly flattened orbits (and as we shall see, comets follow similar or even more elongated paths around the sun), but no planet nor any other satellite in the solar system has such an orbit.

The strange motions of Neptune's satellites are probably not a natural consequence of the planet's formation, but rather, are the result of satellites being captured or the result of some immense gravitational disturbance that disrupted the system at some time in the past, after Neptune and its satellites had formed. Perhaps a near-collision with a massive body caused the upheaval. Some interesting speculations on this will be discussed in the next section, for it has been suggested that Pluto, the ninth planet, was a character in the drama.

FIGURE 9.7. *The Satellites of Neptune.*
This scale drawing illustrates the unusual orbital characteristics of the two confirmed moons, Triton and Nereid.

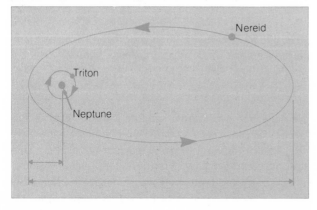

Nereid

Triton

Neptune

Pluto

The ninth and last of the known planets is Pluto, orbiting at an average distance of nearly 40 AU from the sun. From that perspective, the sun would appear as a very bright star, but surely would not produce much warmth. Pluto is the most obscure of all the planets, and we will raise many questions about its nature.

The Discovery: An Inspired Accident

After the triumph of Neptune's discovery, continued analysis of the orbits of Uranus and Neptune seemed to indicate other rather minor discrepancies in their motions. The natural tendency of astronomers, given the recent experience with Neptune, was to speculate that yet another planet, beyond the orbit of Neptune, might be responsible. In this case, the expectation was stronger than the evidence. Nevertheless, several independent calculations, including most notably one carried out in 1915 by Percival Lowell (of Martian canal fame), led to predictions of the location of the supposed ninth planet, and the search was on.

Lowell set out to find his new planet, and spent the last years of his life searching photographic plates for it. The process was complicated by the fact that the predicted position was in a portion of the Milky Way that is crowded with stars; thus the search for a dim object that slowly changed position among all the fixed

stars was tedious and difficult. Lowell was unsuccessful.

In 1930 the American astronomer Clyde Tombaugh (Fig. 9.8), having taken up the search at the Lowell Observatory some time after Lowell died, discovered a tiny dot of light that displayed the expected motion (Fig. 9.9), and he found it only 7 degrees away from the position that Lowell had predicted. The new member of the sun's family was given the name Pluto, after the Roman god of the underworld.

Subsequent analysis of Pluto showed it to be far too small and of too low a mass to have created any noticeable disturbance in the paths of Uranus and Neptune. Reexamination of the data that Lowell and others had used in predicting the existence of a ninth planet showed, furthermore, that there was no evidence for such a planet. The minor discrepancies between calculated and observed positions were all within the uncertainty of the measurements. Apparently Lowell and the others who made the calculations wanted to find evidence for a ninth planet, and so they found it, whereas dispassionate scientific analysis would not have shown that it was there. It is a remarkable coin-

FIGURE 9.8. *Clyde Tombaugh at the Blink Comparator.* This photo shows the discoverer of Pluto as he peers into the eyepiece of the device he used to find the planet on photographic plates. The blink comparator allows two images of the same field of stars to be examined alternately (with the use of a mirror that flips between two positions), so that moving objects stand out.

TABLE 9.3 Pluto

Orbital semimajor axis: 39.44 AU (5,900,000,000 km)
Perihelion distance: 29.57 AU
Aphelion distance: 49.31 AU
Orbital period: 248.8 years (90,700 days)
Orbital inclination: 17°10'12"

Rotation period: 6.387 days
Tilt of axis: 118°

Diameter: 3,000–3,600 km (0.16–0.47 $D_\oplus$)
Mass: 1.3×10^{25} grams (0.002 $M_\oplus$)
Density: 0.5–0.9 grams/cm^3
Surface gravity: 0.024–0.034 earth gravity
Escape velocity: 0.9–1.1 km/sec

Surface temperature: 40 K
Albedo: 0.25–0.36

Satellites: 1

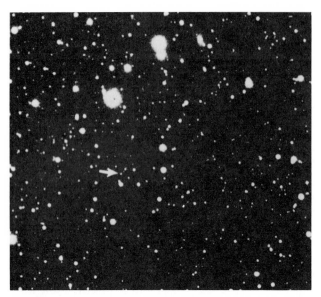 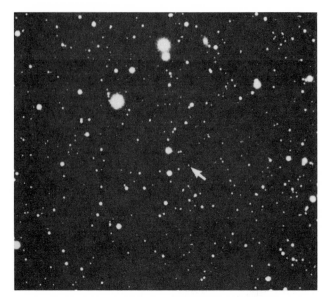

FIGURE 9.9. *The Discovery of Pluto*. These are portions of the original photos in which our ninth planet was discovered in 1930. The position of the planet is indicated by arrows.

cidence, then, that Pluto was found, and especially that it was found not far from the position predicted by Lowell. If it had not been for that prediction, unsound though it may have been, Pluto might not have been discovered for a long time.

Planetary Misfit

The little we know about Pluto shows it to be a nonconformist. Throughout our discussions of the other planets, certain systematic regularities have shown up, and Pluto violates many of them.

First, Pluto's orbit is highly irregular, compared with those of the other planets. It is both noncircular and tilted with respect to the ecliptic. At its greatest distance from the sun, Pluto is nearly 70 percent farther away than at its closest distance, when it actually moves inside of Neptune's orbit for a time. (In fact, Pluto is temporarily the eighth planet from the sun right now, having moved inside of Neptune's orbit in 1978, not to reemerge until 1998.) Pluto's orbit is tilted by 17 degrees to the ecliptic, whereas all the other planetary orbits are tilted by 7 degrees or less.

Pluto also does not seem to fit in with the rest of the planets in terms of physical characteristics. The other outer members of the solar system are giant planets with thick atmospheres, whereas Pluto appears to be small with a low mass. It has been difficult to pinpoint the mass of Pluto; for a long time the only thing known about Pluto's mass was derived from the fact that Pluto does not seem to have any significant effect on the motion of Neptune, which implied that Pluto's mass had to be less than about 10 percent of that of the earth. Pluto's diameter was not established until rather recently, when the technique called interferometry (see chapter 3) was used to discover that it is between 3,000 and 3,600 kilometers.

Spectroscopy has shown that Pluto has methane gas and possibly methane ice as well, so there is a thin atmosphere and perhaps also a frost of methane ice on the surface. Most molecular compounds formed of common elements freeze out into solid form at the low temperature (perhaps 40 K) thought to characterize Pluto.

Brightness variations indicate that Pluto has some surface markings, and is rotating with a period of 6.39 days. The orientation of the rotation axis has been difficult to determine, but recent information indicates a large tilt, probably more than 90 degrees, so that the rotation of Pluto is retrograde. The best estimate currently is that the tilt is 118 degrees.

The small size and apparent low mass of Pluto, combined with its unusual orbit, led scientists to speculate that it was once a satellite of one of the giant

Astronomical Insight 9.2

THE SEARCH FOR PLANET X

The discovery of Pluto was by no means the end of the search for new planets. Quite to the contrary, this event sharpened the interest in continuing the search, and it was carried on for an extended period.

The technique by which Pluto was discovered was tedious but efficient, in that it was thorough. Scientists carried out the task by comparing photographs taken at different times (usually a few days apart), to see whether any of the objects in the field of view moved. This would ordinarily be very difficult to do, particularly in areas of the sky that are crowded with stars, but it was made easier with the aid of a special instrument called a *blink comparator*. This is a microscope for viewing two photographic plates simultaneously, with a small mirror that flips back and forth, providing in rapid succession alternating views of the two plates. If the plates are mounted so that the images of fixed stars are perfectly aligned, then an object that moves between the times of the two exposures will appear to blink on and off at two separate locations while all the fixed objects are steady. To examine a region of the sky using this technique, astronomers therefore mount in the blink comparator two plates taken at different times, then systematically scan them with the movable eyepiece so they can check each object to see whether it is blinking. The entire job could take days or weeks for each plate pair.

The blink-comparator technique, although time-consuming, has the capability of allowing thorough searches of large regions of the sky. The search for a tenth planet, often referred to as Planet X, continued at Lowell Observatory for thirteen years after the discovery of Pluto. Clyde Tombaugh, who found Pluto, was the principal worker in the extended search. Nothing was found.

During and since that time, occasional reports of evidence for a tenth planet have surfaced, but the evidence has always been indirect. Reanalysis of the motion of Neptune has convinced some astronomers that there really may be discrepancies attributable to gravitational perturbations caused by another planet. Other researchers have pointed to the motions of certain comets as indicative of gravitational influence exerted by Planet X. Most of these predictions have suggested a distance from the sun of 50 to 100 AU. The motion of Halley's comet, a bright and very famous object that visits the inner solar system every seventy-six years, has been analyzed by several scientists, who find that its arrival near the sun is usually several days later than it should be according to the laws of motion. Some have attributed this discrepancy to the gravitational influence of a tenth planet, but no one has found such an object at the predicted position. One of the predictions based on the motion of Halley's comet was rather spectacular: the inferred Planet X had a mass three times that of Saturn, and an orbit inclined by 120 degrees to the ecliptic!

Modern information on comets has explained the discrepancy in the motion of Halley's comet another way. As the comet nears the sun and heats up, volatile gases vaporize and stream outward from the heated side, acting like a rocket exhaust to slow the comet's motion.

The routine search for new objects continues today with occasional success, but no new planets have been found. There was a brief flurry of excitement in 1977, when astronomers discovered a sun-orbiting body whose elliptical path carries it almost as far from the sun as Uranus, and as close in as Saturn. It is clearly not a planet, but the exact nature of Chiron, as it is called, is uncertain. It may be an asteroid in a highly unusual orbit, or an old cometary nucleus that never approaches the sun close enough to develop a luminous tail (see chapter 10). Perhaps eventually Planet X will be found, but if so, it will undoubtedly be a dim, cold, distant object, much too faint to be seen without a large telescope.

planets, and that it somehow escaped to follow its own path around the sun. Recall the disorder of the satellites of Neptune; it has been shown that if Pluto were once a satellite of Neptune, a three-way collision among Pluto, Triton, and Nereid could have altered the orbits of the latter two and allowed Pluto to escape into its present orbit around the sun. To do this would have required an extremely special set of circumstances, with the three meeting each other at just the right speed and direction. It could have happened, however; after all, something unusual must have occurred to create the motions of Triton and Nereid. Thus it seemed that a whole set of anomalous motions in the outer solar system could be explained by a single catastrophic event.

This theory has recently suffered a setback, however. Careful examination of photographs of the planet has shown that Pluto has a lump that moves back and forth from one side to the other (Fig. 9.10). Apparently this lump is a satellite very close to the planet, orbiting Pluto with a period of slightly more than six days. To create the distorted appearance of Pluto, the satellite must be relatively large and close to it. Pluto's satellite was given the name Charon. The idea that Pluto has a

FIGURE 9.10. *The Discovery of Charon.*
This is the photo that revealed Pluto's satellite (the bulge at upper right).

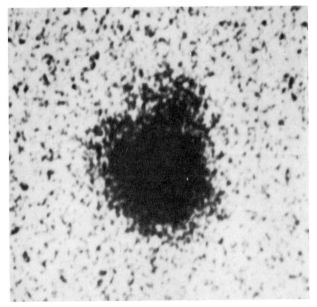

satellite of its own does not square up with the theory that Pluto itself is an escaped satellite of one of the giant planets. We have to drop the latter theory, unless it turns out that Charon split off from Pluto in the cataclysmic process that caused Pluto's escape from Neptune.

The orbital period of Charon is 6.39 days, precisely equal to the rotation period of Pluto. Thus, this is a case where both the satellite and the parent planet are locked in synchronous rotation. To an observer on the surface of Pluto, Charon would appear fixed in position overhead, while the background stars streamed past as a result of the planet's rotation. Charon is unusually large compared with the planet it orbits. This and its proximity have no doubt, helped create sufficiently strong tidal forces to produce Pluto's synchronous rotation. Pluto might be accurately described as a double planet.

The orbital plane of Charon is tilted about 62 degrees with respect to the plane of Pluto's orbit about the sun, and Charon orbits in the retrograde direction (Fig. 9.11). The tentative conclusion that Pluto's rotation axis is tipped 28 degrees (technically 118 degrees, since the spin is retrograde) with respect to its orbital plane is based on the assumption that Charon orbits in the plane of Pluto's equator.

The discovery of Charon gave astronomers a chance to determine the mass of the planet, using Kepler's third law. Pluto's mass is only about 0.0002 of the earth's mass, and this in turn leads to a somewhat low

FIGURE 9.11. *Charon's Orbit.*
This sketch illustrates the orientation of the orbit of Pluto's satellite with respect to the orbit of Pluto itself, which in turn is greatly tilted with respect to the ecliptic. The orbit of Charon is retrograde and tilted 28 degrees away from perpendicular to Pluto's orbital plane.

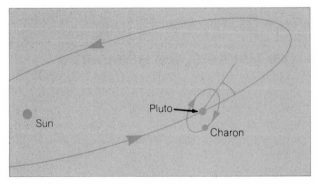

average density, in the range of 0.5 to 0.9 grams/cm^3. Evidently Pluto is not a rocky planet as had previously been assumed, but rather is made largely of ices, much like the satellites of Jupiter and Saturn.

PERSPECTIVE

We have thoroughly digested all the available information regarding the planets. We certainly have many unanswered questions about them, but on the other hand, we have seen a number of systematic trends. Two of the outermost three planets fit the pattern established by the otherouter planets; they are gaseous giants that apparently are similar in many ways to Jupiter and Saturn. The last planet, Pluto, is mysterious, and remains an anomaly to be reckoned with.

Nearly all the pieces of the puzzle are in place. We turn our attention now to the nonplanetary bodies: the asteroids, comets, and meteroids roving through the solar system, to see what they can tell us of its origins. We will then visit the center of power, the sun itself, before finally attempting to unravel the sequence of events that led to the creation of the solar system as we see it today.

SUMMARY

1. Uranus is about fifteen times more massive than the earth, has an average density of 1.2 grams/cm^3, and probably resembles Jupiter and Saturn in structure and composition.

2. The spin axis of Uranus, as well as the orbital plane of its five satellites, is tilted 98 degrees, creating strange seasonal effects.

3. Uranus has five satellites and at least nine thin, dim rings.

4. Neptune was discovered because of its gravitational effects on Uranus, a discovery that vindicated the laws of motion and gravitation developed by Newton.

5. Neptune's two confirmed satellites have unusual orbital properties, possibly created by a major gravitational disturbance at some time in the past.

6. Pluto, the ninth planet, was discovered fortuitously close to the position that Lowell predicted on the basis of suspected irregularities in the orbit of Neptune.

7. Pluto is unusual in that its orbit is highly eccentric and tilted. Also, it is much smaller than the other outer planets, and probably different in structure.

8. Pluto has a relatively large satellite that provides data on the planet's mass and density: these data show that Pluto is probably composed largely of ice.

REVIEW QUESTIONS

1. When Uranus occulted the light from a background star, it took about 1 hour, 40 minutes to pass in front of it. The apparent motion of Uranus as seen from the earth is about 0.0005 seconds of arc per second. Using these numbers, calculate the angular diameter of Uranus.

2. What is the length of the day at the north pole of Uranus? What is the length of the day at the equator when the pole is pointed 90 degrees away from the direction of the sun?

3. Based on what we learned about Jupiter and Saturn, how might we try to determine from remote observations whether Uranus has a magnetic field?

4. At the point in time when Uranus and Neptune are lined up on the same side of the sun, Uranus experiences a gravitational force in one direction from the sun, and another in the opposite direction from Neptune. Calculate how much (by what fraction) this reduces the sun's attraction for Uranus. (hint: you will need data from Appendix 6).

5. Summarize the similarities and differences between Uranus and Neptune.

6. Use Kepler's third law to calculate the semimajor axis of Nereid's orbit about Neptune, given that its orbital period is one year. (Note: you have to use the form

of Kepler's third law that includes the masses of the two bodies.)

7. Contrast the means by which Uranus, Neptune, and Pluto were discovered.

8. Calculate how much the intensity of sunlight on Pluto varies as the planet goes from perihelion to aphelion in its orbit. At perihelion, when Pluto is at its closest approach to the sun, it is still about 30 AU away from it. How does the sun's intensity on Pluto at that time compare with its intensity on the earth?

9. Explain why Pluto has been forced into synchronous rotation with its satellite Charon, whereas the earth has not been similarly forced into synchronous rotation with its moon.

ADDITIONAL READINGS

Allen, D. A. 1983. Infrared views of the giant planets. *Sky and Telescope* 65(2):110.

Beatty, J. K., O'Leary, B., and Chaikin, A. eds. 1981. *The new solar system.* Cambridge, England: Cambridge University Press.

Elliott, J. L., Dunham, E., and Mills, R. L. 1977. The discovery of the rings of Uranus. *Sky and Telescope* 53(6):412.

Harrington, R. S., and Harrington, B. J. 1980. Pluto: Still an enigma after fifty years. *Sky and Telescope* 59(6):452.

Hunten, D. M. 1975. The outer planets. *Scientific American* 233 (3):130.

Tombaugh, C. 1979. The search for the ninth planet. *Mercury* 8(1):4.

Learning Goals

The planets are the dominant objects among the inhabitants of the solar system (except, of course, for the sun), but they are not entirely alone as they follow their clockwork paths through space. The abundant craters on planetary and satellite surfaces have shown us that there must have been a time when interplanetary rocks and gravel were very plentiful, raining down continually on any exposed surface. Today the rate of cratering is much lower than it once was, but some vestiges of the space debris that caused it still remain, orbiting the sun and occasionally becoming obvious to us as they pass near the earth or enter its atmosphere.

There are at least four distinct forms of interplanetary matter, some of which are closely related. In this chapter we will discuss the **asteroids,** myriad rocky chunks up to several hundred kilometers in diameter that orbit the sun between Mars and Jupiter; the **comets,** whose ephemeral and striking appearances have caused us to pause and marvel since ancient times; **meteors,** the bright streaks often visible in our nighttime skies, and the objects that create them; and the **interplanetary dust,** a collection of very fine particles that fill the void between the planets.

Bode's Law and the Asteroid Belt

As we have already seen, some of the early students of planetary motions were concerned with the distances of the planets from the sun. Kepler spent enormous amounts of his time and energy seeking a mathematical relationship that would describe the distances of the planets in terms of geometrical solids. What he finally did discover was something different, a relationship between the distances and orbital periods. Kepler's third law can be used to predict the period of a planet, given its distance from the sun; or its distance, if its period is known. However, the law does not provide any underlying basis for explaining why there are planets at only certain distances from the sun.

In 1766, a German astronomer named J. D. Titius found a simple mathematical relationship that seemed to accomplish what Kepler had set out to do. Titius discovered that if we start with the sequence of numbers 0, 3, 6, 12, 24, 48, and 96 (obtained by doubling each one in order); then add 4 to each and divide by

10; we end up with the numbers 0.4, 0.7, 1.0, 1.6, 2.8, 5.2, and 10.0, corresponding closely to the observed planetary distances in astronomical units. A few years after Titius found this numerological device, it was popularized by another German astronomer, Johann Bode, and eventually became known as Bode's law.

The sequence dictated by Bode's law includes one number, 2.8 AU, where no planet was known to exist. After the discovery of Uranus in 1781, and the recognition that its distance fits the sequence—the next number is $(192 + 4)/10 = 196$, and the semimajor axis of Uranus's orbit is 19.2 AU—there was a great deal of interest in the possibility that a planet existed at 2.8 AU, since Bode's law was doing so well in predicting the positions of other planets.

A deliberate search for such a planet began in 1800, but the discovery of the sought-after object came accidentally. An Italian astronomer named Piazzi noted a new object on the night of January 1, 1801, and within weeks he found from its motion that it was probably a solar system body. When the orbit of the object was calculated, its semimajor axis turned out to be 2.77 AU, very close to the value predicted by Bode's law. The new fifth planet was named Ceres.

A little more than a year after the discovery of Ceres, a second object was found orbiting the sun at approximately the same distance, and was named Pallas. Because of their faintness, both Ceres and Pallas were obviously very small bodies and were not respectable planets. By 1807, two more of these *asteroids,* as they were called, had been found and designated Juno and Vesta. A fifth, Astrea, was discovered in 1845, and in the next decades, vast numbers of these objects began to turn up. The efficiency of finding them was improved greatly when photographic techniques began to be used. Because of its orbital motion, an asteroid will leave a trail on a long-exposure photograph (Fig. 10.1).

Today the number of known asteroids is in the thousands, with about 3,000 of them sufficiently well observed to have had their orbits calculated and logged in catalogues. The total number is probably much more, perhaps 100,000. The term **minor planet** is commonly adopted by modern astronomers, although we will use the traditional name for these objects.

Before we discuss the nature of the asteroids, let us return to our starting point: Bode's law. The discovery of the asteroids seemed to lend a great deal of credence to this mathematical relationship among the planetary distances. There were people who believed that Bode's law reflected some as yet undiscovered physical principle that governed the layout of the solar system. Today the interpretation is rather different. Although there certainly is a regularity in the sequence of planetary distances from the sun (and this regularity must have been dictated by the laws of physics, at the time the solar system formed), astronomers no longer believe that Bode's law itself represents a fundamental physical principle. Bode's law is now regarded as a mathematical coincidence; a numerical sequence that just happens to fit the observed planetary positions. There are, in fact, errors of a few tenths of an AU here and there, and the outermost planets, Neptune and Pluto, do not fit the sequence at all.

The Nature of the Asteroids

It is possible to deduce some of the properties of asteroids, from a variety of observational evidence. Measurements of their brightnesses can lead to size estimates; spectroscopy of light reflected from their surfaces provides information on their chemical makeup; and direct analysis of meteorites that may be remnants of asteroids adds data on their internal properties.

The largest asteroids were the first to have their diameters estimated; the technique was simply to calculate what size the asteroids had to be to reflect the amount of light observed. Much more recently, infrared brightness measurements of both large and small asteroids have provided information on their sizes, since at temperatures of a few hundred degrees asteroids glow at infrared wavelengths. Using the Stefan-Boltzmann law (chapter 3) to calculate the intensity of the emission, astronomers can determine the total surface area of an asteroid.

A few direct measurements of asteroid sizes have also been possible, in cases where these objects have passed sufficiently close to earth that their angular sizes could be directly determined. In recent times, the angular diameters of some asteroids have been successfully measured by the technique of interferometry.

Using these assorted techniques, astronomers have learned that asteroids have a wide range of diameters.

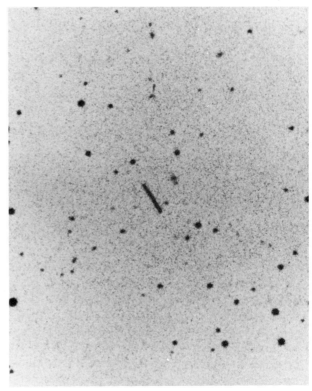

FIGURE 10.1. *Asteroid Motion.*
The elongated image in this photo is the trail made by an asteroid, which moved relative to the fixed stars during a twenty-minute exposure. Most new asteroids are discovered through photos of this type, although the first few were spotted visually.

A few asteroids have diameters as large as several hundred kilometers (Ceres, the largest, is about 1,000 kilometers in diameter), but most are rather small, with diameters of one or two hundred kilometers or less. The largest asteroids are apparently spherical, whereas the smaller ones often are jagged, irregular chunks of rock (Fig. 10.2), varying in brightness as they tumble through space, reflecting light with different efficiencies on different sides. A few asteroids are binary, consisting of two chunks orbiting each other as they circle the sun. The total mass of all the asteroids together is small by planetary standards, amounting to only about 0.04 percent of the mass of the earth.

The compositions derived from spectroscopic analyses are highly varied. The principal ingredients range from metallic compounds to nearly metal-free ones, and from carbon-dominated to silicon-bearing minerals. There are also several classes of asteroids whose

FIGURE. 10.2. *A Portrait of an Asteroid?*
Many asteroids probably bear a general resemblance to this
irregularly shaped object, which is Phobos, one of the tiny
Martian moons.

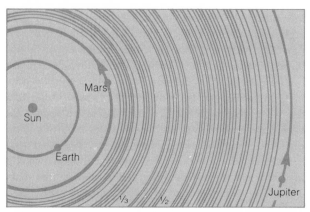

FIGURE 10.3. *Kirkwood's Gaps.*
This schematically shows the orbits of the earth, Mars, and
Jupiter, and a number of possible asteroid orbits. Gaps are
shown at the distances from the sun where asteroids would
have exactly one-half or one-third the orbital period of Jupiter.

composition is not known. Of the largest asteroids in
the main belt between Mars and Jupiter, about three-
quarters are carbonaceous, meaning that they contain
carbon in complex molecular forms. Most of the rest
are composed chiefly of silicon-bearing compounds,
and about 5 percent are metal-rich, the primary metals
being nickel-iron mixtures.

Kirkwood's Gaps: Orbital Resonances Revisited

As increasing numbers of asteroids were discovered
and catalogued throughout the nineteenth century, calculations of their orbits showed remarkable gaps at
certain distances from the sun. One such gap is at 3.28
AU, and another is at 2.50 AU.

The explanation for these gaps was offered by Dan-
iel Kirkwood in 1866, when he realized that these distances correspond to orbital periods that are simple
fractions of the period of Jupiter [Fig. 10.3]. The giant
planet, at its distance of 5.2 AU from the sun, takes
11.86 years to make a trip around it, whereas an asteroid at 3.28 AU, if one existed there, would have a period exactly half as long, 5.93 years. Thus this asteroid
and Jupiter would be lined up in the same way fre-

quently—every time Jupiter made one orbit. The asteroid would therefore be subjected to regular tugs by Jupiter's gravity, and would gradually alter its orbit,
vacating the zone 3.28 AU from the sun. In this manner Jupiter has cleared out several gaps in the asteroid
belt, at this and other distances where the orbital periods would result in regular alignments. The gap at
2.50 AU corresponds to orbits with a period exactly
one-third that of Jupiter. These orbital resonances are
exactly analogous to the ones created by the moons of
Saturn, which are responsible for some of the gaps in
the ring system of that planet [see chapter 8]. Since
Jupiter is the most massive and the nearest of the outer
planets to the asteroid belt, it has by far the greatest
effect in producing gaps, but in principle the other
planets could do the same thing.

The Origin of Asteroids

For a long time the most natural explanation of asteroids was that a former planet broke apart, creating a
swarm of fragments that continued to orbit the sun.
One of the strongest arguments for this theory was
Bode's law. The idea began to lose favor, however,
when it was accepted that Bode's law is not a fundamental physical principle, and it lost more ground
when the total mass of the asteroids was estimated, and
found to be much less than that of any ordinary planet.

Today another rather strong argument against the

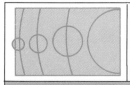

Astronomical Insight 10.1

TROJANS, APOLLOS, AND TARGET EARTH

Not all the asteroids are confined to the main belt between the oribits of Mars and Jupiter. There are exceptions, many of which make up two special categories.

The existence of one of these groups was predicted on the basis of mathematical calculations carried out in the late 1700s by the French scientist J. L. Lagrange. Using Newton's laws of motion, Lagrange showed that a system consisting of two orbiting bodies has certain points where additional objects may orbit in a stable fashion. (Ordinarily, when more than two objects are involved, the system becomes so complex that it is nearly impossible to describe mathematically all the possible motions.) Lagrange showed that if one body orbits another, there are points 60° ahead and 60° behind it, in the same orbit, where additional bodies can stay. In chapter 8 we learned that there is a small satellite occupying the orbit of Dione, one of the larger satellites of Saturn, and that this small satellite stays in position 60° ahead of Dione. There are two such satellites sharing the orbit of Tethys, another moon of Saturn. One is 60° ahead of Tethys in its orbit, and the other is 60° behind. Similar points exist in the orbit of our moon, and there has even been discussion that these might be convenient places to locate permanent manned space stations. (The L–5 Society, a group advocating this idea, takes its name from one of these points, which is called the L–5 point, in honor of Lagrange.)

The same situation exists for the orbit of Jupiter. After a few asteroids were found there in the early decades of this century, concerted searches were made, revealing a large number of asteroids in the Lagrange points, 60° in either direction from Jupiter. There may be hundreds of objects at these two lo-

cations, held in place by the combined gravitational pulls of Jupiter and the sun. These objects are called the **Trojan asteroids,** and they are individually named for classical Trojan and Greek heroes, such as Achilles and Hector.

Another group of asteroids has been identified and named after the first of the class to be discovered, Apollo. These objects are distinct in that their orbits bring them within 1 AU of the sun; thus their paths can cross the orbit of the earth. There are about thirty **Apollo asteroids** known, and a couple of them, Icarus and Eros, have provided useful information for scientists on earth. In 1968, Icarus, which of all the Apollo asteroids comes closest to the sun (0.19 AU), came within 16 million kilometers of Mercury, close enough for Mercury's gravitational pull to noticeably alter its direction, which allowed astronomers to deduce Mercury's mass. Eros passed the earth at a distance of about 23 million kilometers in 1931, giving astronomers a clear view of it through telescopes and allowing them to directly measure its size and shape. Eros was seen to be an irregular chunk of rock, tumbling end over end with a period of about 5.3 hours. This same asteroid passed in front of a bright star in 1975; thus its diameter could be measured precisely. It has an oblong shape, with dimensions of 7 × 19 × 30 kilometers.

Given a long enough time, some of the Apollo asteroids will probably eventually collide with the earth. They are so few in number and the volume of space they occupy is so large, however, that the average time between such collisions is likely to be very long, measured in the millions of years. We can hope, therefore, to escape any truly major impacts in the foreseeable future.

planetary-remnant hypothesis is cited: we know of no reasonable way for a planet to break apart once it has formed, whereas we can easily understand why material orbiting the sun at the position of the asteroid belt could never have combined into a planet in the first

place. It becomes simpler for us to accept the idea that the debris there never was part of a planet, rather than to find a way to imagine the debris first forming a planet and then breaking apart.

The invisible hand of Jupiter's gravity is invoked

again. If, as we suspect, the planets formed from a swarm of debris orbiting the young sun in a disk, no planet could have formed in a location where the pieces of debris could not stick together. Calculations show that once Jupiter formed, its immense gravitational force stirred up the material near its orbit, so that collisions between particles occurred at speeds too great to allow them to stick together. It is as though Jupiter wielded a giant spatula, stirring up the nearby debris and keeping it spread out as a loose collection of rocky fragments. Even today, Jupiter is still at work, keeping the asteroids mixed up and occasionally causing collisions between them that can sometimes break them up. Very recently the *IRAS* satellite has detected rings of dust circling the sun, between the orbits of Mars and Jupiter; that were probably created by such a collision.

The study of meteorites, some of which probably originated as pieces of asteroids that broke apart in collisions, gives us a chance to examine material from the early solar system. It is interesting to note that some of the asteriods apparently have undergone differentiation and developed nickel-iron cores, because some meteorites are almost pure chunks of this material. There must have been sufficient heat inside some of the asteroids to create a molten state, allowing the heavy materials to separate from the rest.

In contrast with the nickel-iron asteroids, some of the other types, particularly the carbonaceous ones, apparently have undergone almost no heating, since they contain high quantities of volatile elements that would have been easily cooked out.

Comets: Fateful Messengers

Among the most spectacular of all the celestial sights are the comets. With their brightly glowing heads and long, streaming tails, along with their infrequent and often unpredictable appearances, these objects have sparked the imagination (and often the fears) of people throughout history. In antiquity, when astrological omens were taken very seriously, great import was attached to the occasion of a cometary appearance (Fig. 10.4). Ancient descriptions of comets are numerous, and in many cases these objects were thought to be associated with catastrophe and suffering.

Included among the teachings of Aristotle was the notion that comets were phenomena in the earth's atmosphere. There was no reliable evidence for this idea, and it is not clear how Aristotle came upon it, but in

FIGURE 10.4. *Calamity on Earth Associated with the Passage of Comets.* This drawing is from a seventeenth-century book describing the universe.

any case it was accepted for centuries to come. However, Tycho Brahe in 1577 was able to prove that comets were too distant to be associated with the earth's atmosphere, because they do not exhibit any parallax when viewed from different positions on the earth. If a comet were really located only a few kilometers or even a few hundred kilometers above the surface of the earth, its position as seen from the earth would change from one location to another. Tycho was able to show that this was not the case, and that therefore comets belonged to the realm of space.

Halley, Oort, and Cometary Orbits

A major advance in the understanding of comets was made by a contemporary and friend of Newton, Edmund Halley. Aware of the power of Newton's laws of motion and gravitation, Halley reviewed the records of cometary appearances, and noted one outstanding regularity. Particularly bright comets seen in 1531, 1607, and 1682 seemed to have similar properties, and Halley suggested that all three were appearances of the same comet orbiting the sun with a 76-year period.

Calculations based on Kepler's third law showed that for a period of 76 years, this object must have a semimajor axis of nearly 18 AU. Halley realized, therefore, that in order for the comet to appear as dominant in our skies as it does, it must have a highly elongated orbit (Fig. 10.5), so that it comes close to the sun at times, even though its average distance is well beyond the orbit of Saturn and nearly as far out as Uranus. Such an eccentric orbit, as a very elongated ellipse is called, had not previously been observed, even though Newton's laws clearly allowed the possibility.

Since Halley's time, searches of ancient reports of comets have revealed that Halley's comet has been making regular appearances for many centuries. The earliest records are from the ancient Chinese astronomers, who apparently observed its every appearance for well over a thousand years, possibly beginning as early as the fifth century B.C.

The most recent visit of Halley's comet was in 1910 (Fig. 10.6), when the earth actually passed through its tail (without any noticeable effects), and it will next be seen in 1986. Unfortunately for us, on its upcoming visit, Halley's comet will be on the far side of the sun from the earth's orbital position, and will not be easily observed from the earth. The best views will be from the southern hemisphere.

FIGURE 10.6. *Halley's Comet as it Appeared in 1910.*

Other spectacular comets have been seen (Fig. 10.8), and a number have rivaled Halley's comet in brightness. Traditionally, a comet is named after its discoverer, and there are astronomers around the world who spend long hours peering at the nighttime sky through telescopes, looking for a piece of immortality.

When a new comet is discovered, a few observations of its position are sufficient to allow computation of its orbit. The results of many years of comet-watching have shown that there are numerous comets whose orbits are so incredibly stretched out that their periods are measured in the thousands or even the millions of years. These comets are only seen once; they return thereafter to the void of space well beyond the orbit of Pluto, where they spend millennia before visiting the inner solar system again.

FIGURE 10.5. *A Cometary Orbit.* This is a rough scale drawing of the orbit of Halley's comet. Most comets actually have much more highly elongated orbits than this, and correspondingly longer periods.

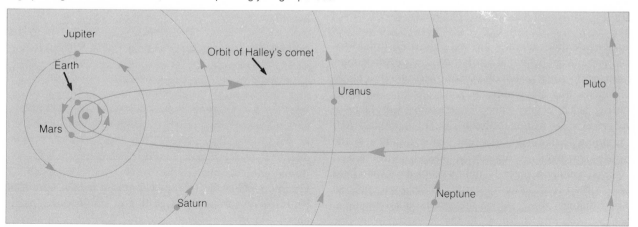

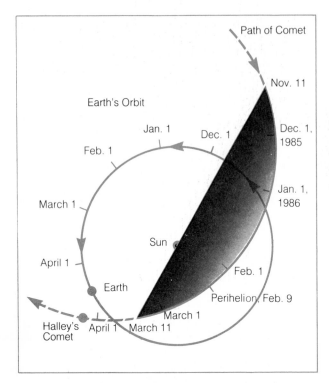

FIGURE 10.7. *The Path of Halley's Comet in 1985–1986.*
This diagram shows the positions of the earth and the comet during the comet's upcoming passage near the sun. The comet will cross the ecliptic in early November 1985, and reach its closest approach to the sun on February 9, 1986. On March 10, it will cross the ecliptic again, so that when it is closest to the earth on April 11 (the position shown here), it will be best seen in the southern hemisphere.

The orbits of comets, particularly these so-called long-period ones, are randomly oriented. Comets do not show any preference for orbits lying in the plane of the ecliptic (in strong contrast with the planets), and about half go around the sun in the retrograde direction.

Consideration of these orbital characteristics, especially the large orbital sizes, led the Dutch astronomer Jan Oort to suggest that all comets originate in a cloud of objects that surrounds the solar system. The **Oort cloud,** as it is now called, is envisioned as being a spherical shell with a radius of 50,000 to 150,000 AU, extending a significant fraction of the distance to the nearest star, which is almost 300,000 AU from the sun.

Occasionally a piece of debris from the Oort cloud is disturbed from its normal path, either by a collision with another object or by the gravitational tug of a nearby star, and it begins to fall inward toward the

FIGURE 10.8. *Comet Kohoutek 1973.*
This was one of the brightest comets in recent years.

sun. Left to its own devices, a comet falling in from the Oort cloud would follow a highly elongated orbit with a period of millions of years, appearing to us as one of the long-period comets when it made its brief incandescent passage near the sun. In many cases, a comet is not left to its own devices, however. Instead it runs afoul of the gravitational pull of one of the giant planets, most often Jupiter. When this happens, the comet may be speeded up, so that it escapes the solar system entirely after it loops around the sun; or it may be slowed down, dropping into a smaller orbit with a shorter period, so that it becomes one of the numerous comets that are seen to reappear frequently.

By invoking the existence of the Oort cloud, we can account for all the observed cometary orbits, and this concept is widely accepted today. We also must consider how the Oort cloud came into being, however. The best guess is that it consists of material left over from the formation of the solar system. If this is true, the comets are no doubt made of very old matter, representing the primordial stuff from which the planets and the sun were created. The formation of the Oort cloud will be discussed at greater length in chapter 12, where the formation and evolution of the solar system as a whole are described.

The Anatomy of a Comet

For most of its life, a comet is just a frozen chunk of icy material, probably consisting of small gravel particles or larger boulders that are embedded in frozen gases. As the comet passes through the outer reaches of its orbit, far from the sun, it does not glow, has no tail, and is not visible from earth.

As the comet approaches the sun, however, it begins to warm up as it absorbs sunlight, and the added heat causes volatile gases to escape. A spherical cloud of glowing gas called the **coma** develops around the solid **nucleus** (Fig. 10.9). Spectroscopic measurements have shown that the gases in the comae of comets are simple molecules such as H_2O, C_2, C_3, CH, CN, CO, and N_2. Other slightly more complex species, such as CO_2, ammonia (NH_3), and methane (CH_4) are probably also present, along with hydrogen molecules (H_2). The observed species glow by a process called **fluorescence.** The molecules absorb light from the sun, which causes them to be excited to high energy levels, and then they emit light as they return to low energy states. A cloud of hydrogen atoms, resulting from the breakup of the molecules in the coma, extends out to great distances from the nucleus (Fig. 10.10). The visible coma may be as large as 100,000 kilometers in diameter, and the halo of hydrogen atoms may extend as much as ten times farther from the nucleus. The solid nucleus is relatively tiny, with a diameter of perhaps a few kilometers.

Sometimes the release of gases from a cometary nucleus is so forceful that it actually alters the course of the comet. If gas is expelled from a specific location in the nucleus (such as the leading side, which is more directly exposed to heating from the sun), it can act like

FIGURE 10.10. *The Halo of Comet Kohoutek 1973.* This is an ultraviolet image made by *Skylab* astronauts; it shows the extent of the hydrogen cloud surrounding this famous 1973 comet. The halo was about a million kilometers in diameter when this photo was taken.

a rocket exhaust. Erratic motions that apparently violate Newton's laws have been observed in several cases, most notably in Halley's comet, whose arrival in the inner solar system is often delayed by this effect.

As a comet nears the sun, the solar radiation and the solar wind force some of the gas from the coma to flow away from the sun, thus forming the comet's tail, which in some instances is as long as 1 AU. There often are two distinct tails (Fig. 10.11): one formed of gas from the coma, usually containing molecules that have been ionized, such as CO^+, N_2^+, CO_2^+, and CH^+; and another formed of tiny solid particles released from the ice of the nucleus. The **ion tail**, the one formed of ionized gases, is shaped by the solar wind, and therefore points almost exactly straight away from the sun at all times. The other tail, called the **dust tail**, usually takes on a curving shape, as the dust particles are pushed away from the sun by the force of the light they absorb. This **radiation pressure** is not strong enough to force the dust particles into perfectly straight paths away from the sun, so the particles follow curved trajectories which are a combination of their orbital motion and the outward push caused by sunlight.

The gases that escape from the nucleus of a comet as it approaches the sun are highly volatile, and would

FIGURE 10.9. *Anatomy of a Comet.* This sketch illustrates the principal features of a comet (not drawn to scale).

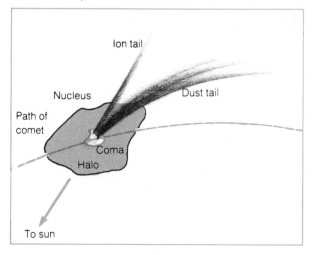

FIGURE 10.11. *Two Tails.*
This photo of Comet West illustrates the distinction between the ion tail, the straight tail at right; and the dust tail, the broad, somewhat curved tail that points more nearly upward in this image.

persed all along the orbital path, so that each time the earth passes through this region, it encounters a vast number of tiny bits of gravel and dust, and we experience a meteor shower.

Meteors and Meteorites

Occasionally one of the countless pieces of debris floating through the solar system enters the earth's atmosphere, creating a momentary light display as it evaporates in a flash of heat created by the friction of its passage through the air. The streak that is seen in the sky is called a **meteor** (Fig. 10.13). Most of us are familiar with this phenomenon (commonly called a shooting star), since it is often possible to see one in just a few minutes of sky-gazing on a clear night. On rare occasions an especially brilliant meteor is seen, possibly persisting for several seconds, and this spectacular event is called a **fireball** or **bolide.** The piece of solid material that creates a meteor is called a **meteoroid.** Most are very small, amounting to nothing more than tiny grains of dust or perhaps fine gravel. A few, however, are larger solid chunks, which are responsible for the bright fireballs.

Occasionally one of the larger meteoroids survives the arduous trip through the atmosphere and reaches the ground intact. Sucn an object is called a **meteorite** (Fig. 10.14), and examples can be found in museums around the world. Meteorites have been the subject of intense scrutiny, for until the last fifteen years or so, they were the only samples of extraterrestrial material scientists could get their hands on.

Throughout history, scientists scoffed at the notion that rocks could fall from the sky, until a meteorite was seen to fall near a French village in 1803, and was found and examined just after it dropped. Such meteorite falls, although rare, are occasionally observed, and have been known to even cause damage (but so far, few injuries).

Primordial Leftovers

Meteorites are old, much older than most surface rocks on the earth. This fact enhances the interest of scientists, who by studying these objects get a glimpse into the early history of the solar system.

Meteorites generally can be grouped into three

not be present in the nucleus if it had ever undergone any significant heating. This tells us that comets must have formed and lived their entire lives in a very cold environment, probably never even getting as warm as 100 K before falling into orbits that bring them close to the sun. If a comet is so easily vaporized, then once it has begun to follow a path that regularly brings it close to the sun, its days are numbered. It may make many round trips, but eventually the comet will dissipate all of its volatile gases, leaving behind nothing but rocky debris. Several cases have been noted where a comet failed to reappear on schedule, but was replaced by a few pieces or perhaps a swarm of fragments (Fig. 10.12). In time, the remains of a dead comet are dis-

FIGURE 10.12. *The Break-Up of Comet West.* This dramatic sequence shows the nucleus of Comet West fragmenting into four pieces.

classes: the stony meteorites, which represent about 93 percent of all meteorite falls; the iron meteorites, accounting for about 6 percent; and the stony-iron meteorites, which are the rarest. These relative abundances were determined indirectly, because the different types of meteorites are not equally easy to find on the ground. Most meteorites found are the iron ones, which, as we just mentioned, represent only a small fraction of those which fall. The stony meteorites look so much like ordinary rocks that they are usually difficult to pick out, and some are burned up in their way through the atmosphere. A particularly good place to search for meteorites is Antarctica, where a thick layer of ice conceals the native rock. Meteorites that fall there are relatively easy to find, and there is little chance for confusion with earth rocks.

The stony meteorites are mostly of a type called **chondrites,** so named because they contain small

spherical inclusions called **chondrules** (Fig. 10.15). These are mineral deposits formed by rapid cooling, which most likely occurred at a time in the history of the solar system when the last solid material was condensing. A few of the stony meteorites are **carbonaceous chondrites** (Fig. 10.16), thought to be almost completely unprocessed since the solar system formed, and therefore representative of the original stuff of which the planets were made. The primordial nature of the carbonaceous chondrites, like the carbonaceus asteroids mentioned earlier, is deduced from their high volatile content, which indicates that they were never exposed to much heat. One particularly fascinating aspect of these meteorites is that in at least one

FIGURE 10.14. *A Meteorite.*
This is a stony meteorite. The black coloring was caused by heating as the object passed through the atmosphere. The light-colored spots are breaks in the fusion crust where interior material is exposed.

FIGURE 10.13. *Two Bright Meteors.*
The streaks of light in this photo are created by a tiny particle entering the earth's atmosphere from space.

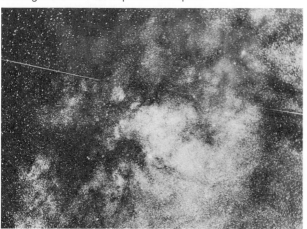

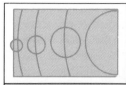

Astronomical Insight 10.2

WHEN IT RAINS METEORS

A meteor shower is an unforgettable experience, well worth the loss of sleep required to get the best view. There are a number of particularly dense showers that occur yearly as the earth passes through the paths of debris that create them. The best time to see a meteor shower is after midnight, when we are on the side of the earth that faces forward as it moves along in its orbit. As the earth plows through the swarm of meteoroids, most of the entries into the earth's atmosphere occur on this leading side.

During a shower, all the meteors seem to approach from a single point, called the **radiant.** This is a simple effect of perspective: the meteoroids are traveling along in parallel paths, but they seem to emanate from a common point as we look in the direction from which they come.

Not all meteor showers are associated with dead comets. Apparently even while a comet lives, making regular appearances, a cloud of debris may be scattered along its orbit, in which case we are treated to the spectacle of a meteor shower whenever the earth passes through the comet's orbit. Hal-

TABLE 10.1. Meteor Showers

SHOWER	APPROXIMATE DATE	ASSOCIATED COMET
Quadrantid	January 3	—
Lyrid	April 21	Comet 1861 I
Eta Aquarid	May 4	Halley's Comet
Delta Aquarid	July 30	—
Perseid	August 11	Comet 1862 III
Draconid	October 9	Comet Giacobini-Zinner
Orionid	October 20	Halley's Comet
Taurid	October 31	Comet Encke
Andromedid	November 14	Comet Biela
Leonid	November 16	Comet 1866 I
Geminid	December 13	

ley's comet is an example: the earth passes through its orbit twice each year, and each of these occasions is marked by a meteor shower.

If the particles following a cometary orbit are not scattered uniformly along it, but are instead concen-

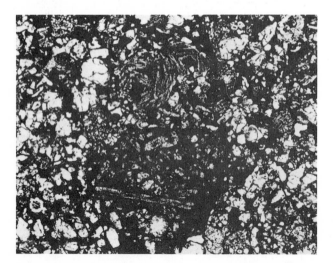

FIGURE 10.15. *Cross-Section of a Chondrite.*
This shows the many chondrules (light patches) embedded within the structure of this type of stony meteorite.

case, complex organic molecules called **amino acids** have been found inside a carbonaceous-chondrite meteorite, showing that some of the ingredients for the development of life were apparently available even before the earth formed.

The iron meteorites have varying nickel contents. Some of them have an internal crystalline structure, which indicates a rather slow cooling process in their early history (Fig. 10.17). This has important implications in terms of their origin, as we shall see.

trated in one region, we will not always see equally active meteor showers when the earth passes through the orbit. The density of the shower will vary, depending on how close we pass to the part of the orbit having the greatest concentration of particles. The Leonid shower, for example, is especially brilliant and intense every thirty-three years, when the earth passes through the most densely populated part of the cometary leftovers that create this shower. In some cases, on the other hand, the particles seem to be uniformly distributed along the orbit, so we see about the same intensity of meteors each time.

Table 10.1 lists some of the more prominent meteor showers seen each year.

A meteor shower. Some 78 meteor trails appear in this exposure.

FIGURE 10.16. *A Carbonaceous Chondrite.*
This example, not of the type in which amino acids have been found, shows a large chondrule (the light-colored spot, upper center), which is about 5 mm in diameter.

Dead Comets and Fractured Asteroids

The origins of the meteoroids that enter the earth's atmosphere can be inferred from what we know of the properties of meteorites, asteroids, and comets. As we have seen, most meteors are caused by relatively tiny particles that do not survive their flaming entry into the earth's atmosphere. During **meteor showers,** when meteors can be seen as frequently as once per second, all appear to be of this type. As noted earlier, these

FIGURE 10.17. *Cross-Section of a Nickel-Iron Meteorite.*
This example shows the characteristic crystalline structure
indicative of a slow cooling process from a previous molten
state. Such meteorites are thought to have once been parts of
larger bodies that differentiated.

was subjected to heavy bombardment as well. The difference, of course, is that the earth has an atmosphere, along with flowing water and glaciation, all of which combine to erase old craters in time. A few traces are still seen, however. There is a very large basin under the Antarctic ice that is probably an ancient impact crater, and a portion of Hudson's Bay in Canada shows a circular shape thought to have a similar origin. A number of other suspected ancient impact craters have been found throughout the world (Fig. 10.18).

Although the frequency of impacts has decreased, there are still rare occasions when major impacts occur. For example, the Barringer crater (Fig 10.19) near Winslow, Arizona, was formed only about 25,000 years ago. The possibility exists that other large bodies could hit the earth; given a long enough time, it is almost inevitable.

Microscopic Particles: Interplanetary Dust and the Interstellar Wind

The empty space between the planets plays host to some very tiny particles, in addition to the larger ones we have just described. There is a general population of small solid particles, perhaps a millionth of a meter in diameter, called interplanetary dust grains. There is also a very tenuous stream of gas particles flowing through the solar system from interstellar space.

The presence of the dust has been known for some time from two celestial phenomena, both of which can be observed with the unaided eye, although only with difficulty. The dust particles scatter sunlight, so that under the proper conditions we can see a diffuse glow where the light from the sun hits the dust. This is analogous to seeing the beam of a searchlight stretching skyward: we only see the beam where there are small particles (either dust or water vapor) that scatter its light, some of it reaching our eyes.

One of the phenomena created by the interplanetary dust is the **zodiacal light** (Fig. 10.20), a faintly illuminated belt of hazy light that can be seen stretching across the sky (along the ecliptic) on clear, dark nights, just after sunset or before sunrise. The second observable phenomenon created by the dust is a small bright spot seen on the ecliptic in the opposite direction from

showers are associated with the remains of comets that have disintegrated and have left behind a scattering of gravel and dust. Therefore it is thought that the most common meteors, those created by small, fragile meteoroids, are a result of cometary debris.

The larger chunks that reach the ground as meteorites may have a different origin. It is likely that asteroids occasionaly collide, the collision sometimes being sufficiently violent to destroy them, and the rubble that is left over is dispersed throughout the solar system. Most meteorites are probably fragments of asteroids. The iron meteorites apparently originated in asteroids that had undergone differentiation, whereas the stony ones came from the outer portions of differentiated asteroids, or from smaller bodies that had never undergone differentiation. The chondrites probably fall into the latter category, since the chondrules reflect a rapid cooling that would have characterized very small bodies. This is why the chondrites are thought to be the most primitive of the meteorites, having undergone no processing in the interiors of large bodies.

From our studies of the other planets and satellites, we know that there was a time long ago when frequent impacts occurred, forming most of the craters seen today. Certainly the earth was not immune; no doubt it

FIGURE 10.18. *Impact Craters on the Earth.* This map shows the locations of major craters thought to have been created by impacts of massive objects.

the sun. This diffuse spot, called the **gegenschein** (Fig. 10.21), is created by sunlight that is reflected straight back by the interplanetary dust, which is concentrated in the plane of the ecliptic. This is analogous to seeing a bright spot on a cloud bank or low-lying

FIGURE 10.19. *Meteor Crater Near Winslow, Arizona.* The impact that created this crater occurred about 25,000 years ago.

mist when we look at it with the sun directly behind us; the bright spot is just the reflected image of the sun, and is the counterpart of gegenschein.

It is possible to collect interplanetary dust particles for direct examination (Fig. 10.22). Usually this is done through the use of high-altitude balloons. However, a surprising new technique has recently been developed that involves scooping sludge (which contains dust particles) off the ocean floor. The earth is constantly being pelted by dust particles (which add about eight tons per day to its mass!), and those particles which fall into the oceans can lie undisturbed on the seabed for long periods of time. Studies of the grains show that they are probably of cometary origin, having been dispersed throughout space from the dead nuclei of old comets.

As we will learn in chapter 18, the space between stars in our galaxy is permeated by a rarefied gas medium. In the sun's vicinity, the average density of this gas is far below that of any man-made vacuum; it amounts to only about 0.1 particle/cm^3 (that is, there is 1 atom, on average, in every volume of 10 cm^3, corresponding to a cube about 1 inch on each side). Because of the motion of the sun in its orbit about the galaxy, the interstellar gas streams through the solar system

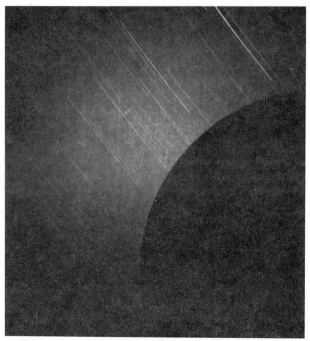

FIGURE 10.20. *The Zodiacal Light.*
This photo shows the diffuse band of light in the plane of the ecliptic; it is caused by the scattering of sunlight from tiny interplanetary dust grains. The zodiacal light is most easily visible about an hour before sunrise or after sunset.

FIGURE 10.21. *The Gegenschein.*
This photo shows the Milky Way (stretching across the upper portion) and a diffuse concentration of light (at lower center) which is the gegenschein. It is created by sunlight reflected directly back to earth by the interplanetary dust, in the direction opposite from the sun.

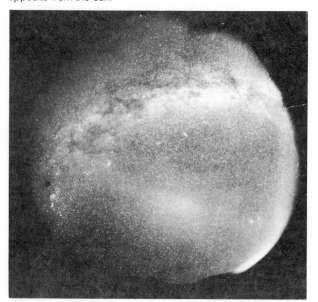

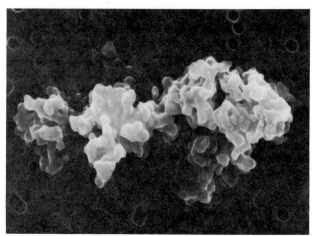

FIGURE 10.22 *An Interplanetary Dust Grain.*
This is a microscopic view of a tiny particle from interplanetary space. The amorphous structure is highly variable from one grain to another.

with a velocity of about 20 kilometers per second. The presence of this ghostly breeze, consisting mostly of hydrogen and helium atoms and ions, was discovered in the early 1970s when observations made from satellites revealed very faint ultraviolet emission from the hydrogen atoms in the gas. The interstellar wind, tenuous as it is, has very little effect on the other components of the solar system, but is nevertheless studied with some interest for what it may tell us about the interstellar medium.

PERSPECTIVE

The interplanetary wanderers discussed in this chapter have given us insight into the history of the solar system, and have told us much about its present state as well. We have found two primary origins of the various objects: comets and asteroids. The former account for most of the meteors and for the interplanetary dust, and the latter are responsible for the meteorites, including the massive bodies that formed the major impact craters in the solar system.

The pieces are nearly all in place. Next we turn our attention inward, to examine the engine that keeps all the machinery in operation.

SUMMARY

1. The asteroids were accidentally and coincidentally discovered near the orbital distances predicted by Bode's law.

2. Thousands of asteroids have been catalogued. They display a variety of sizes (up to 1,000 kilometers in diameter) and compositions (ranging from metals to rocky minerals).

3. Gaps in the asteroid belt are created by orbital resonances with Jupiter, whose gravitational influence was probably also responsible for preventing the asteroids from coalescing into a planet in the first place.

4. Comets are small, icy objects that develop their characteristic comae and tails only when in the inner part of the solar system.

5. Comets apparently originate in a cloud of debris very far from the sun. They occasionally fall inward, either to bypass the sun and return to the distant reaches of the solar system for millennia, or to be perturbed by the gravitational influence of one of the planets, and to thus become periodic comets.

6. When near the sun, a comet ejects gases that glow by fluorescence. Periodic comets eventually lose all of their icy substance in this process, and disintegrate into swarms of rocky debris.

7. A comet may have two tails, one created by ionized gas, and the other made of fine dust particles.

8. A meteor is a flash of light created by a meteoroid entering the earth's atmosphere from space, and a meteorite is the solid remnant that reaches the ground in some cases.

9. Meteorites are either stony, stony-iron, or iron in composition, and they are very old, providing information on the early solar system.

10. Most meteors are created by fine debris from comets, and most meteorites are fragments of asteroids.

11. Interplanetary space is permeated by fine dust particles and an interstellar wind of hydrogen and helium atoms from the space between the stars.

REVIEW QUESTIONS

1. What is the orbital period of an asteroid whose semimajor axis is 2.8 AU?

2. At what distance from the sun would an asteroid have exactly one-tenth the orbital period of Jupiter? Is there a gap in the asteroid belt at that distance?

3. How is the formation of the asteroid belt like the formation of the rings of Saturn?

4. How do the motions of comets differ from those of planets?

5. Summarize the effects of Jupiter on the asteroids, comets, and meteoroids.

6. Summarize the life story of a typical comet.

7. Why are carbonaceous chondrites important clues to the early history of the solar system?

8. Would observers on the surface of Mercury see meteors in the nighttime sky? Would they find meteorites on the surface of Mercury?

9. What does the concentration of the zodiacal light in the ecliptic tell us about the distribution of interplanetary dust in the solar system?

10. What is the orbital period of a Trojan asteroid? How does the period of an Apollo asteroid compare with that of the earth?

ADDITIONAL READINGS

Cassidy, W. A., and Rancitelli, L. A. 1982. Antartic meteorites. *American Scientist* 70(2):156.

Chapman, C. R. 1975. The nature of asteroids *Scientific American* 232(1):24.

Hartmann, W. K. 1975. The smaller bodies of the solar system. *Scientific American* 233(3):142.

Tatum, J. B. 1982. Halley's comet in 1986. *Mercury* 11(4):126.

Van Allen, J. A. 1975. Interplanetary particles and fields. *Scientific American* 233(3):160.

Whipple, F. L. 1974. The nature of comets *Scientific American* 230(2):49.

THE SUN

Learning Goals

BASIC PROPERTIES AND INTERNAL STRUCTURE OF
THE SUN
Hydrostatic Equilibrium and Internal Conditions
The Transport of Energy Inside the Sun
Nuclear Reactions and Energy Production
STRUCTURE OF THE SOLAR ATMOSPHERE
THE SOLAR WIND

SUNSPOTS, SOLAR ACTIVITY CYCLES, AND THE
MAGNETIC FIELD
The 22-year Cycle
The Role of the Magnetic Field
Flares and Other Violence
Solar-terrestrial Relations

Our star, the sun, is rather ordinary by galactic standards. It has modest mass and size, and there are stars as much as a few hundred times larger and a million times more luminous. Its temperature is also moderate, as stars go. In many respects the sun is entirely a run-of-the-mill entity.

Within our solar system, on the other hand, the sun is clearly the king. In mass it outranks even Jupiter by a factor of more than one thousand. It is the only body that glows under its own power (except for the excess infrared and radio emission from Jupiter and Saturn), and its light provides all the illumination by which we see the planets. The sun gives us the warmth necessary for biological activity on earth, and provides all the forms of energy available to us (except nuclear).

In this chapter we will consider the basic properties of our star, placing particular emphasis on its interaction with the planets.

Basic Properties and Internal Structure

The sun is a ball of hot gas (Fig. 11.1). Its density on average is 1.41 grams/cm^3, not much more than that of water, but its center is so highly compressed that the density there is about ten times greater than that of lead. The interior is gaseous rather than solid, because the temperature is very high, around 10 million degrees at the center, diminishing to just under 6,000 K at the surface. At these temperatures, the gas is partially ionized in the outer layers of the sun, and completely ionized in the core, all electrons having been strippped free of their parent atoms.

The sun is held together by gravity. All of its constituent atoms and ions attract each other, and the net effect is that the solar substance is held in a spherical shape. Hot gas exerts pressure on its surroundings, and this pressure, pushing outward, balances the force of gravity, which is pulling the matter inward. This balance is called **hydrostatic equilibrium,** with gravity and pressure equaling each other everywhere. The deeper the layer, the greater the weight of the overlying layers, and the more the gas is compressed. The higher the pressure, the greater the temperature required to maintain the pressure, so we find that the pressure and

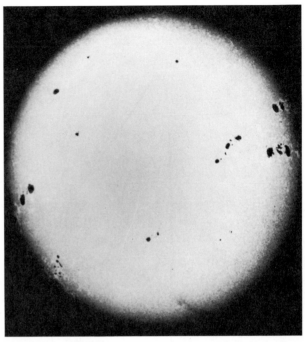

FIGURE 11.1. *The Sun.*
This is a white-light photo, showing numerous sunspots.

temperature both increase as we approach the center of the sun.

The composition of the sun is the same as that of most other stars: about 70 percent of its mass is hydrogen, 27 percent is helium, and the rest is made up of other elements. In the outer layers, where no nuclear reactions have taken place, a greater fraction of the mass is in the form of hydrogen (about 79 percent). Thus the sun's chemical makeup is similar to that of the outer planets, and would resemble the terrestrial planets as well, except that these bodies have lost most of their volatile elements such as hydrogen and helium.

TABLE 11.1 The Sun

Diameter: 1,319,980 km (109.3 D$_\oplus$)
Mass: 1.99 × 10^{33} grams (332,943 M$_\oplus$)
Density: 1.409 grams/cm^3
Surface gravity: 27.9 earth gravities
Escape velocity: 618 km/sec
Luminosity: 3.83 × 10^{33} ergs/sec
Surface temperature: 6,500 K (deepest visible layer)
Rotation period: 25.04 days (at equator)

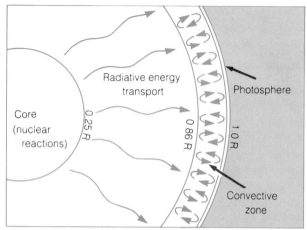

FIGURE 11.2. *The Internal Structure of the Sun.*
This shows the relative extent of the major zones within the sun, except that the depth of the photosphere is greatly exaggerated.

It is apparent that all the components of the solar system formed together, from the same material.

The ultimate source of all the sun's energy is in its core, within the innermost 10 percent or so of its radius (Fig. 11.2). Here nuclear reactions create heat and photons of light at γ-ray wavelengths. It is this light that eventually reaches the surface and escapes into space, but it is a laborious journey. Each photon is absorbed and reemitted many times along the way, gradually losing energy (Fig. 11.3). In the process the photon becomes a visible-light photon, and the energy it

FIGURE 11.3. *Random Walk.*
A photon is continually being absorbed and reemitted as it travels through the sun's interior, and each time it is reemitted, it is in a random direction. Thus, its progress from the core (where it is created) to the surface is very slow.

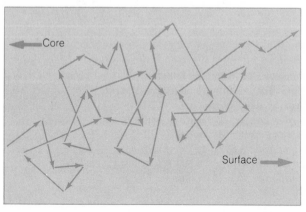

loses heats the surroundings. Because a photon only travels a short distance before it is absorbed, and because when it is reemitted it is in a random direction, progress toward the surface is very slow. It takes an individual photon as long as a million years to migrate from the center of the sun to the surface, even though the light travel time if it were unimpeded would be only two seconds. If the sun's energy source were suddenly turned off, we would not be aware of it for a million years!

Throughout most of the solar interior, the gas is quiescent, without any major large-scale flows or currents. The energy from the core is transported by the radiation wending its slow pace outward, except in the layers near the surface, where convection occurs and heat is transported by the overturning motions of the gas. As we shall see, the bubbling, boiling action in the outer portions of the sun creates a wide variety of dynamic phenomena on the surface.

The sun rotates, and its does so differentially. This is apparent from observations of its surface features, which reveal that, like Jupiter and Saturn, the sun goes around faster at the equator than near the poles. The rotation period is 25 days at the equator, 28 days at middle latitutdes, and even longer near the poles. The differential rotation probably plays an important role in governing variations in the solar magnetic field, which in turn have a lot to do with behavior of the most prominent surface features, the sunspots.

There is evidence that the core of the sun rotates much more rapidly than the surface. The only clue is the presence of very subtle oscillations on the solar surface, which may be wave motions affected by the rapid spin of the interior. The exact rate of internal rotation is not known, but the core probably spins with a period of only a few days, in contrast to the 25-day surface-rotation period. The rapid spin of the solar core is probably a direct result of the collapse and accelerated rotation of the interstellar cloud from which the sun formed. The origin of the sun's rotation is discussed in chapter 12.

Nuclear Reactions

One of the biggest mysteries in astronomy in the early decades of this century involved the sun. The problem was how to account for the tremendous amount of energy it radiates, made particularly perplexing, by geological evidence showing that the sun has been able to produce this energy for at least four or five billion years. Two early ideas, that the sun is simply still hot from its formation, or that it is gradually contracting, releasing stored gravitational energy, were both ruled out, since neither mechanism could possibly supply the energy needed to run the sun for a long enough time.

The first hint at the solution came in the first decade of the 1900s, when Albert Einstein developed his theory of relativity. He showed that matter and energy are equivalent, and that one can be converted into the other by the famous formula $E = mc^2$, where E is the energy released in the conversion, m is the mass that is converted, and c is the speed of light. This mechanism can produce enormous amounts of energy, and physicists began to contemplate the possibility that somehow this energy was being released inside the sun and stars.

In the 1920s, after pioneering work on atomic structure by Max Planck, Niels Bohr, and others, the concept of nuclear reactions began to emerge. Like chemical reactions, nuclear reactions are transformations, except that in this case it is the subatomic particles in the nuclei of atoms that react with each other, rather than the electrons in the outer orbits. There are **fusion** reactions, in which nuclei merge to create a larger nucleus, representing a new chemical element; and **fission** reactions, in which a single nucleus, usually of a heavy element with a large number of protons and neutrons, splits into two or more smaller nuclei. In both types of reaction, energy is released as some of the matter is converted according to Einstein's formula. In the 1920s, Enrico Fermi, Werner Heisenberg, and Wolfgang Pauli explored these possibilities, and by the 1930s, Hans Bethe had suggested a specific reaction sequence that might be operative in the sun's core.

Be the envisioned a fusion reaction in which four hydrogen nuclei (each consisting of only a single proton) combine to form a helium nucleus, made up of two protons and two neutrons. The reaction occurs in several steps: (1) two protons combine to form **deuterium,** a hydrogen isotope that has a proton and a neutron in its nucleus (one of the two protons undergoing the reaction converts itself into a neutron, by emitting a positively charged particle called a **positron**) and another particle called a **neutrino,** which has very unusual properties, (2) the deuterium combines with another proton to create an isotope of helium (^{3}He) consisting of two protons and one neutron; and (3) two of these ^{3}He nuclei combine, forming an ordinary helium nucleus (^{4}He, with two protons and two neutrons in the nucleus) and releasing two protons. At each step in this sequence, heat energy is imparted to the surroundings in the form of kinetic energy of the particles that are produced, and in the second step a photon of γ-ray light is emitted as well.

The net result of this reaction, which is called the **proton-proton chain,** is that four hydrogen nuclei (protons) combine to create one helium nucleus. The end product has slightly less mass than the ingredients, 0.007 of the original amount having been converted into energy. Simple calculations (see chapter 14) show that this reaction can easily supply enough energy to keep the sun running at its present rate for many billions of years.

Nuclear fusion reactions can take place only under conditions of extreme pressure and temperature, because of the electrical forces that normally would keep atomic nuclei from ever getting close enough together to react. Nuclei, which have positive charges because all their electrons have escaped, must collide at extremely high speeds in order to overcome the repulsion caused by their like electrical charges. The speed of particles in a gas is governed by the temperature, and only in the very center of the sun and other stars is it hot enough (around 10 million degrees) to allow the nuclei to collide fast enough to fuse. The high pressure in the sun's core causes nuclei to be crowded together very densely, and this means that collisions will take place very frequently, another requirement if a high reaction rate is to occur.

Structure of the Solar Atmosphere

Observations of the sun's appearance when viewed in different wavelengths of light make it clear that the outer layers are divided into several distinct zones (Fig. 11.4). The "surface" of the sun that we see in visible wavelengths is the **photosphere,** with a temperature ranging between 4,000 and 6,500 K. When we view the sun at the wavelength of the strong line of hydrogen at 6,563 Å, we see the **chromosphere,** a layer above the photosphere whose temperature is 6,000 to 10,000 K. Outside of that is the very hot, rarefied **corona** (best observed at X-ray wavelengths) whose temperature is 1 to 2 million degrees. In between the chromosphere and corona is a thin region called the **transition zone,** where the temperature rapidly rises. Overall, the tenuous gas within and above the photosphere is referred

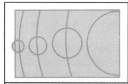

In the early 1960s, measurements of spectral lines in the solar photosphere revealed small Doppler shifts that alternated between blueshifts and redshifts. This seemed to indicate that the sun's surface was rising and falling, a discovery that was for some time quite controversial. The reported Doppler shifts were quite small (less than one one-hundredth of an angstrom; near the limit of detectability), and the idea of a pulsating sun seemed to many scientists unlikely. Although there are stars that undergo much more radical pulsations, creating very noticeable effects (see the discussions of variable stars in chapters 13 and 17), the sun is not prone to such extreme behavior. We like to think of our star as stable and more or less unchanging.

The disputed pulsations would not go away, however, and they became an important phenomenon which scientists set out to explain. The period of a solar pulsation is about five minutes, and the phenomenon is commonly known as the sun's five-minute oscillation. Today there is a widely accepted explanation, in which the sun may be compared to a musical instrument with a resonant cavity.

When waves of any kind are created inside a cavity with reflecting walls, certain wavelengths or frequencies persist while all others die out. The reason is that waves reflected from the walls interfere with incoming waves, the valleys and crests either adding together or canceling each other out. In an enclosed space, the wavelengths that add together constructively are determined by the dimensions of the enclosure; such wavelengths are characteristic of the par-

ticular enclosure and correspond to its so-called resonant frequency. In a musical wind instrument, this frequency determines the pitch of the tone that is emitted. The musician can control the pitch by altering the dimensions of the internal cavity (by opening and closing valves, in most cases).

Inside the sun, there are no walls, but physical conditions that vary with depth can create barriers as real as walls for certain types of waves. It has been deduced that sound waves (compressional waves) can persist in the sun's convective zone, and are confined above and below by the manner in which the sun's temperature varies with depth. Sound waves in a certain range of wavelength are reflected back as they travel farther in or farther out, and this creates a resonant cavity in the outermost portion of the sun for sound waves whose period of pulsation is about five minutes. These waves create a standing pattern of crests and valleys on the solar surface, and at any particular point on the surface, the gas rises and falls with this period.

This phenomenon is quite interesting for its own sake, but is worthy of careful study for another reason as well: the properties of the waves or oscillations of the solar surface can reveal data on conditions in the solar interior, where the waves arise. For example, study of the frequency range of the waves can yield information on exactly how the temperature varies with depth inside the sun, since it is this variation that determines the location of the lower boundary of the resonant cavity. Perhaps even more interesting, analysis of the surface oscillations

to as the solar atmosphere. If we consider the temperature throughout this region, we find that it decreases outward through the photosphere, reaching a minimum value of about 4,000 K. From there the trend reverses itself, and the temperature begins to rise as we go farther out. The chromosphere, immediately above the temperature minimum, is perhaps 2,000 kilometers

thick. Above there the temperature rises very steeply within a few hundred kilometers, to the coronal value of more than a million degrees. Clearly something is creating extra heat at these levels; soon we will discuss where this heat comes from.

First let us discuss the photosphere, the surface of the sun as we look at it in visible light. It is here that

can provide information on the internal rotation of the sun. This is possible because waves traveling over the solar surface in opposite directions are either spread out or squeezed closer together, depending on whether they move in the direction of the sun's rotation or in the opposite direction. This means that waves going in opposite directions have different frequencies, and the difference in frequency depends on the speed of the sun's rotation in the layer where the waves arise. In effect, it is possible to infer the sun's internal rotational velocity at the lower boundary of the resonant cavity in which the waves oscillate.

Using observations of surface oscillations to determine internal rotational speeds of the sun is a complex undertaking, and requires very sophisticated treatment of extensive quantities of data. At present, it has not been possible to carry this effort very far. However, it has been determined from the five-minute oscillations that the sun is rotating more rapidly near the bottom of the convective zone than at the surface, leading scientists to speculate that the rotation near the core is very rapid indeed. If this is true, it could have important implications in terms of the sun's overall structure and evolution. Thus a great deal of significance is being attached to further study of this phenomenon.

Whereas the five-minute oscillation of the sun is now well established and understood, other reported oscillations are not. There is a three-minute oscillation that may be owing to a similar resonant cavity at the level of the chromosphere. Also, there apparently are long-period oscillations that are driven by waves in very deep resonant cavities near the solar core. These controversial oscillations, with periods of 40 and 160 minutes, have the potential of revealing conditions such as the rotation rate very deep inside the sun, and therefore they warrant a great deal of attention.

The detection of long-period oscillations is very difficult, for a variety of reasons. It is important, for example, to observe the sun continuously over many cycles of the oscillations, which means that observation must continue without interruption for several days. This may sound like an impossibility, because of the sun's daily rising and setting, but it is feasible; it can be done from one of the earth's poles during local summer. In fact, solar studies have been carried out from the scientific research station at the south pole. During one of these studies, involving six days of continuous observation of the sun, the 160-minute oscillation was convincingly detected.

Beyond simply detecting the existence of the long-period oscillations, astronomers need to make further refinements if the oscillations are to be exploited as tools for probing the sun's deep interior. An entirely new technology is required, and at present there are several research groups at work developing the necessary instruments. The problem is to find a way to observe very small Doppler shifts in spectral lines at many positions on the sun's surface, and to make these measurements repeatedly at very closely-spaced intervals of time. The velocities involved are very small, only a meter per second or less, so the Doppler shift in a typical visible-wavelength line is only a few millionths of an angstrom. To accurately measure such small shifts is a formidable task, but one that probably will be successfully completed within a few years. When this has been accomplished, a wealth of new information on the internal properties of our local star will be mined.

the density becomes great enough for the gas to be opaque, making it impossible for us to see further into the interior. The sun's absorption lines (Fig. 11.5) are formed in the photosphere, as the atoms in this relatively cool layer absorb continuous radiation coming from the hot interior.

A photograph of the photosphere reveals a cellular appearance called **granulation** (Fig. 11.6). Bright regions, representing areas where convection in the sun's outer layers causes hot gas to rise, are bordered by dark zones where cooler gas is descending back into the interior.

The temperature of the photosphere, roughly 6,000 K, is measured from the degree of ionization in the gas

THE CASE OF THE MISSING NEUTRINOS

Nearly all that we know about the reactions in the sun's core is based on theoretical calculations, and of course astronomers and physicists have been interested in verifying the results of these calculations. We think that the theory is correct because it neatly accounts for the sun's observed energy output, but nevertheless a more direct confirmation has been sought.

One way to do this was suggested by the early work of Enrico Fermi, who postulated the existence of a tiny subatomic particle called the **neutrino,** which, his calculations showed, ought to be released at certain stages in nuclear reactions. This is a strange little particle, having no mass and no electrical charge. Its role is to carry away small amounts of energy, thus balancing the equations describing the reactions. Because of its ephemeral properties, a neutrino hardly interacts with anything at all. Thus, neutrinos produced in the sun's core ought to escape directly into space at the speed of light, in sharp contrast with the photons from the core, which, as we discussed, take a million years or more to get out.

A possible test of the reactions going on inside the sun, therefore, would be to measure the rate at which neutrinos are coming out. Unfortunately, the same properties that allow them to escape the sun so easily also make them very difficult to catch and count. There is an indirect technique, however, based on the fact that certain nuclear reactions can be triggered by the impact of a neutrino. A chlorine atom, for example, can be converted into a radioactive form of argon when it encounters a neutrino. Accordingly, an experiment was set up a few years ago, and is still in operation, in which a huge tank containing a chlorine compound is monitored to see how many of its atoms are being converted into ra-

dio active argon atoms, which in turn indicates how many neutrinos are passing through the tank.

Much to everyone's surprise, neutrinos have not been detected coming from the sun in the expected numbers, and at present no satisfactory explanation has been found. Some of the suggestions are rather startling, such as the idea that the reactions in the sun have actually stopped, as part of some long-term cyclical behavior, or that the true source of the sun's energy is not nuclear reactions at all, but rather a **black hole** in the center (see discussion of these bizarre objects in chapter 16).

Very recently, two glimmers of hope have been offered by scientists concerned with the neutrino puzzle. One idea based on experimental measurements is that the neutrinos are more complex than we realized; they may actually change their state from time to time, meaning they do not interact with chlorine as readily as we previously thought. The other possibility is suggested by the fact that the particular type of neutrinos that react with chlorine are not the type produced directly by the proton-proton chain, but rather are the result of a minor reaction thought to occur in the sun, one that is not responsible for any significant energy production. Thus, our understanding of this particular reaction could be wrong without destroying the entire theory of how the sun produces its energy. Ways are being sought to detect the neutrinos that are produced directly by the proton-proton chain. Such neutrinos are predicted to react with a rare element called gallium. Scientists are making plans to assemble a supply of this material to see whether solar neutrinos can be detected. This will be a difficult undertaking, however, because most of the world's annual production of gallium will be required to provide a big enough target for the elusive solar neutrinos.

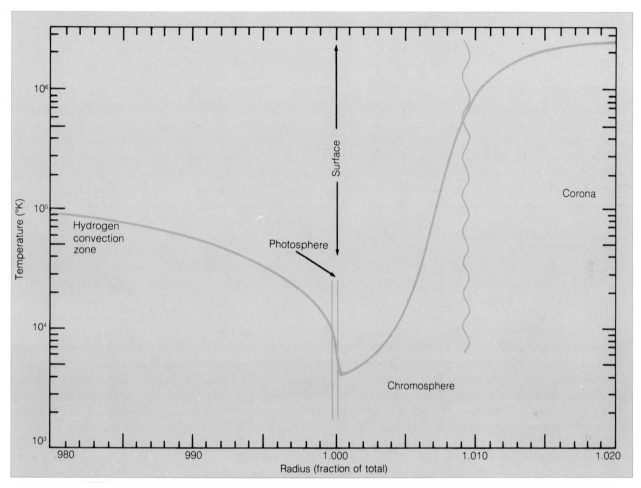

FIGURE 11.4. *The Structure of the Sun's Outer Layers*. This diagram shows the relative heights and temperatures of the convective zone, photosphere, chromosphere, and corona. Heights are expressed in terms of the solar radius, which is roughly 700,000 kilometers.

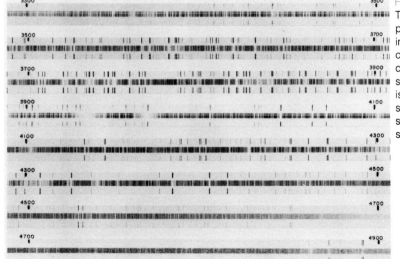

FIGURE 11.5. *Fraunhofer Lines.*
The major absorption lines in the sun's photospheric spectrum are called Fraunhofer lines, in honor of the German scientist who first catalogued many of them in the early nineteenth century. Here we see a photo of a portion of the solar spectrum, the dark band with light lines (this is a negative print). Just above and below the solar spectrum are emission lines made by a special lamp in order to establish the wavelength scale.

Astronomical Insight 11.3

IMITATING THE SUN

Nuclear reactions can produce energy very efficiently, and therefore they have obvious potential benefits for human society, if methods can be found for producing and controlling nuclear energy on earth. To this end, a great deal of effort has gone into developing nuclear power, but the only working reactors today are fission reactors. These have several disadvantages, particularly in that they produce radioactive waste products that are very difficult to store safely.

Fusion reactions can also produce immense quantities of energy, without danger of runaway reactions that could catastrophically destroy the reactor. The sun, after all, runs on power produced by the fusion of hydrogen into helium, both of which are harmless gases. Thus if we could somehow create and control fusion reactions, we might have a permanent solution to the problem of producing sufficient energy to run our society. Unfortunately, the only fusion reactions we have created to date have been the instantaneous ones that occurred in tests of hydrogen bombs.

The difficulty in controlling nuclear fusion reactions is obvious, in view of the conditions required for them to occur inside the sun. The problem is to somehow reproduce, in a controlled environment, the incredible temperatures and pressures of the sun's core. Two techniques for doing this are being developed.

The older of these ideas, dating back to the 1950s, is to contain the superheated gas in a sort of magnetic bottle. Recall that at sufficiently high temperatures, a gas is ionized, and consists entirely of charged particles. Such particles are subject to electromagnetic forces, and can be trapped within a fixed region by a properly shaped magnetic field. A great deal of theoretical work and considerable experimentation have gone into efforts to design a magnetic bottle capable of containing gas under the extreme conditions needed for fusion to occur. The most successful design thus far is a **torus** (a dough-nut-shaped tube), which is twisted into a figure-eight to help keep the particles away from the tube walls. The ionized gas circulates within the enclosed tube, and is kept away from the walls by immense magnetic fields created by electromagnets that surround the tube. This magnetic-confinement technique is being developed primarily at Princeton University. Tests conducted so far have succeeded, for very brief instants, in creating fusion reactions. However, there is still a lengthy development program ahead, for it will be some time before reactions can be sustained and can produce more energy than is required to generate the magnetic fields. A major milestone was passed at the end of 1982, when a new fusion machine at Princeton, the largest constructed so far, passed its first operational test.

The second technique for controlling fusion is being studied mainly at the Lawrence Livermore Laboratories of the University of California. There a device called a **laser** is used to heat material to the required high temperatures, by subjecting the material to extremely intense beams of light. The laser, which was invented in the 1960s, produces a very narrow beam of light, with all the photons at precisely the same wavelength. The power in the beam of light can be immense; lasers have been developed for cutting metal and for performing surgery, for example. The use of lasers to produce fusion reactions involves subjecting small pellets of matter to intense laser beams which instantaneously vaporize the pellets, producing, for a brief instant, the conditions required for fusion to occur. When this technique has been developed to the point where more energy is produced than is required to power the laser, then it can become a useful means of producing energy.

The U.S. government has been supporting fusion research for some time, and, we may hope, will continue to do so. It will take several more years before all the problems and complications are worked out and fusion becomes a viable source of energy. When that day does come, however, it could have major effects on human society, for in the long run our dependence on the earth's limited energy resources could be eliminated.

there and from the use of Wien's law; and the density, roughly 10^{17} particles per cm³ in the lower photosphere, is determined from the degree of excitation, as described in chapter 3. This density is lower than that of the earth's atmosphere, which is about 10^{19} particles per cm³ at sea level.

Most of what we know about the sun's composition is based on the analysis of the solar absorption lines, so the derived abundances represent only the photosphere. We have no reason to expect strong differences in composition at other levels, however, except for the core, where a significant amount of the original hydrogen has been converted into helium by the proton-proton reaction.

Near the edge of the sun's disk the photosphere looks darker than in the central portions. This effect, called **limb darkening,** is caused by the fact that we are looking obliquely at the photosphere when we look near the edge of the disk. We therefore do not see as deeply into the sun there as we do when we look near the center of the disk. The gas we are seeing at the limb is cooler than the deeper-lying gas we see at the disk's center, and thus radiates less.

The chromosphere lies immediately above the temperature minimum. The fact that this region forms emission lines tell us, according to Kirchhoff's laws, that the chromosphere is made of hot, rarefied gas, hotter than that of the photosphere behind it. When viewed through a special filter than allows light to pass through only at the wavelength of the hydrogen emission line at 6,563 Å (Fig. 11.7), the chromosphere has a distinctive cellular appearance referred to as **supergranulation,** similar to the photospheric granulation, but with cells some 30,000 kilometers across instead of about 1,000 kilometers. There is also fine-scale structure in the chromosphere, in the form of spikes of glowing gas called **spicules** (Fig. 11.8). These come and go, probably at the whim of the magnetic forces that seem to control their motions.

The outermost layer of the sun's atmosphere is the corona, which extends a considerable distance above the photosphere and chromosphere. The corona is ir-

FIGURE 11.6. *Solar Granulation.*
This photo, obtained by a high-altitude balloon from above much of the atmosphere's blurring effect, distinctly shows the granulation of the sun's photosphere, which is a result of convective motions in the sun's outer layers.

FIGURE 11.7. *The Chromosphere.*
This photo, taken through a special filter that allows light to pass through only at the wavelength of the bright hydrogen emission line at 6,563 Å, reveals the locations of ionized hydrogen gas in the sun, which are primarily in the chromosphere. The light areas (such as at right) are active regions associated with sunspots, where the chromosphere glows especially brightly at the observed wavelength. The dark streaks are filaments that do not glow strongly in this emission line.

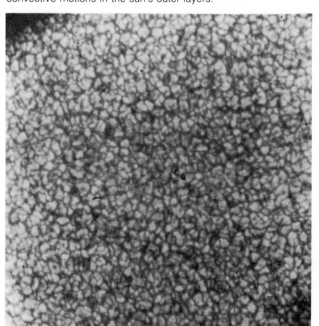

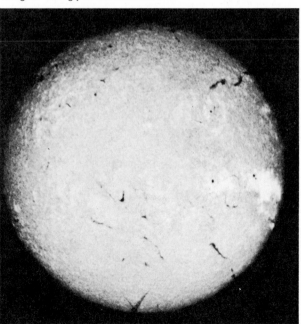

FIGURE 11.8. *Spicules.*
This photo shows the chromosphere's transient features known as spicules. These spikes of glowing gas, which are apparently shaped by the sun's magnetic field, come and go irregularly.

FIGURE 11.9. *The Corona.*
This photo, obtained during a total solar eclipse, shows the type of structure commonly seen in the sun's corona. There are giant looplike features, and an overall appearance of outward streaming. Bits of more intense light from the sun's chromosphere are seen around the edges of the moon's occulting disk. Visible at left is the planet Venus.

great geysers of hot gas that spurt upward from the surface of the sun, taking on an arc-shaped appearance. These are usually associated with sunspots, and both phenomena are linked to the solar activity cycle, to be discussed shortly.

FIGURE 11.10. *An X-Ray Portrait of the Corona.*
This image, obtained by *Skylab* astronauts, shows the sun in X-ray light, which reveals only very hot regions. The bright regions are places where the corona is especially dense, and the dark regions, known as coronal holes, are places where it is much more rarefied. The structure seen here changes with time.

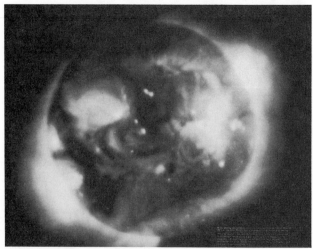

regular in form, patchy near the sun's surface, but with radial streaks at great heights, suggestive of outflow from the sun (Fig. 11.9). The density of the coronal gas is very low, only about 10^9 particles per cm^3. As we mentioned earlier, the corona is very hot, containing highly ionized gas. The source of the energy that heats the corona to such extreme temperatures is not well understood, although a general picture has emerged.

X-ray observations reveal that the corona is not uniform, but instead has a patchy structure (Fig. 11.10). There are large regions that appear dark in an X-ray photograph of the sun; in these areas the gas density is even lower than in the rest of the corona. These **coronal holes,** as such regions are called, are probably created and maintained by the sun's magnetic field. The coronal holes as well as the overall shape of the corona vary with time (Fig. 11.11), showing that the corona is general is a dynamic, active region. Another impressive sign of the corona's dynamic nature is the existence of **prominences** (Figs. 11.12 and 11.13),

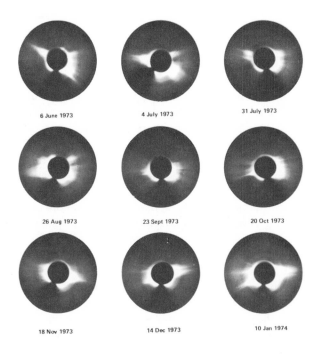

FIGURE 11.11. *Changes in the Corona.*
This series of photos was obtained through the use of a special shutter device that blocked out the solar disk, allowing the corona to be observed without a total solar eclipse. Here we see that the structure of the corona changes quite markedly over a period of a few months.

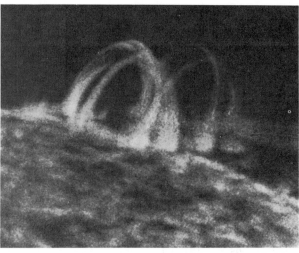

FIGURE 11.12. *A Prominence.*
Here the looplike structures thought to be governed by the sun's magnetic field are readily seen. This photo was obtained by *Skylab* astronauts, using an ultraviolet filter.

The hot outer layers of the sun have provided astronomers with a second major mystery regarding the solar energy budget. In contrast with the mystery of the sun's internal energy source, which has been solved, the mystery concerning the mechanism for heating the chromosphere and corona remains unsolved. It is generally accepted that the heating must come from the boiling and churning of the outer convective layer of the solar interior, but it is not clear how this activity is translated into heat at great distances above the photo-

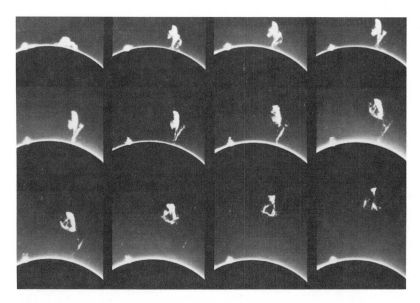

FIGURE 11.13. *An Eruptive Prominence.*
This striking sequence shows an outburst of ionized gas from a prominence on the sun's limb. The photos were obtained through the use of a special device that blocked out the sun's disk.

sphere. For awhile it was assumed that sound waves were responsible, but today it is thought that various kinds of magnetic waves are responsible. There is certainly enough energy in the convective motions in and below the photosphere to account for the heating; the problem is one of understanding how this energy is transported into the higher levels.

Recent satellite observations in ultraviolet and X-ray wavelengths have shown that stars similar to the sun also have chromospheric and coronal zones. Thus if we can understand how the sun operates, we will also gain a deeper understanding of how other stars work.

The Solar Wind

The long, streaming tail of a comet always points away from the sun, regardless of the direction of the comet's motion. The significance of this was fully realized in the late 1950s, when the first U.S. satellites revealed the presence of the earth's radiation belts and the fact that they are shaped in part by a steady flow of charged particles from the sun. The solar wind reaches a speed near the earth of 300 to 400 kilometers per second, and apparently persists to a distance beyond the orbit of Neptune (the *Pioneer 10* spacecraft, on its way out of the solar system, still detected solar wind particles as it crossed the orbit of Neptune in mid–1983). It is this flow of charged particles that forces cometary tails to always point away from the sun.

The existence of the solar wind is evidently a natural by-product of the same heating mechanisms that produce the hot corona of the sun. It was originally thought that particles in this high-temperature region move about with such great velocities that a steady trickle escapes the sun's gravity, flowing outward into space. However, X-ray observations of the sun have shown that the situation is not that simple. It is the solar magnetic field that governs the outward flow of charged particles. The coronal holes, mentioned earlier, are regions where the magnetic-field lines open out into space. Charged particles such as electrons and protons, constrained by electromagnetic forces to follow the magnetic-field lines, therefore escape into space only from the coronal holes. The speed of the solar wind is relatively low close to the sun, but accelerates outward (quickly reaching the velocity of 300 to

400 km/sec already mentioned), after which it is nearly constant. The wind nearly reaches its maximum velocity by the time it passes the earth's orbit, and beyond there it flows steadily outward. It is thought that at some point in the outer solar system, the wind comes to an abrupt halt where it runs into an invisible and tenuous wall of matter swept up from the interstellar medium that surrounds the sun. Perhaps *Pioneer 10* will eventually discover where this boundary is.

Most of the direct information we have on the solar wind comes from satellite and space-probe observations, because the earth's magnetosphere shields us from the wind particles. Solar-wind monitors are placed on board most spacecraft sent to the planets. One striking discovery has been the fact that the wind is not uniform in density, but instead seems to flow outward from the sun in sectors, as though it originates only from certain areas on the sun's surface. This is explained by the X-ray data already mentioned, which indicate that the wind emanates only from the coronal holes. Because the base of the wind is rotating with the sun, the wind sweeps out through space in a great curve, similar to the trajectory of water from a rotating lawn sprinkler (Fig. 11.14).

Occasional explosive activity occurring on the sun's surface releases unusual quantities of charged parti-

FIGURE 11.14. *The Solar Wind.*
This schematic diagram illustrates how ionized gas from the sun spirals outward through the solar system in a steady stream. The solar magnetic field creates sectors of variable density in the wind.

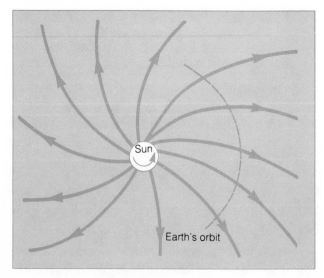

cles, and some three or four days later, when this burst of ions reaches the earth's orbit, we experience disturbances in the ionosphere that can interrupt shortwave radio communications and cause auroral displays. These magnetic storms, as they are often called, are outward manifestations of a much more complex overall interaction between the sun and the earth. At the end of the next section we will discuss the so-called solar-terrestrial relations.

Sunspots, Solar Activity Cycles, and the Magnetic Field

The dark spots on the sun's disk were observed more than three hundred years ago, and were cited by Galileo as evidence that the sun is not a perfect, unchanging celestial object, but rather has occasional flaws. During the centuries since then, observations of the spots (which individually may last for months), have revealed some very systematic behavior. The number of spots varies, reaching a peak every eleven years, and during the interval the spots move steadily from the sun's middle latitudes toward its equator. At the beginning of a cycle, when there is a maximum number of sunspots, most of them appear in activity bands about 30 degrees north or south of the solar equator. During the next eleven years, the spots tend to lie ever closer to the equator, and by the end of the cycle they are

nearly on it. By this time the first spots of the next cycle may already be forming at middle latitutde. A plot of sunspot locations during a cycle clearly shows this effect, and is called a **butterfly diagram** because of the shape the pattern of spots makes (Fig. 11.15).

The sunspots are not totally black, although they appear so when seen against the bright background of the photosphere (Fig. 11.16). Actually they glow rather intensely, but because they are somewhat cooler then their surroundings, they are not as bright. The typical temperature in a spot is about 4,000 K, compared with the roughly 6,000 K temperature of the photosphere. Using Stefan's law, we see that the intensity of light emitted in a spot compared with the surroundings is $(4,000/6,000)^4 = 0.2$; that is, the brightness of the solar surface within a sunspot is only about one-fifth of the brightness in the photosphere outside.

A hint as to the origin of the spots was found when their magnetic properties were first measured. Astronomers accomplished this by using spectroscopy of the light from the spots. The energy levels in certain atoms are distorted by the presence of a magnetic field, which in turn causes the spectral lines formed by those levels to be split into two or more distinct, closely spaced lines. The degree of line-splitting, which is referred to as the **Zeeman effect,** depends on the strength of the magnetic field. Thus the field can be measured from afar by analyzing the spectral lines to see how widely split they are. This technique works for distant stars as well as for the sun.

When the initial measurements of the sun's mag-

FIGURE 11.15. *The Butterfly Diagram.* This is a plot of the latitudes of observed sunspots through several cycles of solar activity. During each cycle, the spots gradually shift their favored locations closer and closer to the solar equator.

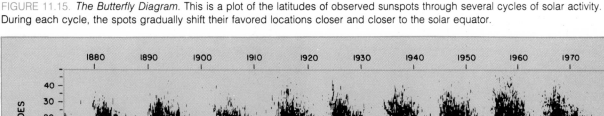

netic field were made in the first decade of this century, it was found that the field is especially intense in the sunspots, about a thousand times stronger than in the surrounding gas. A strong magnetic field creates pressure in a gas, just as high temperature does. Thus a sunspot maintains a balance with its hotter surroundings as the magnetic pressure within the spot counteracts the pressure of the hotter gas around it.

When the magnetic fields of sunspots were measured, they were found to act like either north or south magnetic poles; that is, each spot has a specific magnetic direction associated with it. Furthermore, pairs of spots often appear together, the two members of a pair usually having opposite magnetic polarities. During a given eleven-year cycle, in every sunspot pair the magnetic polarities always have the same orientation. For example, throughout one eleven-year cycle, the spot to the east in each pair will have a north magnetic polarity, and the one to the west a south magnetic polarity. During the next cycle, the polarity of all the pairs will reverse, with the south magnetic spot to the east, and the north magnetic spot to the west. Between cycles, when this arrangement is reversing itself, the sun's overall magnetic field also reverses, with the solar magnetic poles changing places. It is actually twenty-two years before the sun's magnetic field and sunspot patterns repeat themselves, so the solar magnetic cycle is truly twenty-two years long.

Sunspot groups are the scenes of the most violent forms of solar activity, the **solar flares.** These are gigantic outbursts of charged particles, as well as visible, ultraviolet, and X-ray emission, created when extremely hot gas spouts upward from the surface of the sun (Fig. 11.17). Flares are most common during sunspot maximum, when the greatest density of spots is to be seen on the solar surface. Close examination of flare events shows that the trajectory of the ejected gas is shaped by the magnetic lines of force emanating from the spot where the flare occurs. Charged particles flow outward from a flare, some of them escaping into the solar wind. If the flare occurs where the solar wind that hits the earth arises, then the earth gets an extra dosage of solar wind particles several days later, affecting radio communications. The extra quantity of charged particles entering the earth's upper atmosphere also can cause unusually widespread and brilliant displays of aurorae. Apparently flares occur when twisted magnetic-field lines suddenly reorganize themselves, releasing heat energy and allowing huge bursts of charged particles to escape into space.

The combination of all these bits of data on sunspots, magnetic fields, and solar activity cycles has led to the development of a complex theoretical picture, one that successfully accounts for many of the observed

FIGURE 11.17. *A Major Flare.*
The gigantic looplike structure in this ultraviolet photo obtained from *Skylab* is one of the most energetic flares ever observed. Supergranulation in the chromosphere is visible over most of the disk.

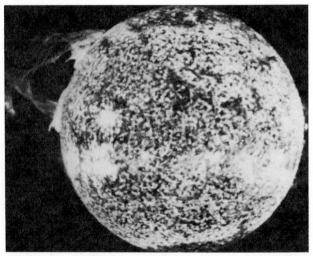

FIGURE 11.16. *Sunspots.*
This is a telescopic view of a group of spots, showing their detailed structure. They appear dark only in comparison with their much hotter surroundings.

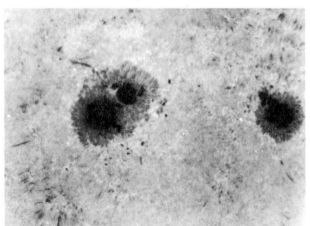

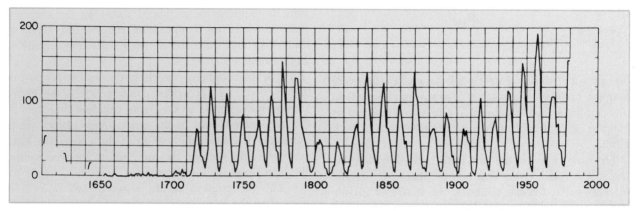

FIGURE 11.18. *The Sun's Long-Term Activity*. This diagram illustrates the relative level of activity (in terms of sunspot numbers) over three centuries. It is clear that the level varies. Note that there is a span of about fifty years (1650–1700), known as the Maunder minimum, when there was little activity. Evidence shows that the sun has long-term cycles that modulate the well-known 22-year period.

phenomena. This theory envisions ropes or tubes of magnetic-field lines inside the sun, connecting its north and south magnetic poles. At some places these tubes become kinked, and loops break through the surface, creating pairs of sunspots with opposing magnetic polarities where they emerge and reenter. Early in the sunspot cycle the magnetic tubes break through the surface at middle latitudes, but later, as the sun's magnetic field is moving toward reversal of the poles, they do so near the equator. This accounts for the latitude-dependence of the spots during a cycle, as shown in the butterfly diagram. When the solar magnetic field reverses itself every eleven years, so do these magnetic ropes; thus when the new cycle begins, the sunspot pairs have their polarities reversed compared with the pairs of the previous cycle.

Records of sunspot counts maintained over centuries have allowed astronomers to study the long-term behavior of the sun. It has been found that the 22-year cycle is not perfectly repeatable, but that there have been longer-term variations in solar activity (Fig. 11.18). Most striking was a prolonged period in the late 1600s when the cycle seemed to stop, with no marked periods of sunspot maxima. This epoch of reduced sunspot activity has been called the **Maunder minimum,** after its discoverer, E. W. Maunder (also the originator of the butterfly diagram). Only time will tell whether the Maunder minimum was part of some much longer-period cycle in the sun's behavior.

The origin of the sun's magnetic field, and its periodic pole reversals, is probably a dynamo, similar in nature to those thought to be at work in the interiors of the planets that have fields. Unlike the terrestrial planets, however, the sun is not rigid, and consequently differential rotation may play a role in creating the instability that causes the dynamo to reverse itself regularly every eleven years. From the sun's behavior we might speculate that the magnetic fields of Jupiter and Saturn, which also rotate differentially, reverse themselves from time to time as well.

The solar activity cycle may have some indirect effects on the climate of the earth. Eleven-year patterns in the occurrence of droughts have been reported, and it seems possible that such patterns are related to the solar cycle, although it is not known how. During the time of the Maunder minimum, the earth's climate was in chaos, with terrible droughts in many areas and particularly severe winters in Europe and North America.

There are other aspects of the relationship between solar activity and the earth's atmosphere. As the solar wind fluctuates in intensity, the earth's magnetosphere varies in extent. We have already noted that solar flares, which occur most often during sunspot maximum, have a significant effect on the earth's ionosphere. The chemistry of the upper atmosphere may also be strongly influenced by variations in the solar ultraviolet emission, which in turn are linked to the solar activity cycle. Solar-terrestrial relations is an important topic; one that merits and is receiving greater attention.

PERSPECTIVE

Our sun, the source of nearly all our energy, is a very complex body. As stars go, it is apparently normal in all respects, so we imagine that other stars are just as complex, even though we cannot observe them in such detail.

We have explored the sun, both in its deep interior and in the outer layers that can be observed directly. We know that nuclear fusion is the source of all the energy, and that the size and shape of the sun are controlled by the balance between gravity and pressure throughout the interior. Somehow heat is transported above the surface, keeping the chromosphere and the corona hotter than the photosphere. Perhaps most intriguing of all is the solar activity cycle and its relationship to the sun's complex magnetic field.

With this examination of the sun, we have completed our survey of the solar system. We are ready to tie together all the diverse pieces of information that we have discussed, and to develop a coherent picture of the system as a whole and the manner in which it formed.

SUMMARY

1. The sun is an ordinary star, one of billions in the galaxy.

2. The sun is gaseous throughout, and hydrostatic equilibrium causes it to be very hot and dense in the core.

3. Energy is produced in the core by nuclear fusion and is slowly transported outward by radiation, except near the surface, where energy is transported by convection.

4. The nuclear reaction that powers the sun is the proton-proton chain, in which hydrogen is fused into helium.

5. Observations of the sun through filters designed to isolate various wavelengths allow us to see distinct levels of the sun's outer layers, each characterized by a different temperature.

6. The outer layers of the sun are the photosphere,

with a temperature of about 6,000 K; the chromosphere, where the temperature ranges from 6,000 to 10,000 K; and the corona, where the temperature is higher than 1,000,000 K.

7. The photosphere, which is the visible surface of the sun, creates absorption lines, whereas the hotter chromosphere and corona create emission lines.

8. The excess heat in the outer layers of the sun is somehow transported there from the convective zone just below the surface; the mechanism for transporting the heat is not well understood.

9. The sun emits a steady outward flow of ionized gas called the solar wind. The wind originates in coronal holes, and is therefore controlled by the solar magnetic field.

10. Sunspots occur in 11-year cycles, which are a reflection of the 22-year cycle of the sun's magnetic field. The spots are regions of intense magnetic fields where flux tubes from the solar interior break through the surface.

11. The solar activity cycle may have important effects on the earth's climate.

REVIEW QUESTIONS

1. Why is lead solid, whereas the sun, many times denser at its core than lead, is gaseous?

2. Why do nuclear reactions occur only in the innermost core of the sun? Explain in terms of hydrostatic equilibrium.

3. Explain why the use of special filters to isolate the wavelengths of certain spectral lines allows us to separately examine distinct layers of the sun.

4. Why are the granules in the sun's photosphere bright, and the descending gas that surrounds them relatively dark?

5. Use Kirchhoff's laws to explain why the photosphere produces absorption lines, but the chromosphere and corona produce emission lines.

6. What would we conclude about the temperature just below the outermost layers of the photosphere if the sun's disk appeared brighter, instead of darker, at the edges?

7. Why does it take about three days for the effects of a solar flare to begin to occur in the earth's magnetosphere?

8. How would the appearance of sunspots be altered if they had weaker magnetic fields than their surroundings?

9. In previous chapters we discussed the effects of solar wind on the planets. Summarize these effects.

10. Describe how energy produced in the sun's core by nuclear reactions heats the chromosphere and corona. In other words, describe how this energy gets from the core to the outer layers, step by step.

ADDITIONAL READINGS

Eddy, J. A. 1977. The case of the missing neutrinos. *Scientific American* 236(5):80.

Gibbon, E. G. 1973. *The quiet sun.* NASA Publication NAS 1.21. Washington, D.C.: NASA.

Gough, D. 1976. The shivering sun opens its heart. *New Scientist* 70:590.

Newkirk, G., and Frazier, K. 1982. The solar cycle. *Physics Today* 35(4):25.

Parker, E. N. 1975. The sun. *Scientific American* 233(3):42.

Wallenhorst, S. G. 1982. Sunspot numbers and solar cycles. *Sky and Telescope* 64(3):234.

Walker, A. B. C. Jr. 1982. A golden age for solar physics. *Physics Today* 35(11):61.

Wilson, O. C., Vaughan, A. H., and Mihalas, D. 1981. The activity cycle of stars. *Scientific American* 244(2):104.

Wolfson, R. 1983. The active solar corona. *Scientific American* 248(2):104.

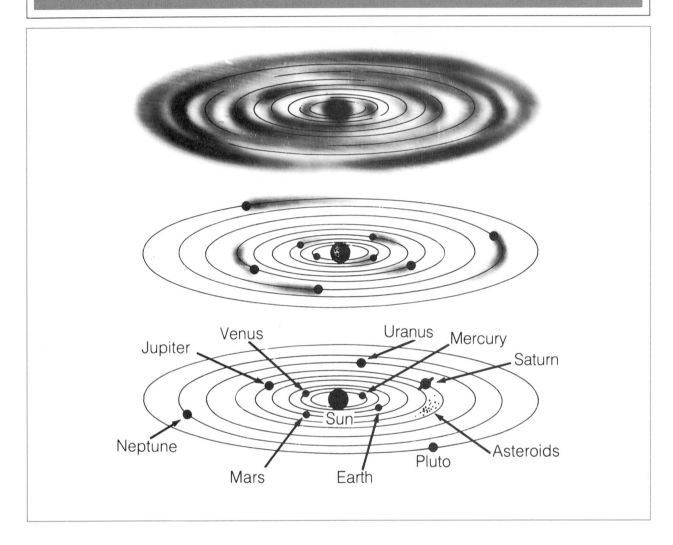

Jupiter Venus Uranus Mercury Saturn

Sun

Neptune Asteroids

Mars Earth Pluto

Learning Goals

12

We have collected a considerable quantity of information about the solar system and its contents. Along the way, we have uncovered various systematic trends, such as the similarities in internal structure and composition among the terrestrial planets, and the similarities among the outer planets, but so far we have said very little about how all of this arose. In discussing the histories of the individual planets, we have referred to some of the important processes thought to have been involved, such as the condensation of the first solid material from a cloud of gas surrounding the young sun, but it remains for us to put together a coherent story of the formation and evolution of the entire system.

Before we describe the modern theory, we will review the overall properties of the solar system that must be accounted for, and then we will have a look at the historical developments and alternative suggestions that scientists have considered over the years.

A Summary of the Evidence

Any successful theory regarding the formation of the solar system must be able to explain a number of facts. In a way this is very much like a mystery story, where the detective (the scientist seeking the correct explanation) has certain clues that reveal isolated parts of the story, from which past events must be reconstructed.

One category of clues in our mystery has to do with the orbital and spin motions of the planets and satellites in the solar system (Figs. 12.1 and 12.2). All the planetary orbits lie in a common plane, all are nearly circular, and all go around the sun in the same direction. Pluto violates these rules to an extent, because its orbit is both elongated and tilted (by 17 degrees) with respect to the ecliptic, but even this misfit goes around the sun in the same direction as the other planets. The spins of nearly all the planets are in this same direction, as are the orbital and spin motions of most of the satellites in the solar system. This motion, which is in the counterclockwise direction when viewed from above the north pole, is said to be **direct,** or **prograde,** motion.

Although we do not believe in any physical basis for Bode's law, we still need to understand what caused the planets to form where they did, and not elsewhere. Another hint having to do with planetary motions is that almost all the planets have small **obliquities,** meaning that their equatorial planes are nearly aligned with the ecliptic plane. Even the sun has its axis of rotation nearly perpendicular to the ecliptic. Only Uranus and Pluto do not fit the pattern.

FIGURE 12.1. *The Co-Alignment of Planetary Orbits.* All nine planets have orbits that are nearly circular and whose planes are nearly parallel. An exception is Pluto, whose orbit is tilted 17 degrees with respect to the ecliptic, and is sufficiently eccentric (that is, elongated) that Pluto is actually closer to the sun than Neptune at times.

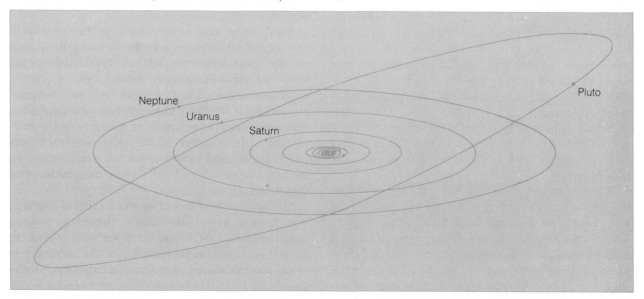

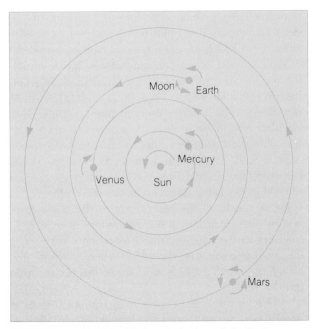

FIGURE 12.2. *Prograde Motions in the Inner Solar System.* The rotations and orbital motions of the planets are generally in the same direction. This sketch of the orbits of the inner four planets shows that Venus is an exception, with retrograde rotation, but that the other three planets and their satellites obey the rule. All the outer planets and nearly all their satellites (except for the Pluto-Charon system) do also.

A second type of clue in the mystery has to do with the natures of the individual planets as well as the systematic trends among the planets. First, we must consider the planetary compositions and seek an explanation for the fact that the inner four, the terrestrial planets, have high densities, indicating that they are made mostly of rock and metallic elements; whereas the outer planets consist primarily of lightweight gases such as hydrogen and helium, as well as ices composed of these and similarly volatile gases. Within each of the two groupings of planets, there are relatively minor differences in chemical composition that must also be accounted for, such as the various isotope ratios we have discussed.

A major mystery for a long time was the slow rotation of the sun. According to many early theories, the sun should be spinning much faster. The difficulty arises in theories that envision the solar system forming from the collapse of a cloud of gas and dust, and these theories, as we shall soon see, are the most successful ones. The laws of physics tell us that the angular mo-

mentum of an object (see chapter 2) must remain constant. This means that if a spinning object shrinks in size, it must spin faster to compensate, thereby maintaining constant angular momentum. Thus the sun, thought to have formed at the center of a collapsing cloud, should have a very rapid rotation rate, rather than the leisurely 25-day period it has. The explanation for this was a long time coming, and for awhile astronomers were side-tracked in their effort to understand the formation of the solar system.

A final category of information that bears on the origin of the solar system is the distribution and nature of the various kinds of interplanetary objects that were described in chapter 10. We learned that asteroids, comets, meteoroids, and interplanetary dust inhabit the space between the planets, and we found that the motions and compositions of these objects reflect conditions that existed very early in the history of the solar system. Somehow this information must fit into our overall picture.

Catastrophe or Evolution?

In discussing theories of the formation of the solar system, we will confine ourselves to those which were suggested after the time of Copernicus, when it was established that the planets orbit the sun. Several primitive ideas that were based on the geocentric notion were outlined in chapter 1.

The first serious theory based on the heliocentric view of the solar system was developed by the French scientist René Descartes, who in 1644 advanced the idea that the solar system formed from a gigantic whirlpool, or vortex, in a universal fluid, with the planets and their satellites forming from smaller eddies. This theory was rather crude, without any clearly specified idea of the nature of the cosmic substance from which the sun and planets arose, but it did account for the fact that all the orbital motions are in the same direction.

The hypothesis of Descartes was the first of a general type known as **evolutionary theories,** theories which hold that the formation of the solar system occurred as a natural by-product of the sequence of events that produced the sun. Evolutionary theories state that no special circumstances were needed to create the planets, other than the fact that the sun formed. Thus there may be planets orbiting other stars.

Immanuel Kant further elaborated on Descartes's idea. In 1755 Kant applied the recently discovered Newtonian mechanics to the problem, and was able to show that a rotating gas cloud would flatten into a disk as it contracted (Fig. 12.3). Kant's theory was called the **nebular hypothesis,** because it invoked formation of the sun and planets from an interstellar cloud, or nebula. In 1796 Pierre Simon de Laplace, a French mathematician, added the notion that as the spinning cloud flattened into a disk, concentric rings of material broke off because of rotational forces, so that at one point the early solar system would have looked very much like the planet Saturn with its rings (Fig. 12.3). Each ring was supposed to have then condensed into a planet. Soon the problem of accounting for the slow solar rotation was recognized, and further development of the evolutionary theories was stymied.

FIGURE 12.3. *The Hypotheses of Kant and Laplace.* Descartes's simple vision of a vortex was refined by Kant, who realized that rotation should cause a collapsing cloud to take on a disklike shape; and by Laplace, who hypothesized that a rotating disk would form detached rings that could then condense into planets.

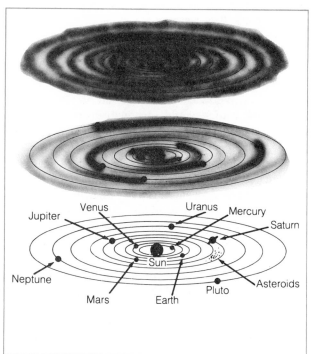

Meanwhile an alternative group of ideas, called **catastrophic theories,** had been suggested. These concepts of the formation of the planetary system envision a solitary sun to begin with, but then invoke some singular, cataclysmic event that disrupted the sun and formed the planets. The first of these ideas was put forth by another Frenchman, Georges Louis de Buffon, who in 1745 suggested that a massive body (which he referred to as a comet, although it was much larger than any cometary nucleus) passed so near the sun that its gravitational pull forced material out of it, this gas then condensing to form the planets (Fig. 12.4).

Buffon's idea was largely ignored until the beginning of the twentieth century, when the difficulty with the sun's slow spin forced scientists to consider alternatives to the evolutionary theories. By 1905, the English astronomers T. C. Chamberlain and F. R. Moulton had suggested an elaboration of Buffon's original idea, in which another star was the object that passed near the sun, creating tidal forces that caused matter from inside the sun to stream out, thereafter condensing into planets.

In modern times, the idea of mass being transferred from one star to another in close double-star systems has been discovered to be very important under certain circumstances (see chapter 14), so the idea of gravitational forces extracting matter from the sun is not out of the question. Further refinements to the bypassing-star hypothesis were soon provided by the British astrophysicists H. Jeffreys and J. H. Jeans.

A disturbing problem with the catastrophic theories was the low probability that the sun could have collided with another star, given the great distances be-

FIGURE 12.4. *The Catastrophe Hypothesis.* Here a by-passing body gravitationally forces material out of the sun; this matter was thought to subsequently form planets.

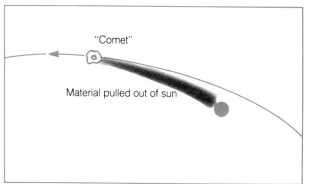

tween stars in our part of the galaxy. Of course, this did not rule out the possibility, for even a low-probability event can occur, but it was sufficiently worrisome to lead to the development of an alternative catastrophic theory, one in which the sun started out as a member of a triple-star system, but was then freed from the gravitational bonds of the other two objects when the three nearly collided. In the process of expelling the sun from the triple-star system, the gravity of the other stars tore some of the solar material out, forming a tail, which later condensed to form the planets. (More recently, a similar three-body interaction has been suggested to explain the strange orbits of Neptune's two satellites and the expulsion of Pluto into its own solar orbit; see chapter 9.) This theory, developed by H. N. Russell, R. A. Lyttleton, and F. Hoyle, replaced the difficulty of forming the sun and the planets together with one of forming the sun and two other stars together. The advantage was that the slow rotation of the sun did not have to be produced during the sun's formation, but could have developed later, when the sun nearly collided with the other stars and was ejected.

By the late 1930s, the catastrophic theories were becoming quite unwieldy, not only because of the special circumstances that had to be assumed, but also because of several fundamental problems. For example, calculations showed that even if some material were pulled out of the sun's interior, it would have so much internal energy (it would be so hot) that it would expand and dissipate into space, rather than condense to form planets. Another difficulty was how to pull material out of the sun, giving the tenuous, extended matter sufficient angular momentum to form the planets, but without also giving it enough speed to escape entirely.

A more modern, but quite definitive, objection to the catastrophic theories has to do with the abundance of the hydrogen isotope deuterium. Deuterium is a form of hydrogen whose nucleus contains a proton and a neutron, in contrast with the normal type of hydrogen whose nucleus consists solely of a proton. It happens that deuterium is destroyed in nuclear reactions, which will inevitably occur if the gas containing deuterium is subjected to high enough temperatures, such as would be the case inside a star. Even in the sun's outer layers, the temperatures are high enough to destroy deuterium. Hence, if the solar system were formed of material pulled out of the sun, we would expect to find no

deuterium in the planets and interplanetary bodies. To the contrary, deuterium is found to be a normal constituent of the solar system, which means it must have been present in the primordial material from which the system formed (in chapter 23 we discuss the origin of deuterium in the universe), and that this material was never part of the sun. This emphatically rules out all catastrophic theories that suggest the planets formed from gas pulled out of the sun or any other star (except possibly a very cool one).

In the 1940s a new theory was proposed by the Soviet astronomer O. Y. Schmidt, and was developed in some detail by H. Alfven, the Swedish astrophysicist who was later to win a Nobel prize for his work on the behavior of ionized gas in the solar magnetic field. The idea was that once the sun had formed, its gravitational field trapped material from the surrounding interstellar medium, which then formed into planets. This general idea, called the **accretion theory** of solar-system formation, never gained much popularity, because the problems with the evolutionary theory began to be solved at about the time the accretion theory was being developed. Most research efforts after this time centered on the evolutionary theory.

Progress was being made in the 1940s on the general question of how a contracting cloud could flatten into a disk and then break up into eddies that could form planets. The German physicist C. F. von Weizsäcker showed that the disk would tend to rotate differentially, meaning that the inner parts would orbit the center faster than the outer regions, and that this would cause the disk to break up into eddies (Fig. 12.5). The work of von Weizsäcker even showed how the relative sizes of the eddies would vary with distance from the center, accounting for both the relative distances of the planets from the sun and their varying sizes from one to the next. More detailed analyses of the stability of rotating disks were later carried out by a number of other scientists, who showed that rather than the rotating disk forming regular eddies, as envisioned by von Weizsäcker, the disk would become clumpy, forming localized regions where the density was high enough to allow material to fall together gravitationally. This process would lead to the disk being broken up into a number of small solid bodies, which in turn could later merge to form the planets.

The breakthrough in understanding the sun's slow rotation came in the early 1960s, with the discovery of

Astronomical Insight 12.1

A DIFFERENT KIND OF CATASTROPHE

The catastrophic theories described in the text have as a common theme the formation of the planets from matter pulled out of the sun by a close encounter with another body. In the more modern versions of these theories, it is the differential gravitational force exerted by the other body that distorts the sun enough to cause some of its substance to break off. Consequently, these theories are often called tidal theories. A radically different tidal theory was proposed in 1960.

The new tidal theory still invoked the disruption of one star by the gravitational attraction of another, but now it was supposed that the sun disrupted a neighbor, rather than the reverse. The premise is that the sun formed as a member of a cluster of stars, and that after it had condensed to its present size, it passed near another cluster member that was just in the process of forming. This embryonic star was still rather extended and cool, so it was easily torn apart by the sun's tidal force, and the material pulled out of it was not too hot to condense into planets.

This idea, first proposed by M. Woolfson, has not been developed as completely as some other theories of the origin of the solar system, and may have faults that are as yet unknown. It does overcome the principal objections to the other catastrophic theories, however. It has failed to gain widespread favor partly because of the success of evolutionary theories.

One interesting aspect of Woolfson's hypothesis is the suggestion that the sun formed as part of a cluster. This was proposed for two reasons: it makes a near-collision between stars more likely; and it provides an opportunity for the young sun to be close to newly forming stars, creating the circumstances necessary for the rest of the scenario to be played out.

There is, however, no evidence that the sun was ever part of a cluster of stars. On the other hand, it is possible that it was (some astronomers even claim that *all* stars form in clusters). A cluster of stars will naturally disperse, given enough time, partly because of the random motions of the stars within it and partly because of the stretching and distortion it may suffer as it orbits the galaxy, occasionally passing through the galaxy's spiral arms. The sun's age is roughly 4.5 billion years, which is sufficient time for any cluster to be dissipated.

Although the Woolfson theory may have few obvious flaws, it is probably doomed to be largely ignored unless some serious problem is found with the evolutionary theory. The principle of Occam's razor prevails, dictating that the simplest solution is most likely the correct one.

the solar wind. Because the sun continuously expels charged particles, interplanetary space is filled with them, and the sun's magnetic field tries to pull them along with it as the sun rotates (Fig. 12.6). This creates a constant drag on the sun (imagine trying to spin a pinwheel under water). This drag, given long enough to act, slows the sun's spin. If the sun was born with a rapid rotation, this **magnetic-braking** process could easily have slowed it to the present rate in the billions of years that have passed. Most likely the density of charged particles in space around the sun was much greater in the early days of the solar system, so the drag created as the sun's magnetic field tugged on the particles was probably greater than it is now, and the braking could have taken place in a relatively short time.

With the understanding of the magnetic-braking process came an understanding of how the solar system formed, and all that remained was further refinement of the idea. There are still many unanswered questions, but the modern theory of solar-system formation is probably correct, as described in the next section.

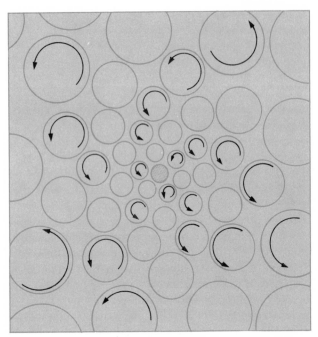

FIGURE 12.5. *Von Weiszäcker's Theory of a Turbulent Disk.* This sketch shows how a rotating disk would form eddies, according to the theory of the German scientist C. F. von Weiszäcker.

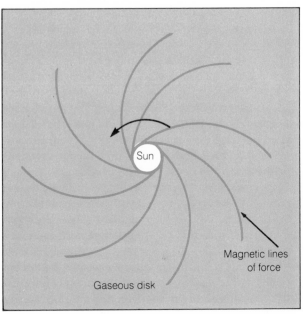

FIGURE 12.6. *Magnetic Braking.* The young, rapidly spinning sun has a magnetic-field structure as shown, the field lines rotating with the sun. The early solar system was permeated with large quantities of debris and gas, some of which was ionized and therefore subject to electromagnetic forces. The sun's magnetic field therefore exerted a force on the surrounding gas, which in turn created drag that slowed the sun's rotation.

A Modern Scenario

The first step in the formation of the solar system was the collapse of an interstellar cloud, one that must have been rotating before it began to fall in on itself. Our galaxy has a large quantity of interstellar material containing both gas and small dust grains (about the size of the interplanetary grains described in chapter 10). The interstellar medium is not uniformly distributed throughout space, but instead is concentrated here and there in large amorphous regions called interstellar clouds. Some of these clouds are sufficiently dense to block out starlight, and they appear as dark regions in photographs of the sky (Fig. 12.7). Even in these areas of relatively high concentration, the material is so rarefied that a quantity of mass equal to that of the sun is spread over a volume up to ten or more light-years across.

The composition of the interstellar material is apparently quite uniform, consisting of about the same mixture of elements as the sun. Thus the sun and other stars are born with a standard composition of almost 80 percent hydrogen (by mass), more than 20 percent helium, and just traces of all the other elements. The planets must also have formed out of material with this composition, yet, as we have seen, their present makeup is quite different from this, especially in the case of the terrestrial planets.

Somehow the extended, tenuous gas in interstellar space had to become concentrated in a very small volume on its way to becoming a star. Gravitational forces were responsible for pulling the cloud together, but there must have been an initial push or a chance condensation to get the process started (see chapter 15 for details). Once the cloud was falling in on itself, the rest of the process leading to the formation of the sun was inevitable, dictated by physical laws.

A combination of theory and observation tells us how the collapse proceeded. The innermost portion of the cloud fell in on itself very quickly, leaving much of the outer material still suspended about the center. The rotation of the cloud sped up as its size diminished, and if the cloud had a magnetic field to begin with

FIGURE 12.7. *A Region of Star Formation.*
The dark patches in this portion of the constellation Ophiuchus are dense interstellar clouds, the kind of environment where cloud collapse and star formation are taking place.

(most do), the field was intensified in the central part as a result of the condensation.

The core of the cloud began to heat up, owing to the energy of impact as the material fell in. It eventually began to glow, first at infrared wavelengths, and finally, after a prolonged period of gradual shrinking, at visible wavelengths. Nuclear reactions began in the center when the temperature and pressure was sufficiently high, and the sun began its long lifetime as a star, powered by these reactions.

The steps described so far regarding the formation of the sun are fairly well understood, and have been observed to be taking place today in many cloudy regions of the galaxy. In a number of stellar nurseries, infrared sources are found embedded inside dark interstellar clouds, indicating that newly formed stars are hidden there, still heating up. The details of planet formation are a little sketchier, however, because we have no way to observe the process as it occurs. All that we know about it has had to be inferred from observations of the end product: the present-day solar system.

As the central part of the interstellar cloud collapsed to form the sun, the outer portions were forced into a disk shape by rotation, just as Kant had shown. At this stage there was an embryonic sun surrounded by a flattened, rotating cloud called the **solar nebula** (Fig. 12.8). The inner portions of the nebula were hot, but the outer regions were quite cold.

Throughout the solar nebula, the first solid particles began to form, probably by the growth of the interstellar grains that were mixed in with the gas. For every element or compound, there is a combination of temperature and pressure at which it "freezes out" of the gaseous form (in direct analogy with the formation of frost on a cold night on earth). The volatile gases require a very low temperature in order to condense, so these materials tended to stay in gaseous form in the inner portions of the solar nebula, but condensed to form ices in the outer portions. The elements that condense easily, even at high temperatures, are called **refractory elements,** and these were the ones that formed the first solid material in the warm inner portions of the solar nebula. This material therefore consisted of rocky debris containing only low abundances of the volatile species.

In due course, rather substantial objects built up, resembling asteroids in size and composition; these are referred to as **planetesimals.** By this time, gravitational instabilities had probably caused the solar nebula to build up eddies or condensations in some regions. Where these occurred, planetesimals had a greater chance of colliding, and in time these objects began to coalesce, forming the larger masses that became planets. In most cases, the planet that was formed rotated in the same direction as the overall rotation of the disk, but in two cases, Venus and Uranus, something happened to change the direction of rotation. Perhaps an unusually large planetesimal collided with each at some point, altering the natural direction of spin.

The scenario just described apparently applies to the terrestrial planets, which therefore seem to have formed in two stages: (1) the condensation of refractory elements, leading to the development of planetesimals; and (2) the accretion of the planetesimals to form planets. The low quantity of volatile elements that characterizes the terrestrial planets was already established when the planets formed, and then was exaggerated by the release of volatile gases that occurred during the planets' early histories, when they underwent molten periods.

There is some uncertainty about the sequence of events that led to the formation of the outer planets.

FIGURE 12.8. *Steps in the Sun's Formation.* An interstellar cloud, initially very extended and rotating very slowly, collapses under its own gravitation. This happens most quickly at the center. As the collapse occurs, the internal temperature rises and the rotation rate increases. Eventually the central condensation becomes hot and dense enough to be a star, with nuclear reactions in its core.

These planets contain a much higher proportion of volatile gases, which is to be expected, given the lower temperatures in the outer portions of the solar nebula. Thus, planetesimals that formed there would have contained higher relative abundances of gases like hydrogen and helium, and so would the planets that later formed through the coalescence of these planetesimals.

An alternative viewpoint has developed, however, which holds that the outer planets formed directly from the gas of the solar nebula, without an intermediate condensation stage. This idea is prompted in part by the extensive ring and satellite systems of the outer planets, which resemble miniature solar nebulae. The theory suggests that these planets simply formed like the sun did, from gravitational collapse of swirling gas clouds. The idea is that large eddies developed in the outer portions of the gaseous disk, and these became sufficiently dense to gravitationally fall in on themselves, forming their own small rotating disk systems in the process (Fig. 12.9). The central portion of each disk then became a planet, whereas the outer portions underwent the same kinds of instabilities as the solar nebula, forming condensations that eventually became satellites (or rings, if the solid fragments that formed were inside the Roche limit, where tidal forces prevented their coalescing into a satellite).

The fact that Uranus has a rotation axis that is tilted 98 degrees with respect to the ecliptic lends support to this idea. Apparently the rotating disk that was to become Uranus was tilted, so that not only the planet but also its system of satellites and rings were inclined equally, something that would be difficult to explain if Uranus was forced into its tilt at some time after the satellites had formed. An additional bit of evidence in favor of this theory for the formation of the outer planets is their rapid rotation rates, which would have developed naturally as the disks collapsed.

The asteroids probably formed as planetesimals (similar to those which eventually created the terrestrial planets), but were prevented from coalescing into a planet by the gravitational effects of Jupiter, as we noted in chapter 10. The comets probably formed farther out, though not at the distance of the Oort cloud, where they now reside. At such a great distance (50,000 to 150,000 AU) from the center of the solar nebula, it is doubtful that any condensation could have

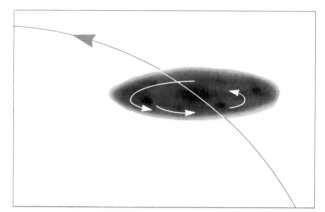

FIGURE 12.9. *Formation of a Giant Outer Planet.*
Here a rotating disk has formed around a condensation in the outer solar system. Lumps in the disk grow to become satellites, except in the innermost portion where tidal forces prevent this from occurring and a ring system forms instead. The planet that is forming at the center rotates rapidly as a result of its contraction from a much larger gas cloud. The entire system of planet, rings, and moons orbits the sun.

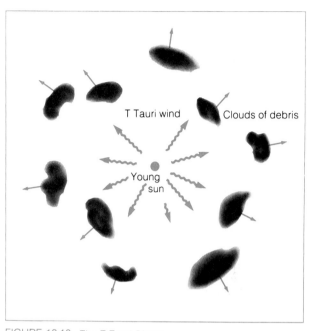

FIGURE 12.10. *The T Tauri Phase.*
A star like the sun is thought to go through a phase very early in its lifetime when it violently ejects material in a high-velocity wind. This wind sweeps away matter left over from the formation process.

occurred, because the density would have been too low. Therefore it is thought that the comets condensed at intermediate distances, probably near the orbits of Uranus and Neptune, and then were forced out to their present great distances by the gravitational effects of the major planets. Both Jupiter and Saturn, lying closer to the sun, would have forced the cometary bodies out farther, by speeding them up a little each time they passed by. Eventually, most of the comets retreated to a great distance, forming the Oort cloud.

Once the planets and satellites were formed, the solar system was nearly in its present state, except that the solar nebula had not totally dissipated. There was still a lot of gas and dust swirling around the sun, along with numerous planetesimals that had not yet accreted onto planets. During this time the sun's magnetic field, pulling at ionized gas in the nebula, was very effective in slowing the sun's rotation, through the magnetic-braking process. Meanwhile, the remaining planetesimals were floating about, now and then crashing into the planets and their satellites. Most of the cratering in the solar system occurred during this period, in the first billion years.

The leftover gas in the solar nebula was dispersed rather early, when the infant sun underwent a period of violent activity and developed a strong wind that swept the gas and tiny dust particles out into space

(Fig. 12.10). Only the larger solid bodies, the planetesimals and the newly formed planets, were left behind. The wind phase has been observed in other newly formed objects called **T Tauri** stars, and is apparently a natural stage in the development of a star like the sun.

Once most of the remaining interplanetary debris was eliminated, either by the T Tauri wind or by accretion onto planetary bodies, the solar system looked much like it does today. The planets have gone their own ways: most of the terrestrial planets have continued to evolve geologically, whereas the outer planets apparently have remained perpetually frozen in their original condition.

Are There Other Solar Systems?

As far as we know, the entire process by which our planetary system formed was a by-product of the sun's formation, requiring no singular, catastrophic event. Thus we might suspect that the same processes oc-

Even in the evolutionary theory that is now widely accepted, the development of the solar system required explosive events. The evolution of the universe itself may be viewed as the result of a series of explosions, including the initial one from which the universe has been expanding ever since. A variety of astronomical evidence (discussed thoroughly in chapter 23) shows that the universe began in a single point, some ten to fifteen billion years ago, and at some later time began to expand outward. The original conditions were so extreme in temperature and density that the birth of the universe must have been an explosive event, commonly called the *big bang*.

For a brief time during the early stages of the universal expansion, conditions were suitable for nuclear fusion reactions to occur, and the primordial substance was converted from a soup of subatomic particles to one of recognizable elements, primarily hydrogen and helium, with almost no trace of heavier species. Eventually, as the universe continued to expand, galaxies and individual stars formed. Just as nuclear reactions form a heavier element (helium) from a light one (hydrogen) in the sun, similar reactions take place inside all stars, gradually enriching the universal content of heavy elements. Stars more massive than the sun undergo more reaction stages, going from the production of helium to that of carbon, and from carbon to other elements. When these stars die, often explosively, their newly formed heavy elements are returned to interstellar space, and are available for inclusion in new stars that form later. Thus, during the five to ten billion years that passed between the time of the big bang and the time of the solar-system formation, the chemical makeup of our galaxy gradually changed, so that about 2 percent of the total mass was in the form of elements heavier than helium.

Since this roughly represents the composition of the sun and the primordial material in the solar system, it might seem that there is no more to the story of the origin of the elements. There are compelling reasons to believe, however, that the formation of the solar system was influenced by at least two local explosive events. Certain atomic isotopes that are present in solar-system material are the result of explosive nuclear reactions that must have taken place in relatively recent times, compared to the time that has passed since the big bang. These isotopes, one a form of aluminum (with 26 instead of the normal 27 neutrons and protons) and the other a form of plutonium (with 244 protons and neutrons instead of 242), are radioactive, with half-lives short enough that they could not have been created much before the solar system formed. The plutonium isotope in question probably formed in a stellar explosion called a *supernova* that occurred a few hundred million years before the solar system formed, whereas the aluminum must have been created in a nearby supernova only a million years or so before the solar system was born.

The second of these explosions may have played a crucial role in triggering the formation of our sun and planetary system. We have said that the solar system formed from an interstellar cloud that collapsed. What we have not discussed, however, is what caused this collapse to occur. Normally an interstellar cloud is quite stable, and it will not fall in on itself unless something happens to compress it. Various mechanisms can bring this about, and one of them is a shock wave from an explosive event such as a supernova. Thus, in a singular event sometimes known as the "bing bang", our solar system may have been born as the direct consequence of the death of a nearby star. We derive from this event not only some minor anomalies in nuclear isotope abundance, but also the very existence of the solar system. It is interesting to note that even in an evolutionary origin as discussed in this chapter, the solar system has been strongly influenced by what can only be described as catastrophic events.

curred countless times as other stars were born, and therefore there may be many planetary systems in the galaxy. Even if we rule out double- and multiple-star systems, in which perhaps all the material was either ejected or included in the stars, there remains a vast number of possible planetary systems.

To detect planets orbiting distant stars is not easy. For a long time there was one star suspected of having an orbiting planet. The gravitational pull of a very massive planet would cause its parent star to wobble a little in position as the planet orbited it. A nearby star called Barnard's star was thought to be wobbling in such a way, but a recent reexamination of the data has shown that there is no clear-cut indication of such a motion. At the moment the case for the existence of a massive planet circling Barnard's star is not well established.

Very recently, infrared observations have led to the detection of a massive planet orbiting the young star T Tauri. We have already mentioned T Tauri, because it is the prototype of a newly formed star undergoing violent activity such as expelling material in a rapid wind. Using the technique called **speckle interferometry,** in which the distorting effects of the earth's atmosphere are eliminated by analyzing how the light waves are displaced and organized on the way through it, astronomers recently discovered that T Tauri has a very nearby companion. The companion was at first thought to be a star, making T Tauri a binary system (these are quite common; see chapter 13). Now, however, estimates of the characteristics of the new object show it to be a giant planet, five to twenty times as massive as Jupiter. This discovery lends further credence to the evolutionary theory, because no longer is the solar system the only example known of a star being accompanied by planets.

The *IRAS* satellite (see Chapter 3) has discovered infrared radiation from the bright A star called Vega, and this radiation may be due to solid material orbiting the star. The data appear to rule out the presence of plant-size objects, but it is possile that the radiation is coming from a cloud of dust or larger particles that could potentially form planets.

While infrared observations show some promise as a technique for detecting other planetary systems. Other kinds of new technology may also provide new information in this area. The *Space Telescope,* for example, will orbit above the earth's atmosphere, and will therefore be free of the distortion and fuzziness that plague ground-based telescopes. It is expected that the *Space Telescope* will allow astronomers to see objects comparable to Jupiter, if such objects orbit nearby stars. There is little hope of directly detecting smaller objects like the terrestrial planets, however.

The clarity of the *Space Telescope* images will also allow measurements of stellar motions to be much more precise than those made from earth; thus tiny oscillations in the positions of stars will be more easily detected. Again, this will probably not permit the detection of small planets like the earth, but could locate Jupiter-like objects. The use of interferometry at ground-based observatories may also provide new hope of detecting small oscillations in star positions indicative of the presence of planets. Astronomers are waiting eagerly for the chance to use these new instruments and techniques in a search for planets in other solar systems. It will be fascinating to learn the results in the coming years.

PERSPECTIVE

In this chapter we have outlined a natural, evolutionary process that seems to account for all the observed properties of the solar system. The orbital characteristics and the compositions of the planets, the slow rotation of the sun, and the nature and motions of the interplanetary material are now understood. Many details have yet to be worked out, but the overall scenario is probably in essence correct.

We have completed our examination of the solar system, and we are ready to move on. The next logical step is to explore the realm of the stars.

SUMMARY

1. The facts that must be explained by a successful theory of the formation of the solar system include the systematics of the orbital and spin motions of the planets and satellites, the contrasts between the terrestrial and giant planets, the slow rotation of the sun, and the existence and properties of the interplanetary bodies.

2. There are two general classes of theories: catastrophic and evolutionary.

3. Catastrophic theories state that some singular event, such as a near-collision between stars, distorted the sun by tidal forces, pulling out matter that condensed to form the planets.

4. Evolutionary theories postulate that the planets formed as a natural by-product of the formation of the sun. These theories, although they are simpler, were not accepted for awhile because of their failure to explain the slow spin of the sun.

5. When magnetic braking was understood in the 1960s, the evolutionary theory became most widely accepted.

6. In the modern theory, the planets formed from condensations in the solar nebula, the flattened disk of gas and dust that formed around the young sun.

7. The terrestrial planets formed from the coalescence of planetesimals; solid, asteroid-like objects that consensed from the hot inner portions of the solar nebula.

8. The giant planets formed from small-scale eddies in the solar nebula that contracted and increased their spins, in parallel with the same processes that occurred in the solar nebula itself.

9. The evolutionary theory leads scientists to suspect that many other solar systems exist around other stars, but so far none has been detected because of the difficulty of observing planets at interstellar distances.

REVIEW QUESTIONS

1. Why are the rotation of Venus and the retrograde orbits of some satellites not explained by the theory of solar-system formation described in the text?

2. We might have mentioned, as another fact that must be explained, the fraction of the total mass of the solar system that is contained in the sun. Using data from tables in the appendixes, calculate the fraction of the total mass of the solar system that is contained in the sun.

3. How might the present-day solar system be different if the sun had been formed with no magnetic field?

4. The terrestrial planets have very low abundances of

volatile elements. They also have lower masses than the giant planets, which have about the same composition as the sun. If you assume that 10 percent of the earth's mass is in the form of iron, what would the mass of our planet be if it had all the elements in the same proportion to iron as in the sun? How would the earth compare to Jupiter in this case? (Hint: you will need to use data from Appendix 2).

5. Summarize the differences between the formation of the terrestrial planets and that of the giant planets in terms of the evolutionary theory described in the text.

6. Compare the differences between the terrestrial and giant planets with the differences between the inner and outer of Jupiter's four Galilean satellites. Explain the contrasts among the Jovian satellites in terms of the evolutionary theory of solar-system formation.

7. Why does the evolutionary theory make it doubtful that a major tenth planet exists?

8. The giant planets rotate rather rapidly in comparison to the terrestrial planets. Why is this so? What forces might be acting to slow their rotations?

9. How much dimmer would Jupiter appear from the earth if it orbited alpha Centauri, the nearest star? The distance to alpha Centauri is about 4.3 light-years, or 270,000 AU.

ADDITIONAL READINGS

Beatty, J. K., O'Leary, B., and Chaikin, A., eds. 1981. *The new solar system.* Cambridge, England: Cambridge University Press.

Cameron, A. G. W. 1975. The origin and evolution of the solar system. *Scientific American* 233(3):32.

Falk, S. W., and Schramm, D. N. 1979. Did the solar system start with a bang? *Sky and Telescope* 58(1):18.

Head, J. W., Wood, C. A., and Mutch, T. A. 1977. Geologic evolution of the terrestrial planets. *American Scientist* 65:21.

Reeves, H. 1977. The origin of the solar system. *Mercury* 7(2):7.

Sagan, C. 1975. The solar system. *Scientific American* 233(3):23.

Schramm, D. N., and Clayton, R. N. 1978. Did a supernova trigger the formation of the solar system? *Scientific American* 237(4):98.

Wetherill, G. W. 1981. The formation of the earth from planetesimals. *Scientific American* 244(6):162.

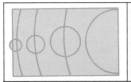

David N. Morrison

Through active involvement in research and in NASA administration, David Morrison has become widely recognized as an authority in the area of planetary exploration. Educated at the University of Illinois and at Harvard University, Dr. Morrison has devoted most of his own research to solid bodies in the solar system such as planetary satellites and asteroids (there is even an asteroid that bears his name). His involvement in solar system exploration goes far beyond that, however, as he is very active in working with NASA to plan for future missions. Currently chairperson of NASA's external advisory Solar System Exploration Committee, Dr. Morrison has also served in various capacities within NASA. Most of his professional career has been spent at the University of Hawaii, where he is now Professor in the Institute for Astronomy and the Department of Physics and Astronomy. Author of many popular articles on astronomy and solar system exploration, Dr. Morrison is serving as President of the Astronomical Society of the Pacific, an organization that encompasses both professional and amateur astronomers.

During the 20 years from the first *Mariner* flyby of Venus to the *Voyager 2* encounter with Saturn, planetary exploration experienced its first Golden Age. More than 40 robot spacecraft probed first toward the moon, Venus, and Mars, then ultimately to every planet known to ancient peoples, from Mercury to Saturn. The United States launched most of these spacecraft, bearing names symbolic of their exploratory missions: *Ranger, Surveyor, Pioneer,* and *Voyager.* The Soviet Union, the other nation to contribute to this era of discovery, focused its efforts on the moon and Venus, but it, too, achieved remarkable successes. Within less than a generation, humans effectively discovered more than two dozen worlds, placing our planet for the first time in its cosmic context.

In spite of this brilliant beginning, planetary exploration has fallen upon hard times. No new NASA missions were initiated between 1977, when the *Galileo* orbiter and probe of Jupiter were authorized, and 1983, when a Venus radar mapping orbiter was begun. This failure of initiative has ensured a major gap in U.S. exploration between 1981 and 1988, and has sacrificed leadership in this field throughout most of this decade to increasingly ambitious European and Soviet planetary programs.

In an effort to rebuild the U.S. planetary exploration program, a high-level NASA advisory group called the Solar System Exploration Committee worked from 1981 through 1983 to redefine our goals, with emphasis on more economical ways to undertake planetary missions.

Our choice of planetary missions has always depended on our technological capability. In the past, improvements in technology have led to more ambitious missions of increasing complexity, such as *Viking, Voyager,* and *Galileo.* However, this "bigger is better" philosophy has conflicted with recent constrained NASA budgets: the planetary community seemed to be pricing itself out of the market. Further, a vicious cycle was developing: the long intervals between flights decreased the opportunities for cost-saving inheritance of hardware and designs from previous missions, while increasing the pressure to include the maximum scientific payload in each flight. The new plan seeks to break this cycle and increase the opportunities for new missions by reducing their individual costs and phasing them together in a more efficient way. One excellent way to control mission costs is to use inherited hardware and software from one mission to the next, and concentrate on what we already know how to do well—flybys, simple planetary orbiters and atmospheric probes—and avoid large-scale technology development. With these restrictions, we will have to defer some extremely interesting missions, such as a Mars rover or sample return. But a great deal of exciting science can still be carried out within a core program of frequent, but modest, missions.

Continued FUTURE EXPLORATION OF THE SOLAR SYSTEM

Economy in spacecraft can best be achieved if we undertake planetary exploration as a continuing program rather than as episodic events. For near-earth missions to the moon, Venus, Mars, and some asteroids, we can take advantage of spacecraft already developed by aerospace companies for commercial application in earth orbit. Even after being adapted for planetary missions, the costs of such spacecraft are modest by comparison with the specialized spacecraft now used for planetary exploration. Missions to the outer solar system, to comets, and to main belt asteroids cannot be carried out with modified earth satellites. Instead, we should develop a new generation of *Mariner/Voyager* type spacecraft. They will be as simple as possible and easily reconfigured from one application to the next. Informally known as *Mariner Mark II*, these spacecraft are presently under study at the Jet Propulsion Laboratory. Atmospheric probes to the outer planets and Saturn's cloud-covered moon, Titan, can be done using the Jupiter probe already designed for the Galileo mission. Finally, computer techniques should simplify the control and operation of spacecraft and the handling of the data they return to earth.

If these cost-saving approaches are adopted, we believe that a vigorous mission program—about one launch per year—can be carried out at an annual cost of only about $350 million in current-year (fiscal year 1984) dollars. This is only one-third of the NASA planetary budget of the early 1970s, which reached nearly one billion dollars

annually (in fiscal 1984 dollars) during the development of *Viking* and *Voyager*.

There are 14 missions in the proposed core plan, conveniently divided into three groups: inner planets, small bodies (comets and asteroids), and outer planets. The first and highest priority mission is one that has been under consideration for many years: a *Venus Radar Mapper (VRM)*. This spacecraft, to be launched in 1988, will fill the greatest gap in our knowledge of the inner planets: the surface topography of Venus. Although hidden from normal view by thick clouds, the surface of Venus can be mapped from orbit with subkilometer resolution by synthetic aperture radar. This mission, which was approved by Congress in 1983, may test a very recent theory that Venus has active volcanoes.

A Mars orbiter, based on a commerical earth satellite, is the second recommended inner planet mission. It should be launched in 1990, and would focus on the surface chemistry and climate of the red planet. The daily and seasonal exchange of water and carbon dioxide among the polar caps, atmosphere and soil could be studied. Later inner planet missions are: a Mars orbiter to investigate the upper atmosphere, a hard-landing Mars surface probe, a lunar polar orbiter, and an advanced Venus atmosphere probe.

The first mission to the small bodies, and the first of the *Mariner Mark II* series, will be a rendezvous with a short-period comet. By flying for months in close formation with the comet, the spacecraft will

detail the comet's cycle of activity as it is heated by the sun. This mission will be a major step beyond the fast flybys of Halley's comet planned for 1986 by Soviet, European, and Japanese scientists. The other small bodies missions in the core program are flybys and orbiters to main belt asteroids and to the smaller earth-approaching asteroids, and a simplified sample return mission to a comet.

We have given first priority in the outer solar system to a probe to the atmosphere of Titan, the remarkable moon of Saturn with its thick nitrogen- and methane-atmosphere and complex organic chemistry. Other missions are a Saturn orbiter and probes to Saturn and Uranus, as well as *Galileo*, which is scheduled for a 1986 launch and 1988 arrival at Jupiter.

While the United States could carry out all of these missions alone, we anticipate and welcome international cooperation. The European Space Agency or individual countries could do some missions, and other missions would lend themselves to joint activity. There is particular interest right now in U.S. and European cooperation on a Saturn orbiter and Titan probe. The planetary science community believes that this proposed mission plan is both realistic and exciting. If these recommendations are adopted, beginning in 1988 we will see a launch rate of planetary missions as high as that which characterized the Golden Age of the 1960s and 1970s, but it will be achieved at half the cost. We will be using new technology to hold down costs, and thereby to stimu-

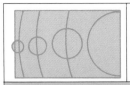

late activity. While we can do nothing to correct the data gap of the 1980s, we can begin now to rebuild our capability and prepare the missions that will restore U.S. leadership in the 1990s.

What I have discussed here is only a core program, limited to individually modest missions and specifically avoiding many new technological challenges. But I do not believe that our objectives should be limited to this core program, even though it seems quite rich by comparison with the currently depressed state of planetary exploration. As a great nation with ambitious long-term goals in space, we can and should be willing to undertake *Viking*-class missions, such as a Mars rover or sample return mission. Only by accepting such major challenges will humans ultimately be able to gain the advantages of a permanent presence in space.

THE STARS

Introduction to Section III

This section is devoted to the stars and their properties. In its four chapters we learn about how stars are observed, how their fundamental properties are deduced from observations, and how stars work. We see what the processes are that control the structure of stars, and we learn how stars live, evolve, and die. We discover that stars are dynamic, changing entities, and only seem fixed and immutable because the time scales on which they evolve are vastly longer than the human lifetime.

The first of the four chapters describes the three basic types of stellar observations, and how the data obtained from these techniques are used to derive fundamental properties of stars. We systematically consider several stellar properties, discuss how each is determined, and summarize the typical values that are found.

Chapter 14 then initiates the discussion of how stars function and evolve. We describe the physical processes that occur in stars, and how these processes govern the stellar parameters previously discussed.

Here we draw upon knowledge of our own sun, whose well-observed properties provide vital information about other stars as well. Although the discussion is self-contained and does not require knowledge of the contents of chapter 11, there are references to that chapter; thus it is recommended that it be read in conjunction with this section of the book.

The final two chapters, 15 and 16, describe the evolution of stars, from their beginnings to their deaths. In chapter 15 we base our discussion on theoretical calculations as well as on observations of star-formation regions and clusters of stars. In chapter 16 we learn about the remnants of stars that have finished their lives, and here we discover some of the most bizarre and exciting objects in the universe.

Having studied the stars, we will then be ready to consider large groupings such as our own galaxy, which contains many billions of stars, and whose own evolution is governed in part by that of the individual stars within it.

Learning Goals

ASTRONOMICAL OBSERVATIONS
Positional Astronomy
Stellar Brightness Measurements
   The Magnitude System
   Color Index and Stellar Temperature
Stellar Spectroscopy and Spectral Classes
BINARY STARS
MEASUREMENT OF STELLAR PROPERTIES
Absolute Magnitudes and Stellar Luminosities

Stellar Temperatures
The Hertzsprung-Russell Diagram
   The Correlation Between Temperature and Luminosity
   Spectroscopic Parallax: A Powerful Distance Tool
Stellar Diameters
Binary Stars and Stellar Masses
Stellar Composition, Rotation, and Magnetic Fields

13

Everything we can learn about a star is contained in the light we receive from it. Fortunately, a lot of information is there, and astronomers have learned how to extract much of it. In this chapter we will first discuss the three basic types of astronomical observations, and then we will learn how astronomers use these techniques to measure fundamental parameters of stars.

Three Ways of Looking at It: Positions, Magnitudes, and Spectra

Nearly all observations of stars fall into one of three categories; measurements of star (1) positions; (2) brightnesses; or (3) spectra. The first two types of observations have rather long histories, positional measurements dating back to the first human observations of the skies, and brightness measurements (although rather crude ones) having been made by Hipparchus in the second century B.C.

Positional Astronomy

The science of measuring star positions is called **astrometry.** Ancient astronomers used simple devices such as quadrants and sextants (Fig. 13.1) to measure angular positions on the sky. Today the principal technique is to photograph the sky and to carefully measure the positions of the star images on the photographic plates. If a number of photographs of a given portion of the sky are taken and measured separately, the results can be averaged to produce a more accurate determination than is possible by measuring a single photograph. We can now measure a stellar position to a precision of less than 0.01 second of arc. When the *Space Telescope* is in operation in the late 1980s, even more accurate measurements will be possible, because the fuzziness caused by the earth's atmosphere will be eliminated.

Modern astronomers make astrometric measurements for a variety of reasons. For example, such measurements are important for the analysis of stellar motions and for what these motions can tell us about the structure of the galaxy (this is discussed in chapter 17). Astrometric data are also very important in accurately

cataloguing stars so that the stars can be observed by other telescopes. (A major task now under way is the preparation of star lists for *Space Telescope* observations; the lists include stars that are much fainter than those listed in current catalogues.) Finally, and perhaps most important, positional measurements can be used to measure distances to stars, making use of the stars' parallax motions.

Recall from chapter 1 that stellar parallax is the apparent shifting in a nearby star's position resulting from the orbital motion of the earth (Fig. 13.2). Ancient and medieval astronomers were unable to detect stellar

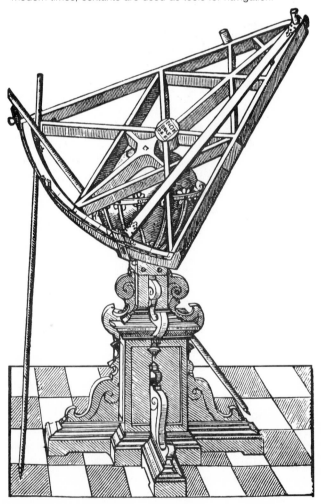

FIGURE 13.1. *A Sextant.*
Instruments of this type were used for centuries for the measurement of star positions relative to each other. Even in modern times, sextants are used as tools for navigation.

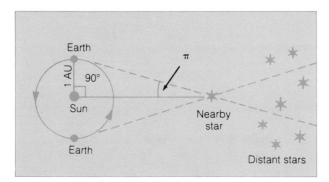

FIGURE 13.2. *Stellar Parallax.*
As the earth orbits the sun, our line of sight toward a nearby star varies enough in direction to make the star appear to move back and forth with respect to more distant stars. The parallax angle π is defined as one-half of the total amount of apparent annual motion. If π is 1 second of arc, then the sun-star distance is 206,265 times the sun-earth distance. This defines the parsec, the distance to a star whose parallax angle is 1 arcsecond.

parallax, leading most of them to reject the idea that the earth orbits the sun. Eventually the heliocentric theory won out, but stellar parallax remained undetected until 1838. The reason stellar parallaxes had defied earlier observers now became obvious: even for the closest stars, the maximum shift in position is much less than 1 second of arc! The stars are simply much farther away than the ancient astronomers had dreamed possible. The annual parallax motion of the star alpha Centauri, the star nearest the sun, is comparable to the angular diameter of a dime as seen from a distance of about two miles, a very small angle indeed.

The successful detection of stellar parallax led to a direct means of determining distances to stars. The amount of shift in a star's apparent position resulting from the earth's motion about the sun depends on how distant the star is; the closer it is, the bigger the shift. The parallax angle π is defined as one-half the total angular shift of a star during the course of a year. This angle, once it is measured, tells us the distance to a star, in a unit of measure called the **parsec.** A star whose parallax is 1 arcsecond is, by definition, 1 parsec away. (The word *parsec,* in fact, is a contraction of *parallax-second,* meaning a star whose parallax is 1 arcsecond.) In mathematical terms, the distance to a star whose parallax is π is $d = 1/\pi$. Thus, if a star has a parallax of 0.4 arcseconds, it is $1/0.4 = 2.5$ parsecs away. Recall that the closest star has a parallax of less

than 1 arcsecond; we conclude therefore that even this star is more than 1 parsec away.

In more familiar terms, a parsec is equal to 3.26 light-years, or 3.08×10^{18} centimeters, or 206,265 astronomical units. The use of parallax measurements to determine distances is a very powerful technique, and is the only direct means astronomers have for measuring how far away stars are. Unfortunately, parallaxes are large enough to be measured only for stars rather close to us in the galaxy. The smallest parallax that can be measured is about 0.001 arcseconds, corresponding to a distance of $1/0.001 = 1,000$ parsecs. Accurate measurements can be made only for somewhat larger parallax angles, corresponding to distances within a few hundred parsecs. The galaxy, on the other hand, is more than 30,000 parsecs in diameter! Clearly, other distance-determination methods are needed if we are to probe the entire galaxy. One very powerful method will be described later in this chapter.

Stellar Brightnesses

The second general type of stellar observation is the measurement of the brightnesses of stars. This was first attempted in a systematic way by Hipparchus, who, more than two thousand years ago, established a system of brightness rankings that is still with us today. Hipparchus ranked the stars in categories called **magnitudes,** from first magnitude (the brightest stars) to sixth (the faintest visible to the unaided eye). In his catalogue of stars, and in all since then, these magnitudes are listed along with the star positions.

The magnitude system has been modernized, so that all astronomers use the same technique for measurement, and so that reliance on subjective impressions has been eliminated. It was discovered in the mid-1800s that what the eye perceives as a fixed *difference* in intensity from one magnitude to the next actually corresponds to a fixed intensity *ratio*. Measurements showed that a first-magnitude star is about 2.5 times brighter than a second-magnitude star, a second-magnitude star is 2.5 times brighter than a third-magnitude star, and so on (Fig. 13.3). The ratio between a first-magnitude star and a sixth-magnitude star was found to be nearly 100. In 1850, the system was formalized by the adoption of this ratio as exactly 100; thus, the ratio corresponding to a one-magnitude difference is the fifth root of 100, or $(100)^{1/5} = 2.512$. Therefore a first-magnitude star is 2.512 times brighter than a star

From the earliest times when astronomers systematically measured star positions, they began to compile lists of these positions in catalogues. Hipparchus developed an extensive catalogue, as did early astronomers in China and other parts of the world, more than two thousand years ago (see chapter 1).

To list stars in a catalogue of positional measurements requires some kind of system for naming or numbering the stars, and a variety of such systems have been employed. The ancient Greeks designated stars by the constellation in which the stars were located and their brightness rank within the constellation. For example, the brightest star in Orion is α Orionis, the next brightest is β Orionis, the next is γ Orionis, and so on. (The constellation names are spelled in their Latin versions, since this system was perpetuated by the Romans and the Catholic church after the time of the Greeks.) Today many stars are still referred to by their constellation rankings, with the Greek alphabet designating the various brightness ranks.

After the time of Ptolemy, when western astronomy went into decline for some 1,300 years, Arab astronomers, occupying northern Africa and southern Europe, carried on astronomical traditions. These people assigned proper names to many of the brightest stars, and these names, such as Betelgeuse (α Orionis), Rigel (β Orionis), and Bellatrix (γ Orionis), are also still in use.

With the advent of the telescope, when many new stars were discovered that were too faint to be seen with the naked eye, catalogues rapidly outgrew the old naming systems. Most catalogues therefore assigned a number to each star, usually in a sequence related to the star's position. Thus, modern catalogues, such as the *Henry Draper Catalog*, the *Boss General Catalog*, the *Yale Bright Star Catalog*, and the *Smithsonian Astrophysical Observatory Catalog*, list stars by increasing right ascension (that is, from west to east in the sky, in the order in which the stars pass overhead at night). In some cases, separate listings are made for different zones of declination (for different strips of sky, separated in the north-south direction). Because each of these catalogues includes a different particular set of stars, each has its own numbering system, and often no cross-reference to other catalogues is provided. Hence a given star may have a constellation-ranked name, an Arabic name, and different numbers in a variety of catalogues, so the astronomer must match up the coordinates to be certain of referring to the same star in different catalogues. Actually, things are not quite so bad as that, since some modern catalogues do provide cross-references to others. Nevertheless, one of the principal tasks of a would-be astronomer is to become familiar with star catalogues and their use.

---

of second magnitude, a sixth-magnitude star is 2.512 × 2.512 = 6.3 times fainter than a fourth-magnitude star, and so on.

Magnitudes are most commonly measured through the use of **photomultiplier tubes;** these are devices that produce an electric current when light strikes them. They are used in many familiar applications, such as in door-openers in modern buildings. The amount of electric current produced is determined by the intensity of light, so an astronomer need only measure the current to determine the brightness of a star (Fig. 13.4), and hence its magnitude.

Once magnitudes could be measured precisely, it was found that stars have a continuous range of brightnesses, and do not fall neatly into the various magnitude rankings. Therefore fractional magnitudes must be used. Deneb, for example, has a magnitude of 1.26 in the modern system, although it was classified simply as a first-magnitude star in the old days. Each of the magnitude categories is found to include a range of stellar brightnesses. This is especially true for the first-magnitude stars, some of which turned out to be as much as two magnitudes brighter than others. To measure these especially bright stars in the modern system

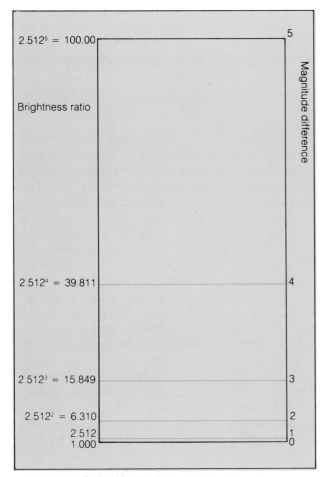

FIGURE 13.3. *Stellar Magnitudes.*
This diagram shows schematically how brightness ratios and magnitude differences are related. To determine the relationship between brightness ratio and fractional magnitudes requires calculations that involve logarithms (see appendixes).

FIGURE 13.4. *Measuring Stellar Magnitudes.*
This is a schematic illustration of photometry, the measurement of stellar brightnesses. Light from a star is focused (often through a filter that screens out all but a specific range of wavelengths) onto a photocell, a device that produces an electric current in proportion to the intensity of light. The current is measured, and, through comparison with standard stars, converted into a magnitude.

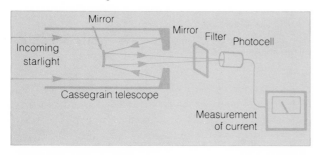

requires the adoption of magnitudes smaller than 1. Sirius, for example, which is the brightest star in the sky, has a magnitude of -1.42. By using negative magnitudes for very bright objects, astronomers can extend the system to include such objects as the moon, whose magnitude when full is about -12, and the sun, whose magnitude is -26 (thus the sun is 25 magnitudes, or a factor of $2.512^{25} = 10^{10}$, brighter than Sirius).

Of course, there are stars too faint to be seen by the human eye, so the magnitude scale must also extend beyond sixth magnitude. With moderately large telescopes, astronomers can observe stars as faint as fifteenth magnitude, and with long-exposure photographs taken with the largest telescopes, they can detect stars as faint as twenty-fifth magnitude. Such a star is nineteen magnitudes, or a factor of $2.512^{19} = 4 \times 10^7$, fainter than the faintest star visible to the unaided eye. The *Space Telescope* is expected to detect stars as faint as twenty-eighth magnitude, which is a factor of $2.512^3 = 15.9$ fainter yet.

The stellar magnitudes we have discussed so far all refer to visible light. As we learned in chapter 3, however, stars emit light over a much broader wavelength band than the eye can see. We also learned that the wavelength at which a star emits most strongly depends on the star's temperature. Therefore by measuring a star's brightness at two different wavelengths, astronomers can learn something about the star's temperature. Filters that allow only certain wavelengths of light to pass through are used for this purpose. Typically, a star's brightness is measured through two such filters, one that passes yellow light, and one that allows blue light to pass, resulting in the measurement of V (for visual, or yellow) and B (for blue) magnitudes (Fig. 13.5). Since a hot star emits more light in blue wavelengths than in yellow, its B magnitude is *smaller* than its V magnitude. For a cool star, the situation is reversed, and the B magnitude is larger than the V (remember that the magnitude scale is backward in the sense that a smaller magnitude corresponds to a greater brightness).

The difference between the B and V magnitudes is called the **color index.** The exact value of this index is a function of the temperature of a star, so stellar temperatures can be estimated simply by measuring the V and B magnitudes. A very hot star might have a color index $B - V = -0.3$, whereas a very cool one might typically have $B - V = +1.2$.

One additional type of magnitude should be mentioned, although it is a very difficult one to measure

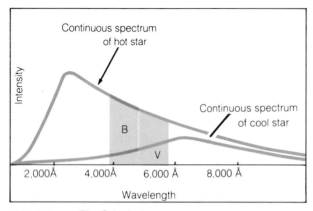

FIGURE 13.5. *The Color Index.*
Here are continuous spectra of two stars, one much hotter than the other. The wavelength ranges over which the blue *(B)* and visual *(V)* magnitudes are measured are indicated. We see that the hot star is brighter in the *B* region than in the *V* region of the spectrum; therefore, its *B* magnitude is *smaller* than its *V* magnitude, and it has a *negative B–V* color index. The opposite is true for the cool star.

directly. This is the **bolometric magnitude,** which includes all the light emitted by a star at all wavelengths. To determine a bolometric magnitude requires ultraviolet and infrared observations, as well as visible, and can best be done with the use of telescopes in space. This is particularly true for hot stars, which emit a large fraction of their light in ultraviolet wavelengths. For cool stars, which emit little ultraviolet light but a lot of infrared radiation, bolometric magnitudes can be measured from earth-based observations with fair accuracy. The bolometric magnitude of a star is always smaller than the visual magnitude, because more light

is included when all wavelengths are considered than when only the visual wavelengths are measured. We will see later in this chapter how bolometric magnitudes are related to other stellar properties, particularly luminosity.

Measurements of Stellar Spectra

The first observations of stellar spectra, made in the mid-1800s (before it was understood how spectral lines are formed), were accomplished through the use of **spectroscopes.** These are simple devices that allow the observer to see a star's light spread out according to wavelength, but not to record it. In the late 1800s the technique of photographing spectra was introduced, which led to a systematic study of stellar spectra. A photograph of a spectrum is called a **spectrogram** (Fig. 13.6).

Among the first astronomers to systematically examine spectra of a large number of stars was the Roman Catholic priest Angelo Secchi, who in the 1860s used a spectroscope to catalogue hundreds of spectra. He found the appearance of the spectra to vary considerably from star to star, although the stars were consistent in one respect: they all showed continuous spectra with absorption lines. The work of Kirchhoff soon explained this to be due to the relatively cool outer layers of a star, which absorb light from the hotter interior. At first it was thought that the differing appearances of stellar spectra were caused by differences in chemical composition of stars, and in one of the early classification schemes, stars were assigned to categories based on their compositions. The basis of the modern classi-

FIGURE 13.6. *A Stellar Spectrum.* A stellar spectrum, like the one shown here, is usually displayed as a negative print. Hence light features are absorption lines, wavelengths at which little or no light is emitted by the star. This point is illustrated by the intensity plot above the spectrum.

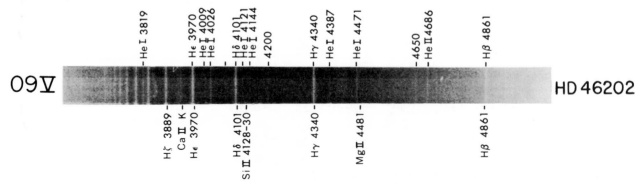

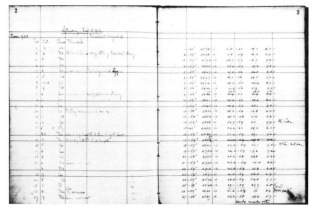

FIGURE 13.7. *Annie J. Cannon.* Cannon, a member of the Harvard College Observatory for almost fifty years, classified the spectra of several hundred thousand stars. At right is shown a page from one of her notebooks. Today she is recognized as the founder of modern spectral classification.

fication system for stellar spectra was founded by a group of astronomers at Harvard University, most notably Annie J. Cannon (Fig. 13.7). Cannon found a smooth sequence of types of spectra, in which the pattern of strong absorption lines changed gradually from one type to the next. Having already assigned letters of the alphabet to the various types, she placed them in the sequence O, B, A, F, G, K, M (Fig. 13.8).

It was later realized that the differing appearances of stellar spectra were caused not by differences in chemical composition, but by differences in temperature. The hotter a star is, the more highly ionized is the gas in its outer layers; the degree of ionization in turn governs the pattern of spectral lines that will form (if this is confusing, review the discussion of ionization in

chapter 3). Therefore Cannon's sequence of spectral types was a temperature sequence: the hottest stars are the O stars, and the coolest are the M stars. She was able to discern subtle differences between spectra, and assigned subclasses to each of the major classes. In this system the sun is a G2 star, being intermediate between types G and K.

In the modern classification system, a few key spectral lines serve to establish the type of a given star (Fig. 13.9). For the O stars, ionized helium, which requires a very high temperature for its formation, is the principal species that reveals the spectral type. For the slightly cooler B stars, it is atomic helium, and for the A stars, atomic hydrogen lines dominate. The F stars

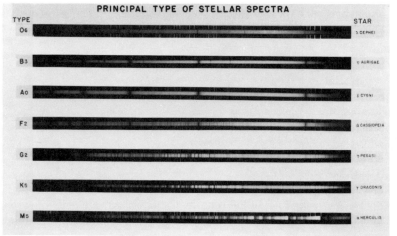

FIGURE 13.8. *A Comparison of Spectra.* Here we see several spectra representing different spectral classes. A few major absorption lines are identified, which are arranged in order of decreasing stellar temperature (top to bottom).

There is some irony to be found in the fact that Harvard University, long a stronghold of the all-male tradition in American colleges, was also the institution that nurtured some of the nation's first leading women astronomers. Today, women are underrepresented among professional astronomers, but were even more so at the turn of the century when the foundations of modern stellar spectroscopy were developed at the Harvard College Observatory.

In 1877, when Edward C. Pickering became director of the observatory, Secchi's work (described in the text), based on visual inspection of stellar spectra through a spectroscope, was the only attempt that had been made to classify stars according to the appearance of their spectra. Henry Draper, an American amateur astronomer, had in 1872 become the first to photograph the spectrum of a star. Upon Draper's death in 1882, his widow endowed a new department of stellar spectroscopy at Harvard. Pickering, as director, hired among his assistants a number of women, several of whom went to work on the problem of classifying spectra of stars.

An innovative technique was used to photograph spectra of large numbers of stars. A thin prism was placed in front of the telescope, so that each star image on the photographic plate at the focus was stretched, (in one direction) into a spectrum. If color film had been used, each stellar image would have looked like a tiny rainbow. Pickering and his group refined this **objective prism** technique to the point where all stars in a field of view as faint as ninth or tenth magnitude would appear as spectra well-enough exposed for classification. Thus it became possible to amass stellar spectra in vast quantities.

The problem of sorting out and classifying the spectra fell to Pickering's associates. One of them, Williamina Fleming, published the first *Draper Catalog of Stellar Spectra* in 1890. In this catalogue, some 10,351 stars in the northern hemisphere were assigned spectral classes A through N, in a simple elaboration of a rudimentary classification scheme adopted earlier by Secchi. A number of Fleming's classes were later dropped.

While the first catalogue was being prepared, a niece of Draper, Antonia Maury, joined the staff and set to work on the analysis of spectra of bright stars. Spectra for these objects could be photographed through the use of thicker prisms (actually, a series of two or three thin prisms), so that the spectra were more widely spread out according to wavelength, which allowed greater detail to be seen. Maury concluded, on the basis of these high-quality spectra, that stars should be grouped into three distinct sequences rather than just one. She discovered that some stars had unusually narrow spectral lines, others had rather broad ones, and a third group had medium-width lines. It later was found that her sequence *c*, the thin-lined stars, are giant stars (the lower atmospheric pressure in these stars causes the spectral lines to be less broadened than in main-sequence stars). Maury's discovery helped Einar Hertzsprung confirm the distinction he had recently found between giant and main-sequence stars, and he thought her work to be of fundamental importance. Unfortunately, Maury's three sequences were not adopted, and subsequent work on spectral classification at Harvard was based on only a single sequence.

Annie J. Cannon, the most important of the Harvard workers in spectral classification, arrived on

have strong hydrogen lines, but also lines caused by certain metallic elements that are ionized once (that is, these atoms have lost just one electron). The G stars have a mixture of ionized and atomic metals, and in the K and M stars these elements are nearly all in atomic form. The cooler M stars also have strong molecular lines.

Having established this classification system around the turn of the century, Cannon then catalogued nearly a quarter of a million stars, a monumental task that

the scene in 1896. She gradually modified the classification system to the present sequence, finding that the arrangement O, B, A, F, G, K, M was a logical ordering, with smooth transitions from one type to the next (at this time, no one knew this was a temperature sequence). Cannon was also able to discern such fine differences in stellar spectra that she established the ten subclasses for each major division that are in use today. In 1901 she published a catalogue of classifications for 1,122 stars, and then embarked on her major task: the classification of more than 200,000 stars whose spectra appeared on survey plates covering both hemispheres. By this time she was so skilled that she could reliably classify a star in a few moments, and most of the job was done in a four-year period between 1911 and 1914. The resulting *Henry Draper Catalog* appeared in nine volumes of the *Annals of the Harvard College Observatory*, the final one being published in 1924. Pickering died in 1919, before the catalogue was complete, and was succeeded as director by Harlow Shapley.

Cannon continued her work, later publishing a major *Extension* of the catalogue, along with a number of other specialized catalogues. She died in 1941. The American Astronomical Society subsequently established the Annie J. Cannon prize for outstanding research by women in astronomy.

The successes of the Harvard women were not confined to stellar spectroscopy. Another major area of interest to Pickering and later to Shapley was the study of variable stars, and Henrietta Leavitt, who joined the group as a volunteer in 1894, played a leading role in this area. Following some early work in establishing standard stars for magnitude determinations, Leavitt by 1905 was at work identifying variable stars from comparisons of photographic plates (she was eventually to discover more than 2,000 of them). In the process, she examined the Magellanic Clouds for variables, and noticed that their periods of pulsation correlated with their average brightnesses. This was a discovery of profound importance, for the period-luminosity relationship for variable stars was to become an essential tool for establishing both the galactic distance scale and the intergalactic scale (see chapters 17 and 20).

The Harvard women, including those discussed here and several whose names are now obscure, were a remarkable group, responsible for a number of major advances in the science of astronomy at a time when its basis in physics was just becoming clear. In this day of awakening recognition of the proper role of women in all areas, it is fitting to consider and appreciate the pioneering work done by this group.

took about five years (although the publication of the results, called the *Henry Draper Catalog*, took place over a much longer period). This catalogue has been a fundamental reference for generations of astronomers, and the system of classification established by Cannon

has, with some modification, been in use since its development.

There are some stars that do not fit neatly into the standard spectral classes, and these are often referred to as **peculiar stars.** In most cases these are stars

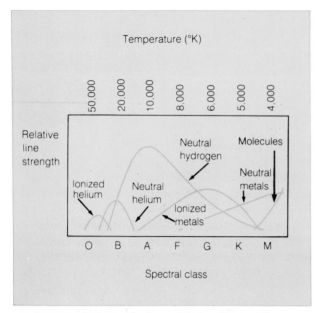

FIGURE 13.9. *Ionization for Stars of Various Spectral Types.* This diagram shows which ions appear prominently in the spectra of stars of different classes. Note that the degree of ionization in the hot stars (left) is much greater than in the cool ones (right).

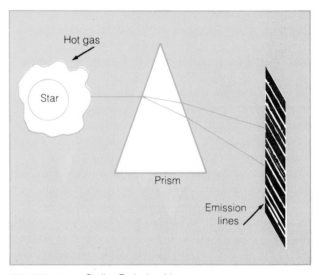

FIGURE 13.10. *Stellar Emission Lines.* Very few stars have emission lines in their visible-wavelength spectra (but many do in ultraviolet wavelengths). When such lines are present, it usually signifies the existence of hot gas above the surface. According to Kirchhoff's second law, a rarefied, hot gas produces emission lines.

whose chemical composition, at least at the surface where the spectral lines form, is unusual. Most stars have the same basic composition (see the list of relative abundances of the elements in the sun, in Appendix 2). Others are unusual because they have emission lines in their spectra, which, according to Kirchhoff's laws, implies that they must be surrounded by hot, rarefied gas (Fig. 13.10). The so-called peculiar stars are probably quite normal, but are in short-lived stages of evolution, so there are not many of them around at any one time.

The **variable stars** represent another kind of unusual star. The majority of these are stars whose brightness fluctuates regularly as they alternately expand and contract (Fig. 13.11). As in the case of the peculiar stars, the variables are normal stars in special stages of their lifetimes, where particular combinations of atmospheric pressure and ionization conditions produce instabilities that cause the pulsations. The most widely known pulsating variable stars are the **δ Cephei** stars, giants whose spectral type varies between F and G as the stars pulsate. As we will see in chapter 17, these stars, as well as the less luminous **RR Lyrae** variables, are very useful tools for measuring distances,

because their luminosities can be inferred from their periods of pulsation. Other pulsating stars include the **Mira** stars, or **long-period variables,** M supergiants that take a year or longer to go through a complete cycle; and a variety of shorter-period variables that are not quite as regular as the δ Cephei and RR Lyrae stars. Some stars are seen to vary erratically, even explosively, and these will be discussed in chapter 16.

Binary Stars

About half of the stars in the sky are members of double or **binary,** systems, where two stars orbit each other regularly. All types of stars can be found in binaries, and their orbits come in many sizes and shapes. In some systems the two stars are so close together that they literally are touching, and in others they are so far apart that it takes hundreds or thousands of years for them to complete one revolution.

Binary systems can be detected by each of the three different types of observations we have discussed. Positional measurements, of course, tell us when two

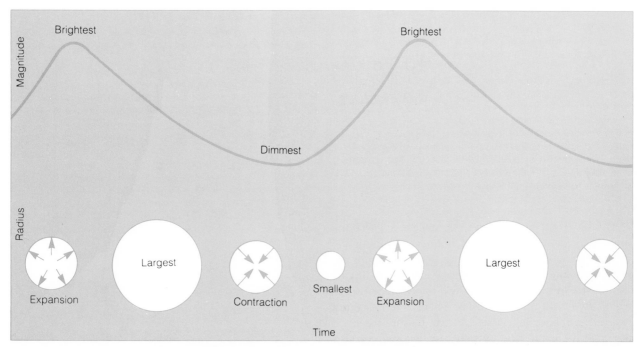

FIGURE 13.11. *A Pulsating Variable Star.* The curved line shows how the brightness (in magnitude units) varies as the star expands and contracts. The series of sketches illustrates how the expansion and contraction phases are related in sequence to the variations in brightness. The surface temperature also varies.

stars are very near each other in the sky. Sometimes this occurs by chance, when a nearby star happens to lie nearly in front of one that is in the background (Fig. 13.12). This is called an **optical binary,** and is not a true binary system, since the two stars do not orbit each other. When a pair of stars is seen close together, and measurements show that the stars are in motion about one another, this is called a **visual binary.** Accurate positional measurements are needed to reveal the orbital motion, because the two stars are very close together in the sky, and because they appear to move very slowly about each other. The distance between the two stars must be many AUs in order for both stars to appear separately as seen from the earth, and therefore the orbital period is many years.

It is possible for binary systems to be recognized even when only one of the two stars can be seen. If positional measurements over a long period of time reveal a wobbling motion of a star as it moves through space (Fig. 13.13), it can be inferred that the star is orbiting an unseen companion. Such systems, detected because of variations in position, are called **astrometric binaries.**

Simple brightness measurements can also tell us when a star, which may appear to be single, is actually part of a binary system. If we happen to be aligned with the plane of the orbit, so that the two stars alternately pass in front of each other as we view the system, then the observed brightness will decrease each time one star is in front of the other. This is called an **eclipsing binary** (Fig. 13.14).

FIGURE 13.12. *An Optical Binary.*
A chance alignment of two stars at different distances may give the appearance of a double star. Here stars A and B appear very close together, as seen from earth, but are in fact widely separated.

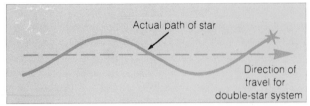

FIGURE 13.13. *An Astrometric Binary.*
In some cases, careful observation of the motion of a star across the sky reveals a curved path, such as the one shown here. Such a motion is caused by the presence of an unseen companion star, so that the visible star orbits the center of mass of the binary system as it moves across the sky.

Finally, spectroscopic measurements can also be used to recognize binary systems, again even in the case where a star appears single. If two stars are actually present, and if they have nearly equal brightnesses, spectral lines from both will appear in the spectrum. If we find a star with line patterns representing two different spectral classes, we can be sure there are two different stars; such a system is called a **spectrum binary.** Even if it is not clear that two different spectra are present (that is, if the two stars have similar spectral classes, or if one is too faint to contribute noticeably to the spectrum), the Doppler effect may reveal the fact that the star is double. Because the stars are orbiting each other, they move back and forth along our line of sight (as long as we are not viewing the system face-on), and this motion produces alternating blueshifts and redshifts of the spectral lines (Fig. 13.15). Binaries in which these periodic Doppler shifts are seen are called **spectroscopic binaries,** and they are more common than spectrum binaries.

A given double-star system may fall into more than one of these categories. For example, a relatively nearby system seen edge-on could be a visual binary, if both stars can be seen in the telescope; it may also be an eclipsing binary, if the two stars alternately pass in front of each other; and it almost surely will be a spectroscopic binary, since the motion back and forth along our line of sight is maximized when we view the orbit edge-on.

Binary stars merit our attention in part because there are so many of them, but even more important, because of what they can tell us about the properties of individual stars. As we will see in the next section, much of our basic information on the nature of stars comes from measurements of binary systems.

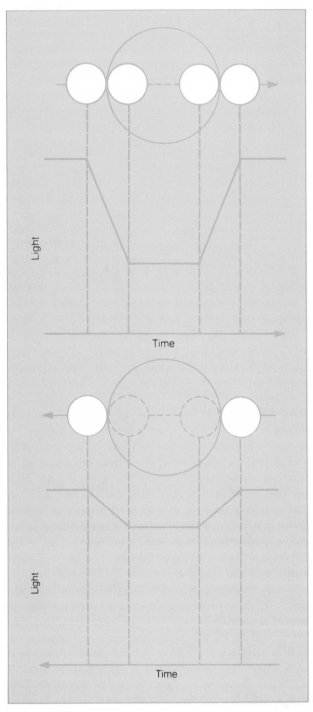

FIGURE 13.14. *An Eclipsing Binary.*
If our line of sight happens to be aligned with the orbital plane of a double-star system, the two stars will alternately eclipse each other, and this causes brightness variations in the total light output from the system, as shown.

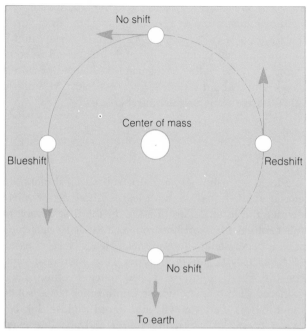

FIGURE 13.15. *A Spectroscopic Binary.*
As a star orbits the center of mass of a binary system, its velocity relative to the earth varies. This produces alternating redshifts and blueshifts in its spectrum, as the star recedes from and approaches us. Only one of the stars is shown here to be moving, although in reality both do.

Fundamental Stellar Properties

Before being able to understand how stars work, astronomers needed to determine their physical properties and how they are related to each other. A number of fundamental quantities characterize a star, and in these sections we will see how the observations described previously can be used to determine those quantities. These include the luminosities, the temperatures, the radii, and above all, the masses of stars. We will also learn about a new, very powerful distance-determination method used to measure the distances to stars.

Absolute Magnitudes and Stellar Luminosities

The luminosity of a star is the total amount of energy the star emits from its surface, in all wavelengths, and it is usually measured in units of ergs per second (see

chapter 2). To make such a measurement, astronomers must take into account the distance to a star and the light received from the star at *all* wavelengths, not just those wavelengths to which the human eye responds. In order to eliminate the effect of distance, astronomers use the **absolute magnitude,** defined as the magnitude a star would have if it were seen at a standard distance of 10 parsecs (Fig. 13.16). This distance was chosen arbitrarily, but is used by all astronomers. A comparison of absolute magnitudes reveals differences in luminosities, because all the effects of distance have been canceled out.

It helps to consider a specific example. Suppose a certain star is 100 parsecs away, and has an **apparent magnitude** (the observed magnitude) of 7.3. To determine the absolute magnitude, we must find out what this star's magnitude would be if it were only 10 parsecs away. The inverse-square law tells us that since the star would be a factor of 10 closer, it would appear a factor of $10^2 = 100$ brighter. Thus, if it were only 10 parsecs away rather than 100, this star would be a factor of 100 brighter, corresponding to exactly 5 magnitudes brighter. Therefore, its absolute magnitude is $7.3 - 5 = 2.3$. By following a similar line of thought, we can derive a star's absolute magnitude also from the measured apparent magnitude and the distance. Other examples are given in the review questions at the end of this chapter and in appendix 9.

FIGURE 13.16. *The Absolute Magnitude.*
We imagine that we can move stars from their true positions to a uniform distance from us of 10 parsecs. The magnitude a star would have at this distance is called the absolute magnitude, and is directly related to the stellar luminosity, since the distance effect has been accounted for.

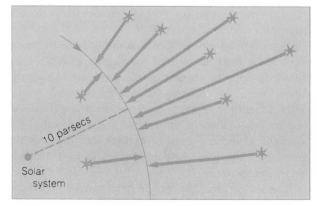

Let us now return to the question of luminosities. By determining the absolute magnitudes of stars, we can compare their luminosities, since the distance effect has been removed. Since we must allow for light emitted by a star at all wavelengths, the bolometric absolute magnitude is the quantity that is used. Thus, if one star's bolometric absolute magnitude is 5 magnitudes smaller than another's, we know that its luminosity is a factor of 100 greater. To express the luminosity of a star in terms of ergs per second, the star is compared with a standard star for which a direct measurement of the luminosity has been made by actually measuring the intensity of light at all wavelengths that reaches the earth, and allowing for distance.

What astronomers find when they determine stellar luminosities is that the values from star to star can vary over an incredible range. There are stars with luminosities as small as 10^{-4} that of the sun and as great as 10^6 that of the sun, a range of 10 billion from the faintest to the most luminous! Luminosity is by far the most highly variable parameter for stars; the other parameters that we will discuss cover ranges of only a few hundred or so from one extreme to the other.

Stellar Temperatures

Earlier in this chapter, we learned that the color of a star depends on the star's temperature, as does the spectral class, so that by observing either, we can deduce the temperature. Recall that the color index, the difference between the blue (B) and visual (V) magnitudes, is a measured quantity that indicates temperature. A negative value of $B - V$ means that the star is brighter in blue than in visual light, and therefore is a hot star. A large positive value indicates a cool star. A specific correlation of color index with temperature has been developed, and is used to determine temperatures from observed values of $B - V$.

More refined estimates of temperature can be made from a detailed analysis of the degree of ionization, which is done by measuring the strengths of spectral lines formed by different ions. This is basically the same as simply estimating the temperature from the spectral class, since in either case the point is that the strength of spectral lines of various ions depends on how abundant those ions are, which in turn depends on the temperature.

The temperature referred to here may be called the surface temperature, although stars do not have solid surfaces. We are really referring to the outermost layers of the atmosphere, where the absorption lines form. This region is called the photosphere of a star, and it actually has some depth (although it is very thin compared with the radius of the star). Stellar temperatures range from about 2,000 K for the coolest M stars to 50,000 K or more for the hottest O stars.

The Hertzsprung-Russell Diagram and a New Distance Technique

We have seen that temperature and spectral class are closely related, and we have learned how astronomers deduce the luminosities of stars. In the first decade of this century, the Danish astronomer Einar Hertzsprung and, independently, the American Henry Norris Russell (Fig. 13.17), began to consider how luminosity and spectral class might be related to each other. Each gathered data on stars whose luminosities (or absolute magnitudes) were known, and found a close link between spectral class (temperature) and absolute magnitude (or luminosity). This relationship is best illustrated in the diagram constructed by Russell in 1913 (Fig. 13.18), now called the **Hertzsprung-Russell,** or **H-R diagram.** In this plot of absolute magnitude (on the vertical scale) versus spectral class (on the horizontal axis), stars fall into narrowly defined regions

FIGURE 13.17. *Henry Norris Russell.*
One of the leading astrophysicists during the era when an understanding of the physical processes of stars was emerging, Russell made many major contributions in a variety of areas.

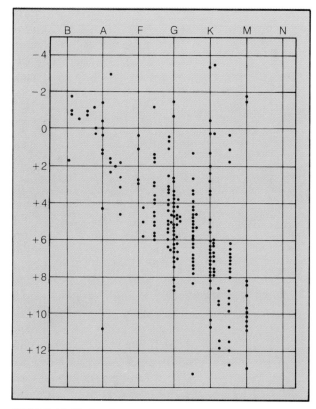

FIGURE 13.18. *The First H-R Diagram.*
This is the first plot showing absolute magnitude versus spectral class, constructed in 1913 by Henry Norris Russell.

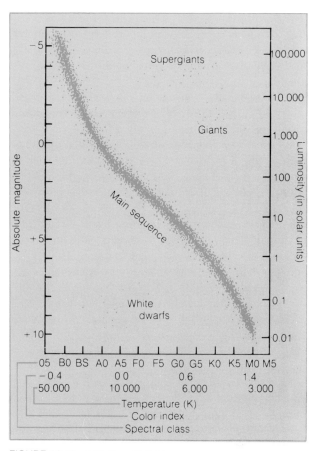

FIGURE 13.19. *A Modern H-R Diagram.*
This diagram shows the locations of a large number of stars, and gives alternative units on both axes for the two general parameters, luminosity and temperature.

rather than being randomly distributed (Fig. 13.19). A star of a given spectral class cannot have just any absolute magnitude, and vice versa.

Most stars fall into a diagonal strip running from the upper left (high temperature, high luminosity) to the lower right (low temperature, low luminosity). This strip has been given the name the **main sequence**. A few stars are not in this sequence, but instead appear in the upper right (low temperature, high luminosity). Since the spectra of these stars indicate that they are relatively cool, their high luminosities cannot be due to greater temperatures than the main-sequence stars of the same type. The only way one star can be a lot more luminous than another of the same temperature is if it has a lot more surface area; recall that the two stars will emit the same amount of energy per square centimeter of surface. Hertzsprung and Russell realized that these extra-luminous stars located above the main sequence must be much larger than those on the main

sequence, and they named these stars **giants** and **supergiants**.

The distinction among giants, supergiants, and main-sequence stars (commonly known as **dwarfs**) has been incorporated into the spectral classification system used by modern astronomers. A **luminosity class** (Fig. 13.20) has been added to spectral type. The luminosity classes, designated by Roman numerals following the spectral type, are I for supergiants (this group is further subdivided into classes Ia and Ib), II for extreme giants, III for giants, IV for stars just a bit above the main sequence, and V for main-sequence stars, or dwarfs (not to be confused with white dwarfs). Thus, a complete spectral classification for the bright summertime star Vega, for example, is A0V, meaning that it is an A0 main-sequence star. The red

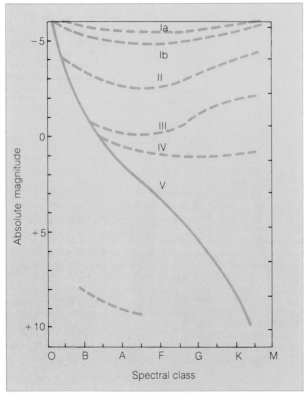

FIGURE 13.20. *Luminosity Classes.*
This H-R diagram shows the locations of stars of the luminosity classes described in the text. A complete spectral classification for a star usually includes a luminosity-class designation, if it has been determined.

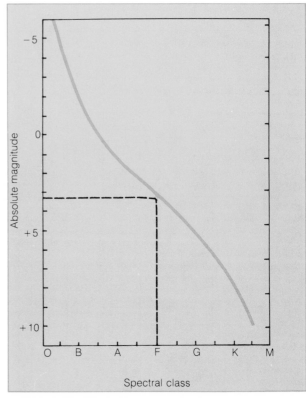

FIGURE 13.21. *Spectroscopic Parallax.*
Knowledge of a star's spectral class (including the luminosity class) allows the absolute magnitude, and hence the distance, to be determined. Here it is shown how the absolute magnitude of an FO main-sequence star is ascertained, leading to an absolute-magnitude estimate of +3.3. The distance can be found by comparing the absolute and apparent magnitudes of the star (this is explained in the text and in the appendixes).

supergiant Betelgeuse, in the shoulder of Orion, has the full classification M2Iab (it is intermediate between luminosity classes Ia and Ib). It is usually possible to assign a star to the proper luminosity class from examination of subtle details of its spectrum. (For example, giant and especially supergiant stars have relatively low atmospheric pressures, which affect the spectral-line widths and the state of ionization of certain elements.

Another group of stars has become known (mostly since the time when Russell first plotted the H-R diagram), that do not fall into any of the standard luminosity classes, but which instead appear in the lower left (high temperature, low luminosity) corner of the diagram. Since these stars are hot but not very luminous, they must be very small, and they have been given the name **white dwarfs.** These objects have some very bizarre properties, which will be discussed in chapter 16.

The H-R diagram can be used to find distances to stars, even stars that are very far away (Fig. 13.21). The idea is really very simple: if we know how bright a star is intrinsically (that is, how much energy it is actually emitting from its surface), and we measure how bright it appears to be, we can determine how far away it is, because we know that the difference between its intrinsic brightness and its observed brightness is due to the distance. The only problem lies in knowing the intrinsic brightness of the star (its luminosity), and this is where the H-R diagram comes in.

Once we determine the spectral class of a star, we can place it on the H-R diagram (as long as we are sure we know the luminosity class, so we know whether it is on the main sequence or is a giant or supergiant). Then we simply read off of the vertical axis the absolute magnitude of the star, which is a measure of the star's luminosity. A comparison of the absolute magnitude with the observed apparent magnitude, therefore, amounts to the same thing as a comparison of the intrinsic and apparent brightnesses of the star, and from such a comparison the distance can be found.

Let us consider a few examples of how the difference between the apparent and absolute magnitudes yields a star's distance. If the difference $m - M$ (the apparent minus the absolute magnitude), which is called the **distance modulus,** is 5, then the star appears 5 magnitudes, or a factor of 100, fainter than it would at the standard distance of 10 parsecs. A factor of 100 in brightness is created by a factor of 10 change in distance, so this star must be 10 times farther away than it would be if it were at 10 parsecs distance; therefore, it is $10 \times 10 = 100$ parsecs away. Similarly, a star whose distance modulus $m - M$ is 10 is 1,000 parsecs away. If $m - M = 15$, then the distance is 10,000 parsecs. It should be obvious that if $m = M$ (that is, $m - M = 0$), then the distance to the star must be 10 parsecs, because this is the distance that defines the absolute magnitude. As a rule of thumb, it helps to remember that for every 5 magnitudes of difference between the apparent and absolute magnitudes, the distance increases by a factor of 10.

This method is very powerful because it can be used for very large distances. We need only to place a star on the H-R diagram so that its absolute magnitude can be determined, and to measure its apparent magnitude. Because this distance-determination technique requires that the spectrum of a star be classified before the star can be placed on the H-R diagram, it is called the **spectroscopic parallax** method. (The word *parallax* is used by astronomers as a general word for distances, even though, technically speaking, no parallax is measured in this case.)

Stellar Diameters

We have learned that a star's position in the H-R diagram depends partly on the star's size, since the luminosity is related to the total surface area. If two stars have the same surface temperature (and therefore the same spectral class), but one is more luminous than the other, we know that it must also be larger. The Stefan-Boltzmann law (see chapter 3) specifically relates luminosity, temperature, and radius; use of this law allows the radius to be determined if the other two quantities are known.

Eclipsing binaries provide another means of determining stellar radius, one that is independent of other properties. Recall that these are double-star systems in which the two stars alternately pass in front of each other, as we view the orbit edge-on. The eclipsing binary is very likely to also be a spectroscopic binary, so the speeds of the two stars in the orbits can be measured from the Doppler effect. We therefore know how fast the stars are moving, and from the duration of the eclipse we know how long it takes one star to pass in front of the other. The simple formula, distance = speed $\times$ time, thus gives us the diameter of the star that is being eclipsed. Even if no information on the orbital velocity is available, the relative diameters of the two stars can be deduced from the relative durations of the alternating eclipses.

Eclipsing binaries provide the most direct means of measuring stellar sizes, but unfortunately, there are not many of them. In most cases, the radii are estimated from the luminosity and temperature, as just described. In a few cases, stellar radii have been measured directly by use of speckle interferometry, a sophisticated technique for clarifying the image of a star by removing the blurring effects of the earth's atmosphere (see chapter 12). Only relatively nearby, large stars can be measured this way, however.

Stars on the main sequence do not vary greatly in radius, ranging from perhaps 0.1 times the sun's radius for the M stars at the lower right-hand end to 10 or 20 solar radii at the upper left. Of course, large variations in size occur as we go away from the main sequence, either toward the giants and supergiants, which may be 100 times the size of the sun, or toward the white dwarfs, which are as small as 0.01 times the size of the sun.

Binary Stars and Stellar Masses

The mass of a star is the most important of all its fundamental properties, for it is the mass that governs most of the others. This point will be discussed at some length in the next chapter.

Unfortunately, there is no direct way to see how much mass a star has. The only way to measure a star's mass is by observing its gravitational effect on other objects, and this is possible only in binary-star systems, where the two stars hold each other in orbit by their gravitational fields. However, binary systems are common, so we have many opportunities to determine masses by analyzing binary orbits.

The basic idea is rather simple, although the application may be quite complex, depending on the type of binary system. Kepler's third law is used, in the form derived by Newton. Remember that if the period P is measured in years, and the average separation of the two stars (the semimajor axis a) is measured in astronomical units, and the masses m_1 and m_2 are measured in units of solar masses, then Kepler's third law is

$$(m_1 + m_2) P^2 = a^3$$

We need only observe the period and the semimajor axis in order to solve for the sum of the two masses. Careful observations of the sizes of the individual orbits

FIGURE 13.22. *Binary-Star Orbits.*
This figure illustrates the terms used in Kepler's third law. Two stars of masses m_1 and m_2 (m_1 is larger than m_2 in this case) orbit a common center of mass, each making one full orbit in period P. The semimajor axis a that appears in Kepler's third law is actually the sum of the semimajor axes of the two individual orbits about the center of mass; this sum corresponds to the average distance between the two stars.

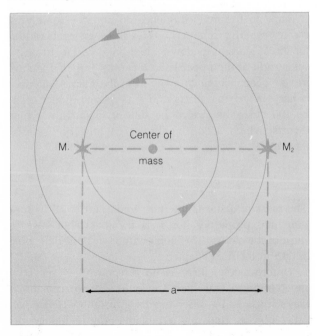

(actually, of the relative distances of the two stars from the center of mass; see Fig. 13.22) also yields the ratio of the masses. When both the sum and the ratio are known, it is simple to solve for the individual masses.

Complications arise when some of the needed observational data are difficult to obtain, however. The period is almost always easy to measure with some precision, but not so the semimajor axis. The main problems are that the apparent size of the orbit is affected by our distance from the binary, and so the distance must be well known if a is to be accurately determined; and that the orbital plane is inclined at a random, unknown angle to our line of sight, so that the apparent size of the orbit is foreshortened by an unknown amount. In some cases it is possible to unravel these confusing effects by carefully analyzing the observations, and in other cases it is not. Even so, some information about stellar masses can be gained, but usually only in terms of broad ranges of possible values.

The masses of stars vary along the main sequence from the least massive stars in the lower right to the most massive at the upper left. The M stars on the main sequence have masses as low as 0.05 solar masses, whereas the O stars reach values as great as 60 solar masses. It is likely that stars occasionally form with even greater masses (perhaps up to 100 solar masses), but as we will see in the next chapter, such massive stars have very short lifetimes, so it is rare to find one.

The giants, supergiants, and white dwarfs have masses comparable to those of main-sequence stars. Hence their obvious contrasts with main-sequence stars in other properties such as luminosity and radius have to be the result of something other than extreme or unusual masses. This is discussed in the next chapter.

For main-sequence stars, there is a smooth progression of all stellar properties from one end to the other. The mass and the radius vary by similar factors, whereas the luminosity changes much more rapidly along the sequence. The mass of an O star is perhaps 100 times that of an M star, whereas the luminosity is greater by a factor of 10^8 or more. There is a **mass-luminosity relation** among main-sequence stars, a numerical expression in which the luminosity of a star is proportional to an exponential power of the mass. In its simplest form, this relation states that the luminosity is proportional to the cube of the mass. More precise versions of the mass-luminosity relation reflect some variation in the exponent along the main sequence.

◄ Jupiter and the
Galilean satellites

◄ Turbulent motions in
the Jovian atmosphere

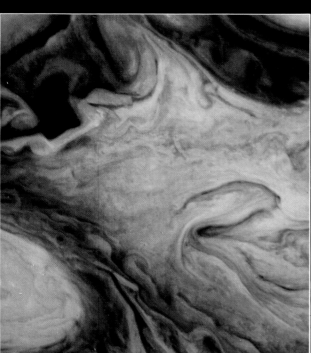

Io against the backdrop of Jupiter ▼

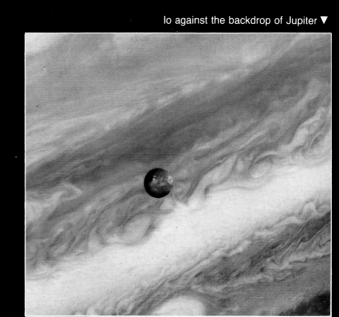

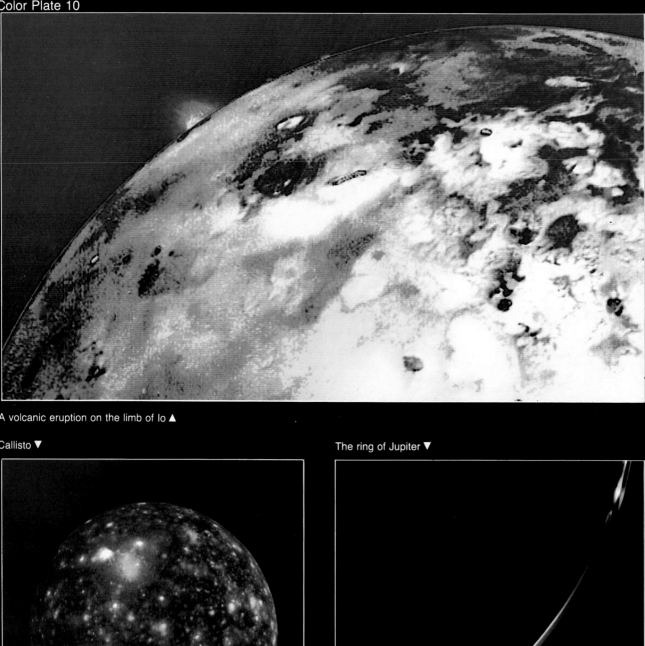

A volcanic eruption on the limb of Io ▲

Callisto ▼

The ring of Jupiter ▼

The rings of Saturn ▲

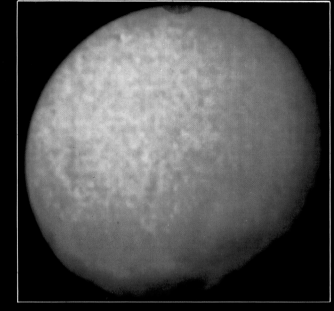

Enceladus ▲

◀ Saturn and some
of the major satellites

Color Plate 12

◀ A false-color
image of saturn

Structure in
Saturn's atmosphere ▼

Ring structure reconstructed
from *Voyager* photopolarimeter data ▼

An ultraviolet false-color image of Comet Kohoutek ▲

False color representing ▲
brightness structure of Comet Bennett

Comet Kohoutek ▼

Dynamics of the sun's corona ▲

Solar active regions seen in extreme ultraviolet light ▲

A solar prominence ▼

◄ The Pleiades

An X-ray image of
a group of O and B stars ►

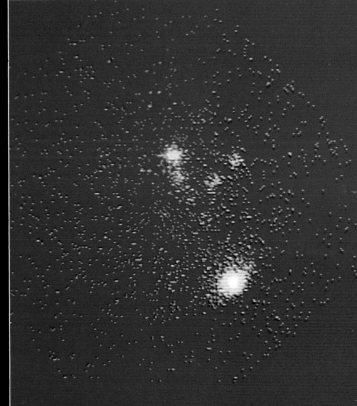

A false-color image
of the red supergiant
Star Betelgeuse, showing
convective structures
on the surface ▼

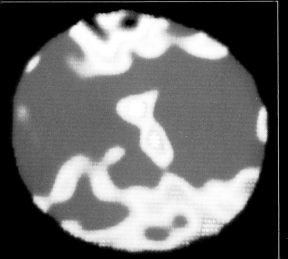

The Cone Nebula ▼

The Dumbbell Nebula, a planetary nebula ▲

A false-color image of the
nebulosity surrounding the
young massive star Eta Carina;
the colors represent
emission by different ions
▼

An X-ray image of the Eta Carina region ▼

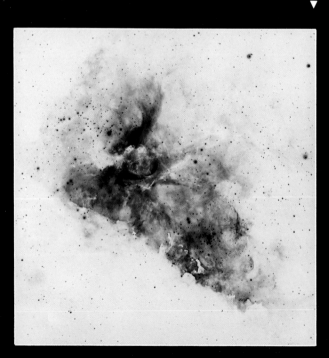

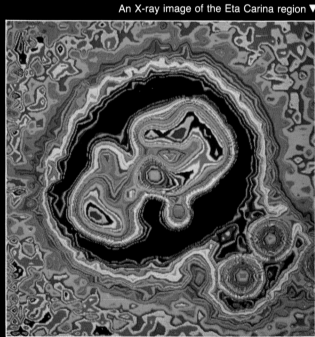

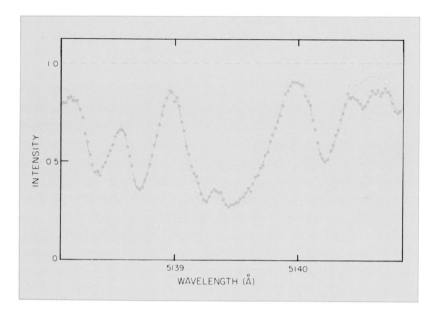

FIGURE 13.23. *Determination of Stellar Composition*. This diagram illustrates a modern technique for measuring the chemical composition of a star. A theoretical spectrum (solid line) is compared with the observed spectrum (dotted and dashed line). The abundances of elements assumed present are altered in the computed spectrum until a good match is achieved.

Other Properties

There are several other properties of stars that can be determined, primarily from analyses of their spectra. Perhaps the most fundamental of these is the composition. Stars are generally made of the same material in the same proportions, but a sophisticated analysis was required to show this. It was necessary first to develop techniques for taking the effects of temperature into account, for these effects, as we mentioned earlier, are dominant factors in controlling the strengths of lines in a stellar spectrum. In modern work on stellar composition, scientists use complex computer programs to calculate simulated spectra with different assumed compositions until they find a match with the observed spectrum (Fig. 13.23).

Another stellar property that can be learned from the spectrum is the rotational velocity at the surface, because the Doppler effect causes broadening of the spectral lines (Fig. 13.24). A spectral line forms over the entire disk of a star, and generally one edge of the disk is approaching the earth, and the other is receding. Hence some of the gas creates a blueshift, and some a redshift (and the gas in the central portions of the disk has little or no shift), so the rate of rotation of the star determines the degree of broadening of the spectral lines.

The last property to be discussed is the magnetic

FIGURE 13.24. *The Effect of Rotation on Stellar Absorption Lines*. Rotation of a star broadens its spectral lines because of the Doppler effect. Here the same line is shown as it would appear in the spectum of a slowly rotating star (top) and one that is spinning rapidly (bottom).

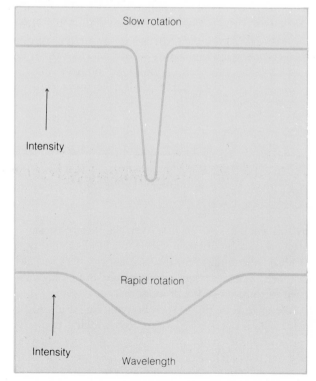

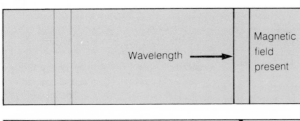

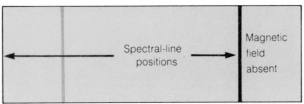

FIGURE 13.25. *The Measurement of a Magnetic Field.*
The presence of a magnetic field causes certain absorption
lines of some elements to be split, in a process called Zeeman
splitting. The amount of separation is a measure of the strength
of the magnetic field.

field of a star. It is difficult to assess the importance of
this property because magnetic fields are often hard to
measure. The Zeeman effect, a process whereby certain
spectral lines are split, owing to the presence of a magnetic field (see chapter 11) allows astronomers in some
cases to determine magnetic-field strength (Fig. 13.25).
However, the splitting of spectral lines can be seen only
in very slowly spinning stars whose lines are very narrow. For the many stars that have broader spectral
lines, we have no reliable method of measuring the
magnetic-field strength.

PERSPECTIVE

In this chapter we have learned the basics of stellar
observations, by discussing the three fundamental types
of observations and the characteristics of stars revealed
by each. Considerable astronomical terminology has
been introduced in the process.

We have gone on to learn how all the basic properties of stars are derived from observational data. We
can categorize stars, classify them; describe them in
any way we wish. Now we are ready to see how stars
work; *why* the various quantities are related the way
they are, and no other way.

SUMMARY

1. There are three basic types of stellar observation:
measurements of positions, brightnesses, and spectra.

2. Positional astronomy (astrometry) has developed
techniques capable of measuring stellar positions to an
accuracy approaching 0.001 arcseconds, which is sufficient to detect proper motions, binary motions in
some cases, and stellar parallaxes for stars up to 1,000
parsecs away.

3. Astronomers generally carry out stellar brightness
measurements by using the stellar magnitude system,
in which a difference of one magnitude corresponds to
a brightness ratio of 2.512.

4. Magnitudes can be measured at different wavelengths, allowing the determination of color indices; or
over all wavelengths, resulting in the determination of
bolometric magnitudes.

5. Stellar spectra contain patterns of absorption lines
that depend on the surface temperatures of stars, and
which therefore can be used to assign stars to spectral
classes that represent a sequence of temperatures.

6. Peculiar stars are those which do not conform to
the usual spectral classes, most often because they have
unusual surface compositions, but sometimes because
they have emission lines.

7. Astronomers can detect binary-star systems on the
basis of positional variations of one or both stars (astrometric binaries), brightness variations (eclipsing binaries), or composite or periodically Doppler-shifted
spectral lines (spectrum or spectroscopic binaries).

8. We determine stellar luminosity through knowledge of the distance to a star and the star's apparent
magnitude. The absolute magnitude, a measure of luminosity, is the magnitude a star would have if it were
seen from a distance of 10 parsecs.

9. Stellar temperature can be inferred from the $B-V$
color index, estimated from the spectral class, or determined from the degree of ionization in the star's outer
layers.

10. The Hertzsprung-Russell diagram shows that the
luminosities and temperatures of stars are closely related, and that stars which do not fall on the main sequence are either larger (as in the case of the red
giants) or smaller (white dwarfs) than those on the
main sequence.

11. We can measure the distance to a star by first determining the star's spectral class, then using the H-R diagram to infer the star's absolute magnitude, and finally comparing the absolute magnitude with the apparent magnitude to yield the distance. This technique is called spectroscopic parallax.

12. We can determine stellar diameters directly in eclipsing binaries, through knowledge of the orbital speed and the duration of the eclipses.

13. Through the use of Kepler's third law, we can derive stellar masses in binary systems, when the period and the orbital semimajor axis are observed.

sition of the sun on the H-R diagram must have a larger radius. If its luminosity is a hundred times greater than that of the sun, how much larger is its radius?

9. Summarize the various reasons it is difficult to determine the masses of a pair of stars in a binary system, even though in the ideal case this can be done.

10. Explain why the spectral lines of a particular element (say, hydrogen) may have quite different strengths in two stars, whereas the abundance of hydrogen may be the same in the two stars.

REVIEW QUESTIONS

1. Does the sun have annual parallax motion, as seen from the earth?

2. How much fainter is a twelfth-magnitude star than an eleventh-magnitude star? How does a fourth-magnitude star compare in brightness with a fifth-magnitude star?

3. Briefly discuss the role of stellar parallax in the development of the heliocentric theory.

4. Explain why the bolometric magnitude of a star is always smaller than the visual magnitude.

5. Explain, in terms of what you learned about ionization in chapter 3, why stars of different temperature have different absorption lines in their spectra.

6. If a certain star has absolute bolometric magnitude $M_{bol} = 1.5$, and another has $M_{bol} = 4.5$, which is more luminous, and by how much?

7. If one star is four times hotter than another, and has twice as large a radius, how does its luminosity compare with that of the other star? (Note: you must review some material in chapter 3).

8. Explain why a G2 star that lies well above the po-

ADDITIONAL READINGS

Probably the best sources of additional information on stars and their properties are general astronomy textbooks, of which there are a number readily available in nearly any library. A few more specific resources are listed here as well.

Abt, H. A. 1977. The companions of sun-like stars. *Scientific American* 236(4):96.

Aller, L. H. 1971. *Atoms, stars, and nebulae.* Cambridge, Ma.: Harvard University Press.

De Vorkin, D. 1978. Steps towards the Hertzsprung-Russell diagram. *Physics Today* 31(3):32.

Evans, D. S., Barnes, T. G., and Lacy, C. H. 1979. Measuring diameters of stars. *Sky and Telescope* 58(2):130.

McAlister, H. A. 1977. Binary star speckle interferometry. *Sky and Telescope* 53(5):346.

Mihalas, D. 1973. Interpreting early-type stellar spectra. *Sky and Telescope* 46(2):79.

Philip, A. G. D., and Green, L. C. 1978. The H-R diagram as an astronomical tool. *Sky and Telescope* 55(5):395.

Struve, O., and Zebergs, V. 1962. *Astronomy of the 20th century.* New York: Crowell Collier and Macmillan.

Upgren, A. 1980. New parallaxes for old: a coming improvement in the distance scale of the universe. *Mercury* 9(6):143.

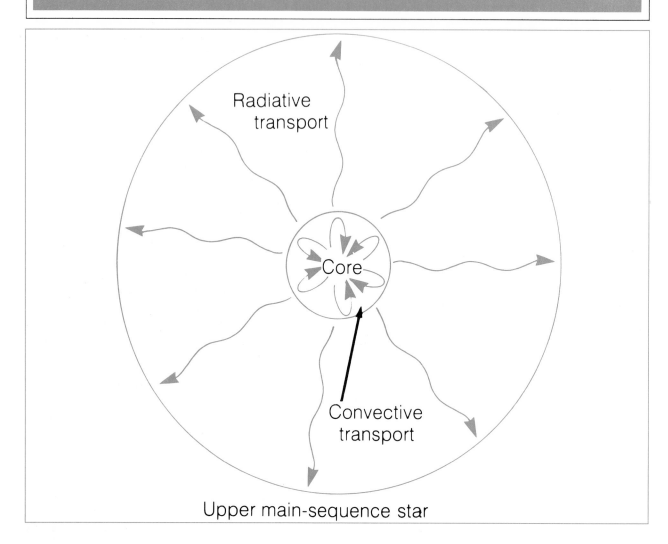

Upper main-sequence star

Learning Goals

THE NATURE OF A STAR
HYDROSTATIC EQUILIBRIUM AND THE ROLE OF MASS
NUCLEAR REACTIONS AND ENERGY TRANSPORT
STELLAR LIFE EXPECTANCIES

HEAVY-ELEMENT ENRICHMENT
STELLAR CHROMOSPHERES AND CORONAE
STELLAR WINDS AND MASS LOSS
MASS TRANSFER IN BINARY SYSTEMS
STELLAR MODELS

We have learned how astronomers determine from observations the physical characteristics of stars. From this we have discovered a great deal about the stars' surface properties, but very little about what goes on inside them, or how they evolve. To understand these secrets, we must apply the laws of physics and calculate theoretically the internal structure and evolution of stars, and the calculations must reproduce the observable properties. If this is done successfully, we can be somewhat confident that the calculations also accurately describe the internal conditions of stars.

What is a Star Anyway?

We have described a star's surface as being a hot, gaseous region, often sufficiently hot that the gas is ionized. Because only absorption lines form in this region, we know that deeper in the star it must be hotter, so that continuous radiation is created there and flows out through cooler layers before escaping into space. The surface layer where the continuous spectrum is emitted and the spectral lines form (called the **photosphere**) has a low density, well below that of the earth's atmosphere at sea level. The density must be very high in the star's interior, however, because the average density (the total mass divided by the volume of the star) is close to that of water, or roughly 10^8 times greater than the surface density. To achieve such a high average density when the outer layers are so rarefied requires a still higher interior density, perhaps 100 times that of water, or about 30 times that of rock.

Despite this, a star is not solid inside. Although the density increases toward the center, so does the temperature, keeping the star in a gaseous state. Temperatures range from 10 to 100 million degrees absolute in the cores of stars. Under such extreme conditions, the gas particles whiz around at very high velocities (typically 10^8 cm/sec, or 1,000 km/sec), and they collide very frequently and with great impact. The result is that *all* electrons are knocked loose, and the gas is fully ionized, meaning that it consists only of bare nuclei and free electrons. This situation is different from that of the surface layer, where the atoms may have lost a few electrons, but most elements still have some electrons orbiting their nuclei.

A star, then, is a spherical ball of gas with density and temperature increasing toward the center. Most stars are made primarily of hydrogen, although as we will see, the composition changes gradually over a star's lifetime.

Hydrostatic Equilibrium and the Central Role of Mass

We may wonder what keeps a star in the state we have described. Why is it spherical? Why do density and temperature increase inward? The answer is that all the gas particles exert gravitational forces on each other, so that the star is held together by its own gravity. This force is always directed toward the center; thus, the star is forced into a symmetric, spherical shape. The fact that gas is compressible explains how gravity creates a state of high density inside.

A gas that is compressed heats up, causing it to exert greater pressure on its surroundings. Thus a star's interior is very hot and the internal pressure is high. If it were not for this pressure, gravity would cause a star to keep shrinking. A balance is struck between gravity, which is always trying to squeeze a star inward, and pressure, which pushes outward (Fig. 14.1). This balance is called **hydrostatic equilibrium,** and it plays a dominant role in determining an object's internal structure.

But what determines the balance; that is, what causes a star to reach a certain state of internal compression and no other? The answer is the mass of the star. The amount of gravitational force is set by the total mass, and this force in turn determines how much pressure is needed to balance gravity. The star will be compressed until this pressure is reached, and then it will become stable. The pressure required to balance gravity dictates the temperature inside the star. Hence a star's mass determines the internal density and temperature, as well as the overall size of the star, since this is a function of mass and density.

Luminosity is also governed largely by the mass. The luminosity of a star is simply the amount of energy generated inside it that eventually reaches the surface and escapes into space, and it is determined by the temperature in the interior. As we have just learned, the temperature is set primarily by the mass.

Thus virtually all the observable properties of a star

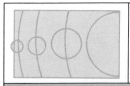

Astronomical Insight 14.1
HENRY NORRIS RUSSELL

During the first decades of the twentieth century, physics underwent a period of extraordinary change and ferment. In what could be called a second Renaissance, the structure of the atom was revealed, the secrets of the photon and the emission and absorption processes in atoms were unraveled, Einstein developed his theories of general and special relativity, and by the late 1930s, the production of energy by nuclear reactions was understood. All these discoveries had applications to stars, and among the scientists who led the way into these new territories, none showed greater insight and breadth than the American astrophysicist Henry Norris Russell. This was the era when astronomy truly became astrophysics, and Russell played a key role in the transition.

Educated at Princeton, Russell received a Ph.D. there in 1900, and was at first quite inactive owing to illness. By 1905, following a three-year stay at Cambridge University in England, Russell took a position at Princeton, resuming what was to become a

lifelong association with that university. In 1912 he was appointed director of the Princeton University Observatory, a post he was to hold for the next thirty-five years. During these years he not only made numerous fundamental breakthroughs in understanding the physics of stars, but also traveled and lectured extensively, keeping scientists at widespread locations abreast of new developments, in a sense acting as mentor to a whole generation of astronomers. His seemingly limitless energy also led him into endeavors that brought him before the public eye (he wrote more than five hundred popular articles, along with a textbook that became a standard for more than thirty years). In an era before television and instant celebrity, he was known to the public by name and by sight throughout the United States.

Russell worked in nearly every area of astronomy, in each instance quickly developing new insights and techniques. He was both an observer and a theorist. His first work involved the development of as-

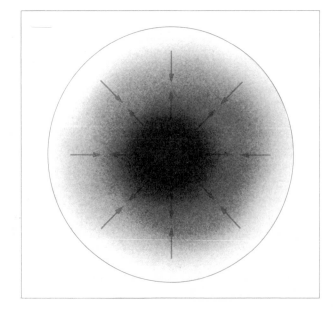

FIGURE 14.1. *Hydrostatic Equilibrium.*
This is a cutaway sketch of a star, showing its spherical shape. The arrows represent the balanced forces of gravity (inward) and pressure (outward), and the shading indicates that the density increases greatly toward the center. Equilibrium is reached when the core becomes sufficiently hot to attain the pressure necessary to counterbalance gravity.

depend on its mass (Fig. 14.2). This explains why the main sequence represents a smooth run of masses increasing from the lower right to the upper left in the H-R diagram: the luminosity and temperature, which define a star's place in the diagram, both depend on mass.

Of course, there are stars that do not fall on the main sequence, yet their masses are not different from those of main-sequence stars. This tells us there must be something other than mass that can influence a star's properties. This other parameter is the chemical composition. As we will see, this varies as a star ages.

trometric methods and the measurement of stellar distances through the trigonometric parallax technique, and his later efforts were devoted to the structure and evolution of stars, with a good deal of research on stellar spectra and the structure of atoms thrown in for good measure.

His principal work can be classified into four general areas. First, he studied double stars, and developed very sophisticated techniques for analyzing the light variations in eclipsing binaries to determine not only stellar radii, but also stellar densities and internal structure. To do this required the development of new mathematical methods, as well as a physical theory of how light is emitted from the surface of a star.

Russell's second major area of research was the study of stellar evolution, an interest that was inspired by the recognition that there exist giant and dwarf stars, and which resulted in his development of the H-R diagram. He was led into this work by his earlier studies of stellar distances, which enabled him to determine absolute magnitudes for a number of nearby stars. Along the way, Russell developed several theories about stellar evolution, each of which was consistent with what was known at the time, but each of which later had to be modified.

Russell's work on stellar spectra was inspired by research on ionization balance done by others, and his own recognition that it should be possible to measure a star's chemical composition through the analysis of its spectral lines. Having developed techniques for doing this, Russell is credited with ascertaining that the primary constituent of the sun (and, by inference, the stars in general) is hydrogen. This was a profound discovery about the nature of the universe, one that later played a major role in studies of its evolution.

Finally, Russell's fourth major area of work, the study of atomic structure (specifically, electron energy levels and the spectra they produce) was an outgrowth of his interest in stellar composition. He found a great lack of laboratory data when he sought to analyze stellar spectra, and accordingly he developed both theoretical and experimental means to supply the needed information.

Although it is true that many of Russell's advances were made through collaboration with other scientists and not in isolation, it is equally true that he was quick to rise to a role of leadership in every instance. If his era represented a new beginning in astronomy, then surely he must be regarded as a true Renaissance man of his time.

The supergiants, giants, and white dwarfs have different chemical makeups than main-sequence stars; in the next section we will learn how the differences arise.

The amount of internal compression, and hence the temperature and luminosity of a star, depend on the average mass of the individual nuclei in the star's core. The heavier the particles that make up the gas, the more tightly they are compressed by gravity, the hotter it gets, and the greater the luminosity. When a star is formed, it consists mostly of hydrogen, but as it ages,

FIGURE 14.2. *The Importance of Mass.* The mass of a star is the single quantity that governs all other stellar properties, for a given composition. The sequence shown here is generally the same, but the details vary for different chemical compositions.

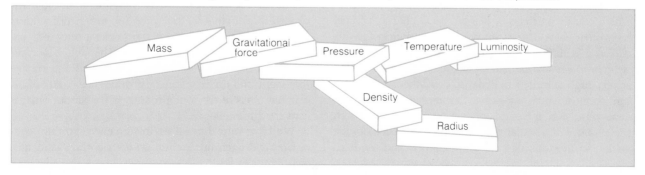

its core material is converted to helium at first, and later possibly to other, even heavier elements. In the process the core heats up and the star becomes more luminous.

The fact that all properties of a star depend on just its mass and composition was recognized several decades ago, and is usually referred to as the **Russell-Vogt theorem**, after the astrophysicists who first stated it. In modern times, it is known that other influences, such as magnetic fields and rotation, must play roles in governing a star's properties as well. Just what these roles are is not yet well understood, however.

Nuclear Reactions and Energy Transport

We have seen that the core of a star must be very hot in order to maintain the pressure required to counterbalance gravity. Since energy flows outward from the core, there must be some source of heat in the interior. If there were not, the star would gradually shrink.

Nuclear fusion reactions provide the only source of heat capable of maintaining the required temperature over a sufficiently long period of time. A particular type of reaction was described in chapter 11, where we discussed the sun's interior. Recall that in atomic fusion reactions, nuclei of light elements such as hydrogen combine to form nuclei of heavier elements. In the process, a small fraction of the mass is converted into energy according to Einstein's famous formula $E = mc^2$: it is this energy, in the form of heat and radiation, that maintains the internal pressure in a star.

Fusion reactions can occur only under conditions of extremely high temperature and density. The nuclear force that holds the protons and neutrons together in a nucleus and which causes fusion to occur only acts over very short distances. The electromagnetic force, which causes particles of like electrical charge to repel each other, acts over a much greater distance; for this reason it is difficult for nuclei in a gas to undergo nuclear fusion reactions. The nuclei must therefore collide at very high velocities in order to combine, so that they can get close enough together despite their electromagnetic repulsion for each other (Fig. 14.3). The high temperature of a stellar core imparts high speeds to the nuclei, so they collide with great energy, and the high

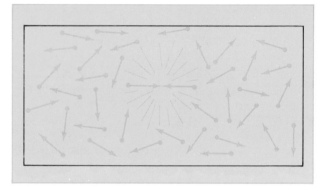

FIGURE 14.3. *Nuclear Fusion.*
The core of a star is a sea of rapidly moving atomic nuclei. Occasionally a pair of these nuclei, most of which are simple protons (hydrogen nuclei), collide with sufficient velocity to merge together, forming a new kind of nucleus (deuterium in this case, composed of one proton and one neutron). Energy is released in the process.

density means that collisions will be frequent. Even so, only occasionally do two nuclei combine in a fusion reaction; a single particle may collide and bounce around inside a star for millions or even billions of years before it reacts with another. There are so many particles, however, that reactions are constantly occurring.

The amount of energy released in a single reaction between two particles is small, about 10^{-5} erg. An erg (defined in chapter 2) is itself a small quantity; a 100-watt light bulb radiates 10^9 ergs per second. Hence a light bulb would require 10^{14} reactions per second to keep glowing, if nuclear reactions were its energy source. A star like the sun emits more than 10^{33} ergs per second, so a tremendously large number of reactions must be occurring in its interior at all times.

The reaction that takes place inside all stars on the main sequence converts hydrogen nuclei into helium nuclei. A hydrogen nucleus consists solely of a proton, whereas a helium nucleus contains two protons and two neutrons. The result of the reaction is that four hydrogen nuclei (that is, four protons) are combined into one helium nucleus; two of the protons must be converted into neutrons in the process. The helium nucleus has less mass than the total of the four hydrogen nuclei that went into the reaction. The difference, 0.007 of the original mass of the four protons, is the amount that is converted into energy.

We have said that all main-sequence stars are converting hydrogen into helium in their cores. The details of the reactions differ from one part of the main sequence to another, however. In stars on the lower portion of the main sequence, including the sun, the dominant process is a rather simple sequence of reactions called the **proton-proton chain.** The more massive stars on the upper main sequence, which have higher internal temperatures, undergo a more complex sequence called the **CNO cycle.** In this sequence, carbon acts as a *catalyst,* meaning that it is necessary to get the reactions going, but is not used up in the reactions. The term *CNO cycle* has been adopted because oxygen and nitrogen also appear in the reaction sequence. The result is the same as before: four hydrogen nuclei are combined into one helium nucleus, and 0.007 of the original mass is converted into energy. The details of the proton-proton chain and the CNO cycle are given in Appendix 10.

Once it is produced in the core, energy must somehow make its way outward to the stellar surface. For most stars, two energy-transport mechanisms are at work at different levels (Fig. 14.4). One of these is **radiative transport,** meaning that the energy is carried outward by photons of light. In the core, where it is very hot, the photons are primarily γ-rays, but as they slowly move outward, being continually absorbed and reemitted on the way, they are gradually converted to longer wavelengths. When the light emerges from the stellar surface, it is primarily in the visible-wavelength region (or the ultraviolet or infrared, if it is a very hot or very cool star).

The second means of energy transport inside stars is **convection,** a process discussed briefly in chapter 11. Convection is an overturning of the gas, as heated material rises and cooled material sinks. The same process causes warm air to rise toward the ceiling of a room, and is responsible for the overturning of water in a pot that is being heated from the bottom.

Whether radiative transport or convection is the dominant energy-transport mechanism in a star depends on the temperature structure (specifically, how rapidly the temperature decreases with distance from the center). Most stars have both a convective zone and a radiative region. In stars like the sun (those on the lower half of the main sequence, of spectral types F, G, K, and M), convection occurs in the outer layers whereas radiative transport is the principal means of energy transport in the interior. For stars on the upper main sequence, the situation is reversed: there is convection in the central core, but not in the rest of the star, where radiative transport is responsible for conveying the energy to the surface.

FIGURE 14.4. *Energy Transport Inside of Stars.* In a star on the upper portion of the main sequence, energy is transported by convection in the inner zone and by radiation in the outer regions. The opposite is true of lower main-sequence stars.

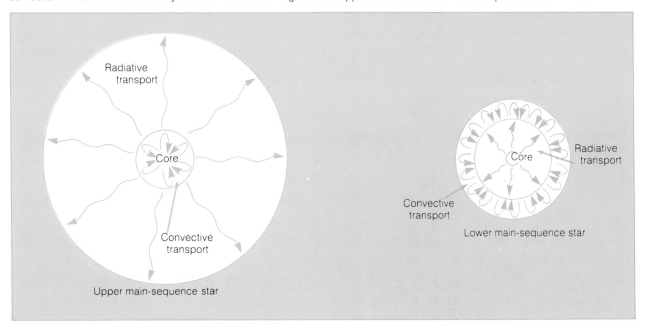

Stellar Life Expectancies

The lifetime of a star is measured by the amount of energy the star can produce in nuclear reactions, and the rate at which the energy is radiated away into space (Fig. 14.5). When all the available nuclear fuel has been used up, the star undergoes major changes in its structure and properties, and it leaves the main sequence (this will be discussed in detail in the next chapter). We can estimate the hydrogen-burning lifetime of a star simply by determining how much mass can be converted to energy in the reactions, and then using the formula $E = mc^2$ to see how much energy will be produced.

Let us consider the sun first. It has a mass of 2×10^{33} grams. Only the innermost 10 percent or so of this mass will ever undergo nuclear reactions, because only in the core are the temperature and density sufficiently high. Therefore we expect only 2×10^{32} grams of the sun to be available for reactions. Only 0.007 of this, or 1.4×10^{30} grams, will actually be converted into energy. From Einstein's formula, we find that $E = mc^2$

$= (1.4 \times 10^{30} \text{ grams}) \times (3 \times 10^{10} \text{ cm/sec})^2 = 1.26 \times 10^{51}$ ergs. This is all the energy the sun can ever produce in its lifetime on the main sequence.

To estimate how long it will take the sun to use up all this energy, we need only to take into account its luminosity, which is the rate at which the energy is produced and radiated away. For the sun, this is roughly 4×10^{33} ergs/sec, so the lifetime is $(1.26 \times 10^{51} \text{ ergs})/(4 \times 10^{33} \text{ ergs/sec}) = 3.15 \times 10^{17}$ sec. This is just about 10^{10} years. Therefore, we can expect the sun to run out of hydrogen fuel 10 billion years after it began burning it. As we learned in chapter 11, the age of the sun is roughly 4.5 billion years, so there are still more than 5 billion years to go before the fuel is expended.

What of other stars, with greater or smaller masses? A star near the top of the main sequence may have 50 times the mass of the sun, but at the same time it uses its energy as much as 10^6 times more rapidly (recall how rapidly the luminosity of a star varies with its mass, as discussed in the last chapter). These numbers lead to an estimated lifetime that is $50/10^6 = 5 \times 10^{-5}$ times that of the sun. This star will last only half a

FIGURE 14.5. *Stellar Lifetimes*. The large bucket on the left represents the large quantity of energy produced by a massive star during its lifetime, and the water gushing out of the hole at the bottom of the container depicts the star's high luminosity, or energy-loss rate. At right is a small bucket representing a low-mass star, which produces much less energy during its lifetime, but which loses the energy so slowly (in other words, the star has such a low luminosity) that it outlives the more massive star by a wide margin.

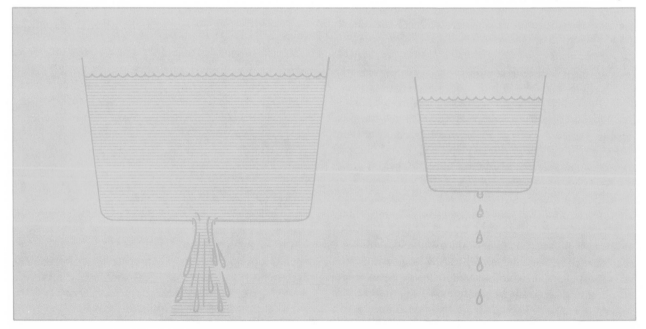

million years! (Actually, in such a massive star, more than 10 percent of the total mass can undergo reactions, and the lifetime is accordingly longer than this estimate; it is a few million years.) By astronomical standards, such massive stars exist only for an instant before using up all their fuel and dying. This is one reason why stars of this type are very rare; only a few may be around at any given moment, regardless of how many may have formed in the past. (As it happens, very massive stars also do not form very often, another reason for their rarity.)

Let us consider a lower-main-sequence star, for example an M star with mass 0.05 solar masses, and luminosity 10^{-4} times that of the sun. Its lifetime is $0.05/10^{-4} = 500$ times greater than the sun's lifetime, or about 5×10^{12} years. This is a very long time, probably greater than the age of the universe. No star with such a low mass has yet had time to use up all its fuel; all such stars ever born are still with us. For this reason, the majority of stars in existence today are low-mass stars (it is also true that these stars form in greater numbers).

Heavy-Element Enrichment

Clearly nuclear reactions have an effect on the chemical composition of a star, because they change one element into another. We have stressed that all stars consist of about the same mixture of hydrogen and helium, with a trace amount of other elements. However, now we find that in the core of a star, hydrogen is gradually converted into helium. Although this may not immediately affect a star's surface composition, which is what astronomers can measure directly from spectral analysis, it does change the internal composition. It is this change that causes a star to evolve, because the core density is altered as hydrogen nuclei combine into the heavier helium nuclei, and because when the hydrogen runs out, the star must make major structural adjustments as its primary source of internal pressure disappears.

These adjustments are discussed in chapter 15; for now, let us consider only the change in abundances of the elements. There are other nuclear reactions that can take place in stars after the hydrogen-burning stage has ended, if the core temperature reaches sufficiently high levels. One such reaction is the conversion of helium into carbon, by a sequence called the **triple-alpha reaction** (helium nuclei are called alpha particles, and in this reaction, three of these combine to form one carbon nucleus). When the triple-alpha reaction takes place, the composition of the star's core changes from helium to carbon.

There are many additional reactions that can occur, if in later stages the core temperature goes even higher (Fig. 14.6). These reactions produce ever-heavier products, so that as a star goes through successive stages of nuclear burning, its internal composition changes from

FIGURE 14.6. *The Enrichment of Heavy Elements*. This diagram schematically illustrates the sequence or element formation that occurs in a stellar core, as one nuclear fuel after another is exhausted. The first step, the conversion of hydrogen into helium, occurs in all stars, but the number of subsequent steps that a star goes through depends on its mass. The heaviest element that can be formed by stable reactions inside stars is iron.

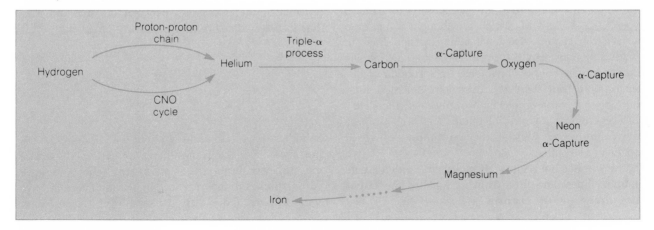

what had originally been predominantly hydrogen to elements as heavy as iron (which has 26 protons and 26 neutrons in its nucleus).

These reactions require ever-higher temperatures, because as the particles become more massive, it takes more energy to keep them moving fast enough to react. Furthermore, the heavier nuclei have greater electrical charges and therefore stronger repulsive forces that tend to keep them apart. For both these reasons, reactions involving the fusion of heavy nuclei require higher temperatures than those involving the fusion of light-weight nuclei such as hydrogen.

It is the most massive stars that go through the greatest number of reaction stages and, as we have just mentioned, these stars live only a short time. In later chapters (16 and especially 19) we will discuss the role played by these massive stars in the enrichment of chemical abundances in the galaxy.

Stellar Chromospheres and Coronae

Let us now consider other processes that may affect a star's structure and evolution. Some very important ones take place near the surface, rather than deep inside.

As we have discussed, in most stars convection occurs at some level in the interior. For stars on the upper portion of the main sequence, it takes place near the center, and there are no directly observable consequences. For stars on the lower portion of the main sequence, however, convection occurs in the outer layers, and there are important effects. Stars of spectral types F, G, K, and M, which are thought to undergo surface convection, also have faint emission lines in their visible-wavelength spectra. These lines are usually almost drowned out by the intense light from the stellar photosphere, but there are ultraviolet emission lines that are easy to observe (Fig. 14.7) because the stellar surface emits little ultraviolet light (remember, these are relatively cool stars, so they emit most strongly at longer wavelengths).

The sun has hot regions above its surface that create emission lines, and these zones are called the **chromosphere** and the **corona** (see chapter 11). The fact that other cool stars have the same kinds of emission lines indicates that they, too, have chromospheres and coronae (Fig. 14.8). The sun's corona is extremely hot, exceeding 10^6 K, some two hundred times hotter than the photosphere. Apparently very high temperatures also exist in the coronae of other stars, although as yet the observations have not been sufficiently extensive to allow precise determinations. The emission lines that best indicate the temperature of a corona lie at very short ultraviolet wavelengths not accessible to any telescopes yet launched.

The presence of chromospheres and coronae seems to be linked with the presence of convection in the outer layers. Stars on the lower portion of the main sequence generally have both phenomena. The kinetic energy of the turbulent motions in the convective layer is transported into higher zones, where it causes heating. As we learned in our discussion of the sun, the precise mechanism for converting the energy of convection into heat is not understood.

What of the red giants and supergiants? These stars can have surface temperatures comparable to the lower-main-sequence stars, but obviously their internal structure is rather different, since they are so much larger. These stars also have convection in their outer layers, however, and indeed they also have chromospheres, although it is not certain that the extremely high temperatures characteristic of coronae are present.

A very important modern area of research, made feasible by the development of ultraviolet telescopes for use in space, has to do with the study of phenomena in stars that also occur in the sun. These phenomena include chromospheres and coronae, and probably activity cycles like the sun's 22-year cycle (see chapter 11). Interest in the connection between the sun and stars has inspired intensive new observing programs.

Stellar Winds and Mass Loss

So far we have said nothing about the possibility of the hot stars having chromospheres or coronae. Some of the upper-main-sequence stars are known to have emission lines in their spectra, indicating that they have some hot gas above their surfaces. Visible-wavelength

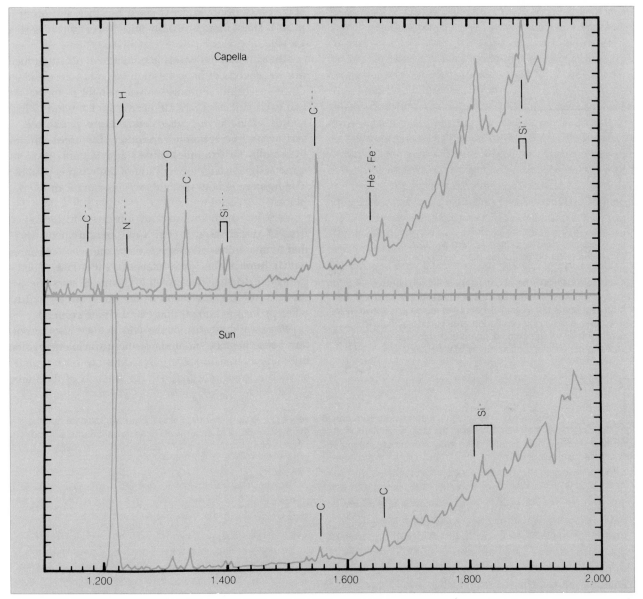

FIGURE 14.7. *Ultraviolet Emission Lines in the Sun and Capella*. This star, a binary consisting of a G giant and an F giant, has a much richer ultraviolet emission-line spectrum than the sun, a G main-sequence star. Some of the emission lines are labled with the identities of the ions that produce them (the number of pluses represents the number of electrons missing).

emission lines have been found in only a few extremely hot or luminous stars, however.

In the late 1960s, with the first observations of ultraviolet spectra, it was found that many hot stars (nearly all the O stars and many of the hotter B stars) have ultraviolet emission lines. Furthermore, there are absorption features that show enormous Doppler shifts, indicating that the stars are ejecting material at speeds as great as 3,000 kilometers per second! The most extreme **stellar winds,** as these outflows are called

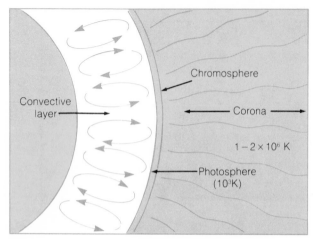

FIGURE 14.8. *A Stellar Chromosphere and Corona.*
Stars in the lower portion of the main sequence all have
chromospheres and coronae, analogous to the sun (see chapter
11 for more details). The photosphere is the region where the
star's continuous spectrum and absorption lines are formed; the
chromosphere is a thin, somewhat hotter region just above; and
the corona is a very hot, extended area outside of that. The
source of heat for the chromosphere and corona is probably
related to convective motions in the star's outer layers.

(Fig. 14.9), occur in the O stars, but they also occur in
most B stars, though usually with lower velocities (Fig.
14.10).

The cause of the winds is not known, although there
are indications of how the high velocities are reached.
Once gas begins to move outward, light from the star
can exert sufficient force to accelerate the wind to high
speeds. This force, called **radiation pressure,** is
very weak (we certainly cannot feel the breeze from a
light bulb, for example), but O and B stars are so luminous that strong acceleration of the wind is possible.
The mystery is how the outflow gets started in the first
place.

Very high temperatures are observed in the winds
from O and B stars. Recent X-ray measurements show
that temperatures of 10^6 degrees or more are common,
which implies that these stars have coronae. This is
rather surprising, because the O and B stars are not
thought to have convection in the outer layers, and no
other process is known that could create coronae.

Whatever the cause of the winds, they have important consequences. Analysis of the ultraviolet emission
lines shows that in some cases stars are losing matter
at such a great rate that a large fraction of the initial

FIGURE 14.9. *Stellar Winds.* A luminous hot star (left) ejects gas at a very high velocity; the lengths of the arrows indicate that the
gas accelerates as it moves away from the star. A luminous cool star (especially a K or M supergiant) is so large that the surface
gravity is very low, and material drifts away at relatively low speeds. In both types of stars, radiation pressure probably helps
accelerate the gas outward.

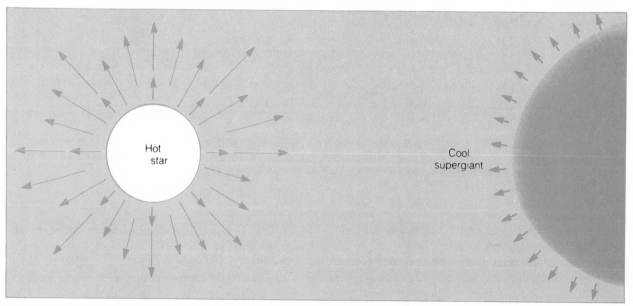

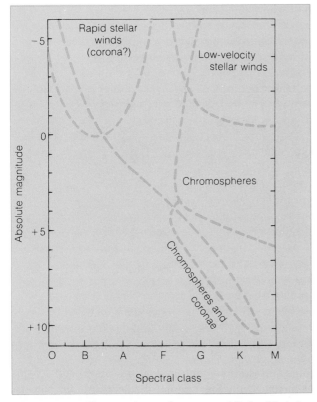

FIGURE 14.10. *Chromospheres, Coronae, and Stellar Winds in the H-R Diagram.*
Here we see that chromospheres exist in most stars of type F or cooler, whereas coronae may be confined to main-sequence stars. Very luminous stars, hot and cool alike, have sufficiently strong winds to result in significant loss of mass.

mass may be lost during their lifetimes. An O star might lose as much as 1 solar mass every 100,000 years; if such a star begins life with 20 or 30 solar masses and lives 1,000,000 years, it could lose 10 solar masses, a significant fraction of what it had to begin with. This has important effects on how such a star evolves, as we will see.

The supergiants in the upper right of the H-R diagram also lose mass through stellar winds (Fig. 14.10), but these winds are of a different nature. The red supergiants are so large that their surface gravity is very low, so that gas in the outer layers is not tightly bonded to the star. Radiation pressure can easily push the gas outward at the relatively low speed of 10 or 20 kilometers per second. The amount of mass lost can be just as great as in the hot stars, however, because these

low-velocity winds are much denser than the high-speed winds from the O and B stars.

Because the red supergiants are relatively cool objects, the gas in their outer layers is not ionized, and even molecular species form there. In addition, small solid particles called dust grains can condense, so that the star becomes shrouded in a cloud. In extreme cases the dust cloud becomes so thick that little or no visible light escapes to the outside. The dust grains become heated, however, and emit infrared radiation, so the star can still be detected with an infrared telescope.

Some of the dust that forms in the outer layers of red supergiants escapes into the surrounding void, and interstellar space gets contaminated in this way with a kind of pollution, an interstellar haze (the interstellar dust is discussed at greater length in chapter 18).

Mass Exchange in Binary Systems

All the processes discussed so far can occur in single stars or in stars that are part of binary systems. There are additional factors that come into play in certain kinds of binaries, but which do not affect single stars.

We mentioned that a star can lose matter into space. If such a star has a companion, then the companion may catch some of the cast-off material and gain mass in the process (Fig. 14.11). In some cases the companion actually helps its neighbor lose matter. There is a point between the two stars where their gravitational forces just balance, and a particle of gas that reaches that point can fall either way, into either star. If one of the stars is swelling up on its way to becoming a red giant, when its outer layers reach this balance point, gas will begin to flow down onto the other star. It is possible for a pair of stars with unequal initial masses to reverse themselves, so that the one that starts out with most of the mass ends up with the least.

Scientists have observed a number of binary systems where this mass-exchange phenomenon is taking place. These systems are characterized by flowing streams of gas swirling around the two stars, creating emission lines with Doppler shifts indicating the motions of the gas. In some of these systems the star that is receiving mass has finished its evolution and has become a stellar remnant such as a white dwarf, neutron star, or black hole (see chapter 16), and the gas falling

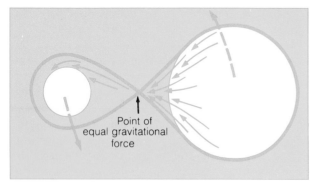

FIGURE 14.11. *Mass Transfer in a Binary System.*
The more massive star in a binary will be the first to evolve and swell up as a red giant. In the process, its outer layers may come sufficiently close to the companion star to be pulled to it gravitationally. As we will learn in later chapters, such a transfer of mass can radically affect both stars.

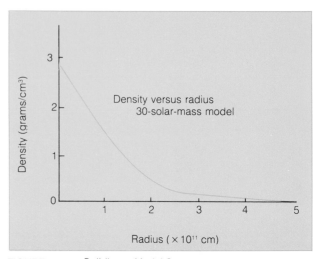

FIGURE 14.12. *Building a Model Star.*
Modern astrophysicists reconstruct what is happening inside stars by using complex computer programs that simultaneously solve large numbers of equations describing the physical conditions. Often the result is plotted, as a convenient way of showing the implications of these calculations.

in creates spectacular effects ranging from nova explosions to intense X-ray emission.

No matter how it happens, the loss or gain of mass by a star has important effects on how the star evolves. We learned that the mass of a star determines all other stellar properties; thus it follows that the other properties will vary if the mass does.

Stellar Models: Tying It All Together

Once astrophysicists have understood all the phenomena that govern a star's internal structure and its outward appearance, they must combine their knowledge of all these processes into a theoretical calculation that will tell them what goes on inside the star, and what will happen to it in time (Fig. 14.12). When the calculations correctly reproduce the observable quantities, it is assumed that they also correctly illustrate the nonobservable aspects, such as the star's interior conditions. We can have reasonable confidence in such calculations, because it can be demonstrated that the resulting solutions are unique; that is, no other solution to the mathematical equations can exist. The most difficult question is whether we know all the correct equations to begin with.

The calculation of a model star consists of simultaneously solving a set of basic mathematical relationships that describe the physical processes, and solving them repeatedly for different depths inside the star. In this way the calculations may proceed from the observed surface conditions to the center, or from assumed central conditions to the surface, with adjustments made until the calculations match the observed properties of the star. To carry out such a calculation requires a substantial amount of time on a large computer.

In order to see how a star evolves, we must compute the model many times over, each time taking into account changes produced in the star that occurred in the previous step in the calculation. For example, a certain amount of hydrogen is converted into helium at each step, and this change in the chemical composition of the core must be considered during the next step. The conversion of hydrogen into helium uses up the hydrogen fuel, so that the star will eventually exhaust its supply, and it causes the density of the core to increase, because helium nuclei are more massive than hydrogen nuclei. These factors cause the star's overall structure to change, and the calculations keep track of the changes, allowing us to trace the evolution. In the next chapters we will discuss the evolution of stars, and how it is determined from a combination of theory and observation.

PERSPECTIVE

We now are acquainted with all the physical processes that affect a star; we know what a star is and what makes it run. We have seen that mass plays a dominant role in dictating all the other properties, although composition, which changes with time, is also important. We are prepared to see what the life story of a star is: how it is born, how it lives, and how it ends its days.

SUMMARY

1. A star is gaseous throughout, with temperature and density increasing toward the center.

2. The balance between gravity and pressure within a star, called hydrostatic equilibrium, governs the internal structure.

3. The mass of a star, along with its chemical composition, govern its central temperature and pressure (through hydrostatic equilibrium), its luminosity, its radius, and its internal structure.

4. Nuclear fusion reactions take place in the core of a star, where the temperature and density are high enough to allow nuclei to collide with sufficient velocity and frequency.

5. In most stars (those on the main sequence) hydrogen is converted into helium in the reactions in the core.

6. The energy inside a star is transported by convection or by radiation. On the lower main sequence (including the sun), radiative transport dominates in the inner parts of the star, and convection operates in the outer layers. On the upper main sequence, convection is dominant in the core, and radiative transport occurs in the outer layers.

7. The lifetime of a star depends on its mass and how rapidly it uses up its nuclear fuel, and decreases dramatically from the lower main sequence to the upper main sequence.

8. Nuclear reactions in stars, and the recycling of matter between stars and interstellar space, result in a gradual increase in the abundance of heavy elements in the universe.

9. Stars in the cool half of the H-R diagram have chromospheres and coronae, in analogy with the sun.

10. Stars in the hot portion of the H-R diagram, and very luminous cool supergiants as well, have stellar winds that can cause the loss of large fractions of the initial stellar masses. The loss of mass can have significant effects on how the stars evolve.

11. In certain binary-star systems, matter is transferred from one star to the other, significantly affecting the structure and evolution of both stars.

12. To determine the interior conditions of a star, astronomers must compute them from known laws of physics and observations of the star's surface conditions.

REVIEW QUESTIONS

1. How do we know that a star must be in a state of balance between pressure and gravity? What would happen if this balance did not exist?

2. Explain in your own words why stellar mass and composition control all other basic properties of a star, such as temperature, luminosity, and radius.

3. Explain how energy is produced by nuclear reactions, and why these reactions occur only in the central core of a star.

4. Using information from chapter 11, explain why radiative energy transport is a very slow process that requires millions of years for a single photon of light to travel from the center to the surface of a star.

5. Estimate the lifetimes of the following stars: (a) a B star, with 15 times the mass of the sun and 4,500 times the sun's luminosity; and (b) a K star, with half the sun's mass and one-eighth its luminosity.

6. Why are very luminous stars very rare?

7. Explain why each nuclear-reaction stage that can occur in a star as heavier and heavier elements are created in its core requires a higher temperature than the preceding stage.

8. If a star has 20 solar masses to begin with and a lifetime of 5 million years, how much of its mass will it lose if it has a stellar wind that removes mass at a rate of 1 solar mass every 500,000 years?

9. Summarize the differences in energy generation, internal structure, and overall properties, between stars

on the lower main sequence and those near the top of the main sequence.

ADDITIONAL READINGS

Aller, L. H. 1971. *Atoms, stars, and nebulae.* Cambridge, Ma.: Harvard University Press.

Hoyle, F. 1975. *Astronomy and cosmology: a modern course.* San Francisco: W. H. Freeman.

Perry, J. R. 1975. Pulsating stars. *Scientific American* 232(6):66.

Shu, F. 1982. *The physical universe.* San Francisco: W. H. Freeman.

Smith, E. P., and Jacobs, K. C. 1973. *Introductory astronomy and astrophysics.* Philadelphia: Saunders.

Snow, T. P. 1981. Dieting for stars, or how to lose 10^{25} grams per day (and feel better, too!). *Griffith Observer* 45(5):2.

Weymann, R. J. 1978. Stellar winds. *Scientific American* 239(2):34.

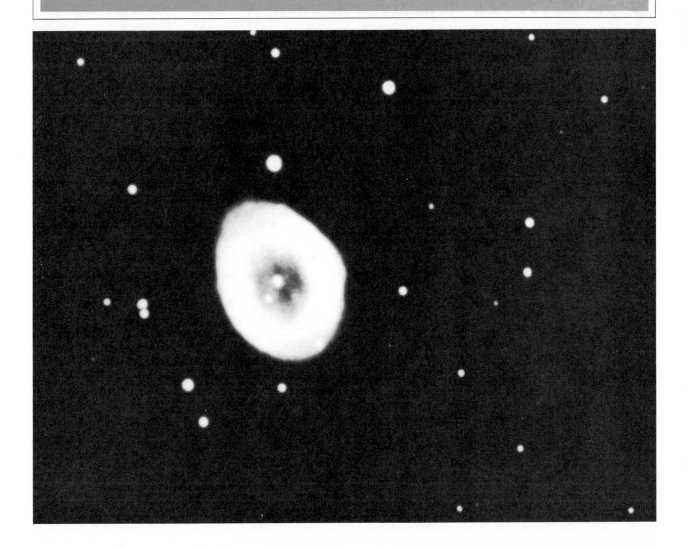

Learning Goals

How can we watch a star evolve? After all, even the most short-lived ones are around for hundreds of thousands of years, whereas human studies of astronomy date back only four or five millennia. Occasionally we catch a star in the act of change, as it makes a transition from one stage in its evolution to another, but clearly we cannot hope to see a star through its whole lifetime, watching it form, live, and die.

We can, however, piece together a story of stellar evolution by examining many stars of different ages, deducing from this the sequence of events that occurs in a single star, and then computing theoretical models that we require to reproduce the observed stellar properties. These models take into account the physical processes that govern stellar structure, and allow us to understand the development sequence that stars go through in their lives.

The Observational Evidence: Stars in Groups

Contrary to ancient teachings that the heavens are immutable, in modern times rapid changes in the skies have been observed on many occasions. We have already learned a little about variable stars (see chapter 13), and we know that "new stars" occasionally show up, in the form of **nova** or **supernova** outbursts (both Tycho Brahe and Johannes Kepler made much of the appearance of supernovae in the heavens of their times). Historically, there was no distinction between what is now known as a nova and what we now define as a supernova. Each manifests itself by the sudden appearance of a starlike object, where before there was nothing obvious. Today important distinctions have been drawn between novae and supernovae, and indeed two types of supernovae have been recognized. Novae are distinguished by their lesser luminosity. The absolute magnitude of a nova is typically -8, whereas a supernova can have an absolute magnitude as great as -19 to -21, corresponding to a factor of hundreds of thousands in luminosity. Novae are only observable within our galaxy; most of our information on supernovae, on the other hand, comes from observations of their appearances in other galaxies. It is estimated that many novae occur in a galaxy per year, whereas supernovae are thought to take place only once every few decades in a galaxy the size of ours. A nova flares up and then decreases in brightness within a few weeks; a supernova may flare up nearly as quickly, but takes many months to become dim again. Both novae and supernovae have emission lines in their spectra, with Doppler shifts indicating that gas is flowing outward, but supernovae show evidence of much higher velocities (up to several thousand km/sec!).

The two kinds of supernovae are distinguished observationally by differences in the way their brightness diminishes, and by spectroscopic contrasts. As we will learn in this and the next chapter, it is thought that the two types, called **type I** and **type II supernovae,** are completely different in origin, and that both are distinct from novae.

While novae and supernovae are both associated with explosive outbursts in stars, there are other kinds of sudden changes in stellar brightness that also serve as direct evidence that stars can evolve on short time scales. A number of cases have been found where formerly dim stars rapidly increase in brightness, and then stay bright. These stars, some of which are called **FU Orionis** objects after the first one discovered, are thought to be very young stars just shedding the clouds of gas and dust that surround them following their formation.

Besides finding stars in states of rapid transition, astronomers draw upon another, more common, phenomenon in deducing the manner in which stars evolve. Within our galaxy, stars are not uniformly distributed, but instead many are located in concentrated regions called **clusters.** By making comparisons of the stars in a cluster, which evolve at different rates because of their differing masses, scientists can gain a wealth of information on how stars change as they age.

A few clusters are sufficiently prominent to be visible to the unaided eye, and others can be viewed with a small telescope or a good pair of binoculars. The Pleiades (Fig. 15.1), a bright cluster to the west of Orion, is perhaps the best-known cluster for northern-hemisphere observers, being easily visible in the evening sky throughout the late fall and early winter. Other clusters easy to find with a small telescope are the Hyades, the double cluster in Perseus (Fig. 15.2), and M13, a fantastic, spherically shaped collection of hundreds of thousands of stars. It is apparent that the stars in a cluster are gravitationally bound to it, each star orbiting the common center of mass.

FIGURE 15.1. *The Pleiades.*
This is a relatively young galactic cluster, and is a prominent object in the late fall and winter.

FIGURE 15.2. *The Double Cluster in Perseus.*
Known as *h* and χ Persei, these are young galactic clusters.

FIGURE 15.3. *The Very Sparse Galactic Cluster NGC 7510.*

There are several distinct types of star clusters. Those containing a modest number of stars (up to a few hundred), located in the disk of a galaxy, are referred to as **galactic,** or **open, clusters** (Fig. 15.3). The Pleiades and the Hyades are examples. Loose groupings of hot, luminous stars are also found in the plane of the galaxy, and are called **OB associations,** being dominated by stars of spectral types O and B. These clusters, in contrast with the other types mentioned here, are probably not gravitationally bound together, but appear grouped simply because the stars have recently formed together. The giant **globular clusters** (Fig. 15.4) are found in a spherical volume about the galactic center (Fig. 15.5), and are not confined to the plane of the galaxy.

There are two important assumptions astronomers make about a cluster of stars: (1) the stars are all at the same distance from us; and (2) the stars formed together so that all are the same age and had the same composition initially. The first of these assumptions means that any observed differences from star to star within a cluster must be real differences, and not a

FIGURE 15.4. *A Globular Cluster.*
This is M92, a prominent example.

false effect created by differing distances from us. Brightness differences, for example, have to be the result of variations in the luminosities of stars within the cluster.

Because distance effects are eliminated when we compare stars within a cluster, it is possible to plot an H-R diagram for a cluster without first determining the absolute magnitudes of the stars. We simply plot apparent magnitude against spectral type, or more commonly, against the color index *B–V*, which is easier to measure for a large number of stars. To do this requires only the determination of the visual *(V)* and blue

FIGURE 15.5. *The Locations of Clusters in the Galaxy.*
The open or galactic clusters are found in the disk of the Milky Way, as are the OB associations, which are always located in spiral arms (see chapter 17). The globular clusters, on the other hand, inhabit a large spherical volume, and are not confined to the plane of the disk.

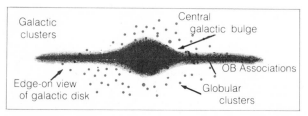

(B) magnitudes of the cluster stars, a rather straightforward observational procedure, and then the construction of a plot of *V* versus *B–V*. The result, called a **color-magnitude** diagram, looks just like an H-R diagram, except that the vertical scale is an apparent-magnitude scale. This is because the *differences* in magnitude of the cluster stars are the same, whether we use apparent or absolute magnitudes.

One advantage of plotting cluster H-R diagrams is that they can be used to determine the distances to clusters. A comparison of a cluster H-R diagram with a standard one (that is, one with an absolute magnitude scale) allows us to see what the absolute magnitude scale for the cluster diagram should be. This in turn tells us the difference between the absolute and apparent magnitudes for the cluster, and from this difference the distance is known. This procedure, known as **main sequence fitting,** depends only on the assumption that the cluster has a main sequence which is identical to that of the standard H-R diagram, a safe assumption as long as the chemical composition of the cluster is not unusual.

The second major assumption made about clusters, that all the stars are the same age, is very important in the study of stellar evolution. The basic premise is that the stars formed together, out of a common cloud of interstellar material, and did not just happen to come together. Binary systems as well as clusters must *form* as binaries or clusters. A corollary is that all stars in a cluster must begin with the same chemical composition, since they form from a common source of material.

The knowledge that all stars in a cluster have the same age and initial composition gives astronomers tremendous leverage in deducing the evolutionary histories of stars. The only major parameter that differs from one star to another within a cluster is mass, so the observed differences among stars in a cluster offer us clues as to how the properties of stars depend on mass. Comparison of H-R diagrams of star clusters of different ages then presents us with a picture of how the different stellar masses evolve.

There are striking differences among H-R diagrams of different clusters (Fig. 15.6). The Pleiades, for example, has an almost complete main sequence, but no red giants. By contrast, M13, a globular cluster, has no stars on the upper portion of the main sequence, but a large number of red giants. A wide variety of intermediate cases can be found.

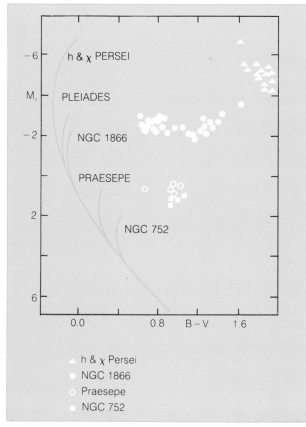

FIGURE 15.6. *H-R Diagrams for Star Clusters.*
Here, plotted on the same axes, are H-R diagrams (based on color index rather than spectral type) of several clusters. The various symbols in the right-hand portion of the diagram distinguish among the giant stars in different clusters.

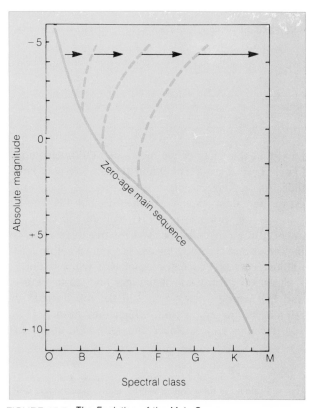

FIGURE 15.7. *The Evolution of the Main Sequence.*
As a cluster ages, the stars on the upper main sequence move to the right. The farther down the main sequence this has happened, the older the cluster.

From considerations discussed in the last chapter, we expect massive stars to evolve most rapidly. These are the stars that start out near the top of the main sequence, and they have such tremendous luminosities that they use up their nuclear fuel rapidly. Tying this together with the observed differences among cluster H-R diagrams leads to the conclusion that the upper-main-sequence stars must evolve into red giants when they have used up all their hydrogen fuel (Fig. 15.7). In a very young cluster, even the most massive stars might still be on the main sequence, whereas in an older cluster, these massive stars will have had time to use up their core hydrogen and become red giants. Hence the Pleiades is relatively young, whereas M13 is very old.

As a cluster ages, more and more of its stars use up their hydrogen and leave the main sequence. The upper main sequence gradually erodes away as its stars move to the right in the H-R diagram, toward the red-giant region. This process creates a definite cut-off point on the main sequence, above which there are no stars. The position of this **main sequence turn-off,** as it is commonly called, indicates the age of a cluster. If, for example, we find a cluster whose main sequence stops at the position of stars like the sun (that is, if it has no stars above G2, or $B-V = 0.6$, on the main sequence), then we know that all stars more massive than one solar mass have burned their hydrogen and evolved toward the red-giant region. In the last chapter we learned that the sun's lifetime for hydrogen burning is about 10 billion years; we conclude, therefore, that such a cluster is just about 10 billion years old. If it were younger, its main sequence would extend farther up, and if it were older, even the sunlike stars would have had time to evolve into red giants, and the main sequence turn-off would be farther down.

Young Associations and Stellar Infancy

Earlier in this chapter we mentioned OB associations, loosely bound groups of O and B stars. We now know that these are very young clusters, probably the youngest in the galaxy, because of the fact that they contain stars on the extreme upper portion of the main sequence. In many cases the stars are moving away from the center and escaping the association, so within a few million years no concentration of stars will remain.

Stars form from interstellar gas and dust; we even find that young clusters may still be embedded in the remains of the cloud from which they formed (Figs. 15.8 through 15.11). Active star formation apparently occurs in relatively dense, obscured regions of interstellar material. The best way to probe these dark and dusty regions is to observe them in infrared wavelengths. One reason is that infrared radiation penetrates much farther through clouds of gas and dust than does visible light (Fig. 15.12); another is that the dust around a newborn star is hot, and it glows at infrared wavelengths. To see infant stars, therefore, we must look for infrared sources buried in dark clouds.

We have found many regions in our galaxy where star formation is apparently taking place. One of the best observed of these is in the sword of Orion within the Orion Nebula, a great, glowing region. A very young cluster of stars is located there, and infrared observations reveal a number of infrared sources embedded within the dark cloud associated with the cluster (Fig. 15.13).

In other, similar regions there are faint variable stars called **T Tauri** stars (see chapter 12 and p. 274). These appear to be newborn stars just in the process of shedding the excess gas and dust from which they

FIGURE 15.8. *A Region of Young Stars and Nebulosity.* This is a portion of the Rosette Nebula, showing numerous dark patches and globules where star formation takes place.

FIGURE 15.9. *A Very Young Cluster with Nebulosity.* This is the cluster M16, a group of stars associated with the gas and dust from which it formed.

FIGURE 15.10. *A Dark Cloud.*
The regions without stars in this photo are portions of the Coalsack, a well-known concentration of interstellar gas and dust in the southern Milky Way.

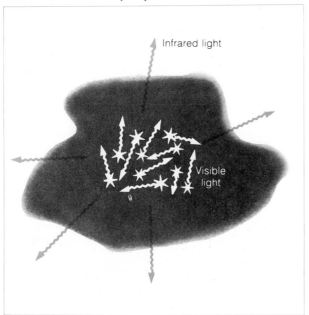

FIGURE 15.11. *A Newborn Star Leaves the Womb.*
This remarkable photo of a dark cloud shows a wispy, glowing trail left behind by a newly formed star that has recently emerged from the cloud.

FIGURE 15.12. *Embedded Infrared Sources.*
This sketch shows a dark interstellar cloud with a few young, infrared-emitting stars inside. The infrared light is able to escape the cloud more readily than visible light, so the best technique for finding young stars is to use infrared telescopes.

formed; in fact, spectroscopic measurements indicate that some material may still be falling into these stars, as though the collapse of the cloud from which they are forming is not yet complete.

Star Formation

From the assorted observations described here, a picture has emerged that shows how stars form. This picture is based on theoretical calculations as well as on observations.

For many years a class of peculiar stars called T Tauri stars has been recognized and studied. These stars, named after the brightest and first-identified example, are characterized by cool-star spectra (most are G stars), and by prominent emission lines. The T Tauri objects are often associated with dense clouds of interstellar matter, and are therefore generally best observed at infrared wavelengths, which penetrate the interstellar dust far better than visible light.

The presence of emission lines has led astronomers to believe that these stars have very active chromospheres, and recent ultraviolet observations have verified that belief. Whereas emission lines in visible wavelengths do not provide much information on the temperature of the gas, there are many good diagnostics of hot gas in ultraviolet wavelengths. Data obtained primarily with the *International Ultraviolet Explorer* have revealed a number of bright ultraviolet emission lines indicative of a very hot gas. Many T Tauri stars have been found by the *Einstein Observatory* satellite to emit X rays as well. There is no doubt that the T Tauri stars have chromospheres, probably heated by energy transported from the outer convective zones to higher levels. This is analogous to the sun's chromosphere, except that the amount of energy involved in the T Tauri chromospheric heating is apparently greater than in the sun.

There are indications, in the Doppler shifts that sometimes are seen in the emission and absorption lines in T Tauri star spectra, that gas is flowing outward at high velocities. Careful study of the Doppler shifts has shown that the situation is not simple, however. Several cases have been found where T Tauri stars are apparently *receiving* gas that is falling in, and certain stars have even shown both infall and outflow at the same time. A more accurate picture probably would be that rather than having an overall wind flowing one way or the other, a typical T Tauri star has gas simultaneously rising and falling, in analogy with the solar prominences whose motions are governed by the sun's magnetic field.

There are other indirect indications that magnetic fields play important roles in T Tauri stars. From observations of bright patches called **Herbig-Haro objects** that are sometimes seen amid the nebulosity near T Tauri stars, it is inferred that gas is ejected at great velocities from the stars, and is beamed along very specific directions. The Herbig-Haro objects appear to be "hot spots" where the high-veloctiy beams strike surrounding material. Very recent radio observations have directly revealed a linear jet or beam of hot gas emanating from one suspected T Tauri star, helping to clarify this interpretation of the Herbig-Haro objects. The best mechanism for explaining these beams of gas is to invoke a magnetic field, with outflowing ionized gas from localized regions on the stellar surface moving along the field lines.

The general consensus on T Tauri stars is that they are very young objects on their way to becoming stars. This conclusion is based on their locations within dense regions of interstellar material, on their apparent instability (the spectra and the gas motions they reveal are often variable with time), and on comparisons of their positions in the H-R diagram with theoretical calculations of the properties of newly forming stars. The T Tauri wind, a phase thought to be undergone by young stars in general, is usually invoked to explain how the early solar system was cleared of leftover debris from the formation of the planets (see chapter 12).

Very recently, the picture of T Tauri stars as young suns was supported by a spectacular discovery. Using a technique called **speckle interferometry** (see chapter 3) in infrared wavelengths, astronomers found that the prototype star T Tauri itself is a binary system. Further analysis of the properties of the dim companion have established that it is not a star at all, but instead is a planet, some five to twenty times as massive as Jupiter. This is striking confirmation of the evolutionary theory of solar-system formation (discussed in chapter 12), and lends credence to the general supposition that planetary systems may be quite common. It is fitting that this discovery should be associated with the "standard" young star T Tauri, although not surprising, since this is the brightest and best-studied example of its class.

 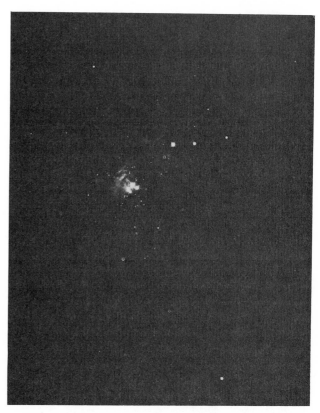

FIGURE 15.13. *The Orion Nebula.*
At left is a normal, long-exposure photo, showing the familiar nebula. At right is a short-exposure photo of the central region of the nebula, taken in infrared light. Here little of the glowing gas can be seen, but the young, hot stars that are embedded with the outer portions are clearly visible

The process begins with the gravitational collapse of an interstellar cloud (Fig. 15.14). This collapse may begin spontaneously, if by chance a portion of the cloud becomes dense enough, or it may be caused by compression due to shock waves created by a nearby supernova explosion. Once cloud collapse starts, gravity takes care of the rest. Calculations show that the innermost portion collapses most quickly, while the outer parts of the cloud are still slowly picking up speed. The temperature in the core builds up as the density increases, but for a long time the heat escapes as the dust grains in the interior radiate infrared light. Eventually, however, the cloud becomes opaque in the center, and radiation no longer can escape directly into space. After this the heat builds up much more rapidly, and the resulting pressure causes the collapse to slow to a very gradual shrinking. Material still falling in crashes into the dense core, creating a violent shock front at its surface. The luminosity of the core, which by this time may have a temperature of 1,000 degrees or more, depends on its mass. If the mass is large; that is, several times the mass of the sun, then the luminos-

ity may be sufficient to blow away the remaining material by radiation pressure. In a lower-mass **protostar,** as the central dense object is called, material continues to fall in. Eventually the density of this material becomes low enough that the protostar can be seen through it. Regardless of mass, a point is reached where a starlike object becomes visible to an outside observer; this stage may be identified with the T Tauri stars. If the obscuring matter is swept away violently, the object may appear to brighten dramatically in a very short time. There are a few cases where a previously rather faint star was found to have brightened suddenly by several magnitudes.

The protostar continues to slowly shrink, growing hotter in its core. The shrinking stops when the temperature becomes high enough for nuclear reactions to begin in the center. This is a major landmark in the process of star formation, for once the reactions have started, the star is on its way to becoming stable (that is, no further shrinking will occur), and it lies on the main sequence, where it will spend most of its life converting hydrogen into helium in its core.

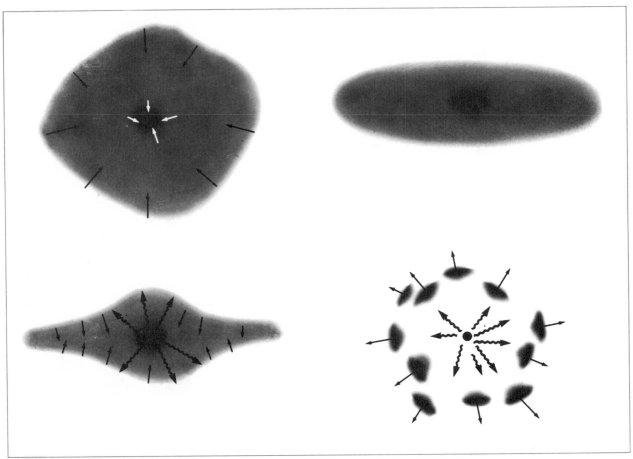

FIGURE 15.14. *Steps in Star Formation.*
A rotating interstellar cloud collapses, most rapidly at the center. At first infrared radiation escapes easily and carries away heat, but eventually the central condensation becomes sufficiently dense that this radiation is unable to escape, and the core becomes hot. After a lengthy period of slow contraction, nuclear reactions begin in the protostar. At some stage, in a process not well understood,

It is convenient to trace on the H-R diagram the path the star takes as it forms (Fig. 15.15). When it is a cold, dark cloud, its position is far to the right and down, well off of the scales of temperature and luminosity appropriate for stars. As the cloud heats up in the core and begins to glow in infrared wavelengths, it moves up and to the left, eventually attaining sufficient luminosity to fall onto the standard H-R diagram, but still off to the right of the main sequence. When it reaches the phase of slow contraction, it moves gradually to the left and down, finally reaching the main sequence when the reactions begin.

The entire process, from the beginning of cloud collapse until the newly formed star is on the main sequence, takes many millions or even billions of years for the least massive stars, and only a few hundred thousand years for the most massive. In a cluster with stars of various masses, it can happen that the most massive stars will form, live, and die before the least massive ones even reach the main sequence. The H-R diagram for such a cluster shows stars located to the right of the main sequence at the lower end (Fig. 15.16).

The Evolution of Stars Like the Sun

Since so much of a star's evolutionary behavior depends on the star's mass, we will discuss examples of different masses separately. In each case, it will prove convenient to refer to a star's location on the H-R dia-

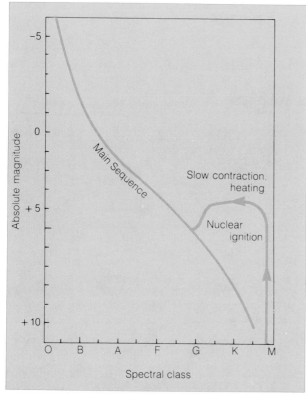

FIGURE 15.15. *The Path of a Newly Forming Star on the H-R Diagram.*
As a protostar heats up, it eventually becomes hot and luminous enough to appear in the extreme lower right-hand corner of the H-R diagram. When the core becomes opaque, the rapid collapse slows to a gradual shinking, and the protostar moves across to the left as it heats. Eventually nuclear reactions begin, and the star moves to the main sequence.

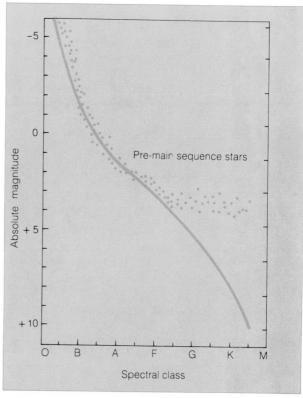

FIGURE 15.16. *Pre-Main-Sequence Stars.*
H-R diagrams of very young clusters often show stars that have not yet reached the main sequence, but are still moving toward it. On the other hand, some of the massive, short-lived stars at the top of the main sequence may have already evolved away from it, having used up their hydrogen fuel.

gram, and how this location changes as the star evolves. We will begin with a star that has a mass like the sun's (Fig. 15.17). When such a star arrives on the main sequence, it is similar to the sun in all its properties. Thus it is a G2 star, with a surface temperature of around 6,000 K and a luminosity of about 10^{33} erg/ sec. Its composition initially is nearly 80 percent hydrogen by mass, and more than 20 percent helium, with only about 1 percent of the mass composed of other elements. It has convection in the outer layers, and it has a chromosphere and a corona.

In the core, the proton-proton chain converts hydrogen into helium. As we learned in chapter 13, this process lasts some ten billion years for a star of 1 solar mass. During this time, the core gradually shrinks as the hydrogen nuclei are replaced by a smaller number of the heavier helium nuclei, and the internal temper-

ature rises. This causes an increase in luminosity, and the star gradually moves up on the H-R diagram. The main sequence, therefore, is not a perfectly narrow strip, but has some breadth since stars on it move slowly upward as their luminosities increase. The starting point, the lower edge of the main sequence, is called the **zero-age main sequence (ZAMS),** because this is where newly formed stars are found (Fig. 15.18).

When the hydrogen in the core is gone, reactions there cease. By this time, however, the temperature in the zone just outside has nearly reached the point where reactions can take place there, and with a little more shrinking and heating of the core, the proton-proton chain begins again, in a spherical shell surrounding the core (Fig. 15.19). At this point the star has an inert helium core, and it is producing all its energy in the hydrogen-burning shell, which steadily

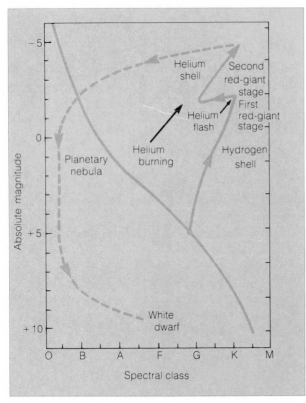

FIGURE 15.17. *Evolutionary Track of a Star Like the Sun.*
This H-R diagram illustrates the track a star of 1 solar mass is thought to follow as it completes its evolution. The path indicated by dashes, following the second red-giant stage, is less certain than the earlier stages.

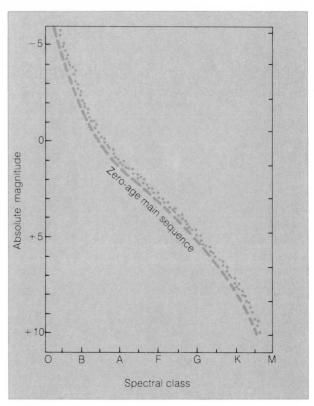

FIGURE 15.18. *The Width of the Main Sequence.*
As stars on the main sequence gradually convert hydrogen into helium in their interiors, the cores shrink a little and get hotter. This in turn increases the luminosity, and the stars gradually move upward on the H-R diagram. As a result, the main sequence, consisting of stars of a variety of ages, is not a narrow strip, but has some breadth.

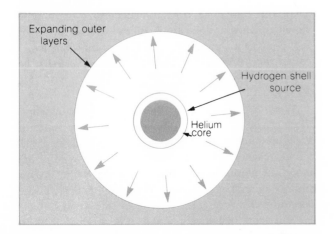

FIGURE 15.19. *The Hydrogen Shell Source.*
After the hydrogen in the core of a star is completely used up, the core, now composed of helium, continues to shrink and heat. This causes a layer of gas outside the core to reach the temperature required for nuclear reactions. A shell source, in which hydrogen is converted into helium, is ignited. The star's outer layers expand and cool, and the star becomes a red giant.

moves outward inside the star, eating its way through the available hydrogen.

The core continues to shrink and heat. This heating enhances the nuclear reactions in the shell, causing it to produce more and more energy. Thus as the source of energy is heated and moves toward the surface, the outer layers of the star are forced to expand, and as this occurs, the surface cools. The luminosity increases because of the increased surface area, so the star moves upward on the H-R diagram (Fig. 15.20). It also moves to the right, because the surface temperature is decreasing. The star becomes a red giant, reaching a size of ten to a hundred times its main-sequence radius. The outer layers are constantly overturning as convection occurs to a great depth (Fig. 15.21).

The helium core becomes extremely dense. At first the temperature is not high enough for any new nuclear reactions to start there (the triple-alpha reaction, in which helium nuclei combine to form carbon, requires a temperature of about 100 million degrees). As the core continues to shrink, the matter there takes on a very strange form. A new kind of pressure gradually takes over from the ordinary gas pressure that has been supporting the core. Electrons have a property that prevents them from being squeezed too close together, and this creates the new pressure. The gas in the stellar core contains many free electrons, and eventually their resistance to being compressed becomes the dominant pressure that supports the core against further collapse. When this happens, the gas is said to be **degenerate.**

A degenerate gas has many unusual properties. One of them is that the pressure no longer depends on the temperature, and vice versa. If the gas is heated further, it will not expand to compensate, as an ordinary gas would. As we will soon learn, important consequences occur if nuclear reactions start in the degenerate region of a star.

We now have a red-giant star, with highly expanded outer layers. Near the center is a spherical shell in which hydrogen is burning in nuclear reactions, and inside of this is a degenerate core containing helium nuclei. The core is small, but it may contain as much as a third of the star's total mass.

During the red-giant phase, the helium core continues to be heated by the reactions going on around it. Eventually the temperature becomes sufficiently high for the triple-alpha reaction to begin. When it does, there are spectacular consequences. Ordinarily when a

FIGURE 15.20. *The Development of a Red Giant.*
As a star's outer layers expand because of the hydrogen shell source, they cool. At the same time, the surface area increases, raising the star's luminosity. The star therefore moves up and to the right on the H-R diagram.

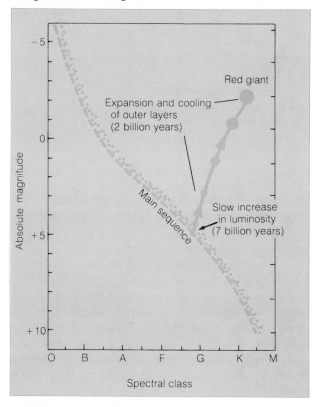

FIGURE 15.21. *The Internal Structure of a Red Giant.*
The stellar core, which actually contains most of the mass of the star, is very small, compared to the huge extent of the outer layers. Most of the volume of the red-giant star consists of relatively rarefied gas, constantly overturning because of convection.

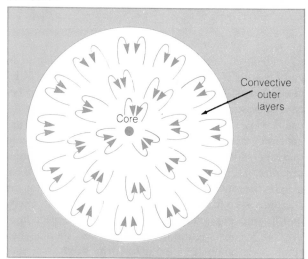

Astronomical Insight 15.2
ELECTRON DEGENERACY

In the text a degenerate gas was described in very crude terms. There it was stated simply that electrons cannot be forced any closer together once a certain state of compression has been reached. It is interesting to examine a little more closely the reasons for this.

We learned, when discussing atomic structure, that an electron in orbit around the nucleus of an atom can only be in certain very specific energy levels. In the core of a star, however, the electrons are all free of atoms, since the extremely high temperature has fully ionized the gas. This usually is taken to mean that the electrons are free to fly around randomly, without regard for fixed energy levels.

This is only an illusion, in a way; under most conditions it only seems that there are no restrictions on the allowed energy state of free electrons. In actuality, when a certain combination of high density and moderately high temperature is reached, the electrons begin to experience certain limitations on their freedom. These limitations result from a combination of two laws of **quantum mechanics,** the branch of physics that describes the motions and forces on acting subatomic particles.

One of these axioms says that even in a gaseous state, electrons (or any other particles) are confined within specific energy levels, regardless of the fact that they are not bound to a nucleus. An energy level in this case means a certain combination of speed, direction, and spin of the electron. The second law holds that no two electrons can occupy the same energy state. The law of physics that makes this statement is called the **Pauli Exclusion Principle,** after the scientist, Wolfgang Pauli, who discovered it.

As a star compresses itself after all the nuclear fuel is depleted, the number of available energy states for the electrons to be in is reduced. A point is reached where all the states have electrons in them. The star cannot collapse any further because to do so would require squeezing more than one electron into the same energy state. This inability of the electrons to be forced any more tightly together creates a physical pressure that is sufficient to hold the star up against gravity, as long as the star's mass does not exceed a certain limit.

Neutron stars, whose cores are made entirely of neutrons, are also subject to the Pauli Exclusion Principle. The neutrons in these stars can be very close together, as close as particles are in an atomic nucleus. In a way, a neutron star is a single, gigantic atomic nucleus, with a density as high as that of nuclear material.

gas is heated, it expands, and this limits how hot it can get, because an expanding gas tends to cool. In a degenerate gas, however, no such expansion occurs, because pressure does not depend on temperature in the usual way. Hence when the reactions begin in the degenerate core of a red giant, the temperature goes up quickly, while the core retains the same density and pressure. The increased temperature speeds up the reactions, producing more heat, in turn accelerating the reactions even further. There is a rapid snowball effect, and in an instant (literally seconds), a large fraction of the core is consumed in the reaction. This spontaneous runaway reaction is called a **helium flash.** Calculations show that there is little or no immediately visible effect of this dramatic event. The overall direction of the star's evolution is changed, however, which would be apparent after thousands of years.

The helium flash quickly disrupts the core, destroying its degeneracy. The core then returns to a more normal state where temperature and pressure are linked, and the triple-alpha reaction continues, now in a more stable, steady fashion. The outer layers of the star begin to retract as the star reverts to a more uniform internal structure, and as they do so, they become hotter. The star moves back to the left on the H-R diagram.

Old clusters, particularly globular clusters, have a number of stars in this stage of evolution, forming a sequence called the **horizontal branch** (Fig. 15.22). This is seen only in clusters old enough for stars as small as one solar mass to have evolved this far; as we learned in previous chapters, this requires an age of ten billion years or more.

In due course, the helium in the core becomes exhausted, leaving an inner core of carbon. Helium still burns in a shell around the core; there could even be an active hydrogen-burning shell farther out in the star at the same time. The star expands and cools, becoming a red giant for the second time. This second red-giant stage is short-lived, however, lasting perhaps one million years. A degenerate core again develops, but

FIGURE 15.22. *The H-R Diagram for a Globular Cluster.* These star clusters are very old, and all the upper-main-sequence stars have evolved to later stages. Here we see a number of stars in or approaching the red-giant stage, and a number on the horizontal branch, where the stars move after the helium flash occurs. Only very old clusters have a horizontal branch. (Besides their great ages, globular clusters also have a relatively low content of heavy elements, and this affects how the stars evolve. See chapter 19).

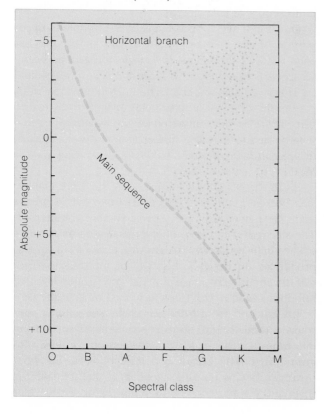

there will not be another dramatic flare-up in its interior. In fact, no further nuclear reactions ever occur in the core of this low-mass star, because the temperature never reaches the extremely high levels needed to cause heavy elements like carbon to react.

The star in its second red-giant stage has two distinct zones (Fig. 15.23): the relatively dense interior, consisting of a carbon core inside a helium-burning shell which in turn is surrounded by a hydrogen-burning shell; and the very extended, diffuse outer layers. The inner portion may contain up to 70 percent of the mass, but occupies only a small fraction of the star's volume.

As the nuclear fuel in the interior runs out, the core shrinks. The details of what happens to the outer layers are not clear, but apparently the star ejects portions of this material in a series of minor outbursts that can be likened to blowing smoke rings. The star gently gets rid of its outer layers.

The evidence that this occurs is primarily observational, since a successful theory of how it happens has not been developed. Astronomers have observed a number of objects that seem to consist of a hot, compact star surrounded by a shell of expanding gas (Fig. 15.24). These shells often have a bluish-green color because of the emission lines by which they glow. Observed through a telescope or on a photograph, the ob-

FIGURE 15.23. *A Star Late in its Lifetime.* Here the core has completed both the hydrogen-burning and helium-burning stages, and is composed of carbon. Outside the core, there may still be an active hydrogen shell source as well as a helium-burning shell, in which case the star is becoming a red giant for the second time.

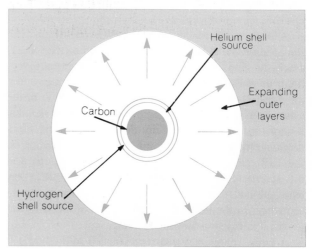

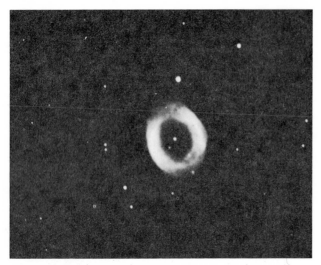

FIGURE 15.24. *A Planetary Nebula.*
This is the Ring nebula, whose striking symmetry makes it a favorite object for viewing through telescopes. The star that ejected the nebula is visible at its center.

jects bear some resemblance to the planets Uranus and Neptune, and thus they are called **planetary nebulae.** Some, such as the Dumbbell nebula and the Ring nebula, are well-known objects visible with a small telescope.

The star that remains in the center of a planetary nebula lies far to the left on the H-R diagram. It has a surface temperature that may be as high as 100,000 K or more, far hotter than any main-sequence star. Evidently, while the shell is being ejected, the star's surface temperature increases as hot innner gas becomes exposed on the surface. In essence such a star is a naked stellar core, left behind when the outer layers are cast off.

Some nuclear reactions may still be going on in a shell inside the star, but they do not last much longer, since the fuel is depleted. The density, already very high, increases as the stellar remnant condenses under the force of gravity. A larger and larger fraction of the star's interior becomes degenerate. As the star shrinks, it stays hot but its luminosity decreases because of its diminishing surface area, and it moves downward on the H-R diagram. Eventually the shrinking stops and the star becomes stable again, but now it is a very small object, so dim that it lies well below the main sequence, in the lower left-hand corner of the H-R diagram. It is called a **white dwarf.**

A white dwarf is a bizarre object in many ways. It

is made of degenerate matter, whose peculiar properties govern its internal structure. It has a mass as great as that of the sun (some are even slightly more massive), yet it is approximately the size of the earth. This means that its density is incredibly high, roughly a million times that of water. A cubic centimeter of white-dwarf material would weigh a ton at the surface of the earth! The properties of white dwarfs and the means by which they are observed are described in the next chapter.

The Middleweights

Let us now consider a star of significantly greater mass than the sun, say 5 to 10 solar masses. Much of this star's evolution is similar to that of the 1-solar-mass star, although there are some major differences. One of the contrasts, of course, is the time scale, because this star will have perhaps 100 times the solar luminosity, and will use up its fuel accordingly faster. Let us escort the star through its development, starting again on the main sequence (Fig. 15.25).

Our middleweight star lands on the middle to upper portion of the zero-age main sequence when it has completed its formation, and nuclear reactions begin. Its spectral type is B8 if its mass is 5 solar masses, and B2 if it is 10 solar masses. The nuclear reaction in its core is the CNO cycle, producing helium from hydrogen. There is no convection in the outer layers, but there is in the core. The surface temperature is roughly between 12,000 K and 20,000 K, depending on the mass.

As in the case of the sunlike star, the more massive star becomes gradually more luminous while on the main sequence, as its core becomes more compressed. The star moves upward a little; this occurs on all parts of the main sequence, making the entire strip appear broadened on the H-R diagram. In the middleweight star the hydrogen in the core is gone (after about a hundred million years) before a shell outside the core is ignited, and for a brief period the star simply contracts and heats, with no reactions taking place. During this time, it turns sharply to the left on the H-R diagram. After only one or two million years, however, hydrogen begins burning again, in a shell outside the core, and the star reverses itself and heads for the red-

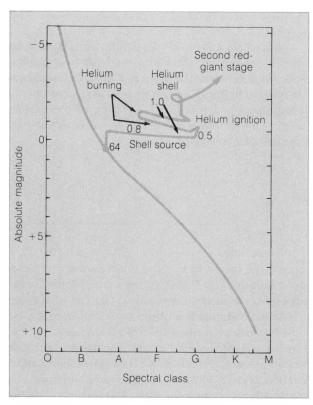

FIGURE 15.25. *The Evolutionary Path of a 5-Solar-Mass Star.*
This star, with its multiple red-giant loops, goes through a more
complicated series of stages than a less massive star. There
are probably additional red-giant stages following those
indicated here, but calculations of the evolution have not yet
been carried that far. The numbers indicate times in units of 10^6
years.

giant region. Within a few hundred thousand years, it
reaches the upper right-hand portion of the H-R diagram. In this case, the core is not degenerate because
the temperature is high enough to prevent a collapse to
the necessary density, so no helium flash occurs. When
the core is hot enough, the triple-alpha reaction quietly
begins, and the star contracts and moves to the left on
the H-R diagram.

This star does not move far to the left, however, and
never really leaves the red-giant region before it develops a new reaction shell outside the core, and expands
and cools once again, reversing itself and heading back
to the right.

What happens next is not clear, but a sketchy picture is emerging. The core temperature continues to increase, and it is possible for new nuclear reactions to

begin. These reactions, called **alpha-capture reactions,** start with carbon and helium nuclei (remember, a helium nucleus, consisting of two protons and
two neutrons, is called an alpha particle). A carbon
nucleus can capture an alpha particle, forming an oxygen nucleus. This in turn can capture another alpha
particle, forming neon, and so on. If the core temperature reaches high enough levels, a complex sequence of
reactions can occur, gradually building up a supply of
heavy elements in the star's core. Each time one fuel
source is exhausted and shell burning begins, the star
expands and moves upward and to the right on the H-
R diagram; each time a new reaction starts in the core,
the star settles down and moves to the left. Thus, a
middleweight star can execute a number of left-to-right
loops in the red-giant region of the H-R diagram. The
number of loops depends on the mass. The situation is
very complicated, however, and theoretical calculations
do not give definitive answers.

One thing that complicates matters is that extremely
luminous red giants lose mass through strong, dense
winds, as described in chapter 14. The rate of mass
loss can be sufficient to significantly alter the mass of
the star within a few hundred thousand years. If a
star's mass is reduced, its evolution is curtailed. The
central density and temperature are lower, which in
turn reduces the number of reaction stages the star
goes through. At some difficult-to-determine point, all
reactions cease. Then the red giant contracts and
moves down and to the left on the H-R diagram, possibly passing through a phase as a planetary nebula.

Whether these middleweight stars end up as white
dwarfs is uncertain. It has been shown theoretically,
and confirmed from observations, that a white dwarf
cannot exist if the star's mass is greater than a certain
limit, about 1.4 times the mass of the sun. If a mass is
any larger, it creates sufficient gravitational force to
overcome the pressure of the degenerate electron gas in
the core, and the star collapses catastrophically. As we
will see, this violent ending definitely occurs in the
most massive stars. For the 5- to 10-solar-mass star,
however, a white dwarf may form if enough mass is
lost to bring the star below the 1.4-solar-mass limit.
Stellar wind and planetary-nebula ejection might reduce the mass far enough, so it is possible that many
stars that start out in the middleweight division end up
as white dwarfs.

However, the white dwarfs formed from the middleweight stars are different from those formed by

lower-mass stars. Recall that for a star of 1 solar mass, the most advanced nuclear reaction that occurs is the triple-alpha process, converting helium into carbon. The 5- to 10-solar-mass star goes well beyond this point, however, so heavier elements are built up in its core. The white dwarf that results, then, may have an interior made of oxygen, neon, or even heavier elements, whereas one produced by a lower-mass star consists of carbon or perhaps only helium.

What of the relatively massive stars that do not shed enough material to become white dwarfs? These stars contract beyond the white-dwarf stage and therefore exceed even the stupendous densities achieved by the white dwarfs. There seem to be two possible end results: the star may contract to another kind of degenerate state, where it can stabilize; or it may collapse indefinitely, never again reaching stability. In either case, the collapse may be violent, with the in-falling matter creating an outburst that we recognize as a type II **supernova explosion.** (Type I supernovae are thought to have entirely different origins, as discussed in the next chapter.)

In the first of these two possibilities, the collapse of the interior is stopped when a new kind of degenerate pressure is created. As the collapse occurs, the protons and electrons are forced so close together that a certain kind of nuclear reaction takes place, in which neutrons are formed. Neutrons, like electrons, can reach a state where they cannot be squeezed any further; this creates a gas pressure that may be sufficient to balance the force of gravity, creating a kind of super white dwarf called a **neutron star.** Like a white dwarf, a neutron star has a strict limit on its mass, around 2 or 3 solar masses. If the stellar core has more mass than this, then the second possibility just mentioned occurs: the collapse never stops. We will return to this case in the next section, where very massive stars are discussed.

The neutron star has even more fantastic properties than a white dwarf. Its diameter is about 10 kilometers, yet it contains more than 1.4 solar masses, and its density far exceeds that of the white dwarf. A cubic centimeter of neutron-star material would weigh a billion tons at the earth's surface!

Observations have confirmed that neutron stars can form during supernova explosions, because there are cases where a neutron star was found in the center of the expanding cloud of debris left over from such an event. One of these apparently formed in historical times, when the supernova that created the present-day Crab Nebula occurred (this was the "guest star" observed by Chinese astronomers in A.D. 1054).

Neutron stars are exceedingly dim objects, very difficult to see directly; for a long time they were thought to be unobservable. The ways in which they have been observed will be described in the next chapter. For now, we have finished our life story of the middleweight star, and we are ready to see how the most massive stars evolve.

The Heavyweights

The most massive stars form in basically the same manner as their less massive relatives, but they do so more quickly (Fig. 15.26). As we learned earlier in this chapter, in a cluster the massive stars can go through the entire evolution from birth to death before the less massive ones even reach the main sequence.

Stars containing 20 or more solar masses start their main-sequence lifetimes as O stars. They have surface temperatures of 30,000 K or more, and luminosities of 10^5 to 10^6 times that of the sun. The dominant nuclear reaction during the main-sequence phase is the CNO cycle. These stars have some excess surface heating, which causes coronae to form, but it is not thought likely that convection occurs in the surface layers. There is convection in the cores of these stars, however.

As in the other cases, these stars undergo gradual core contraction as the hydrogen there is converted into helium. By contrast with the less massive stars, however, the O stars do not move upward on the H-R diagram; instead, they move straight across to the right. The reason is that O stars steadily lose matter through high-velocity winds (described in chapter 14), so that the tendency to increase their core temperature and hence the luminosity is offset by the decreased mass of the overlying layers.

The details of the later stages of evolution for these massive stars are not very clear. Most likely, several stages of nuclear burning take place as the star goes through numerous loops in the red-giant region of the H-R diagram, leading ultimately to the formation of iron in the core. At the same time, the stellar wind may rip away so much of the star's substance that the outer portion of the core is exposed, creating a kind of "peculiar" object called a **Wolf-Rayet star** (Fig. 15.27).

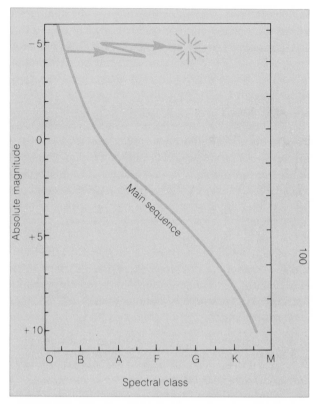

FIGURE 15.26. *Evolution of an O Star.*
The evolutionary track of a very massive star is quite simple: it moves to the right (not up, because mass loss by a stellar wind counteracts the effects of expansion), briefly moves to the left during helium burning, then goes back to the right. The star may go back and forth (rapidly) a few times, but within a total lifetime of around one million years, it exhausts all nuclear fuel and explodes in a supernova.

These are hot stars with very strong winds that have excess quantities of carbon (as a result of the triple-alpha reaction) or nitrogen and oxygen (the products of the CNO cycle) at their surfaces. When the outer layers have been stripped off, we get a direct view of the innards of the star, where material that has recently undergone nuclear reactions is seen. The convection that is taking place in the core helps to bring this processed matter up to the surface.

When iron has been formed in the core of a star, no further nuclear reactions can support the star against the inward force of gravity. Indeed, any additional reactions that could occur would help gravity, because reactions involving elements heavier than iron are all **endothermic,** meaning that more energy is required to create the reaction than is produced. Hence if these reactions start, they actually cool the star's interior, thereby reducing the pressure and allowing gravity to dominate further.

Apparently the following sequence of events can occur. When the core is composed of iron, the reactions stop, and the star begins to collapse, having no means of support (Fig. 15.28). As the collapse begins, the compression of the core causes endothermic reactions to start, and suddenly there is an effective vacuum at the center, as the heat there goes into the reactions. The star now collapses very quickly, in free-fall, and it implodes. It is as though the rug were suddenly pulled out from under the outer layers, and they come crashing down into the center of the star. This causes a massive shock, and the material bounces back violently, in a supernova explosion. The rapid outward expansion of the star during the explosion is driven largely by a flood of neutrinos formed in rapid nuclear reactions. A good deal of matter from the star is dispersed into space as a result of the outburst.

During the supernova explosion, densities and temperatures reach incredible levels for a short time, sufficient to cause the formation of very heavy elements. Most of the atoms heavier than iron in the universe were created in supernova explosions. Of course, some or all of the intermediate-weight elements that had been created in the star's interior during its more placid stages of evolution are also dispersed into space by the explosion, so the process enriches the interstellar gas with a wide variety of elements.

What remains of the original star? Apparently in many cases nothing at all is left, except a cloud of cosmic debris. In other cases, a fraction of the star's final mass remains in the form of a highly condensed remnant. This may be a neutron star, if the remnant mass is below the limit of 2 or 3 solar masses.

If too much matter is contained in this remnant, however, a neutron star cannot form, because the force of gravity is too strong. The star simply continues to collapse; nothing can ever stop it. The resulting object, first envisioned mathematically nearly seventy years ago, is called a **black hole,** and today there is observational evidence that such a thing exists.

Although a black hole literally cannot be seen, there are indirect means by which its presence may be detected. These means, as well as observational data concerning black holes, are described in some detail in the next chapter. For now, we return to the evolution of massive stars.

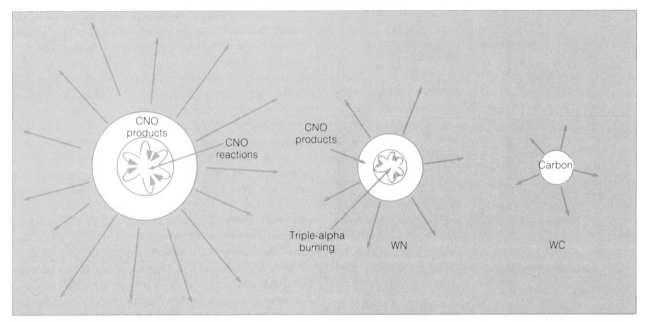

FIGURE 15.27. *Formation of a Wolf-Rayet Star.*
A massive star loses its outer layers through a stellar wind; meanwhile, its core undergoes CNO-cycle reactions, gradually being transformed into helium. When the outer layers have been blown away, the exposed inner region of the star, containing enhanced quantities of nitrogen and oxygen as by-products of the CNO cycle, is a type of Wolf-Rayet star known as a WN star. If the star has progressed to helium burning by the triple-alpha process, then it is a WC star, with an abundance of carbon.

It seems almost certain that all stars starting out with masses in excess of 10 solar masses or so end their lives in supernova explosions. The race between mass loss and the nuclear reactions in the core determines how many nuclear-reaction stages the star undergoes, as well as how much mass it will have left when all the possible reactions are finished. The outcome of the race is difficult to call theoretically. Hence it is difficult to predict exactly which stars will end up as neutron stars and which will form black holes. Some may simply explode, leaving no remnant behind.

FIGURE 15.28. *A Supernova Explosion.*
When the last possible reaction stage has been completed, the star's outer layers collapse inward because the source of internal pressure is gone. The collapse is so violent that the star "bounces back," and explodes as a supernova. The explosion is actually caused by the outward pressure of neutrinos created in rapid nuclear reactions.

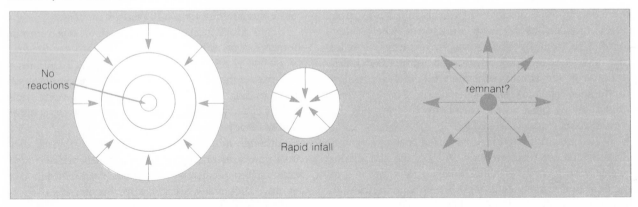

Evolution in Binary Systems

It was pointed out in chapter 14 that some special effects may occur in certain binary systems. If the two
stars are sufficiently far apart, say several astronomical
units or more, then most likely each evolves on its own,
as though the companion were not there. If, however,
they are very close together, then significant interaction
takes place which can drastically alter the normal
course of events (Fig. 15.29).

If either star is massive enough to have a strong wind,
then matter can be transferred to the other star. Even
if no strong wind is present initially, as soon as either
star reaches the red-giant stage, it can swell up sufficiently to dump material onto the other through the
point between them where their gravitational forces are
equal (see chapter 14).

Regardless of how it happens, once mass begins to
transfer from one star to the other, the evolution of one
or both is modified. The one that gains mass speeds up
its evolution, whereas the one that loses it may slow

FIGURE 15.29. *Mass Exchange and Evolution in a Close Binary System.*
The star that is initially more massive evolves faster, and as a red giant it deposits material onto the companion. This speeds up the
evolution of the second star, so that it expands and returns matter to the first star. If by this time the first one is a compact object
such as a neutron star, the consequences of this stage of mass transfer can be violent (see chapter 16).

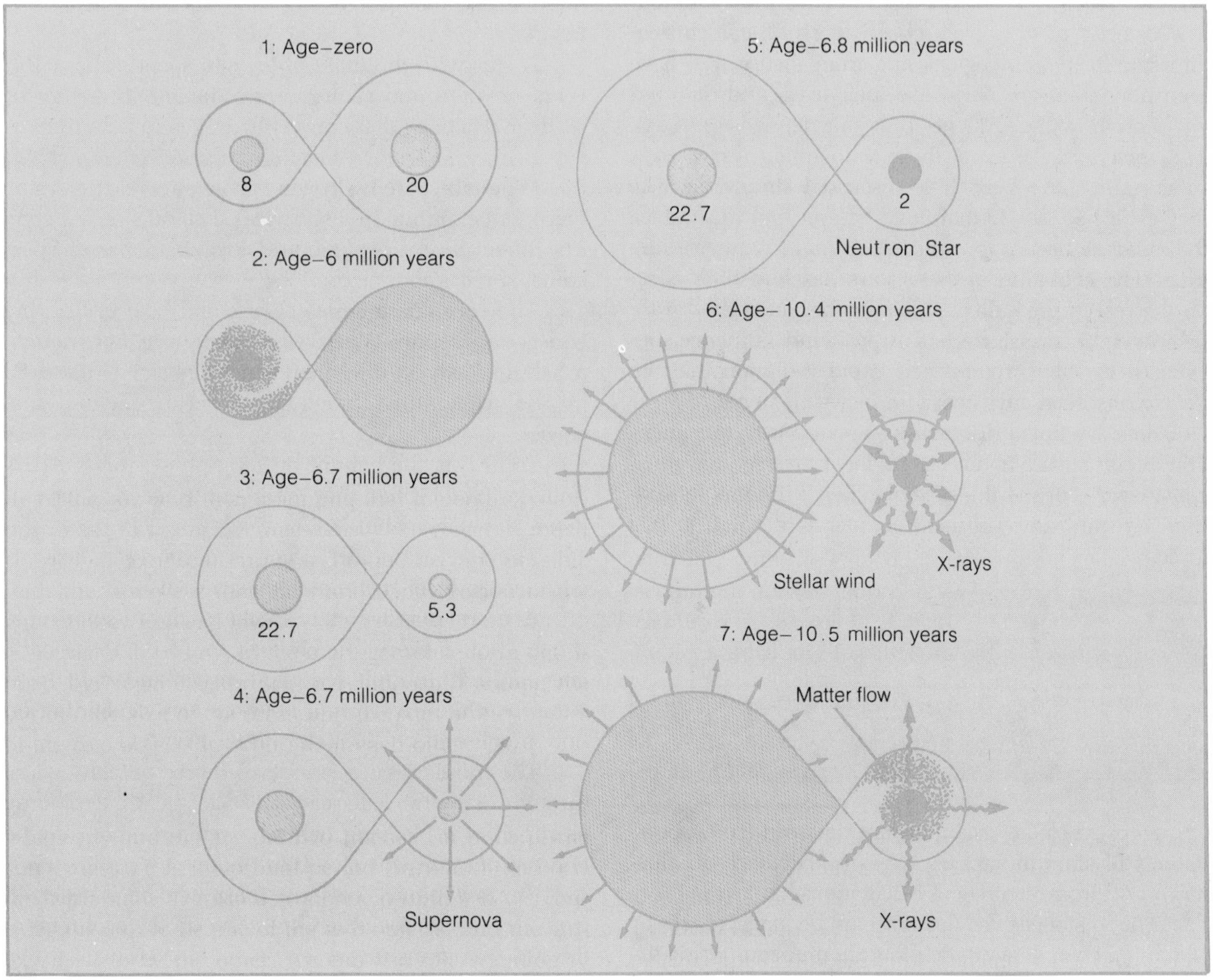

down. The most spectacular observational effects occur when one star has gone all the way through its evolution and has become a white dwarf, neutron star, or black hole, and then it gains new material from the companion star. Special effects, often including X-ray emission, are then observed, and it is under these circumstances that stellar remnants are most easily detected. This will be discussed at some length in the next chapter.

PERSPECTIVE

We have seen how astronomers deduce the evolutionary histories of stars from observations of star clusters and star formation regions and from theory. We have seen how stars are born, live their lives, and die, and we know how the sequence of events depends on mass. The least massive stars, those with much less than 1 solar mass, have been little discussed simply because they evolve so slowly that none has yet had time, since the universe began, to complete its main-sequence lifetime. The evolution of these stars has had little effect on the rest of the galaxy, whereas the much more rare heavyweight stars have had a profound influence, because it is only through their rapid evolution that the galaxy has been enriched with heavy elements.

Before we finish discussing stars and turn our attention to larger-scale objects in the universe, we must consider the properties of the stellar remnants left behind as stars die. The next chapter is devoted to this topic.

SUMMARY

1. Steps in stellar evolution are deduced from observations of stars in various stages (particularly in clusters), and from theoretical calculations.

2. The distances to clusters can be determined by plotting a color-magnitude diagram and comparing the location of the main sequence with that of a standard H-R diagram.

3. The fact that all stars in a cluster have the same age and initial composition means that the observed differences from star to star within the cluster are entirely due to differences in stellar mass. This allows us to determine how stars of different mass evolve.

4. The age of a cluster can be inferred from the location of its main-sequence turn-off point.

5. Stars in the process of forming are often embedded within dense clouds, and are best observed in infrared wavelengths.

6. A star forms from the gravitational collapse of an interstellar cloud, which condenses most rapidly at the center. The new star begins to heat up only after the gas in the core of the cloud becomes opaque, and the resulting protostar continues to slowly shrink until its interior becomes hot enough for nuclear reactions to begin.

7. A star with the mass of the sun spends about 10^{10} years on the main sequence, producing its energy by nuclear reactions that convert hydrogen to helium in its core.

8. When the core hydrogen is gone, the reactions stop there, but continue in a spherical shell outside the core. The outer layers expand and cool, and the star becomes a red giant.

9. The core continues to shrink and heat until it becomes degenerate. When it eventually gets hot enough, a helium flash occurs, after which the star is powered by the triple-alpha reaction, converting helium into carbon.

10. Following the helium-burning phase, the star may undergo instabilities that cause the ejection of one or more planetary nebulae.

11. The star eventually becomes degenerate throughout, and ends its evolution as a white dwarf.

12. A more massive star, in the 5- to 10-solar-mass range, evolves more quickly than the lower-mass stars, goes through multiple red-giant stages, and ends up as either a white dwarf or a neutron star, depending on how much mass it loses during its evolution.

13. The most massive stars evolve very quickly, going through many nuclear-reaction stages before exploding as supernovae, leaving behind a remnant that may be either a neutron star or a black hole.

14. In close binary systems, mass can be transferred from one star to the other, speeding up the evolution of the one that gains mass.

REVIEW QUESTIONS

1. Is it possible for a cluster to contain both hot O stars and cool G stars at the same time?

2. Suppose two stars in the same cluster have identical apparent magnitudes. How do their masses and other properties compare? Could you reach the same conclusion about a pair of stars that are not in the same cluster, but which also have identical apparent magnitudes?

3. Cluster *A* has no stars of spectral types O, B, and A on its main sequence, whereas cluster *B* has stars of all types but O on its main sequence. Which cluster is older? Explain how you made your decision.

4. Why does a protostar not heat up significantly until it becomes opaque?

5. Why can a young star never reach hydrostatic equilibrium until nuclear reactions start in its core?

6. Why is there not a "hydrogen flash" in the core of a star when it forms and the proton-proton chain (or the CNO cycle) begins in its core for the first time?

7. Knowing that a 10-solar-mass star has ten times more total nuclear energy available to it than the sun does, and has about 1,000 times the sun's luminosity, estimate how much shorter the lifetime of the 10-solar-mass star is, compared with the sun.

8. Discuss the role of stellar winds in the evolution of massive stars.

9. Summarize the differences in the evolutions of a 1-solar-mass star and a 10-solar-mass star.

10. It has been discovered that iron is a relatively abundant element in the galaxy, more plentiful than the elements that are a little heavier or a little lighter. In view of what you have learned in this chapter about nuclear reactions in stars, explain why iron is so abundant in the galaxy.

ADDITIONAL READINGS

Bok, B. J. 1981. The early phases of star formation. *Sky and Telescope* 227(2):48.

Cohen, M. 1975. Star formation and early evolution. *Mercury* 4(5):10.

Flannery, B. P. 1977. Stellar evolution in double stars. *American Scientist* 65:737.

Herbst, W., and Assousa, G. E., 1979. Supernovas and star formation. *Scientific American* 241(2):138.

Kaler, J. 1981. Planetary nebulae and stellar evolution. *Mercury* 10(4):114.

Lada, C. J. 1982. Energetic outflows from young stars. *Scientific American* 247(1):82.

Loren, R. B., and Vrba, F. J. 1979. Starmaking with colliding molecular clouds. *Sky and Telescope* 57(6):521.

Seeds, M. A. 1979. Stellar evolution. *Astronomy* 7(2):6.

Shklovskii, I. S. 1978. *Stars: their birth, life, and death.* San Francisco: W. H. Freeman.

Spitzer, L. 1983. Interstellar matter and the birth and death of the stars. *Mercury* 12(5):142.

Sweigart, A. V. 1976. The evolution of red giant stars. *Physics Today* 29(1):25.

Wyckoff, S. 1979. Red giants: the inside scoop. *Mercury* 8(1):7.

Zeilik, M. 1979. The birth of massive stars. *Scientific American* 238(4):110.

STELLAR REMNANTS

16

Learning Goals

16 Our story of stellar evolution is almost complete. We have seen how the properties of stars are measured, and we have learned how stars form, live, and die. The only missing link is what is left of a star after its life cycle is completed.

As we saw in the previous chapter, the form of remnant that remains at the end of a star's life depends on the mass the star had when it finally ran out of nuclear fuel. Three possibilities were mentioned: white dwarf, neutron star, and black hole. In this chapter each of these three bizarre objects is discussed, with emphasis on observable properties.

White Dwarfs, Black Dwarfs

An isolated white dwarf is not a very exciting object, from the observational point of view. It is rather dim (Fig. 16.1), and its spectrum is nearly featureless, except for a few very broad absorption lines created by the limited range of chemical elements it contains (Fig. 16.2). The great width of the lines is caused by the immense pressure in the star's outer atmosphere; pressure tends to smear out the atomic energy levels, because of collisions between atoms. In effect, the energy levels become broadened, and this results in broadened spectral lines when transitions of electrons occur between the levels.

Additional broadening or shifting of spectral lines may occur because of magnetic effects. If the star had a magnetic field before its contraction to the white-dwarf state, the field not only is preserved but also is intensified in the process of contraction. Thus a white dwarf may have a very strong magnetic field, and as we learned previously (chapters 11 and 13), a magnetic field causes certain spectral lines to be split into two or more parts. In a white dwarf, where the lines are already very broad, this splitting (called the Zeeman effect) usually makes the lines appear even broader.

The lines in the spectrum of a white dwarf are always shifted toward longer wavelengths. This is not caused by the star's motion; if it were, it would mean that somehow all the white dwarfs in the sky are receding from us. Instead, a different type of redshift occurs here, one caused by gravitational effects. One of the predictions of Einstein's theory of general relativity is

FIGURE 16.1. *Sirius B.*
The dim companion to Sirius was the first white dwarf to be discovered. The analysis of its mass, as well as of its temperature and luminosity, led to the realization that it is very small and dense.

FIGURE 16.2. *A White Dwarf Spectrum.*
Here we see that white dwarf spectra are without many strong features, and that the spectral lines tend to be rather broad. This white dwarf has a large abundance of carbon, as a result of the triple-alpha reaction.

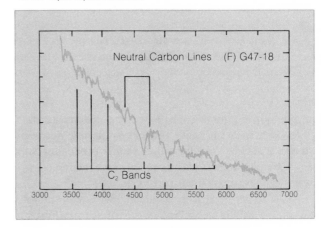

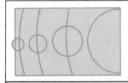

Astronomical Insight 16.1
THE STORY OF SIRIUS B

Sirius, also known as α Canis Majoris, is the brightest star visible in the heavens. It was discovered in 1844 to be a binary, because of its wobbling motion as it moves through space (recall the discussion of astrometric binaries in chapter 13). Its period is about fifty years, and the orbital size about 20 astronomical units; if we analyze these data using Kepler's third law we come to the conclusion that the companion must have a mass similar to that of the sun. At the distance of Sirius (only 2.7 parsecs), a star like the sun should be easily visible, yet the companion to Sirius defied detection for some time.

Part of the difficulty lay in the overwhelming brightness of Sirius itself, whose light would tend to drown out that of a lesser, very close star. Astronomers were so intrigued by the mysterious nature of the companion, however, that they made intensive efforts to find the object, and finally, in 1863, with an exceptionally fine, newly developed telescope, they were able to see it.

It was immediately confirmed that this star was unusually dim for its mass, having an apparent magnitude of 8.7 (the apparent magnitude of the primary star is −1.4) and an absolute magnitude of 11.5, nearly 7 magnitudes fainter than the sun. Spectra obtained with great difficulty (again owing to the brightness of the companion) showed that this object, called Sirius B, has a high temperature, similar to that of an O star.

When Russell constructed his first H-R diagram in 1913, he included Sirius B, finding it to lie all by itself in the lower left-hand corner. The only way for the star to have such a low luminosity while its temperature was so high was for it to have a very small radius, and scientists at the time deduced that Sirius B was about the size of the earth. The incredible density was estimated immediately, and it was realized that a totally new form of matter was involved. The term **white dwarf** was coined to describe Sirius B and other stars of this type, which were discovered soon afterward in other binary systems.

The physics of this new kind of matter took some time to be understood, partly because the theory of relativity had to be applied (the electrons in a degenerate gas move at speeds near that of light). The Indian astrophysicist S. Chandrasekhar was the first to solve the problem of the behavior of this kind of gas, and by the 1930s he was able to construct realistic models of the structure of a white dwarf. From this work came the 1.4-solar-mass limit we referred to in the text.

Modern observational techniques offer the possibility of detecting white dwarfs more easily. For example, even though Sirius B is about 10 magnitudes (or a factor of 10,000) fainter than Sirius A in visible wavelengths, its greater temperature makes it relatively brighter in the ultraviolet. There is even a wavelength, about 1,100Å, below which the white dwarf is actually brighter than its companion. Hence Sirius B and other white dwarfs have been detected with ultraviolet telescopes. The development of more sophisticated space instruments will make it possible to study the properties of many white dwarfs in this way.

that photons of light are affected by gravitational fields. White dwarfs, with their immensely strong surface gravities, provide a confirmation of this prediction. In later sections, we will discuss even more extreme examples of these **gravitational redshifts.**

Left to its own devices, a white dwarf will simply cool off, eventually becoming so cold that it no longer emits visible light. At this point it cannot be seen, and simply persists as a burnt-out cinder, referred to as a black dwarf. The cooling process takes a long time, however; several billion years may go by before the star becomes cold and dark. This may seem surprising, since the white dwarf has no source of energy, only staying hot as long as it can retain the heat that it contained when it was formed. The outer skin of the white dwarf acts like a very efficient thermal blanket, however, since it consists of gas that is nearly opaque. Radiation cannot easily penetrate this outer layer, and because there is no other way for the heat to be transported from the star's interior out into space, the heat takes a long time to filter out.

The thermal blanket created by the outer atmosphere is very thin; most of the volume of the star is filled with degenerate electron gas (Fig. 16.3), as discussed in the last chapter. The atmosphere, consisting of ordinary gas, may be only about 50 kilometers thick (remember, the white dwarf itself is about the size of the earth, with a radius of some 5,000 to 6,000 kilometers). The composition of the interior may be primarily helium, if the star never progressed beyond the hydrogen-burning stage; it may be carbon, if the triple-alpha reaction was as far as the star got in its nuclear evolution before dying; or it may be some heavier element, if the original star was sufficiently massive to have undergone several reaction stages. In any case, the mass of the white dwarf must be less than the limit of 1.4 solar masses.

White Dwarfs and Nova Explosions

Ordinarily, a white dwarf cools off and is never heard from again. There are circumstances, however, in which it can be resurrected briefly for another role in the cosmic drama. It has recently become apparent that white dwarfs are central characters in producing **nova** outbursts (these spectacular events were described in chapter 15).

A typical nova can reach an absolute magnitude in the range of -6 to -9; an otherwise dim star can increase its brightness so much that it might show up briefly as a rather bright object (Fig. 16.4). Usually a nova brightens up to its peak in less than a day, then takes weeks or months to return to its former dimness. Apparently material is ejected by a star during the nova outburst, because spectral lines show Doppler shifts created by matter moving away from the star, and in some cases photographs have revealed expanding nebulae surrounding the star that flared up (Fig. 16.5).

Some stars become novae more than once; several examples are known where two or more nova outbursts, separated by several decades, have occurred in the same star. These **recurrent novae** are reminiscent of another phenomenon, the **dwarf novae.** These stars undergo minor flare-ups, and are called dwarf novae because they do not approach the brightness of a genuine nova; in many ways they appear to be scaled-down versions. The dwarf novae always occur in binary systems where one star is a white dwarf. This seems to be the case for the true novae as well.

To understand how the flare-up occurs, we recall the helium flash that takes place in the degenerate core of certain red giants. When nuclear reactions start in a degenerate gas, they quickly create a violent runaway explosion, because the degenerate gas cannot respond as ordinary gas would. It cannot expand and cool, and thereby hold the reactions down to a moderate, steady rate. Because a white dwarf consists of degenerate gas, it will experience the same kind of runaway explosion as a helium flash, if nuclear reactions are somehow started.

In an isolated white dwarf, there is no way to start new nuclear reactions; all the possible fuel is used up earlier in the star's evolution. In a binary system, however, the companion may shed material when it expands into the red-giant stage, or as a result of a stellar wind, and some of this material can fall onto the white dwarf (Fig. 16.6). This matter forms the fuel for new nuclear reactions, and the impact of its descent onto the surface of the white dwarf can provide the necessary heat. When the temperature gets high enough, reactions begin in the newly accreted matter, and the nature of the underlying degenerate gas ensures that the reactions will quickly flare up, consuming all the new

FIGURE 16.3. *The Internal Structure of a White Dwarf.* The star's interior is degenerate and would cool very rapidly, except that the outer layer of normal gas acts as a very effective insulator, trapping radiation so that heat escapes only very slowly.

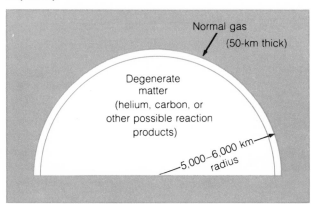

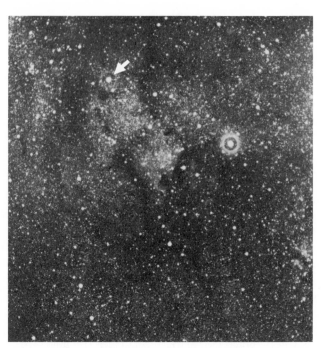

FIGURE 16.4. *Nova Cygni 1975.*
This was one of the brightest novae in recent years. The photo at left shows the region without the nova. At right is the same region photographed near the time of peak brightness, when the apparent magnitude of the object was almost +1.

FIGURE 16.5. *An Expanding Shell Around an Old Nova.*
This is Nova Persei (1901), surrounded by an expanding cloud of gas that was ejected during the outburst.

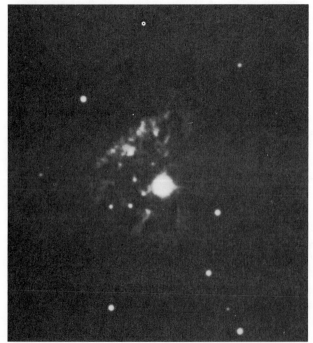

fuel in a very short time. This is the nova outburst, and its magnitude depends on how much material is burned in the brief nuclear conflagration. Apparently in the dwarf novae the outburst occurs after only a small amount of material has accumulated, whereas in a standard nova a great deal more material must build

FIGURE 16.6. *The Nova Process.*
Matter is transferred onto a white dwarf by a red giant companion. Theoretical calculations show that the new material will orbit the white dwarf in a disk, but will eventually become unstable and fall down onto the white dwarf. When it does so, nuclear reactions will flare up, creating a nova outburst.

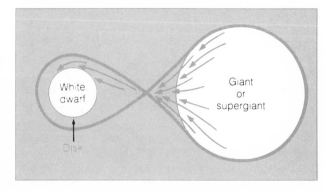

up before the explosion occurs; when it does, however, the outburst is correspondingly more energetic. The explosions recur as more material is subsequently supplied to the white dwarf by its binary companion.

White dwarfs, then, are most prominent in binary systems where mass transfer from a normal companion takes place. We will see later in this chapter that other forms of stellar remnants are also best observed in these circumstances.

Supernova Remnants

We learned in the last chapter that stars too massive at the end of their lifetimes to become white dwarfs will, in most cases, explode in supernovae. This violent demise may leave a stellar remnant (a neutron star or black hole), and it will certainly leave an expanding gaseous cloud called a **supernova remnant.**

A number of supernova remnants are known to astronomers. A few, like the prominent Crab Nebula (Fig. 16.7), are detectable in visible light, but many are most easily observed at radio wavelengths. This is partly because visible light is affected by interstellar dust which can hide our view of remnants, whereas radio waves are largely unaffected by the dust; and partly because the remnants are intrinsically strong radio emitters. The reason for this is not that the remnants are cold, which would cause their peak thermal emission to occur at long wavelengths, as in the case of the outer planets. Here a completely different process, called **synchrotron emission,** produces the radio waves.

Synchrotron emission occurs when electrons move rapidly through a magnetic field. The electrons must have speeds near that of light; as they travel through the magnetic field, they are forced to move in a spiraling path, and they emit photons as they do so (Fig. 16.8). The emission occurs over a very broad range of wavelengths, including some visible, ultraviolet, and even X-ray radiation, but these other wavelengths are often not as easily detected as radio, because the radio emission is the strongest.

Supernova remnants that are detected in visible wavelengths usually glow in the light of several strong emission lines, most notably the bright line of atomic hydrogen at a wavelength of 6,563Å, in the red portion

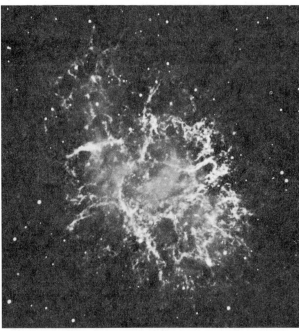

FIGURE 16.7. *The Crab Nebula.*
This is a well-studied supernova remnant, the result of an explosion observed by Chinese astronomers in A.D. 1054. The filamentary structure stands out most clearly in photos such as this, taken through a red filter which reveals directly the regions of ionized hydrogen.

FIGURE 16.8. *The Synchrotron Process.*
Rapidly moving electrons spiral along magnetic-field lines, and emit photons as they do so. The radiation that results is polarized, and its spectrum is continuous but lacks the peaked shape of a thermal spectrum. The electrons must be moving very rapidly (at speeds near that of light); thus whenever astronomers detect synchrotron radiation, they know that a source of large quantities of energy must be present.

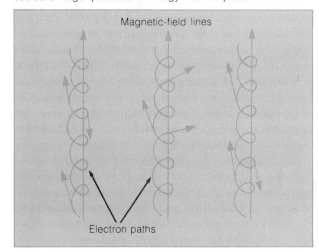

of the spectrum. These lines often show large Doppler shifts, indicating that the gas is still moving rapidly as the entire remnant expands outward from the site of the explosion that gave it birth. Often a filamentary structure is seen, suggestive of turbulence, but probably modified in shape by magnetic fields.

Remnants are visible at the locations of several famous supernova explosions. The most prominent is the Crab Nebula, which was created in the supernova observed by Chinese astronomers in A.D. 1054. Other supernovae seen by Tycho Brahe (in 1572) and by Kepler and Galileo (in 1604) also left detectable remnants, although neither is as bright in visible wavelengths as the Crab. Apparently a supernova remnant can persist for 10,000 years or more before becoming too dissipated to be recognizable.

The energy of a supernova explosion is immense, comparable to the total amount of radiant energy the sun will emit over its entire lifetime. Much of the energy takes the form of mass motions as the remnant expands. This kinetic energy heats the expanding gas to very high temperatures. The entire process has a profound effect on the interstellar gas and dust that permeate the galaxy, as we will learn in chapter 18.

Neutron Stars

We have referred to a neutron star as a stellar remnant composed entirely of neutrons that are in a degenerate state, similar to that of the electrons in a white dwarf. The pressure created by the degenerate neutron gas is greater than that produced by a degenerate electron gas, so a star slightly too massive to become a white dwarf can be supported by the neutrons. Again, however, the star has a limit on its mass. This limit, which depends on other factors such as the rate of rotation, is between 2 and 3 solar masses. Thus there is a class of stars, whose masses at the end of the stars' nuclear lifetimes are between 1.4 and about 2 or 3 solar masses, that become neutron stars.

The structure of a neutron star is even more extreme than that of a white dwarf (Fig. 16.9). All the mass is compressed into an even smaller volume (with a radius of about 10 kilometers), and the gravitational field at the surface is immensely strong. The layer of normal gas that constitutes the neutron star's atmosphere is

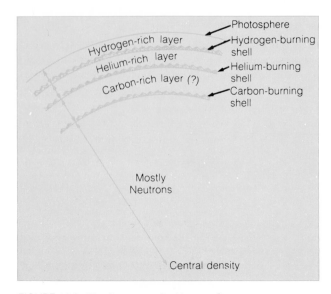

FIGURE 16.9. *The Structure of a Neutron Star.* This diagram shows that the outer regions of a neutron star may consist of thin layers of various elements that were produced by nuclear reactions during the star's lifetime. These outer layers are thought to have a rigid crystalline structure, because of the intense gravitational field of the neutron star.

only centimeters thick, and beneath it may be zones of different chemical compositions resulting from previous shell-burning episodes, each zone only a few meters thick at the most. Inside these surface layers is the incredibly dense neutron gas core, which takes the form of a crystalline lattice. The temperature throughout is very high, but because the surface area is so small, a neutron star is very dim indeed. It may have a magnetic field that is also much more intense than that of a white dwarf, resulting from the compression of the star's original field.

Pulsars: Cosmic Clocks

The properties of neutron stars were predicted theoretically several decades ago, but for a while astronomers did not think these objects could be observed, because of the stars' low luminosities. In 1967, however, an accidental discovery by a radio astronomer led to the establishment of a new class of objects called **pulsars,** which are now believed to be neutron stars. As it turned out, observation of pulsars is only one of two ways to detect neutron stars (the other is described in the next section).

Pulsars are radio sources that flash on and off very

regularly, and with a high frequency (Fig. 16.10). One of the most rapid pulsars is located in the Crab Nebula; it repeats itself thirty times a second. At least two others are known to flash even more rapidly than the Crab pulsar; one of these pulses at the incredible frequency of 642 times per second! The slowest known pulsars have cycle times of more than a minute, on the other hand. In every case, the pulsar is only "on" for a small fraction of each cycle.

When pulsars were discovered, there was a great deal of excitement and a lot of speculation, including the suggestion that they were beacons operated by an alien civilization. Once the initial shock of discovery wore off, however, a number of more natural explanations were offered by astronomers who sought to establish the identity of the pulsars. It was well known that a variety of stars pulsate regularly, alternately expanding and contracting, but none were known to do it so rapidly. Theoretical studies showed that variations as rapid as those exhibited by the pulsars should occur only in very dense objects, denser even than white dwarfs. This led astronomers to think of neutron stars.

Enough was known from theoretical calculations, however, to rule out rapid expansion and contraction of neutron stars as the cause of the observed pulsations. The vibration period of such an object, which is governed by its density, would be even shorter than the observed periods of the pulsars.

A second possibility, that the rapid periods were produced by rotation of the objects, was considered. In order to rotate several times per second, however, a normal star or white dwarf would have to have a surface velocity in excess of the speed of light, a physical impossibility. Such an object would be torn apart by rotational forces before approaching such a velocity. Hence normal stars and white dwarfs were ruled out as candidates to explain pulsars. A neutron star, on the other hand, could rotate several times per second and remain intact.

When the Crab pulsar was discovered, a great deal of additional information became available. This pulsar was found to be gradually slowing its pulsations, something that could best be explained if the pulses were linked to the rotation of the object. The rotation could slow as the pulsar gave up some of its energy to its surroundings.

This discovery cleared up another mystery. The source of energy that powers the synchrotron emission from supernova remnants had been unknown; now it

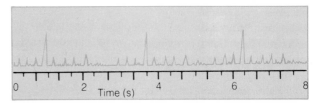

FIGURE 16.10. *Radio Emission From a Pulsar.*
Here, plotted against time, is the radio intensity from a pulsar. The radio emission is weak or nonexistent except for a very brief flash once each cycle (in some cases there is a weaker flash between each adjacent pair of strong flashes).

was suggested that the slowdown of the rotating neutron star could provide the necessary energy, either by magnetic forces exerted on the surrounding ionized gas or by transfer of energy from the pulsar to the surrounding material through the emission of radio waves.

The remaining question was how the pulses were created by the rotation. Evidently a pulsar acts like a lighthouse, with a beam of radiation sweeping through space as it spins. The question was why a rotating neutron star should emit a beam from just one point on its surface.

The most probable explanation has to do with the strong magnetic fields that neutron stars are likely to have. If a neutron star has a strong field, then electrons from the surrounding gas are forced to follow the lines of the field, hitting the surface only at the magnetic poles. The result is an intense beam of electrons traveling along field lines, especially concentrated near the magnetic poles of the neutron star, where the field lines are crowded together. The rapidly moving electrons emit synchrotron radiation as they travel along the field lines, creating narrow beams of radiation from both magnetic poles of the star. If the magnetic axis of the star is not aligned with the rotation axis (Fig. 16.11), these beams will sweep across the sky in a conical pattern as the star rotates. If the earth happens to lie in the direction intersected by one of these beams, then we see a flash of radiation every time the beam sweeps by us.

Although the details are still somewhat vague, this seems to be the best explanation of the pulsars. Since special conditions (that the magnetic and rotation axes must not be aligned, and that the beam must be in our line of sight) are required for a neutron star to be seen from earth as a pulsar, it follows that there should be

Astronomical Insight 16.2

DISCOVERY OF THE PULSARS, AND THE CRAB NEBULA

The radio observatory of Cambridge University, in England, has been used for the last three decades to map the skies in radio wavelengths. This kind of survey creates a vast amount of data, which are routinely processed and which are often not examined in great detail, at least not immediately as they come in.

In 1967, a graduate student named Jocelyn Bell was perusing data from the radio survey when she found an astonishing thing: a radio source that seemed to be blinking on and off very regularly, at a rapid rate. Careful checking showed that this source was always at the same position in the sky, and hence was not some earthly object or a moving vehicle. The regularity of the pulses was so perfect that there arose a great deal of speculation as to whether this was an interstellar beacon left by some intelligent race.

Several additional pulsars, as they were dubbed, were found, and in scientific circles, the search was on for a natural explanation. The concept of the neutron star, first devised several decades earlier, was revived, because it was soon realized that such rapid pulsations could not be produced by anything less dense, even a white dwarf. When a pulsar was then discovered in the midst of the supernova remnant called the Crab Nebula, where a star had exploded recently, the suspicion that neutron stars were responsible for the pulsars approached certainty. The question was settled when the slowdown rate of the pulsar in the Crab was found to be pre-

cisely what was needed to provide the energy of the synchrotron emission, until then a mystery. The Crab pulsar emits pulses at a rate of some thirty times per second, making it one of the fastest of all known pulsars. Apparently this rapid pulse had to do with the star's youth; older neutron stars have had more time to slow down their spin rates as they lose energy to their surroundings. Presumably the more rapid pulsars discovered recently are even younger than the Crab pulsar, but perhaps not; other evidence indicates that they simply have not slowed because they lack surrounding gas to create drag.

The Crab pulsar was identified in visible light by a group of astronomers who built a shutter device and placed it in their telescope in front of a photometer (an instrument for measuring star brightnesses). The shutter was set to open and shut rapidly, at the same rate as the radio pulses from the Crab pulsar. When the astronomers pointed the telescope at a particular dim star near the center of the Crab, and adjusted the shutter, the star suddenly disappeared. This proved that the star was turning itself off and on at the same rate the shutter was opening and closing, and thus showed that the star was definitely the pulsar. It pulses not just in radio, but also in visible and, it was found much later, in X-ray wavelengths as well.

many neutron stars that do not manifest themselves as pulsars. In the next section we will see how some of these nonpulsating neutron stars are detected.

Neutron Stars in Binary Systems

Earlier we made a general statement that stellar remnants are often most easily observed when they are in binary systems. We have already seen that a white dwarf in a binary can flare up violently if it receives

new matter from its companion star. A neutron star reacts similarly in the same circumstances.

If a neutron star is in a binary system where the companion object is either a hot star with a rapid wind or a cool giant losing matter because of its active chromosphere and low surface gravity, some of the ejected mass reaches the surface of the neutron star. Very little of it falls directly down onto the surface; instead, much of it swirls around the neutron star, forming a disk of gas called an **accretion disk** (Fig. 16.12). The indi-

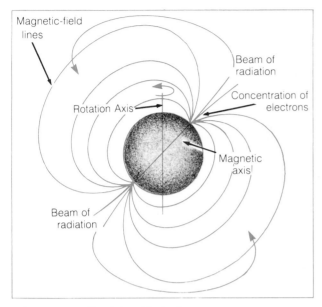

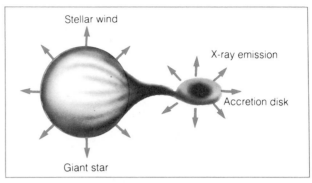

FIGURE 16.12. *An Accretion Disk*.
A giant star, losing mass through a stellar wind, has a neutron-star companion. Material that is trapped by the neutron star's gravitational field swirls around it in a disk, which is so hot (owing to compression) that it glows at X-ray wavelengths. A nearly identical situation can arise when the compact companion is a black hole.

FIGURE 16.11. *The Pulsar Mechanism*.
Here we see a rapidly rotating neutron star, with its magnetic axis out of alignment with the rotation axis. Synchrotron radiation is emitted in narrow beams from above the magnetic poles, where charged particles, constrained to move along the magnetic-field lines, are concentrated. These beams sweep the sky as the star rotates, and if the earth happens to lie in a direction covered by one of the beams, we observe the star as a pulsar.

vidual gas particles orbit the neutron star like microscopic planets, and only fall inward as they lose energy in collisions with other particles. The accretion disk acts as a reservoir, slowly feeding the particles inward, toward the neutron star.

The disk is very hot because of the extreme gravitational forces exerted on it by the neutron star. The gas in the disk is highly compressed, and reaches temperatures of several million degrees. A gas at such a high temperature emits X rays. Hence a neutron star in a mass-exchange binary system is likely to be an X-ray source. A number of such systems have been found in the last decade (Fig. 16.13), with the advent of X-ray telescopes launched on rockets and satellites. Often it is known that an X-ray object is part of a binary system because the Xrays are periodically eclipsed by the companion star. Eclipses are made likely by the proximity of the two stars (mass exchange would not occur unless it were a close binary) and the fact that the mass-

FIGURE 16.13. *An X-Ray Map*.
This shows the locations of X-ray sources discovered by the *Uhuru* satellite, the first major X-ray survey instrument. Most of the sources shown here are binary systems containing evolved, compact companions such as neutron stars or black holes.

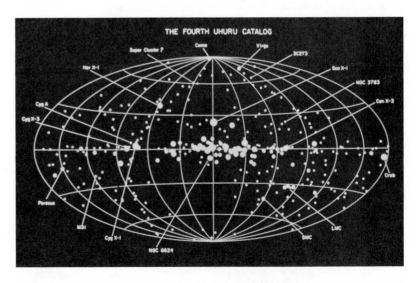

losing companion is likely to be a large star, either an upper-main-sequence star or a supergiant.

The so-called **binary X-ray sources** are among the strongest X-ray–emitting objects known. Many of them are most likely neutron stars, although the possibility also exists that some are black holes, which would similarly produce X-ray emission as material falls inward.

There is a slightly different way in which neutron stars can emit X rays, which also occurs in mass-exchange binaries. If the in-falling material trickles down onto the neutron star in a steady fashion, there is continuous radiation of X rays, as we have just learned. If, however, the material falling in does so sporadically, and arrives in substantial quantities every now and then, a major nuclear outburst occurs each time, in close analogy to the nova process involving a white dwarf. This apparently happens in some cases, producing random but frequent X-ray outbursts (Fig. 16.14). The intensity of the outburst, as in the case of a nova, depends on how much matter has fallen in and been consumed in the reactions. The neutron-star binary systems where this occurs are called **bursters.** One important contrast with novae is that the flare-up of a burster occurs much more rapidly, lasting only a few seconds. Another is that most of the emission occurs only in very energetic X rays, so these objects do not show up in visible light.

Black Holes: Gravity's Final Victory

In the last chapter we saw what happens to a massive star at the end of its life. It falls in on itself, and there is no barrier; neither electron degeneracy nor neutron degeneracy can stop it.

The gravitational field near the surface of a collapsing star grows in strength as the mass of the star becomes concentrated in an ever-smaller volume. This gravitational field has important effects in the near vicinity of the star, although at a distance it remains unchanged from what it was before the collapse. Close to the star, though, the structure of space itself is distorted, according to Einstein's theory of general relativity. Einstein discovered that accelerations caused by changing motion and those caused by gravitational fields are equivalent, and from this it follows that space must be curved in the presence of a gravitational field, so that moving particles follow the same path they

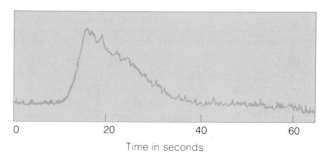

FIGURE 16.14. *The X-Ray Light Curve of a Burster.* This is a schematic illustration of the X-ray intensity from a burster showing how rapidly the emission flares up and then drops off.

would follow if they were being accelerated. Not only physical particles, but also photons of light are affected.

Usually the effects of the curvature of space are not noticeable, except when we engage in very careful observation, or consider very large distances. Later in the text, when we discuss the universe as a whole and its overall structure, we will go into this in more detail. For now, we confine ourselves to local regions where space can be distorted by very strong gravitational fields (Fig. 16.15).

Let us consider a photon emitted from the surface of a star as the star falls in on itself (Fig. 16.16). If the photon is emitted at any angle away from the vertical, its path will be bent over further. If the gravitational field is strong enough, the photon's path may be bent over so far that it falls back onto the stellar surface. A photon emitted straight upward follows a straight path, but it loses energy to the gravitational field, which causes its wavelengths to be redshifted (we already discussed this phenomenon in terms of white dwarfs). When the gravitational field becomes strong enough, the photon loses all its energy, and cannot escape. In other words, the velocity required for the photon to escape from the star exceeds that of light. When this happens the star becomes invisible to us, because no light from the star can reach the earth.

The radius of the star at the time the gravitational field becomes strong enough to trap photons is called the **Schwarzschild radius,** after the German astrophysicist who first calculated its properties some sixty years ago. The Schwarzschild radius depends only on the mass of the collapsing star. For a star of 10 solar masses, it is 30 kilometers. A star of twice that mass would have twice the Schwarzschild radius, and so on.

When a collapsing star has shrunk inside its

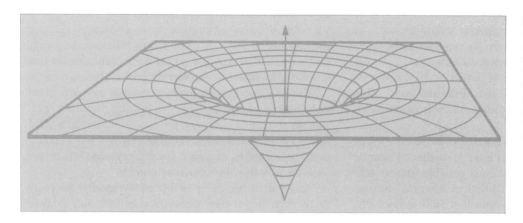

FIGURE 16.15. *Geometry of Space Near a Black Hole.* Einstein's theory of general relativity may be interpreted in terms of a curvature of space in the presence of a gravitational field. Here we see how this curvature varies near a black hole.

Schwarzschild radius, it is said to have crossed its **event horizon,** because we cannot see it or anything that happens to it after that point. We have no hope of ever seeing what happens inside the event horizon, but since we know of no force that could stop the collapse, we assume that it continues. The mass becomes concentrated in an infinitesimally small region at the center, which is called a **singularity,** because mathematically it is a single point.

What happens to the matter that falls into a singularity is a subject for speculation, for no concrete theories have been developed. Among the more fantastic ideas that have been suggested is the possibility that the matter disappears from one place in the universe and

FIGURE 16.16. *Photon Trajectories From a Collapsing Star.* Light escapes in essentially straight lines in all directions from a normal star (left), whose gravitational field is not sufficient to cause large deflections. At an intermediate stage of collapse (center), photons emitted in a cone nearly perpendicular to the surface can escape, but others cannot. Those emitted at just the right angle go into orbit around the star, whereas those emitted at greater angles fall back onto the stellar surface. After collapse has proceeded to within the Schwarzschild radius, no photons can escape.

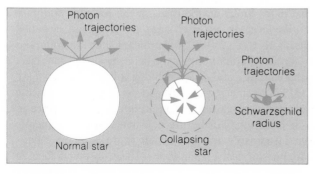

reappears somewhere else, having tunneled through to a different point in space and time. The path it would take has been dubbed a **worm hole,** and the place where it reappears a **white hole.**

Let us return to the collapsed star. The time sense is distorted by the extreme gravitational field and acceleration in such a way that to an observer on earth, the collapse seems to gradually slow, coming to a halt just as the star reaches its event horizon. This slowdown, however, is most significant in the last moments before the star's disappearance, and by then the star is essentially invisible anyway, as most escaping photons are redshifted into the infrared or beyond. The star seems to disappear rather quickly, despite the stretching of the collapse in time resulting from relativistic effects.

Mathematically, a black hole can be described completely by three quantities: its mass, electrical charge, and spin. The mass, of course, is determined by the amount of matter that collapsed to form the hole, plus any additional material that may have fallen in later. The electrical charge, similarly, depends on the charge of the material from which the black hole formed; if the material contained more protons than electrons, for example, the black hole would end up with a net positive charge. Because particles with opposite charges attract each other, and those with like charges repel, it is thought that electrical forces would maintain a fairly even mixture of particles during and after the formation of the black hole, so that the overall charge would be nearly zero. To illustrate this, let us imagine that a black hole was formed with a net negative charge. If there were ionized gas around it afterward (as there likely would be, with some of the matter from the original star still drifting inward), the negative charge of the hole would repel additional electrons, preventing them from falling in, while protons would be accelerated inward. In time, enough protons would be gobbled up to neutralize the negative charge.

The spin of a black hole is not so easily dismissed, however. It stands to reason that if the star were spinning before its collapse, and most likely it would have been, then the rotation would speed up greatly as the star shrank. A high spin rate actually shrinks the event horizon, allowing us to see closer in to the singularity residing at the center. It is even possible mathematically, although it presents a physical dilemma, for the black hole to have so much spin that there is no event horizon, which means that whatever is in the center is exposed to view. A **naked singularity,** as this has been dubbed, would not produce the usual gravitational effects of a black hole, and it would be possible to blunder into one without any forewarning.

For the most part, it is assumed that the spin rate is never so large that a naked singularity can form. In fact, black-hole properties are usually specified by mass alone, and the effects of charge and spin are neglected. It is assumed that our main hope of detecting a black hole is by its gravitational effects, determined entirely by its mass.

Before turning to a discussion of how to find a black hole, we should mention that there may be additional kinds of black holes, formed by processes other than the collapse of individual massive stars. These include "mini" black holes, very small ones postulated to have formed under extreme density conditions early in the history of the universe; and supermassive black holes, thought possibly to inhabit the cores of large galaxies, having formed from thousands or millions of stellar masses coalescing in the center. These other possibilities are discussed in later chapters; for now, we turn to the hunt for stellar black holes.

Do Black Holes Exist?

The mass that goes into a black hole during its formation still exists there, hiding inside the event horizon. Even though no light can escape, gravitational effects persist. The gravitational force of the star is exactly the same as it was before the collapse, except at distances from the hole that are less than the radius of the original star.

Our best chance of detecting a black hole, then, is to look for an invisible object whose mass is too great to be anything else. Even if we do find such a thing, we really have only a circumstantial argument, because the conclusion that it is a black hole relies on the theory that says neither a white dwarf nor a neutron star can survive if the mass is sufficiently great.

The best opportunity for determining the mass of an object occurs when the object is in orbit around a companion star, in which case we can apply Kepler's third law to find the masses. The search for black holes, then, leads us to examine binary systems, where we look for invisible, but massive, objects.

This search has been facilitated by another property of black holes; if new material is pulled into a black hole, the trapped matter gets so highly compressed and heated on the way in that it emits X rays. As we learned earlier, this happens in the case of neutron stars accreting new material, so the emission of X rays by itself does not prove that a black hole is present. We still must determine the mass of the object, to see whether it is too great to be a neutron star.

The skies were first scanned at X-ray wavelengths in the early 1970s, by the *Uhuru* satellite, and later by a number of other X-ray telescopes aboard spacecraft. Many sources of X-ray emission have been found, and the most intense are the X-ray binaries. These are usually known to be binaries because the X-ray emission periodically dips in intensity owing to eclipses by the companion star as it moves in front of the X-ray–emitting object. The orbital period is always only a few hours, and the star that emits the X rays is always invisible. Only the normal companion can be seen. This is often an O or B star with a strong wind, so apparently the matter that falls into the collapsed companion and emits the X rays comes from this wind.

The determination of the masses in a binary system is difficult, if not impossible, if only one of the two stars can be seen, because the data on their velocities and on the inclination angle of the orbits is incomplete. Sometimes enough information can be derived to at least place limits on the mass of the invisible companion, however. In one such case, the X-ray binary called Cygnus X–1 (Fig. 16.17), it appears that the collapsed object has to contain at least 8 solar masses. This object is the leading candidate for being a black hole.

Although the jury is still out on the question of the existence of black holes, the circumstantial evidence in favor is strong. Furthermore, if we follow a principle called **Occam's razor** (which to the scientist means that the simplest explanation of the observed facts is

FIGURE 16.17. *Cygnus X–1.*
This X-ray image obtained by the orbiting *Einstein Observatory* shows the intense source of X-rays at the location of a dim, hot star which is thought to be the normal star that is losing mass to its invisible black-hole companion.

most likely the correct one), accepting that black holes exist is natural, there being no simpler way to explain what happens when a massive star collapses. Most astronomers today have adopted the concept of black holes, and not only believe that they exist, but also consider them an integral part of our universe.

PERSPECTIVE

We have hounded the stars into their graves. We have examined their corpses to see how they decay, and have come away with our minds filled with wonder at the novel forms matter can take. We have seen that the nature of stellar remnants depends entirely on how much mass is left when stars run out of nuclear fuel and collapse, the three possibilities being white dwarfs, neutron stars, and black holes. Stars that die alone, no matter what final form they take, are not likely to be detected again (except for the closest white dwarfs and

the neutron stars that happen to appear to us as pulsars), whereas those which exist in binary systems may be reincarnated in spectacular fashion if mass exchange takes place.

Having learned all we can about stars as individuals, we now are ready to move on to larger scales in the universe, to examine galaxies and ultimately the universe itself.

SUMMARY

1. A white dwarf gradually cools off, taking billions of years to become a cold cinder.

2. If new matter falls onto the surface of a white dwarf, for example in a binary system where the companion star loses mass, then a nova can occur as the degenerate matter of the white dwarf causes a rapid nuclear reaction. Nova outbursts can occur many times in the same white-dwarf binary system.

3. Massive stars are likely to explode as supernovae when all possible nuclear-reaction stages have ceased. The supernova explosion creates an expanding cloud of hot, chemically enriched gas known as a supernova remnant.

4. In some cases a remnant of 2 to 3 solar masses is left behind in the form of a neutron star, which consists of degenerate neutron gas.

5. A neutron star is too dim to be seen directly in most cases, but may be observed as a pulsar (depending on the alignment of its magnetic and rotation axes, and our line of sight), and in a close binary system it may become a source of X-ray emission.

6. Some neutron stars that receive new material in clumps flare up occasionally as X-ray sources called bursters.

7. If the final mass of a star exceeds 2 or 3 solar masses, it will become a black hole at the end of its nuclear-reaction lifetime.

8. The immensely strong gravitational field near a black hole traps photons of light, rendering the black hole invisible.

9. A black hole may be detected by its gravitational influence on a binary companion, or by the X rays it emits if new matter falls in, as in a close binary system where mass transfer takes place. A black-hole binary

X-ray source can be detected only through analysis of the orbits to determine the mass of the unseen object.

REVIEW QUESTIONS

1. Why is the final mass of a star, rather than the mass it begins with, the important criterion for determining what form of remnant the star will leave? Why should the final mass be different from the initial mass for a given star?

2. Using information from chapter 2, compare the surface gravity of the sun with that of a white dwarf having the same mass but only 1/100 of the sun's radius. Make the same calculation for a neutron star that has twice the mass of the sun but only 10^{-6} of the sun's radius.

3. Summarize the techniques by which white dwarfs can be detected from the earth.

4. Calculate (using Wien's law) the temperature at which a cooling white dwarf becomes essentially invisible to us, assuming that this occurs when the wavelength of maximum emission has shifted to the infrared wavelength of 10,000Å.

5. Novae and supernovae, despite the similarity of their names, are completely different phenomena. Explain the differences.

6. Discuss the effect of supernova explosions on the chemical composition of the galaxy.

7. Compare the internal structure of a neutron star with that of a white dwarf.

8. If a certain pulsar has a much longer period than another, which is more likely to be surrounded by a detectable supernova remnant?

9. When a star collapses to become a black hole, the gravitational field close to it becomes very strong. Why does this not affect the orbit of the companion star, in the case where the black hole is a member of a binary system?

10. Both neutron stars and black holes can be detected as X-ray sources under certain circumstances (that is, in mass-transfer binary systems). How can we expect to distinguish between these two kinds of stellar remnants in such cases?

ADDITIONAL READINGS

Anderson, L. 1976. X-rays from degenerate stars. *Mercury* 5(5):2.

Black, D. L. 1976. Black holes and their astrophysical implications. *Sky and Telescope* (50(1):20 (pt. I); 50(2):87 (pt. II).

Gursky, H., and van den Heuvel, E. P. J. 1975. X-ray emitting double stars. *Scientific American* 232(3):24.

Lewin, W. H. G. 1981. The sources of celestial X-ray bursts. *Scientific American* 244(5):72.

Schramm, D., and Arnett, W. 1975. Supernovae. *Mercury* 4(3):16.

Smarr, L., and Press, W. H. 1978. Spacetime: black holes and gravitational waves. *American Scientist* 66:72.

Thorne, K. S. 1974. The search for black holes. *Scientific American* 233(6):32.

Wheeler, J. C. 1973. After the supernova, what? *American Scientist* 61:42.

Andrea K. Dupree

Andrea K. Dupree is an astrophysicist at the Smithsonian Astrophysical Observatory and an associate director of the Harvard-Smithsonian Center for Astrophysics in Cambridge, Massachusetts, where she directs the Solar and Stellar Astrophysics division. Early in her career she investigated the structure and energy balance of the outer solar atmosphere using spectrometers built at Harvard and flown on the Orbiting Solar Observatory *series of satellites and the* Skylab *mission. Her research is currently in the area of stellar and interstellar physics, with emphasis on spectroscopy. A particular interest has been observations and theory of mass loss and atmospheric structure in cool stars, guided by spectra from the* International Ultraviolet Explorer *satellite and ground-based observations. In addition to her research and management activities, Dr. Dupree is also heavily involved in the planning process for future NASA space astronomy instruments, and serves on a number of advisory panels and committees.*

One of the most intriguing and challenging aspects of astrophysics today is stellar activity—the varying emissions from a star that are analogous to the sun's activity cycle. In the last decade, this field has burgeoned, owing to our new views of the sun itself and to new technology both on the ground and in space. These advances allow us to make true quantitative measurements of stellar radiations which are important for theoretical understanding. Moreover, with space observations we can probe the ultraviolet and X-ray emissions from the chromospheres and coronae of stars. We have discovered significant similarities between the sun and stars—with resulting enrichments of both solar and stellar science. A vigorous new field has developed as well: the pursuit of the solar-stellar connection.

The sun is our starting point and, although a rather ordinary middle-aged star, it is rather unique to us. Because of its proximity, we can resolve features on its surface, probe its activity cycle, and in the process try to develop an understanding of the diversity of phenomena that must occur in other stars.

Of course, centuries ago, Galileo Galilei showed that the sun was not the comforting unchanging sphere in the daily sky it had been thought to be, but was instead a blemished one with dark spots marring its visible surface. These spots come and go with some regularity in an eleven-year cycle. In solar physics, a major result of the past decade was the realization of the overriding importance of mag-

netic fields in the sun's large- and small-scale activity. We now know that the spots mark the presence of strong magnetic fields emerging through the solar surface. Spots and their cycles are surface manifestations of the fundamental solar dynamo, amplifying a "seed" internal magnetic field through the interaction of rotation and convection. Associated with the strong surface fields are active regions—dense volumes of hot plasma, shaped by the magnetic fields into giant loops. We observe such loops directly on the sun and find that much radiation emerges from them in the form of emission lines. Enhanced emission from active regions can in fact be used as a tracer of the magnetic field, providing important information on the interactions between ionized gas and magnetic fields, and on the characteristics of the dynamo and subsurface structures as well. We can measure similar emission from other stars, and we can therefore hope to make the same kinds of inferences about their surface and internal behavior.

In other parts of the sun's atmosphere, the magnetic field is open, leading to an easy path of escape for gas in the sun's corona. We believe that much, and perhaps all, of the solar wind comes from these open-field regions. Such a wind acts to brake the rotation of the sun over long periods of time. We can reasonably expect that winds from other stars have played similar roles in their evolutions.

Discoveries about the sun provide clues to the behavior of other stars, but we must be careful be-

cause the sun is only a single dwarf star with its own specific set of physical conditions such as mass, temperature, and chemical composition, and its own history. We may find some surprises. By observing other stars, we can test our understanding of physical processes under varied conditions. Any theory of magnetic activity and the dynamo process, or the coronal-heating question, or the mass-loss mechanism, or any other theory concerning the sun, must stand up against the diversity of physical conditions that exist in other stars.

Cool stars other than the sun exhibit an incredibly wide range of activity. In fact the traditional classification scheme, the H-R diagram, does not suffice to explain the differing levels of emission that are observed. Stars that look identical in their visible-wavelength spectra possess substantial differences in their ultraviolet and X-ray emissions. There must be missing parameters that govern the level of emission. The age of a star may be one such critical characteristic. Surveys of the emission lines from ionized calcium, the best visible-wavelength indicators of stellar chromospheres, have shown that in young dwarf stars the emission is especially intense and the activity cycles are erratic. As a star ages, its activity cycle apparently becomes more stable and its emission level decreases. This decay of activity with age may be related to a star's rotation rate. If mass loss has taken place through a stellar wind, the wind may have acted to brake the rotation. Since it

is rotation that interacts with the convection zone to produce a magnetic dynamo, it is plausible that the magnetic activity would lessen as a spinning star is slowed down. We cannot yet actually resolve spots on a distant star, but we can gain insights into its surface activity by measuring the enhanced emissions known to come from active regions, and using this information to trace magnetic activity and cycles. By monitoring a star over long times, we can measure the periodic variation of its emission-line fluxes, corresponding to the presence and disappearance of its activity patterns, thereby defining the character and extent of its stellar cycle. And, most important, the rotation rate of a star—a fundamental parameter of activity—can be directly measured. Thus it is possible, through long-term monitoring programs, to isolate the relationships between stellar cycles and stellar rotations and ages.

In the search for determinants of activity, we can control the physical variables to some extent by studying stars in a cluster. Here the stars are presumably of the same age and have the same composition. We would then expect that stars at the same position on the cluster H-R diagram should show similar levels of activity. But again, there is a disparity of emission levels in their chromospheric and coronal fluxes, by more than a factor of ten. In such cases, we conjecture that we are seeing the effects of stellar activity cycles, with different stars being observed at different points in their cycles.

Studies of the evolution of stellar activity have consequences for the history of our own sun as well. We believe that a high activity level was experienced by our sun aeons ago—and the stellar studies can help us infer the solar history, to assess its effects on the earth.

Violent forms of stellar activity—transient phenomena and flares—are observed in all kinds of cool stars. In some cases the energies involved in these processes are several orders of magnitude larger than those on the sun. Many cool stars flare quite frequently, with hourly time scales, and in a few binary systems there is evidence of material being transferred during the flare from one star to another.

Some signatures of activity can be much more pronounced in stars other than the sun. The surface fluxes of emission from some stars can be several hundred times the solar value, for example, and the surfaces of the stars can be much more inhomogeneous than that of the sun. The percentage of a star's surface that is covered with spots can be deduced from the ratios of emission lines formed in spots to those formed over the entire surface, and in some cases it is found that as much as 20 percent of a star's surface can be covered. Severe asymmetries of activity—such as stars with spots on one side and a massive wind on the other—have also been found, suggesting magnetic domination of the atmosphere on a much larger scale than we have found on the sun.

Such situations are tantalizing, and we are struggling to understand them in a quantitative way.

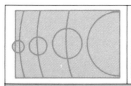
This is difficult, for the complexities are enormous. For example, it is almost inconceivable to develop a theory adequately describing a cool dwarf star with a fully convective core and a spot covering half of its surface, yet this is where the evidence points in some cases. What happens to that magnetic field when the star is rapidly rotating? . . . or when the star has a companion whose wind and magnetic field can interfere and interact? We are working step by step with guidance from the sun, but much of the physics may be an extreme variation of the conditions found in solar physics.

In the future we envision experiments that may let us view the surfaces and extended atmospheres of stars in detail. Telescopes are being planned that are large enough to carry out high-resolution spectroscopy and measure directly the motions of hot plasma expelled from a stellar surface or traveling along a magnetic loop. Satellite interferometers or sophisticated speckle-imaging techniques can enable the surface of a star to be seen directly, with resolution sufficient to allow us to actually see areas of activity.

Despite all the optimism that such advances will be made, however, astronomers must remain receptive to the vagaries of nature. With a confidence based only in solar physics, we may be deluding ourselves as we grope to understand the stellar universe. It will undoubtedly turn out to be a much more complex starry cosmos than we now envision.

Introduction to Section IV

In this section we explore the great conglomeration of stars to which our sun belongs. The Milky Way, a spiral galaxy, is itself a dynamic entity, with its own structure and evolution governed by the complex interactions of stars with each other and with the interstellar gas and dust. Although this system of some 10^{11} individual stars is infinitely more complicated than a single star, and its life story is correspondingly more difficult to unravel, great progress has been made in piecing together the puzzle.

In the first of the three chapters of this section, we examine the overall properties of the Milky Way. We learn how the galaxy's size and shape, as well as the sun's location in the galaxy, were deduced from observations carried out from our position within the great disk, where our view is obscured by interstellar haze. We discuss how the various parameters describing the galaxy were derived from observation, and we learn of some immense and fascinating mysteries that remain.

The second chapter of this section focuses on the diffuse material between the stars. The interstellar medium is so rarefied that a superficial consideration might lead us to deem it unworthy of our time and effort, but we find, to the contrary, that the interstellar medium is a vital part of the galactic life story. Stars form from this material, and in evolving and dying they return their substance to it. The interstellar medium itself is a dynamic, active component of the galaxy, one that in recent years has provided astronomers with several major surprises. In this chapter we survey the general properties of the gas and dust that pervade space, and we pay particular attention to their role in the evolution of the galaxy.

The third and last chapter of this section ties together the disparate data examined in the preceding two chapters, and tells the story of the formation and evolution of our galaxy. We see how the major features of the Milky Way arose from a sequence of developments dictated by simple laws of physics. Although there remain some mysteries, such as the quantity of mass in the far reaches of the galactic halo and the traces of violent activity at the core of the galaxy, most of the story can be told.

Having probed the Milky Way, we will then be ready to move outward to examine and understand the countless galaxies in the universe beyond our own.

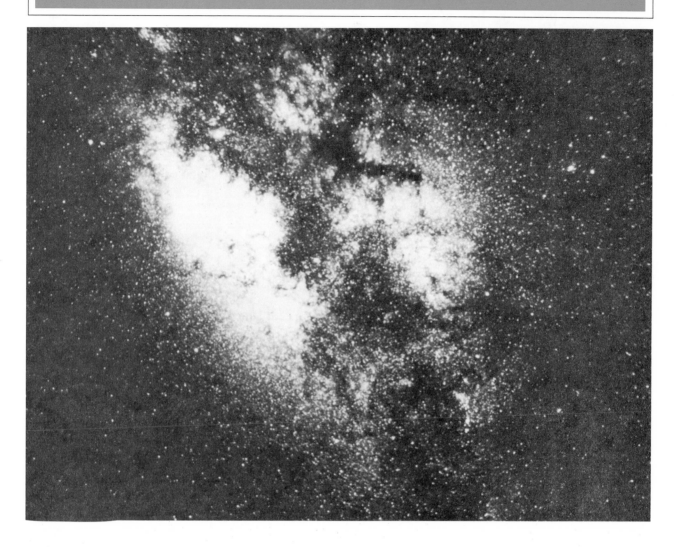

Learning Goals

17

We have discussed stars as individuals, as though they lived in a vacuum, isolated from the rest of the universe. For the purposes of analyzing the structure and evolution of stars, this is a suitable approach, but if we want to understand the full stellar ecology, the properties of individuals must be discussed in the context of the larger environment.

Even with a casual glance at the nighttime sky we can see that the stars generally are grouped rather than randomly distributed. The most obvious concentration is the **Milky Way** (Fig. 17.1), the diffuse band of light stretching from horizon to horizon, clearly visible only in areas well away from city lights. To the ancients the Milky Way was merely a cloud; Galileo with his primitive telescope recognized it as a region having a great concentration of stars. In modern times we know the Milky Way as a galaxy, the great pinwheel of billions of stars to which the sun belongs. The hazy streak across our sky is a cross-sectional, or edge-on, view of the galaxy, seen from a point within its disk.

The overall structure of the Milky Way is like a phonograph record, except that it has a large central bulge, the center of which is called the **nucleus** (Figs. 17.2 and 17.3). Nearly all the visible light is emitted by stars in the plane of the disk, although the galaxy also has a **halo,** a distribution of stars and star clusters centered on the nucleus but extending well above and below the disk. The most prominent objects in the halo are the **globular clusters,** very old, dense clusters of stars characterized by their distinctly spherical shape.

To envision the size of the Milky Way requires a new unit of distance. In chapter 13 we discussed stellar distances in terms of parsecs, and we found that the nearest star is about 1.3 parsecs from the sun. To expand to the scale of the galaxy, we speak in terms of **kiloparsecs** (abbreviated **kpc**), or thousands of parsecs. The visible disk of the Milky Way is roughly 30 kiloparsecs in diameter, and the disk is a few hundred parsecs thick. Light from one edge of the galaxy takes about 100,000 years to travel across to the far edge.

Our galaxy is so large that we must discuss new methods of measuring distance, for even the main-sequence-fitting technique, the most powerful we have mentioned so far, fails when star clusters are too distant and faint to allow determination of the individual stellar magnitudes and spectral types.

Variable Stars as Distance Indicators

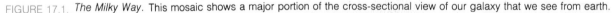

It was stressed in chapter 13 that it is always possible to determine the distance to an object if both the absolute and apparent magnitudes of the object are known. We discussed spectroscopic parallaxes, where the absolute magnitude of a star is derived from its spectral class, and we talked about main-sequence fitting, where a cluster of stars is observed and its main sequence determined, so that it can be fitted to the main sequence in the standard H-R diagram. Now we will learn about a special type of star whose absolute mag-

FIGURE 17.1. *The Milky Way.* This mosaic shows a major portion of the cross-sectional view of our galaxy that we see from earth.

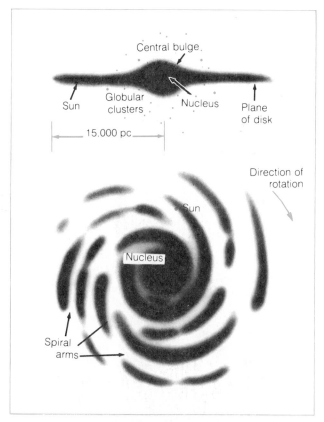

FIGURE 17.2. *The Structure of Our Galaxy.*
These sketches illustrate the modern view of the Milky Way.

FIGURE 17.3. *A Spiral Galaxy Similar to the Milky Way.*
This galaxy (NGC 2997) probably resembles our own, as seen from afar.

nitude can be determined from the other stellar properties, allowing it to be used to measure distances.

In chapter 13 we also briefly mentioned variable stars, including those which pulsate regularly. These stars physically expand and contract, changing in brightness as they do so. One of the first such stars to be discussed is δ Cephei, which is bright enough to be seen with the unaided eye. Following the discovery of δ Cephei in the mid-1700s, other similar stars were found, and these as a class became known as **Cepheid variables,** named after the prototype.

The usefulness of pulsating variables as distance indicators was discovered in 1912, when an astronomer named Henrietta Leavitt (Fig. 17.4) at the Harvard College Observatory studied variable stars located in the Magellanic Clouds, two small galaxies near the Milky Way. Leavitt noticed that the average brightness of these stars was correlated with the period of pulsation; that is, the longer the time between brightenings, the brighter the star. The Magellanic Clouds are so far

away (about 55 kpc) relative to their own dimensions that all the stars there are, for practical purposes, at the same distance from us. This meant that the correlation of period with apparent magnitude was actually a correlation with absolute magnitude (Fig. 17.5), suggesting the possibility that the absolute magnitude could be determined from the period of pulsation.

Subsequent analysis established the numerical relationship between the period and absolute magnitude. Thus the measurement of distance to a variable star becomes a simple matter of determining the period of pulsation by counting the days between times of maximum brightness, then using the established correlation to determine the absolute magnitude. The distance is then found by comparing the absolute and apparent magnitudes, using the standard technique discussed in chapter 13. Because the Cepheid variables are giant stars, they are quite luminous, and can be observed at great distances, even beyond the Milky Way. Therefore these stars are very useful in measuring the scale of our galaxy.

Another type of pulsating star, the **RR Lyrae var-**

Astronomical Insight 17.1
WHAT MAKES A CEPHEID RUN?

Every star has a natural vibration frequency, a period at which it will pulsate if disturbed by an outside force. Ordinarily, however, the pulsations, if started, will die out very quickly, so most stars are quiescent and stable. The pulsations will continue only if some force or energy input is made in synchronization with the vibrations, giving the star a little boost each cycle. A weight on a string behaves the same way; if it is given an initial push and then left alone, it will oscillate for a while, but gradually stop. If, one the other hand, it is given a push each time it reaches a certain point, it will keep oscillating indefinitely, with a regular frequency. This is how a pendulum clock works; a weight or a spring is used to give the pendulum a boost each cycle, and the pendulum in turn maintains a steady tempo.

But what provides the boost to keep a pulsating star expanding and contracting? The answer has to do with the star's outer layers and the way they allow light to pass through. When a Cepheid variable contracts, the density in the atmosphere increases, and helium ions combine with electrons to form helium atoms. These atoms in turn absorb light very efficiently, and in this state the atmosphere acts as a closed valve, keeping radiant energy bottled up inside the star. Heat energy builds up, causing the star to expand again. As it does so, the density in the atmosphere decreases, and the helium atoms become

ionized by the emerging radiation. Light is now free to escape from the star's interior, since there are fewer helium atoms to hold it in, and therefore the pressure that caused the expansion diminishes. Soon enough, the outer layers begin to fall back in, and the star enters a new contraction phase. Helium atoms form again, and again they block the light from the interior, causing a new buildup of pressure and creating a new expansion phase. The cycle continues indefinitely, with the helium acting as a valve, letting light out or keeping it bottled up.

Similar valve mechanisms apply to some of the other types of variable stars, although rather than helium, some other element may play the role of the valve. All stars with just the right combination of temperature and pressure in their outer layers will pulsate because of this valve mechanism, so the variable stars occupy a narrow region in the H-R diagram, where stars have the correct conditions. This area is called the **instability strip,** because all the stars in it are unstable and therefore pulsate.

The average density of a star determines the period of pulsation. The denser a star is, the more rapidly it vibrates. Thus, the range of periods of Cepheid variables represents a range of densities.

iable, also was found to be a reliable distance indicator. These stars have periods of only a few hours, and they all have the same absolute magnitude. Any time we identify an RR Lyrae star by measuring its period, we immediately know its absolute magnitude and hence its distance. The RR Lyrae stars are not as luminous as the Cepheids, but they are still bright enough to be observed at large distances. These stars are often found in globular clusters and are sometimes called **cluster variables.** They played a very important role in the early measurements of the size and shape of our galaxy.

The Structure of the Galaxy and the Location of the Sun

Because the solar system is located within the disk of the Milky Way, we have no simple way of getting a clear view of where we are in relation to the rest of the galaxy. All we see is a band of stars across the sky, which tells us that we are in the plane of the disk. It is not so easy to determine where we are within the disk, with respect to its edge and center.

FIGURE 17.4. *Henrietta Leavitt.*
Leavitt was part of the group of astronomers at Harvard who pioneered the development of spectral classification. In addition, she discovered the period-luminosity relationship for Cepheid variables in the course of studying stars in the Magellanic Clouds.

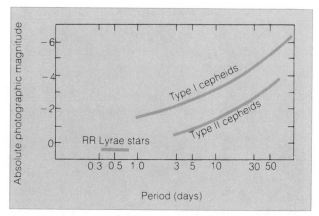

FIGURE 17.5. *The Period-Luminosity Relationship for Variable Stars.*
This diagram shows how the pulsation periods for Cepheid and RR Lyrae variables are related to their absolute magnitudes. The fact that there are two types of Cepheids, with somewhat different relationships, was not recognized at first, and this led to some early confusion regarding distance scales.

The first serious attempts to solve this problem were based on determinations of the density of stars in space around the sun. The method was to choose a random region of the sky and to then count the number of stars of each magnitude there. The number of stars increases with decreasing brightness, because faraway stars appear faint, and in a given direction, there are many more faraway stars than nearby ones. Therefore we can determine the distribution of stars in space by seeing how rapidly their number increases with decreasing brightness (that is, with increasing distance).

In the first decade of the twentieth century, the Dutch astronomer J. C. Kapteyn employed this method, and got a surprising result. The density of stars appeared to fall off in all directions from the sun, implying that our solar system is in the densest portion of the Milky Way (Fig. 17.6). The most logical interpretation was that the sun is at the very center of the galaxy, for it seemed likely that the galaxy should be densest at the core. The thought that the sun was at the center of the known universe made many astronomers uneasy, because in the past, every assumption that the earth held

a special place in the universe had been proved incorrect. It turned out there was indeed a flaw in the work of Kapteyn, although he could hardly have been aware of it. Before the error in Kapteyn's method was found, however, two other studies showed that the sun was probably not at the center of the galaxy.

One of these studies, carried out by Harlow Shapley (Fig. 17.7), made use of the globular clusters, the spherical star clusters found to lie outside the confines of the galactic disk. As noted earlier, these clusters tend to contain RR Lyrae variables, so it was possible for Shapley to determine their distances and hence their locations with respect to the sun and the disk of the

FIGURE 17.6. *The Kapteyn Universe.*
The shaded area in this diagram illustrates roughly the extent of the Milky Way as inferred by J. C. Kapteyn on the basis of star counts. Kapteyn was unable to include stars very far from the sun in the plane of the disk, because interstellar matter obscured the view.

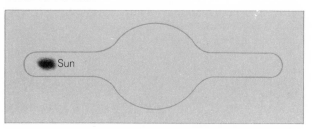

FIGURE 17.7. *Harlow Shapley.*
Shapley's work on the distances to globular clusters was a key step in the determination of the size of the Milky Way. Shapley also played a prominent role in the discovery of other galaxies (see chapter 20). Much of his observational work was done before 1920, when he was a staff member at the Mount Wilson Observatory. He later became director of the Harvard College Observatory.

Milky Way. He found that the globular clusters are arranged in a spherical volume centered on a point several thousand parsecs from the sun (Fig. 17.8), and he argued that this point must represent the center of the galaxy. It would not make physical sense for the glob-

ular clusters to be concentrated around any location other than the center of the entire galactic system.

Shapley's conclusion, first published in 1917, was not widely accepted initially, but other supporting evidence came to the fore in the 1920s. Two scientists, Jan Oort of Holland (already mentioned in connection with the origin of comets; see chapter 10), and the Swede Bertil Lindblad, carefully studied the motions of stars located in the vicinity of the sun. What they found was that these motions could best be understood if the sun and the stars around it were assumed to be orbiting a distant point; that is, there are systematic, small velocity differences between stars, similar to those between runners on a track who are in the inside and outside lanes (Fig. 17.9). It appeared from these studies that the sun is following a more or less circular path about a point several thousand parsecs away, indicating that the center of the galaxy is located at that distant center of rotation. This supported Shapley's view of the galaxy, although there were still uncertainties about how far the sun was from the center.

Let us now return to the work of Kapteyn, for the

FIGURE 17.8. *Shapley's Measurements of Globular Clusters.*
This is one of Shapley's original figures illustrating the distribution of globular clusters in the Milky Way. The sun is at the point where the straight lines intersect, and each circle centered on that point represents an increase in distance of 10,000 parsecs. Note that the distribution of globulars is centered some 10 to 20 kiloparsecs from the sun.

FIGURE 17.9. *Stellar Motions Near the Sun.*
Stars just inside the sun's orbit move faster than the sun, whereas those farther out move more slowly. Analysis of the relative speeds (as inferred from measurements of Doppler shifts) and distances of stars like these led to the realization that the sun and stars near it are orbiting a distant galactic center.

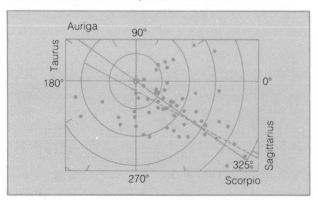

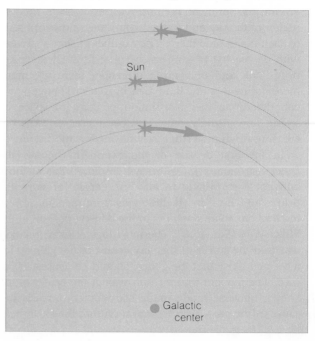

next major discovery was the one that revealed the flaw in his conclusions. In 1930 a study carried out by the American R. Trumpler showed that the galaxy is filled with an interstellar haze that makes stars appear fainter than they would if unobscured. Trumpler made this discovery by examining the apparent brightnesses of star clusters whose distances he knew from their apparent sizes. What he found was that distant clusters appear much fainter than they should, and that the effect increases with increasing distance. This was the first direct evidence that there is a pervasive interstellar medium in the space between the stars; until then, it was known only that interstellar clouds and nebulae existed, as seen on photographs showing dark regions and brightly glowing gas clouds.

Trumpler's discovery explained the contradiction between Kapteyn's results based on star counts, and the picture developed by Shapley and by Lindblad and Oort. Kapteyn's star counts failed to take the interstellar haze into account, so his estimates of stellar distances were erroneous. It was the obscuration by the interstellar medium (see chapter 18) that made the density of stars appear to fall off with distance from the sun.

Consideration of the effects of interstellar material, along with refinements in the assumed relation between pulsation period and absolute magnitude for variable stars, led eventually to a consensus as to the size of the galaxy and the sun's location within it. As mentioned earlier, the disk is some 30 kpc in diameter. The sun is located about 10 kpc from the center, or about two-thirds of the way to the edge. As we will see in later sections, there is still some uncertainty about how far the galaxy extends beyond the sun's position.

Galactic Rotation and Stellar Motions

An important result of the work of Lindblad and Oort was the development of an understanding of the overall motions in the galaxy. Oort's analysis especially was useful in this regard, for he showed not only that the sun and the stars near it are orbiting the distant galactic center, but also that the rotation of the galaxy is differential, meaning that each star follows its own orbit at its own speed. Thus the galaxy in the vicinity of the sun does not act like a rigid disk but like a fluid, with each star moving as an independent particle.

This is not true of the inner portion of the galaxy. There the entire system does rotate like a rigid object; like a record on a turntable. The stars in the inner part of the galaxy are subject to the combined gravitational forces of all the stars around them, and they are not free to follow Keplerian orbits about the galactic center. In the region where rigid-body rotation is the rule, the speeds of individual stars increase with distance from the center, whereas in the outer portion they decrease with distance (Fig. 17.10). In the outer reaches of the galaxy, each star orbits the central portion of the galaxy with little influence from its neighbors, and the stellar orbits are approximately described by Kepler's laws. This is why the orbital speeds decrease with distance from the center in this part of the galaxy. There is an intermediate distance, just inside the sun's orbit, where a transition between rigid-body and Keplerian orbits occurs, and it is here that stars have the greatest orbital velocities. The sun, near this peak position, travels at about 250 kilometers per second, taking roughly 250 million years to make 1 complete circuit about the galaxy. In its 4- to 5-billion-year lifetime, the sun has completed 15 to 20 orbits.

Individual stellar motions do not necessarily follow precise circular orbits about the galactic center. What we have described so far is the overall picture that develops from looking at the composite motions of large numbers of stars. If we look at the individual trees instead of the forest, we find that each star in the great

FIGURE 17.10. *The Rotation Curve for the Milky Way.* This diagram shows how the stellar orbital velocities vary with distance from the center of the galaxy. The fact that the curve does not simply drop off to lower and lower velocities beyond the sun's orbit, as it would if all the mass of the galaxy were concentrated at its center, indicates that there is a lot of mass in the outermost portions of the galaxy.

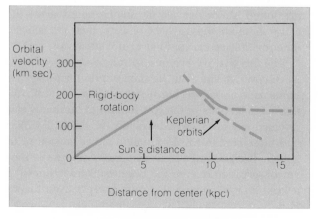

disk has its own particular motion, which may deviate slightly from the ideal circular orbit. These individual motions are comparable to the paths of cars on a freeway, where the overall direction of motion is uniform, but a bit of lane-changing occurs here and there.

In the galaxy, a star's deviation in motion from a perfect circular orbit is called the **peculiar velocity:** in the case of the sun, it is the **solar motion.** The sun has a velocity of about 20 kilometers per second with respect to a circular orbit, in a direction about 45 degrees from the galactic center and slightly out of the plane of the disk. Most peculiar velocities of stars near the sun are comparable, amounting to only minor departures from the overall orbital velocity of about 250 kilometers per second. In chapter 19 we will discuss the **high-velocity** stars, whose orbits greatly deviate from the circular orbits followed by most stars in the sun's vicinity.

Spiral Structure and the 21-cm Line

So far we have spoken of the galactic disk as though it were a uniform, featureless object, but we know that this is not completely accurate. The Milky Way is a **spiral galaxy,** and if we could see it face-on, we would see the characteristic pinwheel shape usually found in galaxies of this type.

There is a common misconception about the spiral structure of the Milky Way (and other spiral galaxies); namely, that there are few stars between the visible spiral arms. In reality, the density of stars between the arms is nearly the same as it is in the arms. The most luminous stars, however, the young, hot O and B stars, are found almost exclusively in the spiral arms. Because these are the brightest of stars, their residence in the arms makes the spiral structure stand out.

The fact that we live in a spiral galaxy was not easily discovered, since our location within the Milky Way limits us to a cross-sectional view. It was not until 1951 that investigations of the distribution of luminous stars revealed traces of spiral structure in the Milky Way, and those studies involved only a small portion of the galaxy. Because interstellar material obscures the view, our observation of even the brightest stars is limited to a local region, about a thousand parsecs from the sun at most.

A major advance in measuring the structure of the galaxy occurred in the same year, when the 21-cm radio emission of interstellar hydrogen atoms was discovered. It had been predicted that hydrogen emits radiation at this wavelength, and the U.S. scientists E. M. Purcell and H. I. Ewen were the first to detect the emission, using a specially built telescope.

One great advantage of the hydrogen 21-cm emission for measuring galactic structure is its ability to penetrate the interstellar medium to great distances. Whereas the brightest stars can be detected at distances of a few hundred parsecs at most, hydrogen clouds in space can be "seen" by radio telescopes from all the way across the galaxy, at distances of several thousand parsecs. Furthermore, hydrogen gas is the principal component of the interstellar medium, and it tends to be concentrated along the spiral arms, so observations of the 21-cm radiation can be used to trace the spiral structure throughout the entire Milky Way (Fig. 17.11).

When a radio telescope is pointed in a given direction in the plane of the galaxy, it receives 21-cm emission from each segment of spiral arm in that direction. Because of different rotation, each arm has a velocity different from the others that are closer to the center or farther out. Therefore what is seen, instead of a single emission peak at exactly 21.1-cm wavelength, is a cluster of emission lines near this wavelength, and the lines are separated from each other by the Doppler effect. By combining measurements such as these with Oort's mathematical analysis of differential rotation, astronomers were able to reconstruct the spiral pattern of the entire galaxy.

The pattern is quite complex, much more so than in some galaxies, where two arms may be seen elegantly spiraling out from the nucleus. The Milky Way consists of bits and pieces of a large number of arms, which give it a definite overall spiral form, but not a smoothly coherent one. As we will learn in chapter 19, the differences between the type of spiral structure seen in our galaxy and the regular appearance seen in others may reflect that the arms have different origins.

The Mass of the Galaxy

Once the true size of the Milky Way was determined, it became possible to estimate its total mass. This can be achieved by measuring the star density in the vicin-

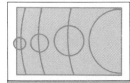

Astronomical Insight 17.2

21-CM EMISSION FROM HYDROGEN

To understand how a hydrogen atom can emit radiation at a wavelength of 21 centimeters, we need to take a closer look at the structure of the atom.

A hydrogen atom consists only of a proton, which forms the nucleus, and a single electron in orbit about it. We have already learned that the electron can occupy a variety of possible orbits, each corresponding to a different energy state, and that a photon of light is absorbed by the electron if the electron moves from a low level to a higher one, or the photon is emitted if the electron drops from a high level to a lower one. The wavelength of the photon is related to the energy difference between the two electron energy levels; the greater the difference, the shorter the wavelength, and vice versa.

It happens that the energy-level structure of the electron is more complicated than previously described. The electron and the proton are both spinning, and the energy of the electron depends on whether the electron is spinning in the same direction as the proton or the opposite direction. If both spin in the same (parallel) direction, then the energy state is slightly greater than when the electron and proton spin in opposite (antiparallel) directions.

As before, the electron can change from one state to the other by either emitting or absorbing a photon. The energy difference is so small, however, that the wavelength of the photon is very much longer than that of visible light. It is 21.1 centimeters.

Hydrogen atoms in space tend to have the electron in the lowest possible energy state, with its spin antiparallel to that of the proton. Occasionally an atom will collide with another, however, causing the electron to jump to the higher state, with the spin then parallel to that of the proton. Following this, the electron spontaneously reverses itself, seeking to return to the lowest energy state, and when it does so, it emits a photon of 21.1-cm wavelength. The probability that the electron will make the downward transition is very low, and the electron may remain in the upper state for as long as 10 million years before spontaneously dropping back down. There are so many hydrogen atoms in space, however, that at any given instant, many photons are being emitted. Hence radio telescopes capable of receiving this wavelength can trace the locations of hydrogen clouds throughout the galaxy and beyond.

---

ity of the sun and then assuming that the entire galaxy has about the same average density. However, a much simpler and more accurate technique became possible with the discovery that the stars in our region of the galaxy obey Kepler's laws.

Kepler's third law, in the more complete form developed by Newton, expresses a relationship among the period, the size of the orbit, and the sum of the masses of the two objects in orbit about each other:

$$(m_1 + m_2)P^2 = a^3,$$

where m_1 and m_2 are the two masses (in units of the sun's mass), P is the orbital period (in years), and a is the semimajor axis (in astronomical units). If we consider the sun to be one of the two objects, and the galaxy itself to be the other, we can use this equation to determine the mass of the galaxy. As we learned ear-

lier, the orbital period of the sun is roughly 250 million years; that is $P = 2.5 \times 10^8$ years. The orbit is nearly circular, with the galactic nucleus at the center, so the semimajor axis is approximately equal to the orbital radius; $a = 10$ kpc $= 2 \times 10^9$ AU. Now we can solve Kepler's third law for the sum of the masses:

$$m_1 + m_2 = a^3/P^2 = 1.3 \times 10^{11} \text{ solar masses.}$$

Since the mass of the sun (1 solar mass) is inconsequential compared with this total, we can say that the mass of the galaxy itself is about 1.3×10^{11} solar masses. The sun is slightly above average in terms of mass, so we conclude that the total number of stars in the galaxy must be $3 - 4 \times 10^{11}$; that is, several hundred billion.

This method, based on Kepler's third law, refers only to the mass inside the orbit of the sun; the matter

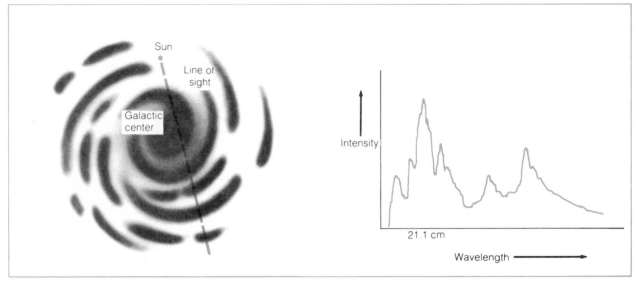

FIGURE 17.11. *21-cm Observations of Spiral Arms.*
At left is a schematic diagram of the Milky Way, showing the direction in which a radio telescope might be pointed to produce a 21-cm emission-line profile like the one sketched at right. There are many components of the 21-cm line, each corresponding to a distinct spiral arm, and each at a wavelength reflecting the Doppler shift between the velocity of that arm and the velocity of the earth.

that is farther out has no effect on the sun's orbit. Thus when we estimate the mass of the galaxy by applying Kepler's third law in this way, we neglect all the mass that lies farther out. So that this problem could be overcome, radio measurements of interstellar gas in the outer portions of the galaxy have recently been used to determine the orbital velocity (and therefore the orbital period) of material in the outer reaches of the galaxy. It was found that the velocity does not decrease as rapidly with distance as had previously been thought, and this in turn led to the conclusion that quite a bit of mass lies beyond the sun's orbit. In a later section, we will discuss the possibility that most of the mass of the galaxy lies in the halo.

The Galactic Center: Where the Action Is

The center of our galaxy is a mysterious region, forever blocked from our view by the intervening interstellar medium (Fig. 17.12). From Shapley's work on the distribution of globular clusters and Oort's analysis of stellar motions, it was known by the 1920s that the center

of our great pinwheel lies in the direction of the constellation Sagittarius. There we find immense clouds of interstellar matter and a great concentration of stars. It is one of the richest regions of the sky to photograph, although the best observations can be made only from the southern hemisphere.

The central portion of the galaxy consists of a more or less spherical bulge, populated primarily by relatively cool stars. The absence of hot, young stars implies that the central region has had relatively little recent star formation, something we will have to deal with in our discussion of the history of the galaxy (chapter 19).

Although we cannot see into the central portion of the nucleus in visible light, this region can be probed at longer wavelengths where the interstellar extinction is not such a problem (Fig. 17.13). Radio and, more recently, infrared observations have revealed some interesting features of the galactic core (Fig. 17.14).

The first clue that something unusual was taking place in the galactic core came from 21-cm observations of hydrogen, which revealed a turbulent mixture of clouds moving about at high speed, and which showed that one of the inner spiral arms, about 3 kpc from the center, is expanding outward at a velocity of more than 100 kilometers per second. This is suggestive

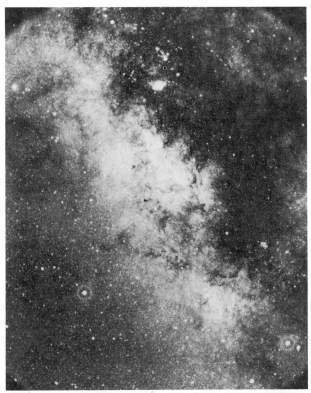

FIGURE 17.12. *The Galactic Center.*
The dark regions in this photo are dust clouds in nearby spiral arms. Because interstellar matter blocks the view, photos such as this do not reveal the true galactic center. but only nearby stars and interstellar matter in the plane of the disk.

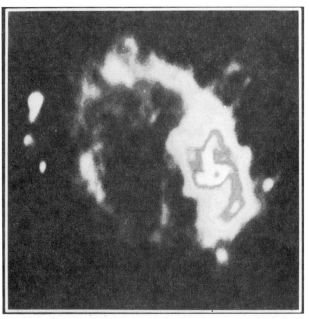

FIGURE 17.13. *A Radio Map of the Galactic Center.*
This image, made at a wavelength of 6 cm (where thermal-continuum emission is measured rather than a spectral feature such as the 21-cm line), shows a tiny, intense spot at the precise location of the galactic nucleus.

FIGURE 17.14. *Activity at the Galactic Center.*
A variety of evidence for energetic motions associated with the center of our galaxy is indicated here.

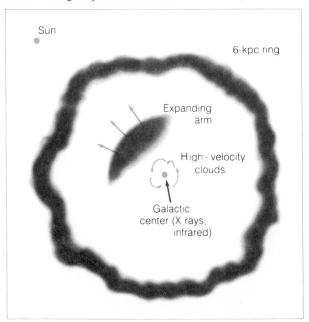

of some sort of explosive activity at the center that forces matter to move outward.

More recently, infrared observations of emission lines from hot gas clouds have shown that interstellar clouds near the center of the nucleus are moving quite rapidly in their orbits, which in turn implies (through the use of Kepler's third law) that they are orbiting a very massive central object.

Additional evidence of violent activity has been found in another portion of the radio spectrum, at the wavelength where interstellar carbon monoxide molecules emit (see chapter 18 for a discussion of molecules in space). Observations of CO have revealed a gigantic ring of interstellar clouds circling the galaxy at a distance of some 6 kpc from the center. It looks as though the ring could have been built up by the expansion of matter away from the galactic center, perhaps as the result of an ancient explosion. The concentration of matter at 6 kpc may be the end result of an earlier episode of the same activity that is responsible for the current expansion of the 3-kpc arm.

Finally, space observations have shown that a very small object precisely at the center of the galaxy is emitting an enormous amount of energy in the form of X rays. This object, whatever it is, is smaller than one parsec in diameter, yet it appears to be responsible for all the violent and energetic activity just described. From the observed stellar velocities, astronomers have estimated the mass of the central object to be roughly 10^6 solar masses. This is only a small fraction of the total mass of the galaxy, but is much greater than the mass of any known object within it, and must therefore represent some new kind of astronomical entity. The best explanation offered by astronomers is that a very massive black hole resides at the core of our galaxy. Its gravitational influence would be responsible for the rapid orbital motions of nearby objects, and matter falling in would be compressed and heated, which would account for the X-ray emission. We do not yet understand how such an object formed, but its presence implies that at some time during the formation and evolution of the Milky Way, great quantities of matter were compressed into a very small volume at its center. Such an intense buildup of matter may have resulted from frequent collisions among stars in the early days of the galaxy, causing vast numbers of them to gradually coalesce into a single, massive object at the center. The prospect that such a beast inhabits the core of our galaxy is quite a bizarre one, yet it may tie in directly with some of the strange happenings that have occurred in the nuclei of other galaxies (see chapter 22).

Globular Clusters Revisited

We have not said much about globular clusters, the gigantic spherical conglomerations of stars that orbit the galaxy. A large globular cluster can be an impressive sight when viewed with a small telescope, and a number of them are popular objects for astronomical photography (Fig. 17.15). More than a hundred of these clusters have been catalogued; no doubt many more are obscured from our view by the disk of the Milky Way.

FIGURE 17.15. *The Globular Cluster M13.*

A single cluster may contain several hundreds of thousands of stars, and have a diameter of ten to twenty parsecs. The mass is usually in the range of several hundred thousand to a few million solar masses. To contain so many stars in a relatively small volume, globular clusters must be very dense, compared with the galactic disk. The average distance between stars in a globular cluster is only about one-tenth of a parsec (recall that the nearest star to the sun is more than one parsec away). If the earth orbited a star in a globular cluster, the nighttime sky would be a spectacular sight, with hundreds of stars brighter than first magnitude.

An H-R diagram for a globular cluster (Fig. 17.16) is rather peculiar looking, compared with the typical diagram for stars in the galactic disk. The main sequence is almost nonexistent, having stars only on the extreme lower portion. On the other hand, there are many red giants, and a number of blue stars that lie on a horizontal sequence extending from the red-giant region across to the left in the diagram. These facts together point to a very great age for globular clusters. As we learned in chapter 15, in a cluster H-R diagram, the point where the main sequence turns off toward the red-giant region indicates the age of the cluster; the lower the turn-off point, the older the cluster. Using this technique to date globular clusters leads typically to age estimates of 14 to 16 billion years, comparable to the accepted age of the galaxy itself. Thus globular

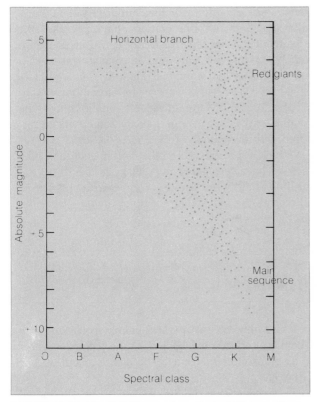

FIGURE 17.16. *The H-R Diagram for a Typical Globular Cluster.* The main sequence branches off at a point near the bottom, indicating the great age of the cluster.

FIGURE 17.17. *A Globular-Cluster X-Ray Source.* Several globular clusters have been found to contain intense X-ray sources at their centers. Here is an X-ray image of a globular, obtained by the *Einstein Observatory,* clearly showing the central source.

clusters are among the oldest objects that exist, and, as we will see, their presence in the galactic halo provides important data on the early history of the galaxy.

In the globular-cluster H-R diagram, the sequence of stars extending from the red-giant region to the left is called the **horizontal branch.** The evolutionary status of these stars has been the subject of a good deal of uncertainty; theoretical models have not yet been developed that can explain how stars evolve to this region of the H-R diagram. It is not even known for certain whether these stars are moving to the left (getting hotter, perhaps following a red-giant phase), or to the right. Most likely these stars have completed their red-giant stages (as discussed in chapter 15) and are moving to the left on the diagram, perhaps on their way to becoming white dwarfs.

Some globular clusters have been found to contain X-ray sources in their centers (Fig. 17.17). In a few cases these are X-ray **bursters,** described in chapter 16 as being neutron stars that are slowly accreting new

matter onto their surfaces, probably in binary systems where the companion star is losing mass. In other cases, however, the X ray source does not have the recognizable properties of a binary system, and may instead be caused by a single object at the center of the cluster. This could be a giant black hole, formed as stars in the cluster collided and fell together in the core.

A Massive Halo:

The halo of the Milky Way galaxy has traditionally been envisioned as a very diffuse region, populated by a scattering of dim stars and dominated by the giant globular clusters. There was little or no evidence of any substantial amount of interstellar material, and the halo was assumed not to contain a significant fraction of the galaxy's mass.

Some of these ideas are changing as a result of very recent discoveries, and now it is believed that the halo of the galaxy may be much more extensive and massive than suspected (Fig. 17.18). We mentioned earlier that radio observations revealed an unexpectedly high

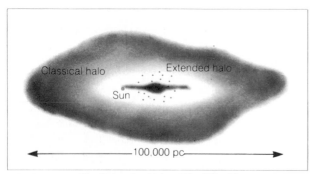

FIGURE 17.18. *The Extensive Halo of the Milky Way.*
This illustrates the shape and scale of the very extended
galactic halo, whose presence has been inferred from the
shape of the galactic rotation curve (see Fig. 17.10) and from
absorption lines formed by interstellar gas.

amount of mass in the outer portions of the disk, indicating that the galaxy is more extensive than previously
thought. In addition, other evidence (to be described in
later chapters) led astronomers to suspect that many
galaxies have large quantities of matter in their halos.
Finally, in the early 1980s, ultraviolet observations revealed a large amount of very hot, rarefied interstellar
gas in the halo of our galaxy, probably extending to
distances of several thousand parsecs above and below
the plane of the disk. This gas is very turbulent, with
clouds traveling at speeds of several hundred kilometers per second, and is highly ionized, indicating a
temperature of at least 100,000 K or more.

At the present time, we cannot say much about the
origin or total quantity of material in the halo of our
galaxy. It is quite possible that in addition to the gas,
there are enough stars, too dim to be detected, to add
significantly to the total mass of the galaxy. Some astronomers believe that as much as 90 percent of the
galaxy's mass may be in the halo. The *IRAS* satellite,
due to complete its infrared survey of the sky in 1984,
will probably tell us whether or not the halo is heavily
populated by cool, dim stars. The *Space Telescope*, to
be launched in the late 1980s, may also help answer
this question.

PERSPECTIVE

We have discussed the anatomy of the galaxy, particularly its overall structure and motions. The solar system
is located some 10 kpc from the center of a flattened,
rotating disk containing a few hundred billion stars.
Radio and infrared observations have revealed the
structure of the disk: it consists of spiral arms, delineated by bright, hot stars and interstellar gas, and a
central nucleus that contains relatively dim, red stars
along with a mysterious, energetic gremlin that stirs up
the core region. Many questions remain about the
structure and extent of our galaxy; some will be answered with the launch of the *Space Telescope.*

We have yet to talk about the workings of the Milky
Way; we have seen what our galaxy is like, but have
said little about why. Before we endeavor to do so, we
will discuss the interstellar medium, a vital part of the
galactic ecology.

SUMMARY

1. The Milky Way is a spiral galaxy, consisting of a
disk with a central nucleus and a spherical halo, where
the globular clusters reside. The disk is about 30 kpc
in diameter.

2. Distances within the Milky Way can be determined
from the measurement of variable-star periods, and application of the period-luminosity relation.

3. The true size of the Milky Way and the sun's location within it were difficult to determine because the
view from our location within the disk is obscured by
interstellar extinction.

4. Star counts seem to indicate that the sun is in the
densest part of the Milky Way, but measurements of the
distribution of globular clusters and analysis of stellar
motions show that the sun is some 10 kpc from the
center. The discrepancy was resolved when it was discovered that interstellar extinction affects the star
counts.

5. The inner part of the galactic disk rotates rigidly,
whereas the outer parts are fluid, with each star following its own individual orbit, approximately described by Kepler's laws.

6. Individual stellar orbits in the sun's vicinity are
generally in the plane of the disk, and are nearly circular. Deviations from perfect circular motion by stars
are called peculiar velocities, and in the case of the
sun, the solar motion.

7. The spiral structure of our galaxy is most easily
and directly measured through radio observations of
the 21-cm line of hydrogen atoms in space. The spiral

pattern is complex, with many segments of spiral arms.

8. The mass of the galaxy, determined by the application of Kepler's third law to the sun's orbit, is roughly 10^{11} solar masses. This technique does not take into account any mass that resides in the halo, or in the galactic plane outside of the sun's position.

9. A variety of evidence indicates that there is chaotic, energetic activity associated with the core of our galaxy. The data show a compact, massive object existing there, and the best explanation is that it is a massive black hole.

10. The globular clusters that inhabit the halo of the galaxy are very old, and therefore provide information on the early history of the galaxy.

11. There may be large quantities of mass in the galactic halo, in the form of interstellar gas or dim, cool stars, or both. It appears possible that as much as 90 percent of the mass of the galaxy is in the halo.

REVIEW QUESTIONS ∘

1. Recalling what you learned in chapter 13 about stellar parallaxes as distance determinators, calculate the distance to a star whose parallax angle is $\pi = 0''.001$ (this is about the smallest angle that can be measured with current techniques). Compare the distance to such a star with the diameter of the galactic disk, and discuss the practicality of using parallax measurements to probe the structure of our galaxy.

2. Suppose a Cepheid variable is found to have a period of ten days. The period-luminosity relation shows that the star's average absolute magnitude is $M = -4$. Its average apparent magnitude is $m = +16$. What is its distance? How does this compare with the diameter of the galactic disk?

3. Why wasn't the period-luminosity relation for Cepheid variables discovered from observations of these stars in our own galaxy?

4. What would have been the result of Kapteyn's star-count method if there had not been any interstellar extinction?

5. Compare Shapley's reasoning that the center of the galaxy should be the place about which the globular clusters orbit, with that of Aristarchus, who determined for similar reasons that the sun and not the earth is at the center of the solar system.

6. Why do the most rapid circular orbits around the galaxy occur at the distance from the center where rigid-body rotation gives way to fluid rotation?

7. Summarize the reasons that the 21-cm line of hydrogen is the best tool we have for tracing the spiral structure of the galaxy.

8. Suppose astronomers analyzed the orbit of a star lying 20 kpc from the galactic center, in order to determine the mass of the galaxy. How would you expect the result to differ from what was found from analysis of the sun's orbit?

9. Summarize the evidence that there may be a massive black hole at the center of the Milky Way.

10. We have learned that the globular clusters are very old objects, comparable in age to the galaxy itself. What does this imply about the chemical composition of the stars in these clusters?

ADDITIONAL READINGS

Bok, B. J. 1972. Updating galactic spiral structure. *American Scientist* 60:708.

— — —. 1981. The Milky Way galaxy. *Scientific American* 244(3):92.

Bok, B. J., and Bok, P. 1981. *The Milky Way.* Cambridge, Ma.: Harvard University Press.

de Boer, K. S., and Savage, B. D. 1982. The coronas of galaxies. *Scientific American* 247(2):54.

Geballe, T. R. 1979. The central parsec of the galaxy. *Scientific American* 241(1):52.

Herbst, W. 1982. The local system of stars. *Sky and Telescope* 63(6):574.

Kraft, R. P. 1959. Pulsating stars and cosmic distances. *Scientific American* 201(1):48.

Sanders, R. H., and Wrixon, G. T. 1974. The center of the galaxy. *Scientific American* 230(4):66.

Seeley, D., and Berendzen, R. 1978. Astronomy's great debate. *Mercury* 7(3):67.

Weaver, H. 1975 and 1976. Steps towards understanding the large-scale structure of the Milky Way. *Mercury* 4(5):18, p. 1; 4(6):18, p. 2; 5(1):19, p. 3

THE INTERSTELLAR MEDIUM

Learning Goals

18 The very rarefied matter between the stars, diffuse though it is, constitutes one of the most important elements in the galactic environment. Stars form from it, and as they age and die, they return their substance to it. As we have seen, the interstellar haze limits our observations of the galactic disk to the nearest few hundred parsecs, allowing us to view only a very small fraction of the entire galactic volume.

Despite its obvious importance, the interstellar medium is so tenuous, by earthly standards, that it verges on nonexistence. Even in the densest interstellar clouds, the density of particles is less than one-trillionth of the earth's sea-level atmospheric density. Only the most sophisticated artificial vacuum pumps can even approach the natural vacuum of space.

The medium is so pervasive, however, and the volume of space so large, that a significant fraction of the total mass of the galactic disk is contained in this form. There are two distinct types of interstellar material, which are thoroughly mixed together throughout space: tiny solid particles called interstellar dust grains; and interstellar gas particles, which may be atoms, ions, or molecules, depending on the temperature and density. The grains are relatively few in number, and constitute only about 1 percent of the total mass in the interstellar medium.

FIGURE 18.1. *A Region of Interstellar Clouds.*
This photo shows a region near the star rho Ophiuchi (itself buried in nebulosity at the center of the photo). The area contains a rich assortment of interstellar clouds and nebulae. At bottom is a globular cluster, much more distant than the interstellar clouds seen here.

Interstellar Dust

The fact that there are interstellar dust grains was evident long ago. Photographs of the Milky Way show extensive, irregular dark regions where dense dust-bearing clouds completely hide the stars behind them (Fig. 18.1). Knowledge of the presence of these clouds is ancient, but it was not recognized until Trumpler's work in 1930 that there is a general distribution of dust grains throughout the spaces in between the obvious clouds. Infrared data from the *IRAS* satellite have shown that interstellar space is permeated by wispy, thin clouds containing dust.

Even though the dust grains represent only a tiny fraction of the total mass in the interstellar medium, they have a very important influence on the starlight that passes through space. There are two basic effects: (1) the obscuration of light from distant stars, commonly referred to as **interstellar extinction;** and (2)

the polarization of starlight. Both effects provide important information about the nature of the dust grains themselves.

The extinction of starlight by interstellar dust particles is the result of a combination of absorption and **scattering,** a process in which the light essentially bounces off of the dust grains. When a photon is scattered, its wavelength remains fixed but its direction is altered. The scattering process is more effective for short wavelengths of light than for longer wavelengths (Fig. 18.2), so a distant star appears redder in color than it otherwise would, because the red light that it emits reaches us more easily than the blue. For a similar reason, the sun appears red when it is near the horizon and its light must traverse a long path through the earth's atmosphere; the fine particles in the air scatter the blue light but allow the red to come through relatively unhindered. Because the extinction of starlight is so much more severe at short wavelengths (Fig. 18.3), ultraviolet telescopes cannot probe the great distances that visible-light telescopes can.

It is possible to determine how much dust lies in the

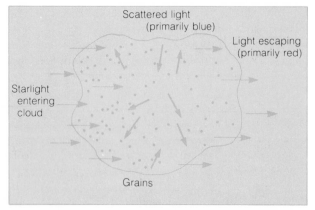

FIGURE 18.2. *The Scattering of Starlight by Interstellar Grains.* The shorter-wavelength photons are more readily scattered by grains; thus it is the longer wavelengths that more easily pass through a cloud. This means that stars seen through interstellar clouds appear reddened.

direction of a given star by measuring how much redder the star appears than it should. Recall the color index B–V, the difference between the blue and visual magnitudes of a star (see chapter 13). The redder a star is, the greater the value of B–V (don't forget: if a star is red in color, it is brighter in visual than in blue light, so the V magnitude is smaller than the B magnitude). The effect of interstellar extinction is to make the value of B–V greater than it would otherwise be.

FIGURE 18.3. *The Interstellar Extinction Curve.* This diagram shows how the extinction (obscuration) of starlight varies with wavelength. The extinction increases steadily toward short wavelengths, and there is a pronounced peak centered near 2,200 Å, in the ultraviolet.

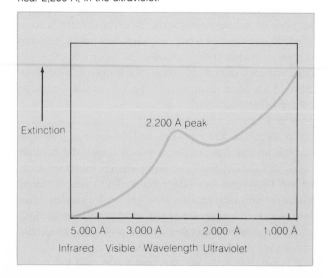

To estimate the amount of dust in the line of sight to a distant star, therefore, simply requires a comparison of the star's B–V color index with the value it would have without extinction. The latter quantity is determined from the spectral class of the star. Remember that B–V and spectral class both measure the same stellar property, namely the temperature, so all stars of a given spectral class therefore have the same intrinsic value of B–V.

Theoretical considerations allow astronomers to determine the average size of the grains, from the variation of extinction with wavelength (Fig. 18.3). The fact that the grains scatter blue light more effectively than red, and the way in which the scattering efficiency varies as a function of wavelength lead to the conclusion that there are two distinct types of grains, both very small: (1) those with an average diameter of about 5×10^{-5} cm (or about 5,000 Å, comparable in size to the wavelength of visible light); and (2) a group of even smaller grains, whose diameters are only about 10^{-6} cm (or about 100 Å), which create extremely strong ultraviolet extinction.

The polarization of starlight tells us something about the shape of the grains. To say that light is polarized means that the waves have a preferred orientation. The interstellar dust creates this effect by selectively absorbing light of one orientation, so that what gets through to us is the light with the orientation perpendicular to this (Fig. 18.4). In doing this, the dust is mimicking the effect of the polarizing lenses that are commonly used in sunglasses.

The only way the interstellar dust grains can produce such an effect is if they are not spherical in shape, but instead are elongated. They also must be aligned in some organized fashion, so that a majority of the grains between the earth and a given star are parallel to each other. Their alignment is probably caused by an interstellar magnetic field, which has a constant orientation over large regions of space. Thus the fact that the grains produce polarization tells us not only something about their properties, but also something about the galaxy; namely, that it has an overall magnetic field. It appears that the field tends to parallel the spiral arms, at least in the vicinity of the sun.

Little has been learned about the detailed shapes or the chemical composition of the grains (Fig. 18.5). Theoretical work indicates that a variety of common substances, including silicates, oxides, graphite, and iron-bearing minerals, could cause the observed extinc-

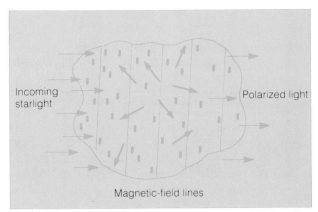

FIGURE 18.4. *The Polarization of Starlight.*
Elongated dust grains, aligned by a galactic magnetic field, tend to scatter photons whose orientations are not parallel to the grains. Thus the light that emerges from an interstellar cloud is polarized.

tion. There is a pronounced peak of extinction at the ultraviolet wavelength of 2,200 Å (Fig. 18.3), which is best explained as being a result of absorption by graphite (a form of carbon), so it is probable that some of the grains contain this substance.

FIGURE 18.5. *An Interstellar Grain?*
No direct information is yet available on the compositions and the detailed shapes of interstellar grains. This sketch illustrates a popular model. For a sobering look at what the grain shapes might really be like, however, see the photo of an interplanetary grain in chapter 10 (Fig. 10.22).

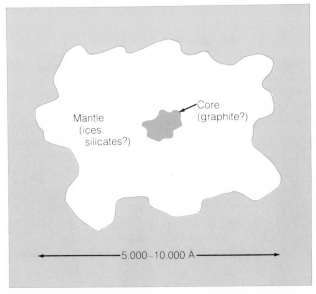

The origin of the interstellar grains is not well understood. There seem to be several possible processes, all of which may contribute to their existence. Tiny solid particles can form from a gas through condensation, if the proper combination of pressure and temperature is reached. As we learned in chapter 12, this probably occurred early in the formation of the solar system. Therefore we expect that some interstellar grains are born in the vicinity of newly formed stars, and are then expelled into space, perhaps by a stellar wind. It also appears likely that some grains form by condensation in the outer layers of cool supergiant stars, where the proper conditions also seem to exist. In addition, it is clear that grains can form in the expanding material in nova and supernova explosions, and in matter expanding outward in planetary nebulae. The formation of interstellar grains is a by-product of the formation and evolution of stars, and is therefore an important part of the large-scale recycling of matter between stellar and interstellar forms.

Observation of Interstellar Gas

Interstellar gas can be detected by several different techniques. As in the case of dust, one manifestation is in the form of rather obvious clouds that can be seen on photographs of the sky. Rather than being dark clouds, however, the visible gas regions are bright concentrations of material that is hot enough to glow. Interstellar gas is also observed in the radio portion of the spectrum, as we learned in the previous chapter when discussing the 21-cm line of hydrogen. As we will see, there are other types of radio emission lines from the interstellar gas as well.

Perhaps the best method for measuring the properties of interstellar gas, however, is to observe the absorption lines that interstellar gas forms in the spectra of stars (Fig. 18.6). As light from a distant star travels through space on its way to earth, it encounters atoms, ions, and molecules along the way, each of which can absorb light at specific wavelengths. The result is that the spectrum of the star has extra absorption lines in it, in addition to the ones created in the star's own atmosphere. It is possible to distinguish the interstellar from the stellar absorption lines by several criteria: the interstellar lines are usually much narrower than those formed in the star's atmosphere; they may occur at a

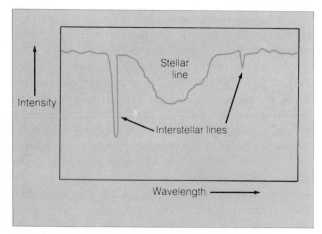

FIGURE 18.6. *Interstellar Absorption Lines.*
This sketch shows the contrast in width between absorption lines formed in a star's atmosphere and those formed in interstellar gas.

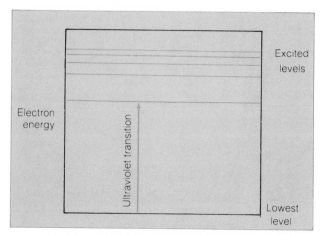

FIGURE 18.7. *The Formation of Interstellar Absorption Lines.*
This energy-level diagram illustrates why most elements have their interstellar absorption lines at ultraviolet wavelengths. In the rarefied interstellar gas, collisions between atoms are rare, and the electrons nearly always stay in their lowest possible energy levels. For most elements, a relatively large amount of energy is needed to cause a transition from the lowest level to a higher one; therefore, only ultraviolet photons can be absorbed, because only they have sufficient energy.

different Doppler shift; and they may represent a different degree of ionization than occurs in the gas in the stellar atmosphere. The first interstellar absorption lines were discovered accidentally just after the turn of the century, but it was not until the 1920s that their interstellar origin was firmly established.

Because of the low density of the interstellar gas, collisions between particles are very rare, and as a result, nearly all the ions and atoms have their electrons in the lowest possible energy level. For most elements, the absorption lines that can be formed when the electrons are in this lowest state lie at ultraviolet wavelengths (Fig. 18.7); only a few relatively rare elements, such as calcium and sodium, have interstellar absorption lines in visible wavelengths. It was not until the 1970s, when ultraviolet telescopes were launched on satellites, that the most common species could be directly observed in space (Fig. 18.8).

The chemical composition of the interstellar medium is nearly identical to that of the sun and other stars, which is no surprise, since the stars form from this material. As we will see in the next section, however, some elements tend to be in the form of grains rather than gas, so the interstellar gas itself has a somewhat different distribution of the elements than the sun.

Clouds and Nebulae

The interstellar material is not uniformly distributed, but instead tends to be patchy, with most of the matter contained in relatively dense concentrations called **in-**

terstellar clouds. In the line of sight to a faraway star, there are likely to be several clouds, their presence being revealed to us by the extra redness they impart to the color of the star and by the interstellar absorption lines they form in the star's spectrum. These most common of interstellar clouds are not usually hot enough to glow nor thick enough to show up as dark patches in the sky like the dense clouds found in some

FIGURE 18.8. *Ultraviolet Interstellar Absorption Lines.*
This is a portion of a spectrum obtained by the *Copernicus* satellite. Several stellar lines and a few interstellar lines are seen here. The interstellar hydrogen feature is so wide because of the great abundance of this element in space.

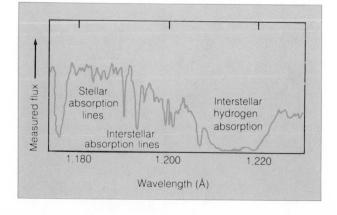

regions, and for this reason they are called **diffuse clouds.** Such clouds were observed directly by *IRAS*.

In comparing the properties of different types of interstellar clouds, we find it convenient to think in terms of density; the number of particles per cubic centimeter. At sea level on the earth, the atmospheric density is about 2×10^{19} molecules/cm^3. In a typical diffuse interstellar cloud, the number is more like 10 to 100/cm^3, a factor of 10^{17} or 10^{18} less! The densest clouds, the dark ones that allow no light to pass through, have densities of about 10^6 particles/cm^3 at most, still a factor of 10^{13} less than the density of the earth's atmosphere. In a diffuse cloud the temperature of the gas is typically in the range of 50 to 100 K, whereas in a dark cloud it is as low as 10 to 20 K.

Diffuse clouds are thought to be generally less than a parsec thick, although in many cases they may stretch for many parsecs in one dimension, as though they were thin sheets of matter, rather than being more or less spherical. The total mass of a diffuse cloud is difficult to assess because of its unknown extent, but is often comparable to the mass of the sun. From this we conclude that these clouds are not the type from which massive stars or star clusters are formed.

The composition of diffuse clouds is dominated by hydrogen, with all other elements present in much smaller quantities. The hydrogen is primarily in atomic form, except in the central portions of the clouds, where the density may be sufficient to allow the formation of molecular hydrogen (designated H$_2$). The abundances of some of the elements are lower than the accepted "cosmic" values. It is as though some of the elements, among them iron, calcium, titanium, and other metallic species, are missing from the interstellar gas. The likely explanation is that these materials are preferentially in the solid dust grains, which gives us some indirect indication of what the grains are made of.

As we have pointed out, diffuse clouds are essentially transparent, but there are conditions under which they are visible in photographs. If a hot star is embedded within a cloud, its radiation ionizes and heats the gas to the point where it glows. Such a cloud is known as an **emission nebula** (Figs. 18.9 and 18.10), nebula being a general word for any interstellar cloud that is dense or bright enough to show up in photographs. In the case of a hot star embedded within a diffuse cloud, the emission occurs when atoms are ionized and then combine again with electrons to form new atoms. When this happens, the electron usually starts out in a high energy state, and then quickly drops down to the

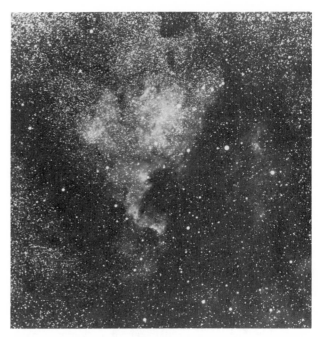

FIGURE 18.9. *An Emission Nebula.*
This is the North American Nebula, so named because of its distinctive shape.

FIGURE 18.10. *Bright Nebulae.*
This shows how reflection and emission nebulae are formed. Light from a hot star is reflected off of a background cloud containing dust. The cloud has a bluish color, because short-wavelength photons are scattered most effectively. The stellar radiation also ionizes gas that is very close to the star, creating a region around it that glows at the wavelengths of certain emission lines. This type of emission nebula is also known as an H II region.

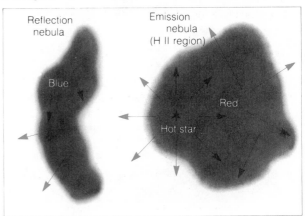

lowest level, emitting a photon of light each time it makes a downward jump. Because these nebulae are dominated by ionized hydrogen, they are called **H II regions,** the H standing for hydrogen, and the Roman numeral II indicating the ionized state (astronomers generally use this type of notation to indicate the degree of ionization of a gas). The strongest line of hydrogen has a wavelength of 6,563 Å, in the red portion of the spectrum, so H II regions stand out especially well in photographs taken with red-sensitive film. A particularly prominent H II region surrounds the stars of the Trapezium in the sword of Orion, and is commonly called the Orion Nebula.

If a diffuse cloud happens to lie just behind a hot star, it may be visible to us as a **reflection nebula** (Fig. 18.10). In this case it is the dust, rather than the gas, that causes the cloud to glow. What we see is the light that has been scattered in our direction by dust grains (Fig. 18.11). This light is always blue, for the same reason the daytime sky is blue: the tiny particles scatter blue light more effectively than longer wavelengths. The best-known reflection nebula is located in the young star cluster called the Pleiades, and it lends photographs of this group of stars an ethereal quality.

FIGURE 18.11. *A Reflection Nebula (NGC 7129).*

Dark Clouds and the Molecular Zoo

In many ways the most fascinating of the interstellar objects are the dark clouds, whose densities and dimensions are sufficient to block out all the light of stars behind them. Dark-cloud regions are highly concentrated in the plane of the Milky Way, and have a variety of amorphous sizes and shapes (Fig. 18.12). Often bright H II regions are located within and among them, for the areas where dark clouds are common are also the sites of star formation, and therefore many young, hot stars are mixed in.

The masses of dark clouds are comparable to, or, in the case of large cloud complexes, far in excess of, a single star mass, ranging up to hundreds of thousands of solar masses in some instances. Thus it is quite possible for massive stars and star clusters to form from material in this type of cloud, and it is in dark-cloud regions that the most active star formation occurs.

As we have already noted, the temperature in the interior of a dark cloud is extremely low, perhaps only 10 to 20 K, and the density is in the range of 10^4 to 10^6 particles/cm^3. Under these conditions, atoms can combine into molecular form, and a wide variety of species have been detected at radio wavelengths.

The discovery of the molecular nature of the gas in dark clouds was relatively recent, the first species (the simple molecule OH) having been detected in 1963. The reasons these observations were not made sooner

FIGURE 18.12. *A Molecular Cloud.*
The dark regions seen here near the bright star omicron Persei (left center) emit a rich spectrum of radio lines produced by interstellar molecules.

are that radio astronomy, then still a young science, was concentrated in other areas (such as measurements of 21-cm radiation), and that more sophisticated radio-astronomy techniques had to be developed.

Because dark clouds are so dense, it is not possible to observe them in the same manner as diffuse clouds. No stars can be seen through dark clouds, so there is no way for us to measure absorption lines formed there. Molecules have special properties, however, that produce radio emission lines, and these lines can be observed.

A molecule is a combination of atoms in a single particle, held together by chemical bonds. These bonds represent a merging of energy states; some of the electrons orbit the entire combination of atoms, acting as the glue that holds them together. The electrons still have fixed energy levels, and there is more complex structure as well. The atoms act as though they are held together with springs, in the sense that they can vibrate, and different frequencies of vibration correspond to different energy states. A molecule can also rotate on its axis, and different rotation speeds also correspond to different energy levels. The vibration and rotation states are close together, so the wavelengths of light corresponding to transitions between them are long. In most cases the rotational levels are so closely spaced that the differences between them correspond to radio wavelengths. Thus if a molecule jumps from one rotational state to a lower one (that is, if it slows its rotation), it emits a photon at a radio wavelength. Each kind of molecule has its own characteristic set of rotational energy levels, so each has its own particular set of radio emission lines. Molecules have the same sort of fingerprints in the radio spectrum that atoms and ions do in visible and ultraviolet light. (Molecules, having such complex structure, also have spectral lines at visible and ultraviolet wavelengths, but these lines cannot be easily observed in dark clouds, for the reasons already stated.)

In a dark cloud, the density is sufficiently high that molecules collide every now and then, and when they do they can be excited to high rotational energy levels. Once a molecule has been kicked into a rapid rotation state, it will soon slow (in a single, instantaneous jump) to a lower state, emitting a radio photon in the process. Thus a dark cloud is constantly releasing radio emission at a variety of fixed wavelengths corresponding to the types of molecules inside. As we learned in discussing the 21-cm line, radio lines are not affected by interstellar extinction, so the emission from a dark cloud escapes easily into space.

Most of the molecules that exist in dark clouds emit at wavelengths of a few millimeters or centimeters. Because the emission from a distant cloud is greatly attenuated (owing to the inverse-square law) by the time it reaches the earth, very sensitive receivers are required to detect it. The technology needed for detecting the emission has been developed only within the last two decades. Every time a new advance in receiver sophistication is made, a number of new molecules are immediately detected.

More than seventy molecules have been discovered in space so far (see appendix 11). Most are rather simple species, the most common being diatomic molecules, which consist of only two atoms. Of these, the dominant one is molecular hydrogen (H_2) which, for a subtle reason, does not emit radio waves, and has been detected in dark clouds only in active regions where it is excited to emit infrared radiation. The next most common molecule is carbon monoxide (CO), and following this the most plentiful species include OH (called the hydroxyl radical), water vapor (H_2O), and an assortment of other simple combinations of abundant elements.

Some rather complex large molecules have been detected as well, the current grand champion being $HC_{11}N$, a species consisting of a total of thirteen atoms, eleven of them carbon. There may be even larger molecules in space, most likely other species containing long chains of carbon atoms, which form easily under the conditions that prevail inside dark clouds.

All the molecules detected so far are made up of combinations of just a few of the most common elements, primarily hydrogen, carbon, nitrogen, oxygen, and sulfur. There probably are molecules containing other elements, but because of their lower abundances, these species are difficult to detect. The second most common element after hydrogen is helium, but helium is inert, meaning that it cannot easily combine with other atoms to form molecules. Helium is definitely present in interstellar clouds, but only in atomic form.

Perhaps the most interesting of all the species detected in dark clouds are the **organic molecules,** those containing certain combinations of carbon atoms that are also found in living material. Their existence shows that when stars form in dark clouds, the ingredients for life are already present. We learned in chapter 10 that amino acids have been found in meteorites,

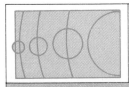

Astronomical Insight 18.1

THE CONFUSION LIMIT

A very interesting yet confounding problem has been recognized recently by researchers involved with radio observations of interstellar molecules. As receiver technology improves and we are able to detect ever weaker radio emission lines, more and more of these lines are found. The reason is that complex molecules tend to have very complicated energy-level structure, so that rather than having just a few strong radio emission lines, they have hundreds or thousands of very weak ones. As a result, the entire portion of the radio spectrum where molecular lines occur is crowded with weak features, and it has already begun to happen that lines from different molecules coincide so closely that they cannot be distinguished from one another.

As the technology continues to improve, this sort of coincidence will occur more and more often.

Eventually, as extremely weak lines become detectable, the point may be reached where the spectrum is so crowded that the characteristic patterns of individual molecules will be lost in the bewildering array of overlapping lines, and we will not be able to identify any new molecules. Hence the true limits on the size and complexity of interstellar molecules may never be known, because for practical reasons we will not be able to identify them, no matter how sensitive the radio receivers.

This potential limit on our ability to detect the largest molecules is fundamental and insurmountable, having to do with the physical nature of the universe out there, rather than our abilities to develop technology here. The term **confusion limit** has been applied to this situation, and there are other examples, such as discussed in chapter 22.

which indicates that these molecules were present before the planets formed. We can speculate that these relatively complex organic molecules formed even before the first solid matter in the solar system.

Interstellar Violence and the Role of Supernovae

Now that we have examined the principal components of the interstellar medium, let us take another look at its overall properties. We have found a diverse collection of ingredients, which can be characterized by a wide variety of temperatures and densities.

When the first spectroscopic observations of the interstellar gas were made more than fifty years ago, astronomers noticed from Doppler shifts of interstellar absorption lines (Fig. 18.13) that some diffuse clouds are moving through space at rather large velocities, up to 100 kilometers per second or even more. Since then its has been found that the motions are often directed

away from groups of bright stars, as though the gas were being expelled by some force originating in the stars.

When 21-cm observations of hydrogen gas became possible, similar motions were detected. A variety of rapidly moving clouds were discovered, and in some regions there appeared to be expanding loops or shells of gas (Fig. 18.14), some of them even showing up in carefully exposed visible-light photographs (Fig. 18.15). Some of these regions were associated with known supernova remnants, where it was clear that the gas was still expanding from the original explosion (Fig. 18.16). In other cases, however, nothing was found at the center of expansion except perhaps a group of young stars. Whatever the cause, the picture developed from these observations was one of an interstellar medium in turmoil, with random, high-speed motions throughout.

In the 1970s new evidence of energetic phenomena in space was uncovered, in two forms. Data obtained from X-ray satellite and rocket experiments showed that the interstellar medium in the plane of the galaxy emits X rays. At about the same time, ultraviolet spec-

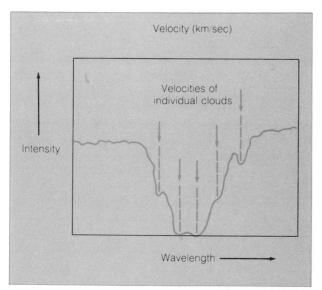

FIGURE 18.13. *The Effect of Motions in the Interstellar Medium.* Interstellar absorption lines often show structure such as this, caused by Doppler shifts of clouds moving at different velocities.

troscopic observations revealed previously undetected ions in space that form only at very high temperatures. The combination of these X-ray and ultraviolet data led astronomers to the conclusion that the space between cool clouds is filled with a very hot gas, whose temper-

FIGURE 18.15. *Barnard's Loop.* This photo of Orion shows a tenuous loop of glowing gas encircling the region of the Belt and Sword, where interstellar clouds and young stars are found. Evidently the loop, named for its discoverer (E. E. Barnard), is composed of gas ejected long ago by the stars in the nebulosity near the center.

FIGURE 18.14. *A 21-cm Picture of the Interstellar Medium.* This cross-section of the disk of the Milky Way shows many looplike structures of hydrogen gas, probably created by old supernova explosions.

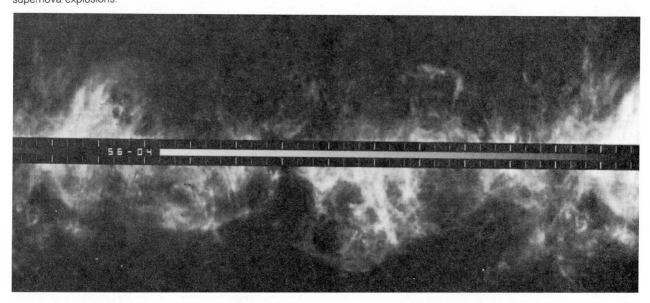

It has been known for a long time that the earth is constantly being bombarded by rapidly moving subatomic particles from space, which have been given the name **cosmic rays.** Consisting of electrons or atomic nuclei, these particles can be detected in several ways, the most common of which is to trace their tracks in special liquid-filled containers called bubble chambers. The cosmic-ray particles are electrically charged, and their passage through such a container creates a trail of fine bubbles. From the length and shape of the tracks, we can infer the charge and speed of the particles. Cosmic-ray velocities are usually near the speed of light.

The earth's magnetic field deflects most cosmic rays, particularly the less massive and energetic ones, so that they reach the ground most easily near the poles. From the point of view of life on earth, this is probably a good thing, for cosmic rays are suspected of causing genetic mutations, and could be harmful in other ways if allowed to reach the earth's surface unimpeded. (On the other hand, mutations caused by cosmic rays are thought to have assisted evolutionary processes in the development of life on earth.)

Extensive observations of cosmic rays show that most are either electrons or nuclei of atoms, with all the electrons stripped free. Most of the latter are single protons, the nuclei of hydrogen atoms, but some are nuclei of heavy elements, including metallic spe-cies known to be formed only in supernova explosions. The majority of cosmic rays that reach the earth originate in the sun, but many, particularly the more massive and energetic ones, come from all directions in the plane of the galaxy, and seem to fill its volume more or less uniformly. Apparently they are trapped by the galactic magnetic field and cannot escape into intergalactic space. Because the paths of cosmic rays are curved and deflected by the galactic magnetic field, it is impossible for us to know where cosmic rays come from. A strong possibility, however, is that they are released in supernova explosions. The total energy contained in cosmic rays darting about the galaxy is very large, and only supernovae seem capable of producing the necessary energy.

The distribution of cosmic-ray nuclei of various elements is generally similar to the chemical abundances found in stars and the interstellar material, although there are some differences, most of which are attributed to nuclear reactions that can occur when the rapidly moving nuclei strike interstellar dust grains. Nuclear reactions of this sort are responsible for the production of certain elements in space, such as the lightweight species, beryllium.

Cosmic rays have little effect on our everyday lives, but they have the potential of telling us quite a bit about the energetic activities that take place in interstellar space.

ature is a million degrees or more. The density of the hot gas between the clouds is very low, even by interstellar standards; it is 10^{-4} to 10^{-3} particle/cm^3, the equivalent of 1 particle in every cubic volume 5 centimeters on a side!

We now know that the interstellar medium not only is diversified, but also is being constantly disturbed, which creates the observed high-speed motions and high temperatures. All this activity requires a substantial amount of energy input; a quiet pond does not spontaneously form ripples and eddies unless something stirs it up.

The agitation of the interstellar medium, it is now understood, is primarily the result of supernova explosions. These occur in the galaxy at a frequency of one every few decades, and each time, a vast amount of energy is released into space. Material expands outward from the site of the explosion, sweeping up a shell of gas that continues to expand, and creating a cavity of very rarefied, extremely hot gas inside the shell. This rarefied, superheated gas is then observable through either synchrotron or thermal radiation in nearly all portions of the spectrum (Fig 18.17). As we have seen, interstellar clouds can be compressed by the

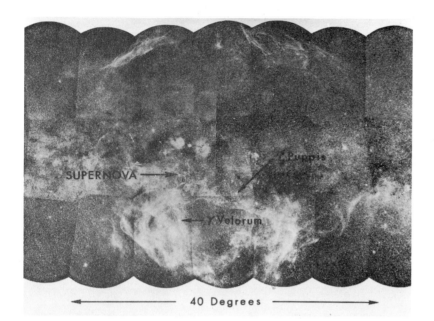

40 Degrees

FIGURE 18.16. *An Extensive Supernova Remnant.*
Here we see, spread over a large region of the sky, a large supernova remnant known as the Gum Nebula. Filamentary structures are obvious, and measurements of interstellar absorption lines show the presence of rapidly moving clouds.

outburst, triggering star formation, and clouds can be accelerated to high velocities as well.

The net effect of repeated supernova explosions in the galaxy is that a large fraction of its volume has become filled with the hot gas that is left behind by the expanding shells, and much of the cloud material has been accelerated, so that high-velocity clouds exist here and there, particularly in loops or fragments of spherical shells (Fig. 18.18).

Another mechanism that has the same effect, and which contributes to this picture, is the expansion of stellar winds. We learned in chapter 14 that hot, blue stars as a rule have high-velocity winds; these are streams of gas flowing outward at speeds as great as 3,000 kilometers per second. Such a wind will act in some ways like a supernova explosion, sweeping up material around the star, and creating an expanding shell (Figs. 18.19 and 18.20). Although a supernova re-

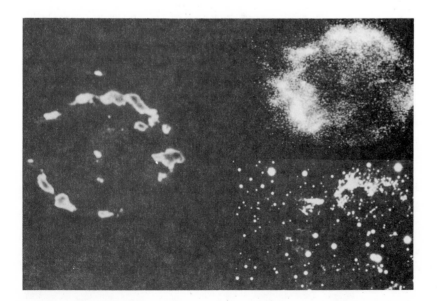

FIGURE 18.17. *X-Ray, Optical, and Radio Images of a Supernova Remnant.*
At upper right is an X-ray image of Cas A, a supernova remnant. Below that is an optical photo, and at left is a radio map. The X-ray and radio images are similar, because both kinds of radiation are produced by the synchrotron process. The visible-light image looks different, because it is in the form of emission lines.

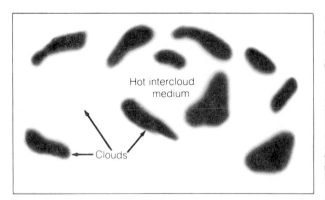

FIGURE 18.18. *A Modern View of the Interstellar Medium.* Recent X-ray and ultraviolet data have led to a picture in which the interstellar clouds are embedded in a very hot (1 to 2 million degree) intercloud gas. The hot gas is created and maintained by expanding supernovae and stellar-wind bubbles.

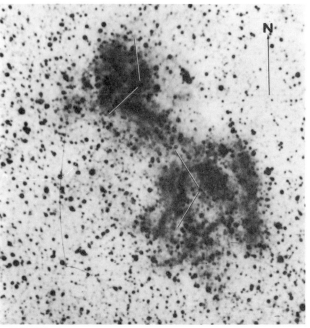

FIGURE 18.19. *Circumstellar Bubbles.* Two Wolf-Rayet stars are shown here, each surrounded by a roughly spherical shell of glowing gas created by the strong winds from the stars. This is a negative print, so stars and glowing gas appear dark against a light background.

leases far more energy in a brief moment, a stellar wind may last several million years, eventually injecting a comparable amount of energy into the surrounding gas. In a cluster of young stars, the winds can combine to evacuate a large bubble around the entire cluster, closely simulating the effect of a supernova explosion.

In one recently discovered case, a gigantic shell was found surrounding a vast volume in the constellation Cygnus (Fig. 18.21). The total energy required to create that structure is estimated to be comparable to that of hundreds or even thousands of supernovae. It is not clear how this "superbubble" formed. Possibly a very

FIGURE 18.20. *Formation of an Interstellar Bubble.* A supernova explosion or a strong stellar wind occurs in the midst of a uniform interstellar medium (left). As the resultant shock waves expand, a spherical cavity is formed, with a concentration of swept-up gas surrounding it. The cavity, or bubble, is filled with very hot, rarefied gas. In the more normal case where the surrounding interstellar medium is initially patchy or cloudy, the bubble formed by the explosion will not be as symmetric as shown here.

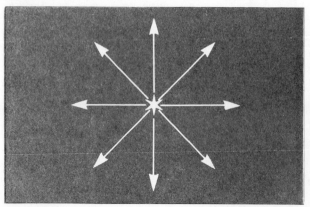

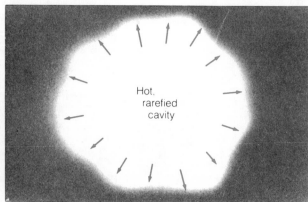

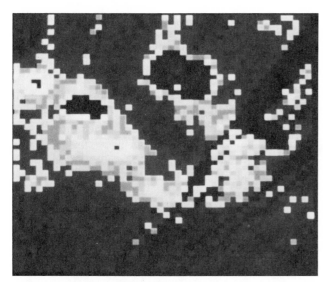

FIGURE 18.21. *The Superbubble in Cygnus.*
This is an X-ray map of a large portion of the constellation Cygnus, showing a gigantic loop of gas that is so hot that it emits X rays. This structure has been interpreted as being a huge bubble created by numerous supernovae that occurred in the same general location. The superbubble is roughly 600 by 450 parsecs in size.

large star cluster at its center experienced a number of supernovae whose expanding shells combined to form the single huge structure seen today.

The hot, tenuous interstellar gas is known to extend well above and below the plane of the galactic disk. Perhaps the galaxy as a whole has a wind emanating outward, a composite of the expansion of all the supernova explosions and stellar winds that have occurred within it. In any case, we can no longer think of the interstellar void as a cold, desolate place where nothing ever happens.

PERSPECTIVE

If this chapter had been written ten years ago, it would have had a much different flavor. The interstellar medium would have been viewed as a quiescent background against which all the activities of stellar evolution are played out. Instead, we now know that the material between the stars plays a dynamic role in the

affairs of the galaxy, with the stars providing the necessary energy as they live and die, and the interstellar material in turn creating the conditions necessary for the births of new generations.

The rich interplay between stars and interstellar matter is the dominant theme of galactic ecology. Its role will be brought to the fore again in the next chapter, as we discuss the formation and evolution of the Milky Way.

SUMMARY

1. The interstellar medium, although extremely tenuous, contains a significant fraction of the mass of the galactic disk.

2. Interstellar grains cause extinction and reddening of starlight, because they scatter and absorb photons, and do so most efficiently at short wavelengths.

3. Analysis of the interstellar extinction created by dust grains leads to the conclusion that many of the grains are about 5,000 Å in diameter, and that there is a second group, responsible for the strong extinction at ultraviolet wavelengths, that are much smaller in size, with diameters of about 100 Å.

4. The grains also cause polarization of starlight, implying that they are elongated in shape and that they are aligned by a galactic magnetic field.

5. The gas in space was first detected through absorption lines in the spectra of distant stars. Most of these lines lie at ultraviolet wavelengths.

6. The interstellar medium is patchy, containing cold clouds embedded in a hot, more rarefied, intercloud medium. Most of the mass in space is in the clouds, and most of the volume is filled with the hot gas.

7. Interstellar clouds may have a variety of densities and temperatures. The so-called diffuse clouds have densities of about 1 to 100 particles/cm^3 and temperatures of 50 to 100 K. These clouds are observable through absorption-line measurements, or, if the clouds are near a hot star, they may appear as emission or reflection nebulae.

8. The dark clouds, with temperatures of 10 to 20 K and densities up to 10^6 particles/cm^3, consist of gas that is primarily molecular in form. The most common molecules are hydrogen (H_2) and carbon monoxide (CO).

A large number of different species have been detected, primarily through their radio emission lines.

9. The interstellar medium displays various forms of energetic activity: there are random, often rapid, cloud motions; the gas between clouds is extremely hot; and here and there are gigantic loop and ring structures suggestive of expanding shells. All this activity can be attributed to energy injected into the interstellar medium by supernova explosions and stellar winds.

8. Why are the dark clouds not observed through interstellar absorption lines?

9. The overall chemical composition of the interstellar medium is thought to be the same everywhere, yet the material is observed in quite different forms in diffuse clouds, dark clouds, and the hot medium between clouds. Summarize the distribution of the elements in each of these components of the interstellar medium.

10. Explain why most of the evidence for violent, energetic activity in the interstellar medium was not discovered until recently.

REVIEW QUESTIONS

1. Explain how absorption lines formed in the interstellar medium can be distinguished from those formed in a star's atmosphere.

2. Explain why we need ultraviolet telescopes in order to observe the interstellar absorption lines of most elements. Why does the same argument not apply to absorption lines formed in stellar atmospheres?

3. How is it that most of the mass in interstellar space is in clouds, yet the clouds fill only a small fraction of the volume?

4. Suppose a B0 star is observed to have a $B–V$ color index of $B–V = 0.20$. From measurements of nearby stars, we know that the normal $B–V$ color index for a star of this type is $B–V = -0.30$. What is the color excess for this star?

5. How can astronomers distinguish between a cool star that is intrinsically red in color and a hotter star that looks just as red because of interstellar dust?

6. Compare the composition of interstellar grains with that of the terrestrial planets, and discuss the similarities in the formation processes for both (see chapter 12).

7. Explain why it is primarily the diffuse clouds, and not the intercloud gas or the dark clouds, that produce the 21-cm emission with which the galaxy is mapped. (Hint: you will have to refer to chapter 17 for information on the formation of the 21-cm line.)

ADDITIONAL READINGS

Blitz, L. 1982. Giant molecular cloud complexes in the galaxy. *Scientific American.* 246(4):84.

Chaisson, E. J. 1978. Gaseous nebulas. *Scientific American* 239(6):164.

Chevalier, R. A. 1978. Supernova remnants. *American Scientist* 66:712.

Heiles, C. 1978. The structure of the interstellar medium. *Scientific American.* 238(1):74.

Herbig, G. H. 1974. Interstellar smog. *American Scientist.* 62:200.

Herbst, E., and Klemperer, W. 1976. The formation of interstellar molecules. *Physics Today* 29(6):32.

Jura, M. 1977. Interstellar clouds and molecular hydrogen. *American Scientist* 65:446.

Kaler, J. B. 1982. Bubbles from dying stars. *Sky and Telescope* 63(2):129.

Knacke, R. F. 1979. Solid particles in space. *Sky and Telescope* 57(4):347.

McCray, R. A., and Snow, T. P. 1979. The violent interstellar medium. *Annual Review of Astronomy and Astrophysics* 17:213.

Miller, J. S. 1974. the structure of emission nebulas. *Scientific American* 231(4):34.

Snow, T. P. 1976. The interstellar medium: much ado about nothing? *Griffith Observer* 40(9):2.

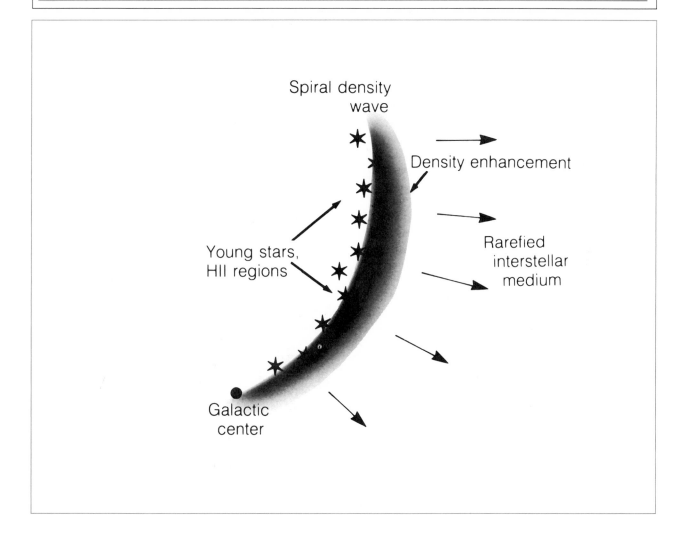

Spiral density wave

Density enhancement

Rarefied interstellar medium

Young stars, HII regions

Galactic center

Learning Goals

STELLAR POPULATIONS AND ELEMENTAL GRADIENTS
Metal Deficiency and Population II
Normal Composition and Population I
Abundance Gradients in the Galaxy
Stellar Motions and the Distribution of Populations
STELLAR CYCLES AND CHEMICAL ENRICHMENT
Nuclear Reactions and Chemical Transformation
The Correlation of Composition With Age

THE FORMATION AND MAINTENANCE OF SPIRAL ARMS
Spiral Density Waves
The Effect of Gravitational Disturbances
GALACTIC HISTORY
Beginnings In a Spherical Cloud
Galactic Collapse and Population II Stars
Disk Formation and Stars of Normal Composition
Continuing Chemical Enrichment

19

The galaxy we observe today is the end product of some 10 to 20 billion years of evolution. A variety of processes have shaped it, and many have left unmistakable evidence of their action. In this chapter we discuss the major influences on the galaxy, along with the developments they have brought about. We have laid the groundwork that will enable us to understand one of the most elegant concepts in the study of astronomy: the interplay between stars and interstellar matter, and the long-term effects of this cosmic recycling on the evolution of the galaxy.

Stellar Populations and Elemental Gradients

To begin this story, we focus on the overall structure of the galaxy (Fig. 19.1), and the types of stars found in various regions. We learned earlier that most of the young stars in the galaxy tend to lie along the spiral arms; indeed, it is the brilliance of the hot, massive O and B stars and their associated H II regions that delineates the arms, making them stand out from the rest of the galactic disk. We also noted that in the nuclear bulge of the galaxy, as well as in the globular clusters and the halo in general, there are few young stars and relatively little interstellar material.

Careful scrutiny has revealed a number of distinctions between the stars that lie in the disk (particularly those in spiral arms), and the stars in the nucleus and halo. Analysis of stellar chemical compositions has shown that stars in the halo and nucleus tend to have relatively low abundances of heavy elements such as metals, whereas stars in the vicinity of the sun (and in the spiral arms in general) have greater quantities of these elements. Nearly all stars are dominated by hydrogen, of course, but in the halo stars the heavy elements represent an even smaller trace than in the spiral-arm stars. The relative abundance of heavy elements in a halo star may typically be a factor of 100 below that found in the sun. If iron, for example, is 10^{-5} as abundant as hydrogen in the sun, it may be only 10^{-7} as abundant in a halo star.

The galaxy has two distinct classes of stars, those in the halo and nuclear bulge, and those in the spiral arms (Fig. 19.2). The groups have been designated as **Population II** (Pop. II) for the stars in the halo, and **Population I** (Pop. I) for those in the spiral arms. The sun is thought to be typical of Pop. I, and its chemical composition is usually adopted as representative of the entire group.

Almost every time an attempt is made to classify astronomical objects into distinct groups, the boundaries between them turn out to be a little indistinct. There are usually intermediate objects whose properties fall between categories, and the stellar populations are no exception. For example, the stars in the disk of the galaxy that do not fall strictly within the spiral arms have intermediate properties between those of Pop. I and Pop. II, and are often called simply the **disk population.** Furthermore, there are gradations within the two principal groups, and we speak therefore of ex-

FIGURE 19.1. *A Spiral Galaxy Similar to the Milky Way.* This vast conglomeration of stars and interstellar matter is following its own evolutionary course, just as individual stars do.

The discovery of stellar populations was actually based on observations of another galaxy, rather than the Milky Way, and the circumstances that led to the discovery were rather unusual. During World War II, following the Japanese attack on Pearl Harbor in late 1941, cities on the west coast of the United States were often kept dark at night, to reduce the danger of bombardment. This was accomplished through enforced blackouts, during which all outside lights were turned off and all windows were covered so that no light got out. As a result of these blackouts, major cities on the west coast experienced some of the darkest nighttime skies of this century. This was a boon to local astronomers.

Many scientists were preoccupied with war-related research efforts, and were not available to take advantage of the excellent observing opportunities. At the Mount Wilson Observatory in the mountains overlooking Pasadena, California, however, there

was a German-born astronomer named Walter Baade, a naturalized U.S. citizen restricted to non-sensitive research because of missing citizenship papers. Baade made extensive use of the 100-inch telescope at Mount Wilson during the blackouts in Los Angeles, and was able to obtain extraordinarily fine photographs of galaxies neighboring the Milky Way.

One of these galaxies is the Andromeda Nebula, a large spiral thought to be similar in many respects to our own galaxy. Baade's images of Andromeda were so fine and clear that he was able to make out individual stars and to assess thier properties. From colors and locations of the stars, he realized that there were systematic differences between different parts of the galaxy, and he developed the notion of two distinct stellar populations. Only later was the idea applied to the Milky Way, after careful observation revealed the same distribution of stellar properties.

treme or intermediate Pop. I or Pop. II objects. It is much more accurate to view the stellar populations as a smooth sequence of properties represented by very low heavy-elements abundances at one end, and sun-like abundances at the other.

FIGURE 19.2. *The Distribution of Populations I and II.* This cross-section of the galaxy shows that Pop. II stars lie in the halo and central bulge, whereas Pop. I stars inhabit the spiral arms in the plane of the disk. Intermediate population stars are distributed throughout the disk.

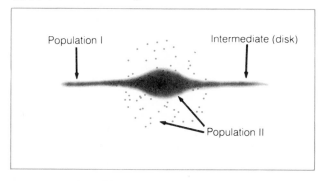

The different stellar chemical compositions found in various parts of the galaxy provide astronomers with very important clues as to the history of the galaxy. Differences that occur gradually over substantial distances are referred to as **gradients,** and the variations in stellar composition from one part of the galaxy to another are called **abundance gradients.** There is a gradient of increasing heavy-element abundances from the halo to the disk (Fig. 19.3). Within the disk itself, there is a similar gradient from the outer to the inner portions (this does not include the stars in the nuclear bulge, which, as we pointed out, tend to have low abundances of heavy elements).

Most of the stars near the sun belong to Pop. I; they are disk stars orbiting the galactic center in approximately circular paths, like the sun. There are a few nearby Pop. II stars here and there, and they are distinguished by a number of properties in addition to their compositions. The most easily recognized is that their motions depart drastically from those of Pop. I stars. As a rule, Pop. II stars do not follow circular paths in the plane of the disk, but instead follow highly elliptical

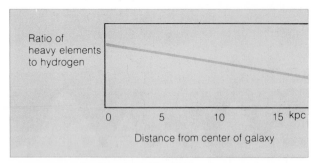

FIGURE 19.3. *An Abundance Gradient.*
This diagram shows how the abundance of a heavy element
(relative to hydrogen) varies with distance from the center of the
galaxy. More stellar generations have lived and died in the
dense regions near the center than farther out, so nuclear
processing is more advanced in the central region.

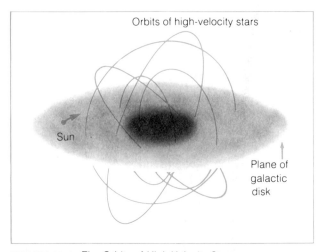

FIGURE 19.4. *The Orbits of High-Velocity Stars.*
These stars follow orbits that intersect the plane of the galaxy.
When such a star passes near the sun, it has a high velocity
relative to us, because of the sun's rapid motion along its own
orbit.

orbits that are randomly oriented, much like cometary
orbits in the solar system. Thus, most of the Pop. II
stars that are seen near the sun are just passing
through the disk from above or below it. The sun and
the other Pop. I stars in its vicinity move in their orbits
at speeds of around 250 kilometers per second, so the
sun moves very rapidly with respect to the Pop. II stars
that pass through its neighborhood in a perpendicular
direction. We therefore classify these Pop. II objects as
high-velocity stars (Fig. 19.4). High-velocity stars
were discovered and recognized as a distinct class even
before the differences between Pop. I and Pop. II were
enumerated. It was only later that these objects were
equated with Pop. II stars following randomly oriented
orbits in the galactic halo.

The systematic differences between the stellar populations arose because different conditions prevailed at
the times when they formed. In this regard it is important to note that globular clusters, which are among the
oldest objects in the galaxy, are representative of Pop.
II; whereas young, newly formed stars are representative of Pop. I.

Stellar Cycles and Chemical Enrichment

The fact that stars in different parts of the Milky Way
have different chemical makeups is readily explained
in terms of stellar evolution. Recall (from chapter 14)
that as a star lives its life, nuclear reactions gradually
convert light elements into heavier ones. The first step
occurs while the star is on the main sequence, during
which time hydrogen nuclei in the star's deep interior
are fused into helium. Later stages, depending on the
mass of the star, may include the fusion of helium into
carbon, and possibly the formation of even heavier elements by the addition of more helium nuclei. The most
massive stars, which undergo the greatest number of
reaction stages, also form heavy elements in the fiery
instants of their deaths in supernova explosions.

All these processes work in the same direction: they
act together to gradually enrich the heavy-element
abundances in the galaxy. Material is cycled back and
forth between stars and the interstellar medium, and
with each passing generation a greater supply of heavy
elements is available. As a result, stars formed where
there has been a lot of previous stellar cycling are born
with higher quantities of heavy elements than stars
formed where little previous cycling has occurred. This
explains why the very old Pop. II stars have low heavy-
elememt abundances they formed out of material that
had not yet been chemically enriched.

The Pop. I stars, such as the sun, are those which
condensed from interstellar material that had previously been processed in stellar interiors and supernova
explosions. Therefore, as a rule, they formed in moderate to recent times in the history of the galaxy, and
are not as ancient as Pop. II objects.

Besides elemental-abundance variations with age,

there can also be variations with location. As noted in the preceding section, there is a distinct abundance gradient within the disk of the galaxy, the stars nearer the center having higher abundances of heavy elements than those farther out. This is a reflection of enhanced stellar cycling near the center, rather than an age difference. The central portion of the disk is where the density of stars and interstellar material is highest, and therefore in this region there has been relatively active stellar processing, because star formation has proceeded at a greater rate than in the less dense outer portions of the disk. Over the lifetime of the galaxy, more generations of stars have lived and died in the central region than in the rarefied outer reaches.

The Care and Feeding of Spiral Arms

In chapter 17 we described the structure of the galaxy and how its spiral nature was discovered and mapped. But we said nothing about how the spiral arms formed, nor why they persist. Both questions are important, for it is clear that if the arms were simply streamers trailing along behind the galaxy as it rotates, they would wrap themselves up tightly around the nucleus, like string being wound into a ball (Fig. 19.5). Because the galaxy is old enough to have rotated at least forty to fifty times since it formed, something must be preventing the arms from winding up, or they would have done so long ago.

Maintaining the spiral arms, a distinct question

from that of forming them in the first place, is a very complex business, and it is not yet fully understood how this is done. The first successful theory appeared in 1960, and is still being refined. The essence of this theory is that the large-scale organization of the galaxy is imposed on it by wave motions. We understand waves to be oscillatory motions created by disturbances, and we know that waves can be transmitted through a medium over long distances while individual particles in the medium move very little. In the case of water waves, for example, a floating object simply bobs up and down as a wave passes by, whereas the wave itself may travel a great distance. Here we are distinguishing between the wave and the medium through which it moves.

The waves that apparently govern the spiral structure of our galaxy are not transverse waves, like those in water, but are compressional waves, similar to sound waves and certain seismic waves (the P waves; see chapter 4 for an elaboration on the different types of waves). In this case the wave pattern consists of alternating regions of high and low density. When a compressional wave passes through a medium, the individual particles vibrate back and forth along the direction of the wave motion. There is no motion perpendicular to that direction, in contrast with water waves.

The theory of galactic spiral structure that invokes waves as a means of maintaining the spiral arms is called the **spiral-density-wave theory.** This hypothesis supposes that there is a spiral wave pattern centered on the galactic nucleus, creating a pinwheel shape of alternating dense and relatively empty regions (Fig. 19.6). The density waves have more effect on the interstellar medium than on stars, so the spiral arms

FIGURE 19.5. *The Wind-Up of Spiral Arms.*
If spiral arms were simple streamers of material attached to a rotating galaxy, they would, within a few hundred million years, wind up tightly around the nucleus. The galaxy is much older than that, so some other explanation of the arms is needed.

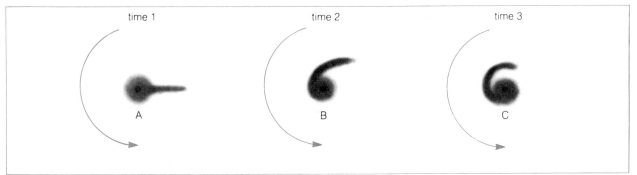

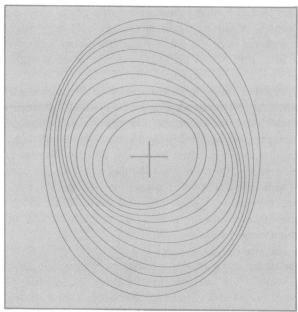

FIGURE 19.6. *The Density-Wave Theory.*
This depicts the manner in which circular orbits are deformed into slightly elliptical ones by an outside gravitational force. The nested elliptical orbits are aligned in such a way that there are density enhancements in a spiral-shaped pattern. This pattern rotates at a steady rate, and does not wind up more tightly.

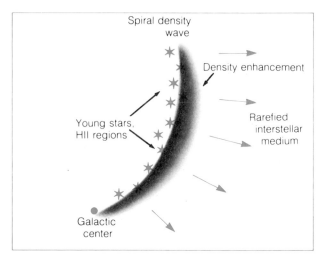

FIGURE 19.7. *The Effect of a Spiral Density Wave on the Interstellar Medium.* As the wave moves through the medium, material is compressed, leading to enhanced star formation. The young stars, H II regions, and nebulosity along the density wave are what we see as a spiral arm.

are characterized primarily by concentrations of gas and dust. These in turn lead to concentrations of young stars, because star formation is enhanced in regions where the interstellar material is compressed (Fig. 19.7).

In its simplest form, the wave pattern is double; that is, there are just two spiral arms emanating from the nucleus, on opposite sides. There are galaxies that have such a simple spiral structure, but many, including the Milky Way, are more complicated. The density-wave theory allows the possibility of more spiral arms, if the waves have shorter "wavelength," or distance, between them.

The waves rotate about the galaxy at a fixed rate that is constant from inner portions to the outer edge. Thus, although the outer portions of the arms appear to trail the rotation, in fact they move all the way around the galaxy in the same time period that the inner portions do. The waves are essentially rigid, in strong contrast with the motions of the stars. Just as in the case of water waves, the motion of the spiral density waves is quite distinct from the motions of individ-

ual particles (that is, stars) in the medium through which the waves travel. In the Milky Way, each star orbits the galaxy at its own speed, which is faster than that of the density waves. This means that as a star circles the galaxy, every so often it will overtake and pass through a region of high density, as it penetrates a spiral arm. Because their motion is slowed slightly when they are in a density wave, stars tend to become concentrated in the arms, just as cars on a highway become jammed together at a point where traffic flow is constricted.

We have noted that the main reason the spiral arms stand out is that hot, luminous young stars are found exclusively in the arms because star formation is enhanced there, where the interstellar gas is compressed. According to the density-wave theory, the most active stellar nurseries should be located on the inside edges of the arms, where stars and gas catch up with and enter the compressed region. This is difficult to check observationally for our own galaxy, but seems to be true of others.

The most luminous stars have such short lifetimes that they evolve and die before having sufficient time to pull ahead of the spiral arm where they were born; therefore, we find no bright, blue stars between the arms. Less luminous stars, those which live billions in-

A DIFFERENT KIND OF SPIRAL ARM

Recently an alternative to the spiral-density-wave theory has been proposed to explain the existence and persistence of the spiral arms. In this picture, the differential rotation of the galaxy, along with the great extent of some regions of star formation, play key roles.

There are regions in the galaxy where very large complexes of interstellar gas and dust are found. Here star formation seems to take place by a sequential process, in which the stellar winds and supernova explosions of massive, short-lived stars give rise to new generations in the near vicinity, as shock waves compress the adjacent clouds. Thus star formation, followed by violent star death, causes on interstellar coud to gradually be devoured as its material is unverted into stars.. In several cases these hotbeds of star formation are many tens of parsecs in extent. On this size scale, the differential rotation of the galaxy distorts the star-forming region. The part of the vast cloudy region nearer the galactic center is pulled ahead of the outer part, and the entire complex is stretched into a curved segment that resembles a portion of a spiral arm. In addition to its shape, such a region has other characteristics of spiral arms, most notably a concentration of interstellar matter and hot, young stars.

In due course a giant dark cloud is consumed by star formation, the luminous stars in the cloud die out, and the entire region fades into obscurity. A spiral-arm segment created in this way is therefore only a temporary feature of the galaxy.

If such segments were continually being formed and dissipated, at any given moment a galaxy would have plenty of spiral structure. The arms would not be permanent, but instead would be constantly coming and going. This is especially likely in galaxies with rather chaotic spiral structure, consisting of many bits and pieces of arms rather than a small number of smooth, complete ones in a simple overall pattern. From what we know of the structure of the Milky Way, this process may well be occurring in our galaxy.

stead of only millions of years, can survive long enough to orbit the galaxy several times, and these stars become spread almost uniformly throughout the galactic disk, between the arms as well as in them. The sun, for example, is old enough to have circled the galaxy some fifteen to twenty times, and therefore has passed through spiral arms and the intervening gaps on several occasions. It is only coincidence that the sun is presently in a spiral arm; it is not necessarily the arm in which the sun formed.

The initial formation of the density wave that is the cause of the spiral arms is a separate question, one on which there is no general agreement. One idea is that gravitational effects of nearby galaxies can disturb a disk-shaped galaxy in such a way that the wave motion is started, in a manner comparable to dropping a rock into a pond and initiating wave motions (Fig. 19.8). In this case, the factor that creates the density waves would be the differential gravitational force, which tends to stretch and distort the galaxy that is subjected to it. The part of the galactic disk nearest the disturbing galaxy would feel a stronger gravitational force than the portions farther away, and as a result, stresses could be placed on the disk that cause oscillatory wave motion.

Another possible mechnism for creating spiral density waves, with no help from neighboring galaxies, occurs in certain galaxies that have noncircular disks. In these **barred spiral galaxies** (see chapter 20), the asymmetric shape of the disk can create the gravitational disturbance needed to initiate spiral density waves.

Aside from the problem of explaining how the arms form in the first place, another area of current research

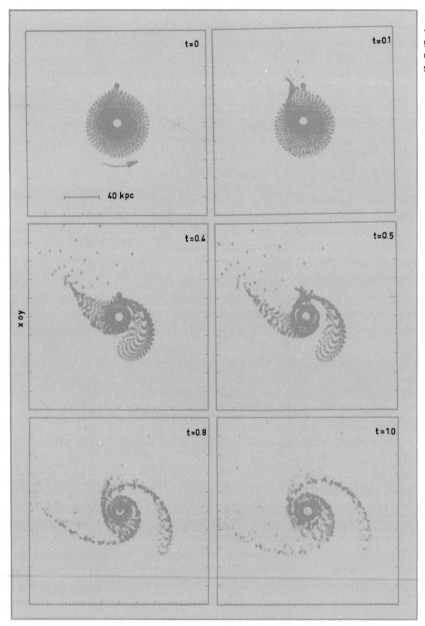

FIGURE 19.8. *The Creation of Spiral Structure.* This sequence of computer-generated models shows how a rotating disk forms spiral density waves when subjected to a gravitational force.

related to spiral arms is aimed at understanding the effects of a massive galactic halo. As we learned in chapter 17, evidence is mounting that the halo of our galaxy may contain as much as 90 percent of the total mass; if so, the gravitational effects of all this material would have some influence on the nature of the spiral arms.

Galactic History

We are now in a position to tie together all the diverse information on the nature of the Milky Way, and to develop from that a picture of its formation and evolution. The pertinent facts that must be explained

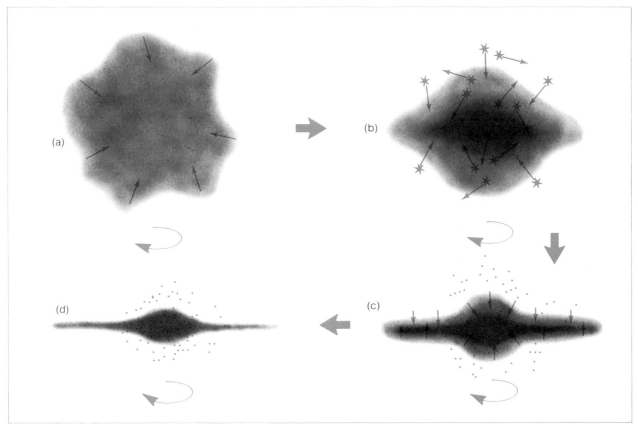

FIGURE 19.9. *Formation of the Galaxy.*
The pregalactic cloud, rotating slowly, begins to collapse (a). The stars formed before and during the collapse have a spherical distribution and noncircular, randomly oriented orbits (b). The collapse leads to a disk with a large central bulge, surrounded by a spherical halo of old stars and clusters (c). The disk flattens further, and eventually forms spiral arms (d). Stars in the disk have relatively high heavy-element abundances, having formed from material that had been through stellar nuclear processing.

include the size and shape of the galaxy, its rotation, the distribution of interstellar material, elemental-abundance gradients, and the dichotomy between Pop. I and Pop. II stars in terms of composition, distribution, and motions. The task of fitting all the pieces of the puzzle together is made easier by the fact that we can reconstruct the time sequence, knowing that as a rule stars with high abundances of heavy elements were formed more recently than those with lower abundances.

The oldest objects in the galaxy are in the halo, which is dominated by globular clusters, but which contains a large (but unknown) number of isolated, dim, red stars as well. Estimates based on the main sequence turn-off in the H-R diagrams for globular

clusters indicate ages of 14 to 16 billion years, and we conclude that the age of the galaxy itself is comparable.

The spherical distribution of the halo objects about the galactic center demonstrates that when they formed, the galaxy was round. Evidently the progenitor of the Milky Way was a gigantic spherical gas cloud, consisting almost exclusively of hydrogen and helium. Very early, perhaps even before the cloud began to contract (Fig. 19.9), the first stars and globular clusters formed in regions where localized condensations occurred. These stars contained few heavy elements, and were distributed throughout a spherical volume, with randomly oriented orbits about the galactic center. Eventually the entire cloud began to fall in on itself, and as it did so, star formation continued to occur, so that many stars were born with motions directed towards the galactic center. These stars assumed highly elongated orbits, accounting for the motions observed today in Pop. II stars.

Apparently the pregalactic cloud originally was rotating, for we know that as the collapse proceeded, a disklike shape resulted. Rotation forced this to happen, just as it did in the case of the contracting cloud that was to form the solar system (see chapter 12). Rotational forces slowed the contraction in the equatorial plane, but not in the polar regions, so material continued to fall in there. The result was a highly flattened disk. Stars that had formed before the disk took shape retained the orbits in which they were born; stars are unaffected by the fluid forces (such as viscosity) that caused the gas to continue collapsing to form a disk.

Stars that formed after the disk developed had characteristics unlike those of their predecessors, in at least two respects: they contained greater abundances of heavy elements, because by this time some stellar cycling had occurred, enriching the interstellar gas; and they were born with circular orbits lying in the plane of the disk. These are the primary traits of Pop. I stars.

Since the time of disk formation, there have been additional, but relatively gradual, changes. Other generations of stars have lived and died, continuing the chemical-enrichment process (particularly in the inner regions of the disk) and creating the chemical-abundance gradient mentioned earlier. Apparently the enrichment process was once more rapid than it is today, because the present rate of star formation is too slow to have built up the quantities of heavy elements that are observed in Pop. I stars. At some time in the past, probably just when the disk was forming, there must have been a period of intense star formation, during which the abundances of heavy elements in the galaxy jumped from almost zero to nearly the present level. A large fraction of the galactic mass must have been cooked in stellar interiors and returned to space in a brief episode of stellar cycling whose intensity has not been matched since.

We have just about recounted all the events that led to the present-day Milky Way, except for the formation of the spiral arms. It is not known when this took place, but it was probably soon after the formation of the disk itself. We see relatively few disk-shaped galaxies without spiral structure, which leads to the conclusion that a disk galaxy does not exist long in an armless state.

It is difficult to guess what will become of the galaxy. It appears that some sort of balance has been reached in the grand recycling process between stars and the interstellar material, so that the interstellar medium is being replenished by evolving and dying stars about as rapidly as it is being consumed by stellar births. Therefore we do not expect the interstellar material to gradually disappear, terminating star formation, as it apparently has in certain other types of galaxies.

PERSPECTIVE

We have now seen our galaxy in a new light, as a dynamic entity. Individual stars orbit its center in individual paths, yet the overall machinery is systematically organized. The majestic spiral arms rotate at a stately pace, while stars and interstellar matter pass through them. We have come to understand the active processes of stellar cycling and chemical enrichment, which still occur today. In a constant turnover of stellar generations, the deaths of the old give rise to the births of the new, while the violence of the death throes energizes a chaotic interstellar medium. The lessons we have learned in examining the Milky Way will be remembered as we move out into the void and probe the distant galaxies.

SUMMARY

1. Stars in the galactic halo have low abundances of heavy elements, and are referred to as Population II stars, whereas those in the spiral arms of the disk have "normal" compositions, and are called Population I stars.

2. There are gradients of increasing heavy-element abundance from halo to disk and from the outer to the inner portions within the disk.

3. Population II stars have randomly oriented, highly elliptical orbits, whereas Population I stars have nearly circular orbits that lie in the plane of the disk. Population II stars, when passing through the disk near the sun's location, are seen as high-velocity stars.

4. The variations in heavy-element abundance from place to place within the galaxy reflect variations in stellar age: very old stars were formed before stellar nuclear reactions produced a significant quantity of heavy elements, and thus have low abundances of these elements; younger stars were formed after the galactic composition was enriched by stellar evolution.

5. Spiral arms are probably density enhancements produced by spiral density waves, which rotate about the galaxy while stars and interstellar material pass through them. Because the interstellar gas is compressed in these density waves, star formation tends to occur there, and this in turn explains why young stars are found predominantly in the spiral arms.

6. The existence and characteristics of Population I and II stars can be explained in a picture of galactic evolution that begins with a spherical cloud of gas that has little or no heavy elements at first. Population II stars formed while the cloud was still spherical or just beginning to collapse, but while it still lacked significant quantities of heavy elements. Population I stars formed later, after the cloud collapsed to a disk, and after stellar evolution produced some heavy elements.

7. The spiral arms formed after the disk was created by the collapse of the original cloud. The spiral density waves that maintain the arms probably started as a result of gravitational disturbances, possibly caused by nearby galaxies.

REVIEW QUESTIONS

1. Explain why the globular clusters have the lowest heavy-element abundances of any objects in the galaxy.

2. The disk population of stars, which uniformly fill the disk of the galaxy without being confined to the spiral arms, have lower heavy-element abundances than extreme Population I stars, which have formed recently in the spiral arms. Explain how this has come about.

3. Suppose a star is observed to have its spectral lines shifted to the red. The strong line of hydrogen whose rest wavelength is 6,563Å is found to lie at a wavelength of 6,567.38Å. What is the line-of-sight velocity of

this star (see chapter 3), and is it a Population I or Population II star?

4. Explain why the evolution of massive stars, rather than that of the much more common low-mass stars, has been the principal contributor to the enrichment of heavy elements in the galaxy.

5. Why are O and B stars in the galaxy found only in the spiral arms?

6. Explain the contrast between the orbital motions of stars circling the galaxy in the disk and the motions of the spiral density waves that are thought to be responsible for the spiral arms.

7. Assuming the mass of the galaxy is 2×10^{11} solar masses, use Kepler's third law to determine the orbital period of a globular cluster whose semimajor axis is 20 kpc (remember to convert this into astronomical units).

8. How was the formation of our galaxy similar to the formation of the solar system (as described in chapter 12)? Does the solar system have the equivalent of a halo?

9. How do we know that the interstellar material that pervades space was, at some time in the past, completely processed through stellar interiors?

10. Summarize the various roles played by supernovae in the evolution of our galaxy.

ADDITIONAL READINGS

Bok, B. J. 1981. Our bigger and better galaxy. *Mercury* 10(5):130.

Bok, B. J. and Bok, P. 1981. *The Milky Way*. Cambridge, Ma.: Harvard University Press.

Burbidge, G., and Burbidge, E. M. 1958. Stellar populations. *Scientific American* 199(2):44.

Iben, I. 1970. Globular cluster stars. *Scientific American* 223(1):26.

Larson, R. B. 1979. The formation of galaxies. *Mercury* 8(3):53.

Shu, F. H. 1973. Spiral structure, dust clouds, and star formation. *American Scientist* 61:524.

_____. 1982. *The physical universe*. San Francisco: W. H. Freeman.

Weaver, H. 1975 and 1976. Steps towards understanding the spiral structure of the Milky Way. *Mercury* 4(5):18, p. 1; 4(6):18, pt. 2; 5(1):19, pt. 3.

Ben M. Zuckerman

Dr. Zuckerman has been one of the leaders of the revolution in astronomy that has been represented by the great advances in infrared and especially radio techniques over the past several years. His major research thrust has been in the observation of emission lines from molecules in dark interstellar clouds and in the gas surrounding cool stars that have ejected matter, and with collaborators he has made many "first" detections of molecular species. For this work Dr. Zuckerman has received several major awards. He has also maintained an abiding interest in the possibility of extraterrestrial intelligence, has participated in radio searches for interstellar signals, and was co-editor of the book Extraterrestrials: Where Are They? (1982; the Pergamon Press, New York). Now a professor of astronomy at UCLA (following a lengthy period at the University of Marland), Dr. Zuckerman is becoming involved in infrared astronomy, particularly in the application of speckle interferometry techniques to infrared wavelengths.

The opening of the radio and infrared spectral domains has resulted in a great flood of discoveries during the past few decades. I think that, without a doubt, radio astronomy has supplied us with more surprises during these years than any other area of astronomy. These discoveries include the quasars (discussed in chapter 22), the pulsars, and the great interstellar molecular clouds. Some objects, such as the molecular clouds, are too cold to emit anything but radio and infrared radiation and must, therefore, be studied primarily at these long wavelengths. But quasars and pulsars, although generally classified as belonging to the realm of high-energy astrophysics, are also very effectively investigated at low energies.

Radio and infrared astronomy hold important keys to our understanding of the galaxy as a whole and of important stellar processes which in turn affect the evolution of the galaxy. Radio data tell us the distribution and characteristics of interstellar gas; the very structure of the galaxy is best revealed to us through 21-cm maps, and indirect evidence for an extensive halo is found in rotation curves derived from radio data. Infrared observations tell us about cool objects in the galaxy, such as star-forming regions, young stars, and low-mass stars. Because we have no other way of directly detecting very low-mass stars, we do not know how many of them there are, relative to hotter and more easily observed stellar types. This textbook stresses that most of the stars in the galaxy are low-mass cool stars,

but we really do not know to what extent these objects dominate. Furthermore, the suggestion that the galaxy has an extensive yet invisible halo may be tested with infrared observations. It is possible that this unseen yet massive halo, if it is really there, is in the form of very cool stars.

The importance of star-formation and star-forming regions is discussed below, in terms of what radio and infrared data can tell us about stellar evolution. From the point of view of the galaxy as a whole, however, there is a different emphasis: we would like to know how *common* the process is, how rapidly the galactic population of stars is replenishing itself. In order to assess this, we must be able to carry out a sensitive census of star-formation regions, and these are typically cool dust-obscured clouds that are invisible to all but infrared telescopes. The only way to obtain the required census is to survey the sky at the appropriate wavelength.

In addition to the evidence for a massive galactic halo, another major surprise brought about by radio and infrared astronomy has been the developing evidence for a massive, energetic object at the heart of the Milky Way. Radio spectroscopic observations show the presence of rapidly-moving clouds near the galactic nucleus. At the same time, infrared data show a hot spot at the very center, and infrared spectroscopic data show that there are interstellar clouds very close to that position that are moving rapidly around an unseen central mass. Analysis of the motions of these clouds has

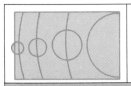

led to the conclusion, described in Chapter 17, that a great deal of mass exists in a very small volume at the heart of the Milky Way.

Many important discoveries in the births, lives, and deaths of stars are also made by radio and infrared astronomers, who work hand in glove to study two of the most important classes of low-energy sources in the Milky Way galaxy: (1) the molecular clouds and their "children", the protostars and young stars that are embedded in these clouds; and (2) dying red giant stars which are accompanied by powerful outflowing winds of dust and molecules. Relative to radio astronomy, infrared astronomy has been hampered by the earth's atmosphere and a more primitive technological base. New ground-based, airborne, and space telescopes and detector systems are helping infrared astronomers to overcome these problems, however. The most notable new facility is NASA's outstandingly successful *IRAS* satellite, which is mapping the mid- and far-infrared sky. In any event, our current picture of the birth and death of stars is a result of the marriage of infrared and radio astronomy.

One of the "hottest" areas of pre-main sequence research involves the study of very energertic winds that appear to emanate from many (possibly all?) young stars. Since star formation involves (we think!) the gravitational accretion of matter, that is, *infall,* it came as a great shock to most of us when infrared and radio spectroscopists discovered that young stars also produce high-velocity *outflowing* winds. These winds are much

more powerful, a factor of a million or more, than the solar wind. They are even more powerful, when compared to the radiative luminosity of the underlying star, then the winds around red giants and planetary nebulae (discussed below). We still do not understand the basic physical mechanism that powers the winds from pre-main sequence stars. One conceivable possibility, at least for solar-mass stars, is that rapid differential rotation and convection in the young stars generates magnetic fields that, via magnetic waves, drive material off of the surface. These winds are probably important in at least two fundamental ways: they terminate mass accretion for young stars (so that all stars do not become massive O stars), and they feed energy and momentum into the surrounding molecular clouds. This keeps the clouds "stirred up" and prevents them from quickly collapsing in a great rush of star formation.

Another hot area is the search for protoplanets and more mature solar systems. Recently, via a new technique of high spatial-resolution infrared astronomy called *speckle interferometry* (mentioned in chapter 12 of the text), two colleagues and I have discovered that the young star T Tauri has a low-luminosity infrared companion (separated from the primary star by about 100 A.U.). Although most astronomers would bet that the infrared companion is a low-mass protostar, some have argued that it is really a giant protoplanet. Even more recently *IRAS* has discovered excess far-infrared radiation originating from the direction of the

bright star Vega. This is being interpreted as due to a cloud of solid particles orbiting around Vega, perhaps similar to the material that formed planets in the solar system so long ago. If large new ground-based infrared telescopes are built during the next decade, then it may be possible, via infrared speckle interferometry, to discover similar dusty clouds around other stars. Indeed this technique may enable us to directly detect very low-luminosity brown dwarfs orbiting about nearby stars (these are dwarfs that are intermediate between giant planets and very low-luminosity stars).

At the other end of the stellar life cycle, radio and infrared astronomers have discovered that a fair number of infrared giant stars are expelling enormous amounts of mass into the interstellar medium. If the sun were to lose mass at the same rate as the most extreme examples of these infrared giants, then the sun would be completely gone in only 10,000 years! This mass loss is probably crucial in determining the ultimate fate of most intermediate-mass stars (2–8 solar masses). That is, in the absence of mass loss, all stars that begin their lives with greater than 1.4 solar masses would die as supernovae as they collapse into neutron stars or black holes. Since around 50 percent of all stars begin their lives with more than 1.4 solar masses, there would then be very many supernovae around! In fact, however, most intermediate-mass stars apparently manage to expel sufficient mass to drop below the 1.4 solar mass limit and,

therefore, expire peacefully as planetary nebulae and white dwarfs rather than as supernovae. Most of this mass loss apparently takes place in a relatively short time when the star is an extreme red giant or infrared giant star. That is, the outflowing material is cool and mainly molecular, containing embedded dust grains that have condensed in the flow. It is the cool dust and the molecules that are readily observable by radio and infrared astronomers.

Spectroscopic studies of this gas and dust indicate that it is often very rich in carbon atoms that have been produced via the triple-alpha process deep within the red-giant stars. This expelled carbon eventually finds its way into the giant molecular clouds mentioned above, where some of it, someday, is then incorporated into a new generation of stars, and, perhaps, planets and people and scrambled eggs.

Radio and infrared astronomers are busy designing and constructing many large new ground- and space-based telescopes and associated detector systems. The next decade promises to be rich in discoveries of the secrets of the Milky Way. We can only speculate about what the future will bring. Infrared telescopes still have a long way to go in sensitivity before they reach levels comparable to those achieved already in other wavelengths, and we can anticipate many new discoveries as the technology catches up. We can expect to learn more and more about the populations of cool objects in the galaxy, we may learn quite a bit about the composition and nature of interstellar dust grains, we will certainly increase our knowledge of physical conditions inside dense interstellar clouds, and we might eventually find unambiguous evidence for the existence of other planetary systems. I believe that infrared astronomy has more galactic news in store for us than any other wavelength band, and I am looking forward to hearing the bulletins as they come in over the next several years.

EXTRAGALACTIC ASTRONOMY

Introduction to Section V

We are ready now to move out of the confines of our own galaxy, to explore the universe beyond. The Milky Way is just one of billions of galaxies in the cosmos, and we will find that although our own system is typical of a certain class of these objects, there is a wide variety of shapes, sizes, and peculiarities associated with other galaxies.

Chapter 20, the first chapter of this section, describes the observational properties of galaxies beyond the Milky Way and the distribution of these galaxies in space. We rely on what was learned in the preceding section, for many aspects of the Milky Way's structure and evolution are characteristic of other galaxies, particularly spirals. We do not discuss the evolution of individual galaxies in great detail; we assume that the forces that have shaped the Milky Way are at work in other situations as well. In our examination of the distribution of galaxies, which will bring us to our first discussions of the universe as a whole, we pay particular attention to clusters of galaxies, and the possibility that these in turn are organized into larger associations. In the process we learn that galaxies within clusters interact with each other in ways that modify their individual properties and also influence the nature of the clusters in which they reside. One question that is of critical importance in our later discussions of the universe as a whole has to do with the uniformity of the distribution of matter. The observational evidence regarding this question is cited in chapters 20 and 21, and the implications are discussed in chapter 23.

In chapter 21, we direct our attention away from individual objects and focus on the universe itself. We describe the two profound observational discoveries that were made concerning the universe: the expansion of the universe and the cosmic background radiation that fills it. Both are legacies of the fiery origin of

the universe, and their observed properties teach us something about its early history. Here we see clearly for the first time that the universe itself is a dynamic, evolving entity. It had a beginning, has been changing since, and will continue to develop.

With the perspective gained from chapter 21, in the next chapter we turn our attention back to individual objects, whose properties can best be appreciated in the context of an expanding universe. A variety of peculiar galaxies, and especially the quasi-stellar objects, have fantastic properties that were discovered by astronomers only because the nature of the universal expansion was already known. These objects, on the frontiers of the observable universe, can potentially tell us a great deal about its history. It is important to realize that very distant objects are seen only as they were long ago, because of the light-travel time. This is useful because it allows us to probe the early history of the universe, but is also a hindrance because it makes it difficult for us to compare distant objects with nearby ones. The quasi-stellar objects are a fundamental component of the early universe, and therefore the mysteries they present to astronomers are among the most important now under study.

The last chapter of this section describes the current state of cosmology, the science of the universe as a whole. Here we tie together all the diverse information gained from the preceding chapters and assess the overall structure of the universe, and we especially emphasize the question of its future. If there is a single premier challenge to modern astronomy, it is this, and we are on the verge of knowing the answer.

Having completed this section, we will be nearly finished with our introduction to astronomy. We are left only with the question of life in the universe, and that is reserved for the final, brief section.

GALAXIES UPON GALAXIES

Learning Goals

We have spoken of our galaxy as one of many, a single member of a vast population that fills the universe. Given all that we have learned about our ordinary position in the cosmos, this is no surprise. Having traced our painful progression from the geocentric view to the realization that we occupy an insignificant planet orbiting an ordinary star in an obscure corner of the galaxy, we should be surprised if we found that our galaxy held any kind of unique status in the larger environment of the universe as a whole. It does not.

Despite the apparent inevitability of this idea, the actual proof that our galaxy is not alone was some time in coming, and arrived only after considerable debate and controversy. The so-called **nebulae,** dim, fuzzy objects scattered throughout the sky, have been known since the early days of astronomical photography, but it was not until the mid-1920s that they were demonstrated to be galaxies, rather than more nearby objects such as gas clouds or star clusters. The proof of their galactic nature was announced in 1924, when the American astronomer Edwin Hubble (Fig. 20.1) reported that he had found Cepheid variables in the prominent Andromeda galaxy (until then known as the Andromeda nebula, since its true nature was not understood), and used the period-luminosity relation to show that this nebula was entirely too distant to be within our own galaxy.

The Hubble Classification System

Having established that nebulae are truly extragalactic objects, Hubble began a systematic study of their properties. The most obvious basis for establishing patterns among the various types was to categorize them according to shape. Hubble did this, designating the spheroidal nebulae **elliptical galaxies** (Fig. 20.2), as

FIGURE 20.1. *Edwin Hubble.*
Hubble's discovery of Cepheid variables in the Andromeda Nebula led to the unambiguous conclusion that this object lies well beyond the limits of the Milky Way, and must therefore be a separate galaxy. Hubble later made important discoveries about the properties of galaxies and what they tell us about the universe as a whole (see chapter 21).

FIGURE 20.2. *An Elliptical Galaxy.*
These smooth, featureless galaxies are probably more common than spirals.

FIGURE 20.3. *A Spiral Nebula.*
This is NGC 6946, an example of a face-on spiral galaxy. During the early years of the twentieth century, there was considerable controversy over the true nature of objects like this.

FIGURE 20.4. *An Edge-On Spiral.*
This view clearly shows the thin layer of interstellar material that resides in the plane of the disk.

distinct from the **spiral galaxies** (Figs. 20.3 and 20.4). Within each of the two general types, Hubble established subcategories on the basis of less dramatic gradations in appearance. The ellipticals displayed varying degrees of flattening, and were sorted out according to the ratio of the long axis to the short axis (Fig. 20.5), with designations from EO (spherical in shape) to E7 (the most highly flattened). The number following the letter E is determined from the formula $10(1 - b/a)$, where a is the long axis and b is the short axis, as measured on a photograph.

Hubble's classification of the spirals was based on the tightness of the arms and the compactness of the nucleus (Fig. 20.6). The types ranged from Sa (tight spiral, large nucleus) to Sc (open arms, small nucleus). The Milky Way is probably an Sb in this system, intermediate in both characteristics, although this is difficult to determine. A very recent study indicates that an Sc designation may be more appropriate.

Hubble recognized a variation of the spiral galaxies in which the nucleus has extensions on opposing sides, with the spiral arms emanating from the ends of the

FIGURE 20.5. *The Shapes of Ellipticals.* The numerical designation (following the letter E) is given by the formula $(1 - b/a) \times 10$, where b is the short axis and a the long axis of the galaxy image. Here are three examples: the EO galaxy has $a = b$ (that is, it is circular in shape); the E3 galaxy has $a = 1.43b$; and the E7 has $a = 3.33b$. E7 is the most highly elongated of the elliptical galaxies.

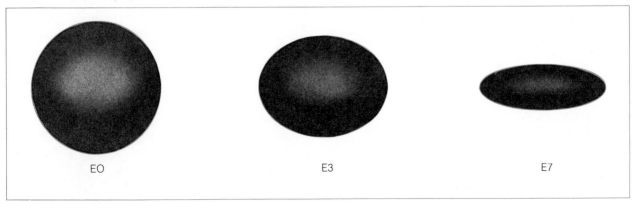

EO   E3   E7

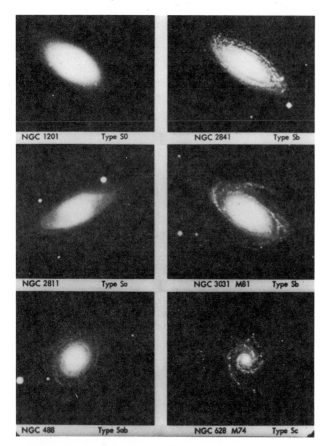

FIGURE 20.6. *An Assortment of Spirals.*
This sequence shows spiral galaxies of several subclasses.

FIGURE 20.7. *A Barred Spiral.*
This is M83, an example of a spiral galaxy with a barlike structure through the nucleus.

FIGURE 20.8. *Barred Spirals.*
Roughly half of all spiral galaxies have central elongations, or bars, from which the spiral arms emanate.

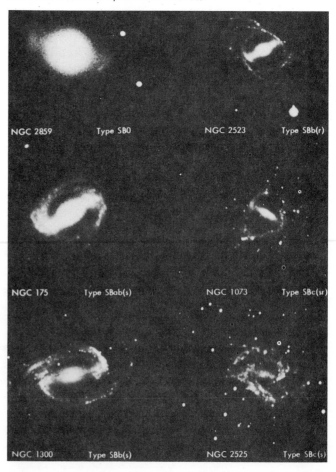

extensions. These he called **barred spirals** (Fig. 20.7), and assigned them the designations SBa through SBc, using the same criteria as before in establishing the a, b, and c subclasses (Fig. 20.8).

Following Hubble's original work on galaxy classification, astronomers recognized an intermediate class called the **S0 galaxies.** These appear to have a disk shape, but no trace of spiral arms.

Hubble arranged the types of galaxies in an organization chart that has become known as the **tuning-fork diagram** (Fig. 20.9). Because there are two types of spirals, Hubble chose not to force all the types into a single sequence, but instead split the sequence into two branches. It was thought for a time that this diagram represented an evolutionary sequence, but later studies showed that all types of galaxies contain old stars, so it became clear that one type of galaxy does not evolve into another. The tuning-fork diagram

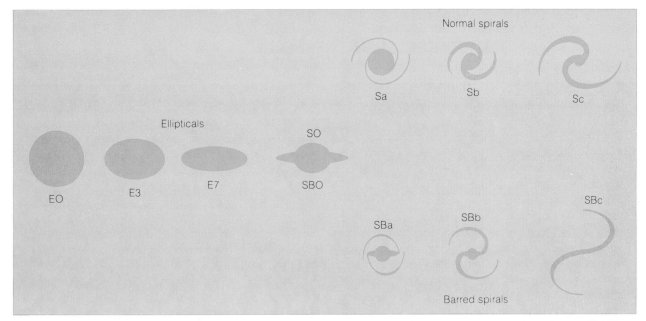

FIGURE 20.9. *The Tuning-Fork Diagram*. This is the traditional manner of displaying the galaxy types, originally devised by Hubble. For quite some time, this was thought to be an evolutionary sequence, although the imagined direction of evolution was reversed at least once. Now it is known that galaxies do not, in the normal course of events, evolve from one type to another.

is not an age sequence after all, and the differences in galactic type must be explained in some other way.

Most of the galaxies listed in catalogues are spirals, which are about evenly divided between normal and barred spirals. Only about 15 percent of the listed galaxies are ellipticals, and a comparable number are SO galaxies. The remaining few percent are called **irregular galaxies;** they do not fit into the normal classification scheme. Because there probably are many small, dim elliptical galaxies that are usually not sufficiently prominenet to be listed in catalogues, it seems likely that ellipticals actually outnumber spirals in the universe. This certainly is the case in dense clusters of galaxies, as we will see.

Although the irregular galaxies are, by definition, misifts, it has proved possible to find some systematic characteristics even in their case. Most have a hint of spiral structure although they lack a clear overall pattern, and these have been designated as **type I irregulars.** The rest, a small group, simply do not conform in any way to the normal standards, and are assigned the classification of **type II irregulars** or **peculiar galaxies** (Fig. 20.10). The Magellanic Clouds, our nearest neighbor galaxies, are both type I irregulars.

FIGURE 20.10. *A Peculiar Galaxy*.
Not all galaxies fit the standard classifications. This is M82, a well-known example of a galaxy with an unusual appearance. For a while it was thought that the nucleus of this galaxy is exploding, but more recent analysis has indicated that the odd appearance is simply the result of a very extensive region of interstellar gas and dust surrounding the center of a spiral galaxy that is being viewed edge-on.

The countless indistinct objects sprinkled across the heavens became one of the leading issues in astronomy in the early decades of the twentieth century. The nebulae inspired an intense controversy, which reached its peak around 1920. By this time, substantial evidence had been collected supporting the opposing viewpoints: that the nebulae were local objects within our galaxy on one hand, and that they were distant galaxies or "island universes" on the other. One of the strongest arguments for the former hypothesis was based on an observation that turned out to be erroneous. There were reports, based on photographs, that some of the spiral nebulae were rotating; different orientations were seen on photographs taken at different times. Although it is no surprise that nebulae rotate, rotation would not be measurable in an object the size of a galaxy, which would take many millions of years to spin once. There would be no hope of detecting differences of orientation in times of a few months or years.

The view that nebulae were local objects was supported by Harlow Shapley, who became the chief spokesman for it. Recall that Shapley was a key figure in establishing the size of the Milky Way and the sun's location within it (see chapter 17). He initially overestimated the size of our galaxy by a substantial amount, and concluded that the Milky Way was large enough to include the Magellanic Clouds within it. These nearby galaxies were, in his view, local objects. This was important, for the Magellanic Clouds clearly were conglomerates of individual stars, rather than gas clouds. If they were proven to lie outside of our galaxy, this would lend weight to the hypothesis that the nebulae were more distant, but similar, objects also consisting of vast collections of stars.

Because of the evidence favoring the hypothesis that nebulae were local objects, the proponents of the island-universe view had difficulty developing strong counterarguments. There was disagreement as to the size of the Milky Way, and those who favored a smaller size than that proposed by Shapley

Measuring the Properties of Galaxies

In order to probe the physical nature of galaxies, we must first know their distances, because without this information, we cannot deduce such fundamental parameters as mass and luminosity.

We have already mentioned one technique for distance determination that can be applied to some galaxies: the use of Cepheid variables. These stars are luminous enough to be identified as far away as a few million parsecs (that is, a few **megaparsecs,** abbreviated **Mpc**), which means we can measure the distances to the Andromeda Nebula and several other neighbors of the Milky Way. This technique is not adequate, however, for probing the distances of most galaxies; thus other methods had to be developed.

Recall from our discussions of stellar distance deter-

mination (chapter 13) that we can always find the distance to an object if we know the object's apparent and absolute magnitudes. This is the basis of the spectroscopic-parallax method, the main-sequence-fitting technique, and even the Cepheid variable period-luminosity relation. A general term for any object whose absolute magnitude is known from its observed characteristics is **standard candle,** and an assortment of these are used in extending the distance scale to faraway galaxies.

The most luminous of stars are the red and blue supergiants, those which occupy the extreme upper regions of the H-R diagram. These can be seen at much greater distances than Cepheid variables, and therefore are important links to distant galaxies. The absolute magnitudes of these stars are inferred from their spectral classes, just as in the spectroscopic-parallax technique. However, these stars are so rare that there is substantial uncertainty in our assuming that they contended to believe that the nebulae, like the Magellanic Clouds, were external objects. Another important part of the controversy hinged on the nature of the bright points of light that occasionally flared up within nebulae; some people argued that these were ordinary novae, in which case they could be near enough to earth to still be within our galaxy, while others argued that these were supernovae, whose brilliance was so great that they had to be at extragalactic distances to appear as faint as they did.

The controversy reached a peak in 1920, when Shapley publicly debated H. D. Curtis, a leading proponent of the island-universe hypothesis, before the National Academy of Sciences. The debate was separated into two parts: a discussion of the size of the Milky Way; and an exchange of arguments on the nature of nebulae. The outcome of the debate was not a clear-cut victory for either point of view, although there appears to have been a consensus that Shapley's arguments were better presented and more thoroughly backed by observational evidence.

Finally, in 1924, the question was settled. Edwin Hubble, working with the recently completed 100-inch telescope at Mount Wilson, was able to make out individual stars in a few of the most prominent nebulae, and he identified some Cepheid variables within them. By applying the standard period-luminosity relation to these variables, Hubble demonstrated beyond any doubt that these nebulae were well outside of our galaxy, even if Shapley's size estimate was accepted. For once and for all it was established that the nebulae are distant galaxies, true island universes.

Subsequent examination of the counterevidence showed an error in the earlier report that some of the nebulae could be seen to rotate. The reported differences in orientation had actually been smaller than the uncertainties of the measurements, so the perceived rotation of the nebulae was not real. As for Shapley's overestimate of the size of the Milky Way, this was found to have been caused largely by his neglect of interstellar extinction, unknown at the time of his work on the size of the galaxy. When this oversight was corrected following the work of Trumpler in 1930 (see chapter 17), Shapley's estimated size of the galaxy shrank to the point where the Magellanic Clouds were definitely outside, and they too were recognized as external galaxies in their own right.

---

form to a standard relationship between spectral class and luminosity. In a variation on this technique, it is simply assumed that there is a fundamental limit on how luminous a star can be, and that in a collection of stars as large as a galaxy, there will always be at least one star at this limit. Thus to determine the distance to a galaxy, we need only to measure the apparent magnitude of the brightest star in the galaxy, and to assume the star's absolute magnitude is at the limit, which is about $M = -8$. This technique extends the distance scale by about a factor of 10 beyond what is possible with Cepheid variables; that is, to 10 or more megaparsecs.

Other standard candles come into play at greater distances. These include supernovae (Fig. 20.11), which at peak brightness always reach about the same absolute magnitude. However, astronomers must exercise care in making measurements, because there are at least two distinct types of supernovae, with different absolute magnitudes. Supernovae can be observed at distances of hundreds of Mpc, and are therefore very useful distance indicators, the major drawback being that we can measure distances only to galaxies that happen to have supernovae occur in them. The ultimate standard candle, useful to distances of thousands of Mpc, is the brightest galaxy in the cluster of galaxies. Astronomers assume that the brightest galaxy in a cluster, as in the case of the brightest star in a galaxy, always has about the same absolute magnitude. (Actually, experience has shown that the brightest galaxy in a cluster may not be so standard from one cluster to another, but that the second-brightest galaxy is a better standard candle.)

Another technique based on the assumption that all objects in a given class are similar makes use of bright H II regions in galaxies. In this case it is assumed that the largest of these regions in galaxies are roughly the same size from one galaxy to another. The angular size

JUNE 9, 1950   FEB. 7, 1951

FIGURE 20.11. *A Supernova in a Distant Galaxy.*
Before the distinction between novae and supernovae became clear, these flare-ups contributed to the controversy over the nature of the nebulae. Now that these occurrences in galaxies are known to be supernova explosions, comparisons of their apparent and assumed absolute magnitudes provide distance estimates.

of the H II regions in a galaxy can then be used to estimate the distance to the galaxy.

It is important to keep in mind how uncertain these techniques are. To assume that all objects in a given class are identical in basic properties such as luminosity or size is always a risky business, especially when such assumptions are applied to objects as distant as external galaxies or as rare as the brightest star in a galaxy. There is little else that can be done, however, so we must simply recognize the inherent limitations in accuracy and take them into account. The uncertainties in distance determinations carry over to our measurements of other properties of galaxies.

The masses of nearby galaxies can be measured in the same manner as the mass of our own galaxy: by applying Kepler's third law to the orbital motions of stars or gas clouds in the outer portions. All that is required is to measure the orbital velocity at some point well out from the center, and to determine how far from the center that point is (which in turn requires knowledge of the distance to the galaxy). Then Kepler's third law leads to

$$M = a^3/P^2,$$

where M is the mass of the galaxy (in solar masses), a is the semimajor axis (in AU), and P is the period (in

years) of the orbiting material at the observed point. (See the more complete discussion in chapter 17 to remind yourself how this equation was developed.)

This technique has its difficulties. It is hard to isolate individual stars in distant galaxies and then to measure (using the Doppler shift) their velocities. Also, both the distance and the orientation of the galaxy must be known before the true orbital velocity and semimajor axis can be determined. The orientation can usually be deduced for a spiral galaxy, since it has a disk shape whose tilt can be seen, and the distance can be estimated through the use of one of the methods just outlined. In most cases the orbital velocities are measured at several points within a galaxy (Fig. 20.12), from the center out as far as possible, and the data are plotted on a **velocity curve** (Fig. 20.13), which is simply a diagram showing the variation of orbital velocity with distance from the center. This is useful for several reasons: it provides a means of checking whether the observations go sufficiently far out in the galaxy to reach the region where the orbiting material follows Kepler's third law; it helps in determining the orientation of the galaxy; and it ensures that the velocities have been measured as far out as possible from the center. The most effective means of obtaining velocity data on the outer portions of a spiral galaxy is to measure the 21-cm emission from hydrogen, which can be detected at greater distances from the center than visible light from stars can be observed.

The technique of measuring rotation curves as just described can best be applied to spiral galaxies, where there is a disk with stars and interstellar gas orbiting in a coherent fashion. In elliptical galaxies there is no such clear-cut overall motion, and a slightly different technique must be used. The individual stellar orbits are randomly oriented within an elliptical galaxy, so there is significant range of velocities within any portion of the galaxy's volume. This range of velocities, called the **velocity dispersion,** is greatest near the center of the galaxy, where the stars move fastest in their orbits, and is smaller in the outer regions. Furthermore, the greater the mass of the galaxy, the greater the velocity dispersion (at any distance from the center), so a measurement of this parameter can lead to an estimate of the mass of a galaxy. The velocity dispersion is deduced from the widths of spectral lines formed by groups of stars in different portions of a galaxy (Fig. 20.14); the greater the internal motion within the region that is observed, the greater the widths of

MAPS OF NGC 784

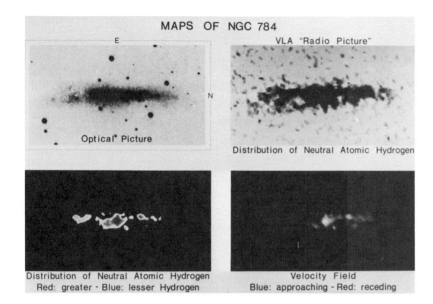

E

N

Optical Picture

VLA "Radio Picture"

Distribution of Neutral Atomic Hydrogen

Distribution of Neutral Atomic Hydrogen
Red: greater - Blue: lesser Hydrogen

Velocity Field
Blue: approaching - Red: receding

FIGURE 20.12. *Determination of a Galactic Rotation Curve from 21-cm Observations.* These are images of NGC 784, a spiral galaxy seen nearly edge-on. At upper left is an optical photo, and at upper right is a radio map showing the distribution of interstellar hydrogen gas. The image at lower left illustrates the density of gas implied by the radio emission, and the image at lower right shows the relative velocites of gas in different parts of the galaxy. Here the lightest areas are receding from us, and the intermediate gray areas are approaching. From data such as these, astronomers measure the variation of velocity with distance from the center of a galaxy. This rotation curve, in turn, is used to estimate the mass of the galaxy.

the spectral lines, owing to the Doppler effect (caused by the fact that some stars are moving toward the earth and others away from it).

Kepler's third law can sometimes be used in an entirely different way in determining galactic masses. There are double galaxies here and there in the cosmos, orbiting each other exactly like stars in binary sys-

FIGURE 20.14. *The Velocity Dispersion for an Elliptical Galaxy.* There is no easily measured overall rotation of an elliptical galaxy, so the mass cannot be estimated from a rotation curve. Instead the average velocities of stars at a known distance from the center are used. Light from a small area of the galaxy's image is measured spectroscopically. The random motions of the stars in the part of the galaxy that is observed broaden the spectral lines by the Doppler effect, and the amount of broadening is a measure of the average velocity of the stars in the observed region. At a given distance from the center of the galaxy, the higher the average velocity, the greater the mass contained inside that distance.

FIGURE 20.13. *A Schematic Velocity Curve for a Spiral Galaxy.* The dashed curve shows the idealized case where the stars in the outer portions of the galaxy move as individuals orbiting a central object, whose mass can therefore be determined through the use of Kepler's third law. The solid line indicates what an actual rotation curve for a spiral galaxy usually looks like; the failure of the curve to drop off indicates that the galaxy has a great deal of mass beyond its visible limits.

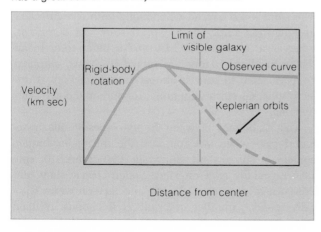

Velocity
(km sec)

Limit of
visible galaxy

Rigid-body
rotation

Observed curve

Keplerian orbits

Distance from center

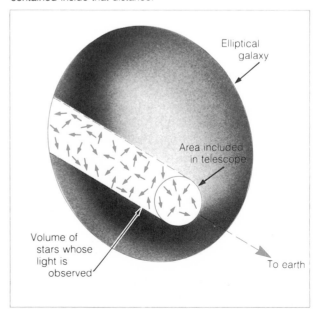

Elliptical
galaxy

Area included
in telescope

Volume of
stars whose
light is
observed

To earth

tems. In these cases, Kepler's third law can be applied, leading to an estimate of the combined mass of the two galaxies. The uncertainties are even more severe than in the case of a double star, however, because the orbital period of a pair of galaxies is measured in hundreds of millions of years. Thus the usual problems of not knowing the orbital inclination or the distance to the system are compounded by inaccuracies in estimating the orbital period, something that is usually well known for a double star. Still, this technique is useful, and has one major advantage: it takes into account *all* the mass of a galaxy, including whatever part of it is in the outer portions, beyond the reach of the standard velocity curve or velocity-dispersion measurements. Galactic masses estimated from double systems are generally much larger than those based on measurements of internal motions within galaxies, possibly indicating that most galaxies have extensive halos containing large quantities of matter. As noted in chapter 17, there is independent evidence that our own galaxy has a massive halo, perhaps containing as much as 90 percent of the total mass.

Once the distance to a galaxy has been established, its luminosity and size can be deduced directly from the apparent magnitude and apparent diameter. Both quantities are found to vary over wide ranges, with luminosities as low as 10^6 and as high as 10^{12} times that of the sun, and diameters ranging from about 1 to 100 kpc.

Generally, elliptical galaxies display a wider range of luminosities and sizes than spirals. The latter tend to be more uniform, with luminsoities usually between 10^{10} and 10^{12} solar luminosities, and diameters between 10 and 100 kpc. The smallest elliptical galaxies are called **dwarf ellipticals.** These may be very common, but because they are too dim to be seen at great distances, we can only say for sure that there are many of them near our own galaxy.

Another class of galaxies that may be unexpectedly common are those embedded in so much interstellar material that they cannot be seen in visible wavelengths. Early results from the *IRAS* satellite, which is mapping the sky at far-infrared wavelegths, indicate a vast number of infrared-emitting galaxies.

It appears that galaxies of a given type tend to be fairly uniform in other properties, just as stars of a given spectral type are the same in other ways. One quantity often used by astronomers to characterize galaxies is the **mass-to-light** (M/L) ratio, which is simply the mass of a galaxy divided by its luminosity, in solar units. Values of the mass-to-light ratio typically range from 50 to 200. Any value larger than 1 means that the galaxy emits less light per solar mass than the sun; that is, such a galaxy is dominated in mass by stars that are dimmer than the sun. A value of M/L = 50, for example, means that 50 solar masses are needed in order to produce the luminosity of one sun. Even the smallest values of M/L that are observed for galaxies are much larger than 1.

Elliptical galaxies tend to have the largest mass-to-light ratios, consistent with our earlier statement that these galaxies are relatively deficient in hot, luminous stars. Spirals, on the other hand, which contain some of these stars, have lower M/L values. It is worth stressing again, however, that even the low values of M/L for these galaxies are much greater than 1, indicating that spirals also are dominated in mass by low-luminosity stars (with some contribution by nonluminous interstellar matter). Hence a spiral galaxy, with all its glorious bright blue stars, actually has far more dim red ones.

The colors of galaxies also can be measured, with the use of filters to determine the brightness at different wavelengths. In general, the spirals are not as red as the ellipticals, again indicating that the latter galaxies contain a higher fraction of cool, red stars. There are also color variations within galaxies; for example, in a spiral, the central bulge is usually redder than the outer portions of the disk where most of the young, hot stars reside.

Both the mass-to-light ratios and the colors of galaxies are indicators of the relative content of Population I and Population II stars. Recall that Pop. I stars tend to be younger, and included in this group are all the bright blue O and B stars. Pop. II objects, by contrast, are old, and include only red, relatively dim stars. Therefore a red overall color along with a high mass-to-light ratio imply that Pop. II stars are dominant, whereas a low value of M/L and a bluer color mean that some Pop. I stars are mixed in. Thus, elliptical galaxies seem to consist almost entirely of Pop. II objects, whereas spirals contain a mixture of the two populations.

This dichotomy between the two types of galaxies is found also when we compare the interstellar-matter content of spirals and ellipticals. Photographs of spirals, especially edge-on views, often clearly show the presence of dark dust clouds, and face-on views typically reveal a number of bright H II regions. Neither dark clouds nor H II regions shows up on photographs of elliptical galaxies. Radio observations of the 21-cm line of hydrogen bear this out: emission is usually present in spirals, but is rarely seen in ellipticals.

The Origins of Spirals and Ellipticals

In attempting to explain the differences between spiral and elliptical galaxies, astronomers have suggested several theories, none of which is fully developed at present.

The weight of all the evidence cited in the previous sections is that spiral galaxies are dynamic, evolving entities, with active star formation and recycling of material between stellar and interstellar forms, whereas elliptical galaxies have reached some sort of equilibrium in which these processes are not taking place at a significant rate. We know from the fact that both types of galaxies contain old stars that there are no systematic age differences between the two types; we cannot simply conclude that the ellipticals are older and have run out of gas.

One of the earliest suggestions to explain the differences between the two kinds of galaxies was that rotation is responsible. We learned in the previous chapter that our own galaxy is thought to have formed from a rotating cloud of gas that flattened into a disk as it contracted. The rotation of the cloud was the cause of the disk formation, so perhaps elliptical galaxies are the result of contracting gas clouds that did not rotate rapidly enough to form disks. It is not clear how this would account for the lack of interstellar material and star formation, however.

One problem with this idea is that elliptical galaxies can and do rotate, yet they have not formed disks. Rotation alone therefore cannot be the full explanation. Apparently the key is whether or not a disk forms before most of the gas in the contracting cloud has been consumed by star formation (Fig. 20.15). If all the gas is converted into stars quickly, before the collapse has proceeded very far, then the result is an elliptical galaxy. If, on the other hand, the initial rate of star formation is not so great, then the cloud has time to form a disk while it still contains a large quantity of interstellar gas and dust. The reason that the timing is so

important is that stars, once formed, will act as individual particles and will continue to orbit the galaxy without forming a disk. In contrast, gas acts as a fluid and will settle into a disk. (The essence of this is whether or not there is viscosity that will dissipate energy and allow the material to sink into the plane of a disk; stars do not encounter each other frequently enough to create a fluid viscosity, but gas particles do.)

Although these ideas represent progress toward understanding why some galaxies are ellipticals and others are spirals, there remain substantial questions. It is not clear why the initial rate of star formation should differ from one galaxy to another, for example, although it has been suggested that it may have to do with the density of the cloud from which the galaxy contracts.

Clusters of Galaxies

Although galaxies may be considered the largest single objects in the universe (if indeed an assemblage of stars orbiting a common center can be viewed as a single object), there are yet larger scales on which matter is organized. Galaxies tend to be located in clusters rather than being distributed uniformly throughout the cosmos, and these clusters in turn have an uneven distribution (Fig. 20.16). Concentrations of clusters are referred to as **superclusters,** which may be the largest things in the universe.

Clusters of galaxies range in membership from a few (perhaps half a dozen) to many hundreds or even thousands. Just as stars in a cluster orbit a common center, so are galaxies in a group or cluster gravitationally bound together, following orbital paths about a central point. In a large, rich cluster, the frequent encounters between galaxies have, over time, caused the members to take on a smooth, spherical distribution; whereas in a small group like the one to which the Milky Way belongs, the arrangement of individual galaxies is more haphazard, creating an amorphous overall appearance.

The Local Group

The Milky Way belongs to a small cluster of galaxies known as the **Local Group.** This cluster consists of about thirty members arranged in a random distribu-

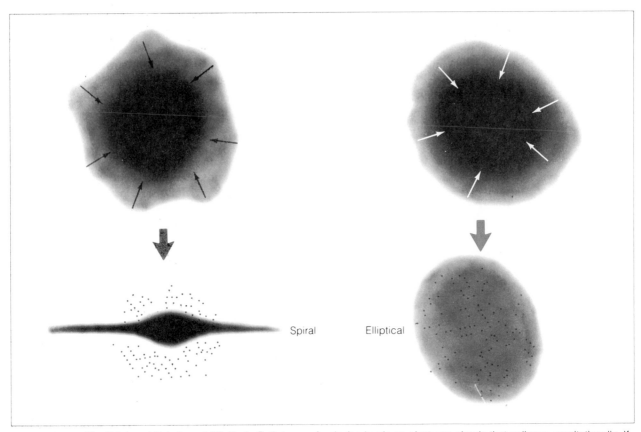

FIGURE 20.15. *The Origins of Spirals and Ellipticals.* Both types of galaxies begin as giant gas clouds that collapse gravitationally. If collapse to a disk occurs before all the gas is converted into stars, the result is a spiral galaxy. If, on the other hand, the gas is entirely used up in star formation before a disk forms, then the result is an elliptical galaxy. The rate of star formation relative to the rate of collapse to a disk is probably determined by the initial density of the cloud, and perhaps is influenced by the rotation rate of the cloud.

FIGURE 20.16. *A Portion of a Cluster of Galaxies in the Constellation Hercules.*
Many different galaxy types are seen here.

FIGURE 20.17. *The Andromeda Galaxy.*
This is a large spiral, comparable to the Milky Way, and the most distant object visible to the unaided eye. Studies of this galaxy have been very useful in helping us better understand the structure and evolution of the Milky Way.

FIGURE 20.18. *Maffei 1 and 2.*
This infrared photo reveals two large galaxies (the fuzzy images, upper right and lower right) that for awhile were considered possible members of the Local Group. More extensive analysis, however, has shown that they are not members.

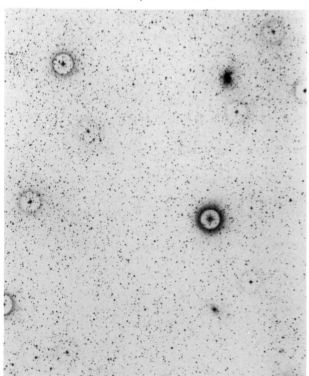

tion. Despite the relative proximity of these galaxies to us, it has been difficult to ascertain their properties in some cases, because of obscuration by our own galactic disk. It is not even possible to say with certainty how many members the Local Group has.

Among the member galaxies are three spirals, two of which, the Andromeda galaxy (Fig. 20.17) and the Milky Way, are rather large and luminous. These are probably the brightest and most massive galaxies in the cluster. Two large galaxies discovered in the 1970s through infrared observations (Fig. 20.18) were for awhile thought to belong to the Local Group, which would have significantly altered the constitution of the cluster, but more recent evidence indicates that both are too distant to be members.

Most of the other members are ellipticals, many of them dwarf ellipticals (Fig. 20.19). There may be ad-

FIGURE 20.19. *The Sculptor Dwarf Galaxy.*
This loose conglomeration of stars lies about 83 kpc from the sun. Such a small, dim galaxy would not be noticed if it were much farther away. The number of systems of this type in the Local Group leads us to assume that they may be very common in the universe.

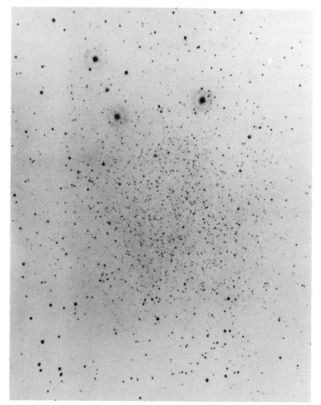

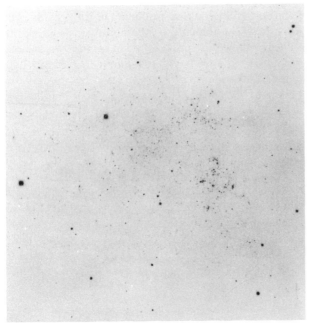

FIGURE 20.20. *IC1613*.
This is an irregular galaxy, a member of the Local Group.

ditional members of this type, undetected so far because of their faintness. Four irregular galaxies (Fig. 20.20), two of them being the large and small Magellanic Clouds (Fig. 20.21), are also found within the Local Group. In addition, there are globular clusters, which are probably distant members of our galaxy, but

FIGURE 20.21. *Both Magellanic Clouds*. The large cloud is at left, and the small cloud is at right.

which lie so far away from the main body of the Milky Way that they appear isolated.

The Local Group is about 800 kpc in diameter, and has a roughly disklike overall shape, with the Milky Way located a little off-center. Beyond the outermost portions of the Local Group there are no conspicuous external galaxies for a distance of some 1,100 kpc.

The Magellanic Clouds and the Andromeda galaxy (also known commonly as M31, its designation in the widely cited Messier Catalog) have been particularly well studied, because of their proximity and prominence and because of what they can tell us about galactic evolution and stellar processing. The large and small Magellanic Clouds appear to the unaided eye as fuzzy patches, easily visible only on dark, moonless nights. They lie near the south celestial pole, and can therefore be seen only from the southern hemisphere. Their name originated from the fact that the first Europeans to see them were Ferdinand Magellan and his crew, who made the first voyage around the world in the early sixteenth century.

The Magellanic Clouds are considered to be satellites of the Milky Way, having lesser masses and following orbits about our galaxy, taking several hundred million years to make each circuit. Lying between 50 and 60 kpc from the sun, both are type I irregulars. They contain substantial quantities of interstellar matter and are quite obviously the sites of active star formation, with many bright nebulae and clusters of hot, young stars (Fig. 20.22). Measurements of the colors of these galaxies, and of the spectra of some of their brighter stars, indicate that they have somewhat lower heavy-element abundances than Pop. I stars in our galaxy. This seems to indicate that the Magellanic Clouds have not undergone as much stellar cycling and recycling as the Milky Way. These and other type I irregulars may generally be viewed as galaxies in extended adolescence, not yet having settled down into mature disks. In the case of the Magellanic Clouds, the reason for the unrest is probably the gravitational tidal forces exerted by the Milky Way.

The Andromeda galaxy is the most distant object visible to the unaided eye, lying some 700 kpc from our position in the Milky Way. All that the eye can see is a fuzzy patch of light, even if a telescope is used, but when a time-exposure photograph is taken, then the awesome disk and spiral arms stand out. The Andromeda galaxy is so large, extending over a full degree across the sky, that full portraits can be obtained only

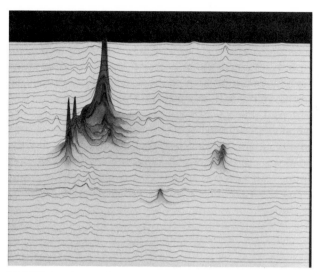

FIGURE 20.22. *An infrared map of the Large Magellanic cloud.* This image, depicting far-infrared data from the *IRAS* satellite, shows an intense peak at the position of the Tarantula nebula, a well-known site of a recent star formation.

FIGURE 20.23. *A Portion of a Rich Cluster of Galaxies.* Most of the objects in this photo are galaxies. This cluster lies in the constellation Coma Berenices.

by using relatively wide-angle telescope optics (most large telescopes have extremely narrow fields of view).

The Andromeda galaxy is probably very similar to our own galaxy, and thus has taught us quite a bit about the nature of the Milky Way. The two stellar populations were first discovered through studies of stars in Andromeda, and the effects of stellar processing on the chemical makeup of different portions of a galaxy are better determined for Andromeda than for our galaxy. This galaxy and other members of the Local Group have also been very important in the development of distance-determination techniques for more distant galaxies and clusters.

Rich clusters: Dominant Ellipticals and Galactic Mergers

In contrast with small clusters such as the Local Group, many clusters of galaxies contain hundreds or even thousands of members (Fig. 20.23). The density of galaxies in such a cluster is relatively high, and therefore there are numerous close encounters between galaxies as they follow their individual orbits about the center. When two galaxies pass close to each other, they exert mutual gravitational forces that can have profound effects. One cumulative result of many such encounters is that the overall distribution of galaxies in the cluster

becomes smooth and more or less spherical, with the greatest density being toward the center. By contrast, small groups or clusters rarely reach this state, but instead retain a more irregular appearance.

The central regions of large, rich clusters are highly dominated by elliptical and SO galaxies (Fig. 20.24), more so than in small groups or isolated galaxies. Near the center of a rich cluster, some 90 percent of the galaxies may be ellipticals or SOs, whereas among noncluster galaxies about 60 percent are spirals. This contrast is probably a direct result of the frequent near-collisions between galaxies in dense clusters. When two galaxies have a close encounter, the tidal forces they exert on each other stretch and distort them. Under some circumstances, the interstellar matter in the two galaxies can be pulled out and dispersed (Fig. 20.25). The outer regions, such as the halos, can also be stripped away. The effect is similar to what happens to rocks in a tumbler: the galaxies gradually are ground down into smoothly shaped remnants. A spiral galaxy, subjected to these cosmic upheavals, may assume the form of an elliptical. In time, most of the spirals in a cluster, particularly those in the dense central region, may be converted into ellipticals and SOs.

The frequent gravitational encounters between galaxies in a rich cluster have another interesting effect: they cause a buildup of galaxies at the center of the cluster. When two galaxies orbiting within a cluster

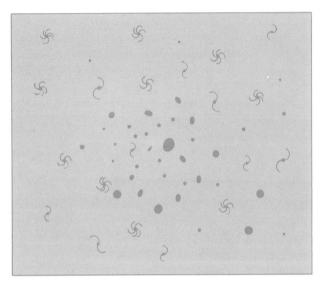

FIGURE 20.24. *Galaxy Types in a Rich Cluster.*
In dense, highly populous clusters of galaxies, nearly all the members in the central portion are ellipticals, and there is often a giant, dominant elliptical at the center. Only in the outer portion are many spirals seen.

come together, one will always gain speed and move to a larger orbit, while the other (always the more massive of the two) will lose speed and drop closer to the center. Thus a gradual sifting process, analogous to the differentiation that occurs inside some planets as the heavy elements sink toward the core, gradually builds up a dense central conglomeration of galaxies at the heart of the cluster. There these galaxies may actually merge, the end result being a single gigantic elliptical galaxy (Figs. 20.26 and 20.27), which continues to grow larger as new galaxies fall in.

Another distinctive characteristic of some rich clusters of galaxies is the existence of a very hot gaseous medium filling the spaces between the galaxies. This so-called **intracluster gas** was discovered through X-ray observations (Fig. 20.28), which show that the temperature of the gas is as high as a hundred million degrees, much hotter than even the highly ionized gas in the interstellar medium in our galaxy. This observation raised the possibility that the general intergalactic void is filled with such gas, although if it is, the density outside of clusters must still be rather low, or

FIGURE 20.25. *A Collision Between Galaxies.*
This shows a pair of galaxies undergoing a near-collision. At right is a computer simulation of their interaction, showing how the present appearance was created.

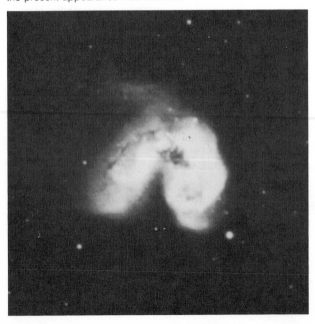

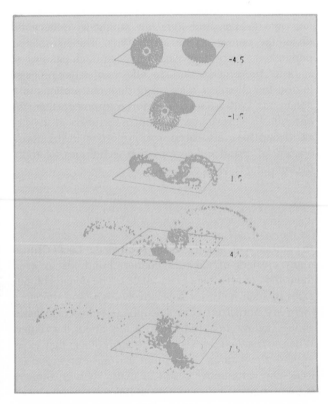

FIGURE 20.26. *A Giant Elliptical Galaxy.*
This is M87, a well-known example of a dominant central galaxy in a rich cluster. This particular galaxy is discussed further in chapter 22.

FIGURE 20.28. *Intracluster Gas in the Virgo Cluster.*
This is an X-ray image from the *Einstein Observatory*, showing emission from the entire central portion of this cluster of galaxies. The X rays are being emitted by very hot gas that fills the space between galaxies.

the X-ray data would have revealed its presence (the significance of a possible general intergalactic medium is discussed in chapter 23).

A different type of X-ray measurement indicates that the hot intracluster gas originates in the galaxies them-selves, rather than entering the cluster from the inter-galactic void. Spectroscopic measurement made at X-ray wavelengths have revealed that the gas contains iron, a heavy element, in nearly the same quantity (rel-ative to hydrogen) as in the sun and other Pop. I stars.

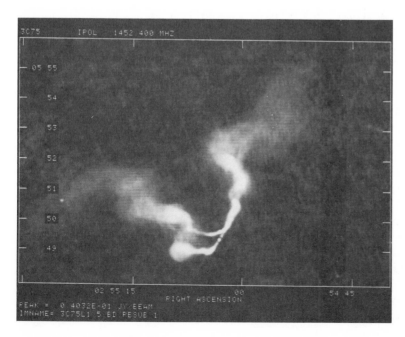

FIGURE 20.27. *A radio image of a galactic merger.*
Here we see a radio map of a giant elliptical galaxy with two bright nuclei (the bright points at lower center, each having a pair of wispy gaseous jets emanating from it; these jets are discussed in chapter 22). Evidently this galaxy is in the process of forming from the merger of two galaxies whose centers have not yet quite combined.

Such a high abundance of iron could only have been produced in nuclear reactions inside of stars. Therefore this intracluster gas must once have been involved in part of the cosmic recycling that goes on in galaxies, as stars gradually enrich matter with heavy elements before returning it to space. How the gas was then expelled into the regions between galaxies is not clear, but it may have been swept out during near-collisions between galaxies, or it may have been ejected in galactic winds created by the cumulative effect of supernova explosions and stellar winds.

Measuring Cluster Masses

There are two methods for measuring the mass of a cluster of galaxies, and both are quite uncertain. This is a crucial problem, as we will see in chapter 23, because of the importance of knowing how much mass the universe contains.

The simpler and more straightforward of the two methods is to estimate the masses of the individual galaxies in a cluster, using techniques described earlier in this chapter, and then add them up. In many cases, particularly for distant clusters where it is impossible for us to measure the rotation curves or velocity dispersions of individual galaxies, the only way we can estimate their masses is to measure their brightnesses and then use a standard mass-to-light ratio to derive their masses. This technique is inaccurate because it depends on our knowing the distance, and because it assumes that the galaxies adhere to the usual mass-to-light ratios for their types. It also does not take into account any matter in the cluster that may lie between the galaxies.

The second method is similar to the velocity-dispersion technique used to estimate masses of elliptical galaxies (Fig. 20.29). The mass of a cluster is estimated from the orbital speeds of galaxies in its outer portions; the faster these galaxies move, the greater the mass of the cluster. This method has the advantage that it measures all the mass of the cluster, whether the mass is in galaxies or between them, but it has the disadvantage that the necessary velocity measurements, particularly for a very distant cluster, are difficult. Furthermore, the technique is valid only if the galaxies are in stable orbits about the cluster; if the cluster is expanding or some of the galaxies are not really gravitationally bound to it, then the results are incorrect. In rich clusters, at least, the smooth overall shape and distribution

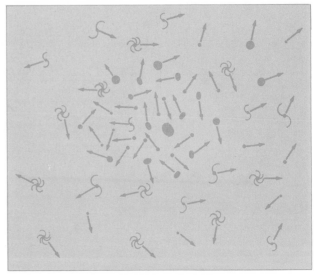

FIGURE 20.29. *Galaxy Motions Within a Cluster.*
Galaxies move randomly within their parent cluster. Analysis of the average galaxy velocities in a large cluster, where the overall distribution of galaxies is uniform, leads to an estimate of the total mass of the cluster, just as the velocity-dispersion method is used to measure masses of elliptical galaxies.

gives the appearance of a bound system, so this technique is probably valid. It may not be applicable to small clusters such as the Local Group, however. This method always leads to an estimated cluster mass that is much greater than that derived by adding up the masses of the visible galaxies, probably because the latter technique neglects intracluster gas and extended galactic halos.

Superclusters

We turn now to consider the overall organization of the matter in the universe. We noted earlier that clusters of galaxies may represent the largest scale on which matter in the universe is organized, but that there is some evidence for higher-order grouping, even in the case of the Local Group (Fig. 20.30). It appears that clusters of galaxies are concentrated in certain regions, commonly referred to as **superclusters.**

The reality of superclusters has been difficult to establish, and for awhile many astronomers were not convinced. Today, however, there are few doubts that clusters of galaxies tend to be grouped, although there

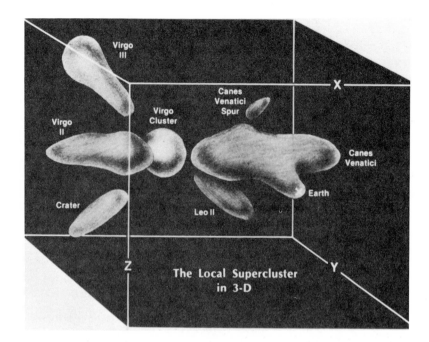

FIGURE 20.30. *The Local Supercluster.* This is an artist's concept of the cluster of galaxy clusters to which the Local Group belongs.

is still disagreement as to the significance of superclusters. The best evidence lies in the distribution of rich clusters (some 1,500 of these have been catalogued), which clearly tend to congregate in certain regions, with relatively empty space in between.

The uncertainty that remains has to do with whether the groupings are random occurrences, or whether they reflect a fundamental unevenness in the distribution of matter in the universe. This is usually tested by comparing the apparent clustering of clusters with what would be expected from coincidence if they were actually distributed in a random fashion (after all, in any random scattering of objects, there will always be occasional accidental groupings). Mathematical experiments of this type appear to show that the observed superclusters could be random concentrations of clusters, and therefore may not have profound significance in terms of the universe as a whole. On the other hand, very recent observations appear to show that a large number of clusters of galaxies may be linked into an enormous filamentary supercluster spanning a major portion of the sky. Previously unrecognized because it crosses the plane of the Milky Way, where our view of distant galaxies is obscured, this supercluster, if real, is far too large to have been the result of random groupings of clusters. This may show that there is a basic unevenness in the distribution of matter in the universe, a possibility whose significance is discussed in chapter 23.

The Origins of Clusters

The basic problem of galaxy formation bears a great deal of resemblance to that of star formation: we know that collapse of an initial cloud will occur only if a sufficiently high density of matter exists in a volume small enough that the mutual gravitational attraction of the gas particles overcomes the particles' natural tendency to fly about as free individuals. Studies of the conditions required for spontaneous gravitational collapse tell us that in the early days of the universe, matter was probably not dense enough (or, conversely, was too hot) to spontaneously form clouds of galactic mass that could collapse. We encountered a similar problem in discussing the formation of individual stars, and in that case we found possible solutions in cloud fragmentation or interstellar turbulence, which can squeeze a cloud and force it to collapse when it otherwise would not do so. Astronomers have suggested similar solutions to the problem of galaxy formation, but there is considerable controversy over the question of whether or not the early universe was turbulent.

Astronomical Insight 20.2

SEARCHING FOR HOLES IN SPACE

One of the fundamental properties of the universe is its homogeneity, or uniformity from one location to another. We discuss in this chapter the hierarchy of galaxies, clusters, and superclusters, but with the tacit assumption that the universe as a whole is more or less uniformly filled with these objects. As we will see in chapter 23, the assumption of homogeneity is commonly made in the construction of models of the universe, but it is an assumption that is sometimes questioned. If the cosmos is not uniformly filled with matter, if the distribution is somehow unbalanced or uneven, it may have important implications in terms of the formation and history of the universe.

A great deal of effort is made by astronomers to measure the distribution of matter. The technique most often used is to choose one or more specific, representative regions of the sky and to carefully search there for all galaxies brighter than some predetermined limit. Distances to all the detected galaxies are then estimated (usually using the Hubble expansion law described in chapter 21), so that a three-dimensional picture can be developed of each sampled region of the sky. From this it is possible to determine the overall distribution of galaxies, and to discover whether any large-scale clumpiness or unevenness exists.

From a study of this type, the astronomical world recently received news of a large portion of space that was apparently devoid of visible matter. In the course of an extensive survey of faint galaxies, a group of astronomers found a region on the sky some 35° across where there appeared to be no galaxies at distances between 240 million parsecs (Mpc) and 360 Mpc. At that distance, 35° of angular diameter corresponds to a linear dimension of 180 Mpc, so it appeared that there was a void more than 100 Mpc across in all dimensions, centered some 300 Mpc away.

Statistical studies show that random distributions of galaxies should produce voids, or empty zones, no larger than about 20 Mpc across. Gravitational effects, in which clumps in the distribution of matter are enhanced by the attraction of galaxies for each other, are thought capable of producing voids up to 35 Mpc in diameter. An empty zone as large as 100 Mpc across would have profound implications, for at the present time there is no known way to produce such a hole in space, unless the universe is fundamentally inhomogeneous. Therefore the report of a hole in space immediately stirred up a great deal of interest, and a number of astronomers began independent studies of the region of the sky where the void was thought to exist.

The original announcement of the existence of the void was based on the discovery of three smaller regions in the same distance range that were empty of galaxies, and the assumption that these three small voids were connected. This had seemed a reasonable assumption, because it was quite unlikely that three unrelated voids in the same general portion of the sky would lie at the same distance from us. Several astronomers set out to search more of the region of the proposed large void, however, to test the assumption that no galaxies would be found in the volume of space between the three small voids.

It was not long before a few faint galaxies were found in the region of the proposed void, and in the distance range previously thought to be empty of galaxies. Apparently there are some galaxies between the voids originally discovered. The evidence for the presence of a hole in space large enough to imply a fundamental unevenness of the distribution of matter has been considerably weakened by these discoveries.

The more recent discovery that superclusters of galaxies may be gigantic filamentary or sheet-like structures, with few galaxies or clusters in the spaces between indicates that significant voids may exist, and that the distribution of matter in the universe may be inhomogeneous. This entire question is at the focus of intense current research efforts.

The Orion Nebula ▶

◀ An infrared image
of two protostars
embedded in a dark cloud

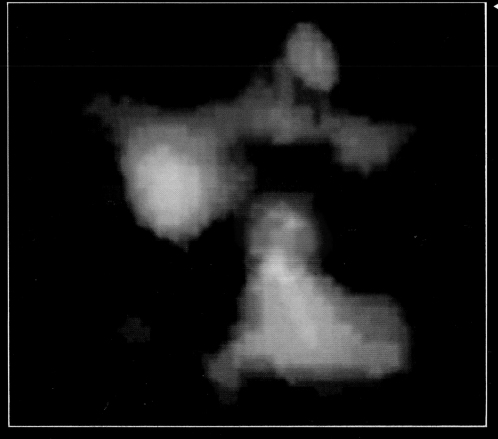

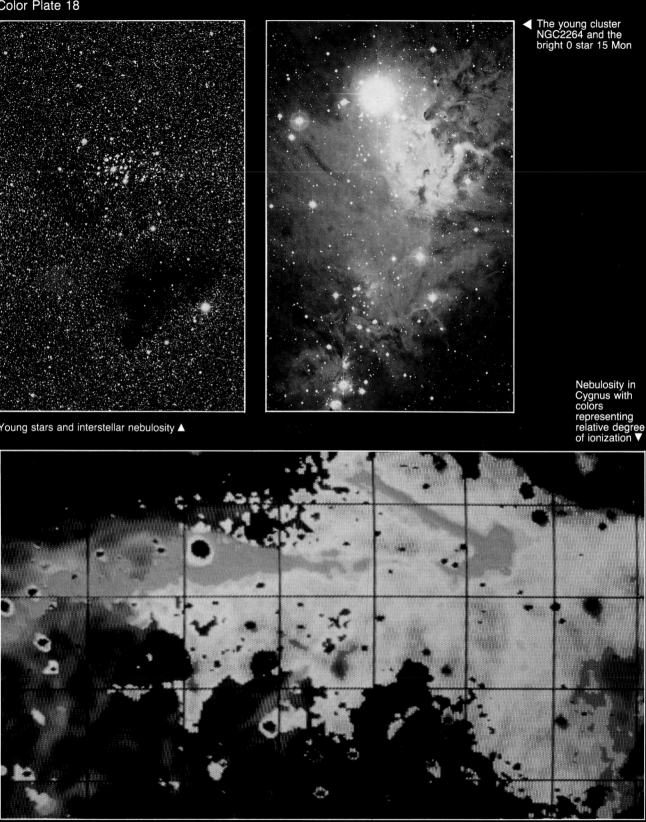

The young cluster NGC2264 and the bright 0 star 15 Mon ◄

Young stars and interstellar nebulosity ▲

Nebulosity in Cygnus with colors representing relative degree of ionization ▼

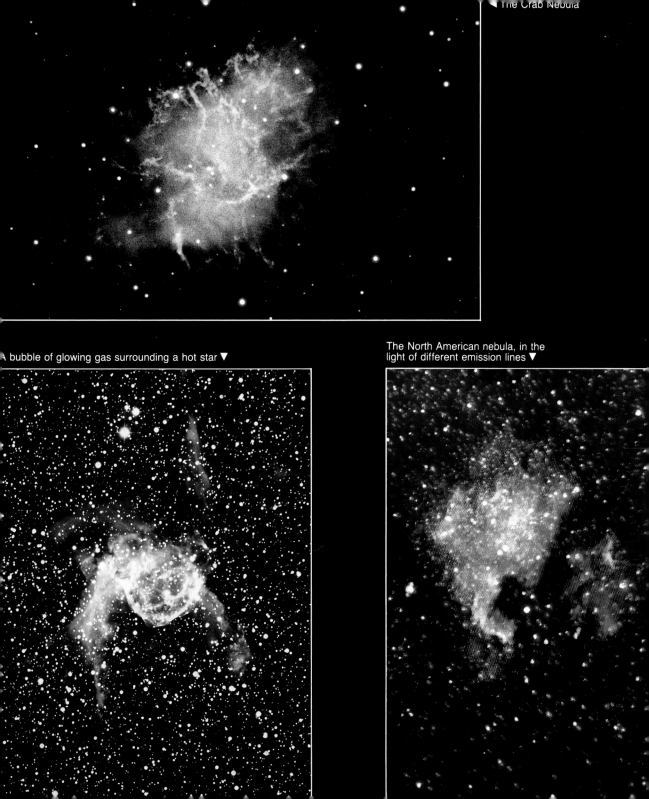

The Crab Nebula

A bubble of glowing gas surrounding a hot star ▼

The North American nebula, in the
light of different emission lines ▼

Color Plate 20

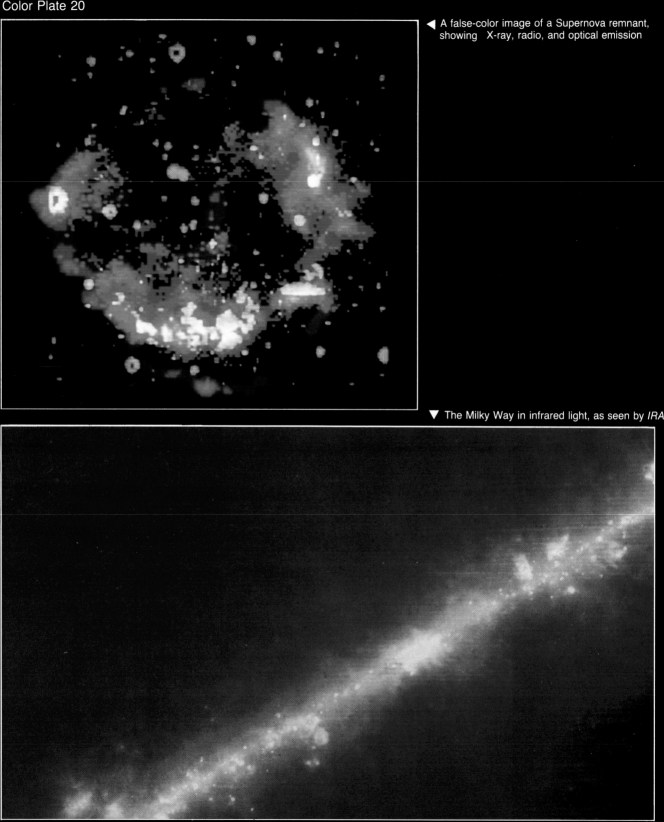

◀ A false-color image of a Supernova remnant, showing   X-ray, radio, and optical emission

▼ The Milky Way in infrared light, as seen by *IRA*

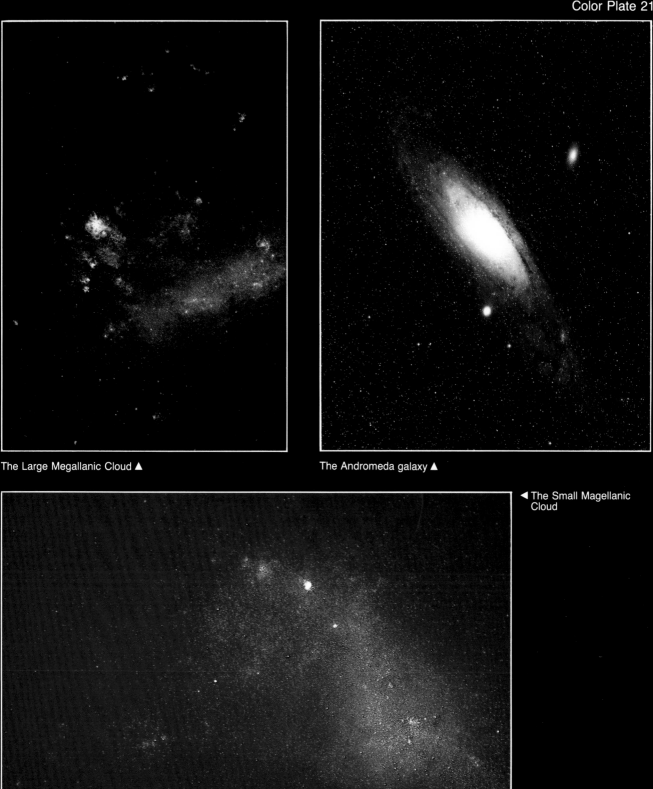

The Large Megallanic Cloud ▲

The Andromeda galaxy ▲

◄ The Small Magellanic Cloud

Color Plate 22

◀ The spiral
galaxy NGC253

The spiral galaxy NGC2997 ▲

A false-color image of a barred spiral galaxy ▼

The Sombrero galaxy ▼

Centaurus A ▲

A false-color radio image of a
radio galaxy with trailing side lobes ▼

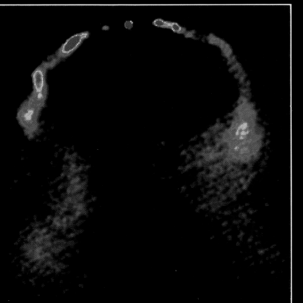

A false-color image of a
remote cluster of galaxies ▼

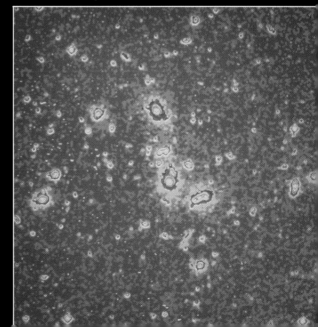

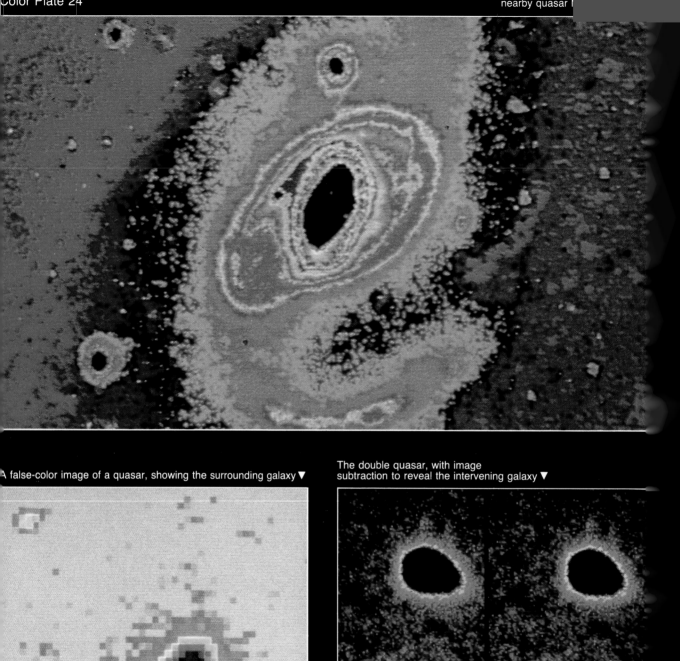

A false-color image of a quasar, showing the surrounding galaxy ▼

The double quasar, with image subtraction to reveal the intervening galaxy ▼

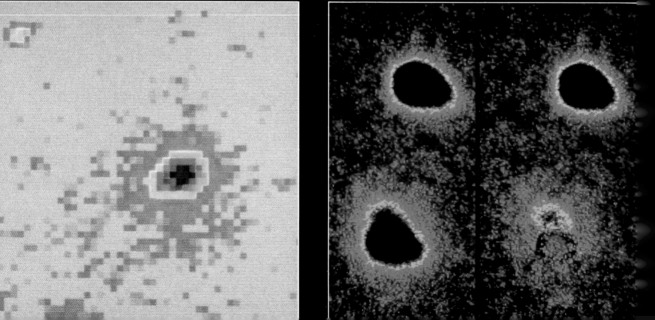

Another suggestion is that massive stars formed early in the history of the universe (calculations show that relatively small condensations could have led to star formation even before galaxies formed), and that the supernova explosions resulting from their deaths could have created sufficiently strong ripples in the universal gas to cause it to fragment and collapse into galaxies.

Whatever the initial cause, the clouds that began to collapse were often much more massive than individual galaxies. As in the case of star formation, when these overly massive condensations developed, they soon fragmented (Fig. 20.31), because as their densities increased, smaller volumes of the gas were able to collapse on their own. The original giant cloud (whose mass was probably about 10^{15} solar masses) broke apart into many small ones, each on its way to becoming an individual galaxy. The entire group was still held together gravitationally, so the galaxies that formed remained in orbit about a common center, and the result was a cluster of galaxies.

One alternative theory attempts to show that the preponderance of elliptical and SO galaxies in rich clusters is the result of the cluster-formation process, rather than the subsequent conversion of spirals into ellipticals and SOs through collisions, as described earlier. The suggestion is that when the original large cloud fragmented, the individual clouds that formed were actually smaller than galaxies. These small clouds then orbited a common center, so that for awhile there existed a cluster of gas clouds. These clouds, especially in the dense central portions of the cluster, occasionally collided with each other and merged together, building up masses comparable to those of galaxies. When a large number of clouds merged, the result most often was a smooth elliptical or SO galaxy. When a small number, perhaps only two or three, combined, the result usually was a flattened, rotating cloud that became a spiral galaxy. This happened much more often in the outer parts of clusters and in small ones such as the Local Group, where the density of clouds was low.

It is quite likely that the situation is actually more complex than the present theories indicate, and that a variety of processes have been involved in the formation of clusters of galaxies and their evolution to the present state. This is a central part of the problem of understanding the early development of the universe, and one that will continue to receive a great deal of attention.

PERSPECTIVE

We have, at last, completed our tour of the universe, having discussed nearly all the forms of organization that matter can take. We still must deal with a few specific kinds of objects that have not yet been described, but the overall picture of the universe in its present state is now more or less complete.

We are ready to tackle questions having to do with the nature of the universe itself, its overall properties and its dynamic nature, and we will do so before examining some of the peculiar objects that are clues to the past.

FIGURE 20.31. *Fragmentation of a Primordial Cloud.* At left is a gigantic cloud, containing the mass of many galaxies. As it collapses, it breaks up into fragments, which then collapse individually, forming galaxies. This is one scenario for the origin of galaxies; in another, the individual clouds that result from fragmentation then merge to form galaxies of varying size and mass.

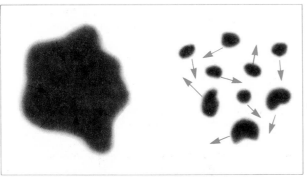

SUMMARY

1. Galaxies are categorized by shape and they fall into two general classes: spirals and ellipticals. There are also a substantial number of SO galaxies (disk-shaped, but with no spiral structure) and irregular galaxies.

2. Distances to galaxies are measured using a variety of standard candles, such as Cepheid variables, extremely luminous stars, supernovae, and galaxies of standard types.

3. Masses of spiral galaxies are determined through the application of Kepler's third law to the outer portions, where the orbital velocity and hence the period are determined from a rotation curve. For elliptical galaxies, the internal velocity dispersion is used. Galactic masses can also be determined from the application of Kepler's third law to binary galaxies.

4. Galactic luminosities and diameters vary quite a lot among elliptical galaxies, but not so much among spirals.

5. All galaxies are dominated by Population II stars. Spirals tend to contain a greater proportion of Population I stars, have substantial quantities of interstellar matter, and generally seem to be in a state of continuous evolution and stellar cycling. Ellipticals, on the other hand, have few or no Population I stars, contain little or no interstellar matter, and generally do not seem to have active stellar cycling at the present time.

6. Spiral galaxies appear to originate from rotating gas clouds that flatten into disks before all the gas is used up in star formation, whereas ellipticals seem to result when star formation consumes all the gas before collapse to a disk occurs.

7. Many galaxies are members of clusters, rather than being randomly distributed throughout the universe.

8. The Milky Way is a member of a cluster called the Local Group, which contains about thirty galaxies.

9. The nearest neighbors to the Milky Way are the Magellanic Clouds, both type I irregulars.

10. The Andromeda galaxy, a huge spiral of type Sb, is similar to the Milky Way in size and general properties; these two galaxies are among the most prominent members of the Local Group.

11. The dominance of elliptical galaxies in rich clusters is probably the result of the conversion of spirals into ellipticals by tidal forces from other galaxies and by drag created by intracluster gas. In many rich clusters there is a giant elliptical galaxy at the center that probably formed from the merging of several galaxies that settled there as a result of collisions.

12. The mass of a cluster can be determined from the sum of the masses of individual galaxies or from the internal velocity dispersion of the galaxies in the cluster.

13. Clusters of galaxies tend to be grouped into aggregates called superclusters, but these may be random concentrations rather than fundamental inhomogeneities of the universe.

14. Clusters of galaxies formed as a result of an uneven distribution of matter at some point early in the history of the universe, but it is not known how this clumpiness originally developed. Theories range from a turbulent early universe to one that fragmented into clouds which then merged to form galaxies.

REVIEW QUESTIONS

1. A supernova of one type has an absolute magnitude of $M = -21$ at peak brilliance. If such objects can be observed to apparent magnitudes as faint as $m = +24$, how far away can supernovae of this type be used as distance indicators?

2. Explain how our knowledge of the distance to a faraway galaxy depends on how well we know the sun-earth distance.

3. In a spiral galaxy with a mass-to-light ratio of 50, it takes 50 solar masses to produce each solar luminosity of energy that the galaxy emits. What does this tell us about the type of star that is most common in such a galaxy?

4. Would you expect elliptical galaxies to have the same proportion of heavy elements as spiral galaxies? Explain.

5. Contrast the evolution of an elliptical galaxy with that of the Milky Way.

6. If a dwarf elliptical galaxy has an absolute magnitude of $M = -15$, and could be detected to an apparent magnitude as faint as $m = +20$, how far away can these galaxies be found? Compare this distance with the diameter of the Local Group, and with the distance to the Virgo cluster, a moderately large cluster some 15 Mpc distant.

7. Discuss the similarities between a rich cluster of galaxies and a globular cluster of stars.

8. Why is it inaccurate to estimate the mass of a cluster of galaxies by using standard mass-to-light ratios for the individual members?

9. Suppose the temperature of the intracluster gas in a rich cluster of galaxies is 100 million degrees (10^8 K).

At what wavelength does this gas emit most strongly? (Review the discussion of Wien's law in chapter 3.)

10. Based on what you learned about the Jeans criterion for gravitational collapse (in chapter 15), discuss the parallel between the problem of the formation of individual stars and that of the formation of individual galaxies.

ADDITIONAL READINGS

de Boer, K. S., and Savage, B. D. 1982. The coronas of galaxies. *Scientific American* 247(2):54.

Geller, M. J. 1978. Large-scale structure of the universe. *American Scientist* 66:176.

Gorenstein, P., and Tucker, W. 1978. Rich clusters of galaxies. *Scientific American* 239(5):98.

Groth, E. J.; Peebles, P. J. E.; Seldner, M.; and Soneira, R. M. 1977. The clustering of galaxies. *Scientific American* 237(5):76.

Hirshfeld, A. 1980. Inside dwarf galaxies. *Sky and Telescope* 59(4):287.

Hodge, P. W. 1981. The Andromeda galaxy. *Scientific American* 244(1):88.

Larson, R. B. 1977. The origin of galaxies. *American Scientist* 65:188.

Mitton, S. 1976. *Exploring the galaxies.* New York: Scribner's.

Rubin, V. C. 1983. Dark matter in spiral galaxies. *Scientific American* 248(6):96.

Sandage, A., Sandage, M., and Kristian, J., eds. 1976. *Galaxies and the universe.* Chicago: University of Chicago Press.

Strom, S. E., and Strom, K. M. 1977. The evolution of disk galaxies. *Scientific American* 240(4):56.

Talbot, R. J., Jensen, E. B., and Dufour, R. J. 1980. Anatomy of a spiral galaxy. *Sky and Telescope* 60(1):23.

Toomre, A., and Toomre, J. 1973. Violent tides between galaxies. *Scientific American* 229(6):38.

van den Bergh, S. 1976. Golden anniversary of Hubble's classification system. *Sky and Telescope* 52:410.

Learning Goals

21 A recurrent theme throughout our study of astronomy has been the dynamic nature of celestial objects. Everything we have studied in detail, from planets to stars to galaxies, has turned out to be in a constant state of change. Now that we are prepared to examine the entire universe as a single entity, we should not be surprised to see this theme maintained. That the universe itself is evolving, with a life story of its own, has been accepted by most astronomers, but there was doubt in the minds of some until very recently. In this chapter we will study the evidence.

Hubble's Great Discovery

Even before it was established that nebulae were definitely galaxies, a great deal of effort went into observing them. Their shapes were studied carefully, and their spectra were analyzed in detail. Typically the spectrum of a galaxy resembles that of a moderately cool star, with many absorption lines (Fig. 21.1). Because the spectrum represents the light from a huge number of stars, each moving at its own velocity, the Doppler effect causes the spectral lines to be rather broad and indistinct. Nevertheless, it is possible to analyze these lines in some detail. One of the early workers in this field, V. M. Slipher, discovered before 1920 that nebulae tend to have large velocities, almost always directed away from the solar system. The spectral lines in most cases were found to be shifted toward the red, and this tendency was most pronounced in the faintest nebulae. Slipher measured velocities as great as 1,800 kilometers per second.

Following his work, others continued to study spectra of nebulae, with heightened interest after Hubble's demonstration in 1924 that these objects were undoubtedly distant galaxies, comparable to the Milky Way in

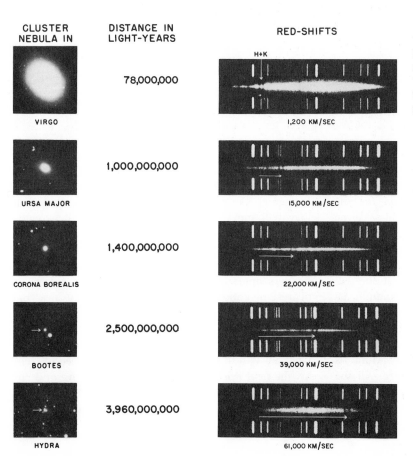

CLUSTER NEBULA IN	DISTANCE IN LIGHT-YEARS	RED-SHIFTS
VIRGO	78,000,000	1,200 KM/SEC
URSA MAJOR	1,000,000,000	15,000 KM/SEC
CORONA BOREALIS	1,400,000,000	22,000 KM/SEC
BOOTES	2,500,000,000	39,000 KM/SEC
HYDRA	3,960,000,000	61,000 KM/SEC

FIGURE 21.1. *Spectra of Galaxies.*
These examples illustrate the broad absorption lines characteristic of spectra of large groups of stars such as galaxies. It was noticed before 1920 that the spectral lines tend to be shifted toward the red in galaxy spectra, indicating that galaxies as a rule are receding from us.

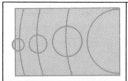

The pioneering observations that led to the discovery of the universal expansion were carried out with a 100-inch telescope at the Mount Wilson Observatory, which overlooks Pasadena, California. This telescope has a remarkable history of involvement with major discoveries, and has ushered in many of the fundamental changes in our perspective on the universe that have taken place in this century.

Named the Hooker Telescope after a Los Angeles businessman who donated funds for the purchase of the primary mirror, the 100-inch instrument has a history dating back to 1908, when the mirror blank was acquired from a French glassworks. Techniques for pouring and annealing such a large mirror were at that time not well developed, and the blank arrived in Pasadena with many small bubbles embedded in it. It was deemed unsuitable, and the glassworks agreed to try to make a better one, but war in Europe intervened. The astronomers at Mount Wilson therefore reexamined the existing glass, and decided after some debate that it should prove adequate; most of the bubbles were buried

deeply enough not to distort the shape of the surface or to weaken the mirror. Grinding and polishing operations began in 1910, and were to take five years.

The shaping of the mirror was undertaken by the leading designer and builder of large telescopes of the day, G. W. Ritchey, and the design and construction of the dome were overseen by George Ellery Hale, director of the Mount Wilson Observatory. Hale, a solar astronomer, was a pioneer in the development of large telescopes, and on no less than four occasions he oversaw the establishment of the then-largest telescope in the world. The 200-inch telescope on Mt. Palomar is named in his honor. One of his primary hopes for the 100-inch instrument was that it would allow the sun's magnetic field to be measured through the use of the Zeemann effect (described in chapter 11), which requires very high-quality spectroscopic data.

The 100-inch telescope finally went into operation in 1918, a full decade after the arrival of the mirror blank from Europe. New and profound discoveries were immediately made, as it became possible for

size and complexity. Hubble and others estimated distances to as many nebulae as possible, by use of the Cepheid variable period-luminosity relation for the nearby ones, and other standard candles such as novae and bright stars for more distant nebulae.

In 1929 Hubble made a dramatic announcement: the speed with which a galaxy moves away from the earth is directly proportional to its distance. If one galaxy is twice as far away as another, for example, its velocity is twice as great. If it is ten times farther away, it is moving ten times faster. The implications of this relationship between distance and velocity are enormous. It means that the universe itself is expanding, its contents rushing outward at a fantastic pace. All the galaxies in the cosmos are moving away from each other (Fig. 21.2).

To envision why the velocity increases with increas-

ing distance, we find it useful to resort to a commonly used analogy. Imagine a loaf of bread that is rising. If the dough has raisins sprinkled uniformly through it, they will move farther apart as the dough expands. Suppose the raisins are one centimeter apart before the dough begins to rise. One hour later, the dough has risen to the point where adjacent raisins are two centimeters apart. A given raisin is now 2 cm from its nearest neighbor, but 4 cm away from the next one over, and 6 cm from the next one, and so on. The distance between any pair of raisins has doubled. From the point of view of any one of the raisins, its nearest neighbor had to move away from it at a speed of 1 cm/hr, the next raisin farther away had to move at 2 cm/hr, the next one had a speed of 3 cm/hr, and so on. If we let the dough continue to rise to the point where the adjacent raisins are 3 cm apart, we find that all the

the first time to observe individual objects within the numberous nebulae. Among the early users of the telescope was Henry Norris Russell, whose groundbreaking work on double stars and on stellar structure and evolution was based largely on observations made with it. Harlow Shapley also used the instrument, although much of his significant work on globular-cluster distances and the size of the galaxy was done with the older 60-inch telescope on Mount Wilson. The measurement of the sun's magnetic field was successfully accomplished in the early years of the 100-inch telescope's operation.

If that had been all that was accomplished with this telescope, it would have been sufficient to ensure its place in astronomical history. It was really only the beginning, however. Another project that began early in the lifetime of the 100-inch instrument was the observation of nebulae, and Hubble's great discovery and analysis of variable stars in the Andromeda galaxy were accomplished with this telescope. Thus the same instrument played a key role both in the determination of the properties of our own galaxy and in the confirmation of the existence of other galaxies, which led to a truly fundamental change in our outlook on the universe.

Hubble continued his work on nebulae, and in 1929 he announced his discovery of the expansion of the universe. Again, the 100-inch telescope had ushered in a new era in astronomy. Later breakthroughs achieved with this instrument included Baade's discovery of stellar populations (described in chapter 19), based on observations of the Andromeda galaxy; pioneering work on the nature of interstellar gas by T. R. Dunham and P. Merrill; analysis of interstellar extinction by J. Stebbins and A. E. Whitford; and, as long ago as 1920, interferometric measurements of stellar angular diameters by A. A. Michelson and F. G. Pease.

The 100-inch telescope was finally surpassed in size in 1948, when the 200-inch Hale Telescope went into operation. By that time industrial and residential expansion of Los Angeles had created both atmospheric pollution and a bright nighttime sky at Mount Wilson, and the 100-inch telescope no longer enjoyed such a fine observing site. Today the Hooker Telescope is still in operation, but it is generally used for observations of relatively bright stars. The addition of sophisticated electronic detectors has helped to preserve the instrument's ability to analyze the light of faint objects.

distances between pairs have tripled over what they were to start with, and the speeds needed to accomplish this are again directly proportional to the distance between a given pair. The farther away a raisin is to begin with, the farther it must move in order to maintain the regular spacing during the expansion, and therefore the greater its velocity. In the same way, the galaxies in the universe must increase their separations from each other at a rate proportional to the distance between them (Fig. 21.3), or they would become bunched up. By observing the velocities of galaxies, astronomers are keeping watch on the raisins in order to see how the loaf of bread is coming along.

It is important to realize that it does not matter which raisin we choose to watch; from any point in the loaf, all other raisins appear to move away with speeds proportional to their distances from that point. Thus we do not conlude that our galaxy is at the center of the universe; from *any* galaxy, it would appear that all others are receding.

Following Hubble, a number of other astronomers extended the study of galaxy motions to greater and greater distances, with the same result: as far as the telescopes can probe, the galaxies are moving away from each other at speeds proportional to their distances. Expansion is a major feature of the universe we live in; one that must be taken into account as we seek to understand its origins and its fate. The data showing universal expansion can be displayed on a plot of velocity versus distance (as was done by Hubble; see Fig. 21.4), or on a plot of redshift versus apparent magnitude (Fig. 21.5), since increasing apparent magnitude (that is, decreasing brightness) indicates increasing distance.

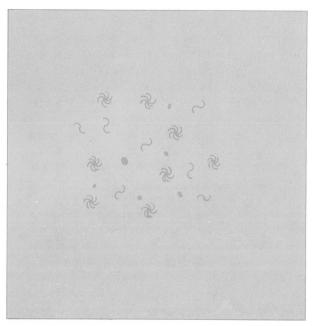

FIGURE 21.2. *The Expanding Universe.* This shows a number of galaxies at one time (left), and again at a later time (right). All the spacings between galaxies have increased because of the expansion of the universe.

The galaxies within a cluster have their own orbital motions, which are separate from the overall expansion (Fig. 21.6). As the universe expands, the clusters move apart from each other, while the individual galaxies within a cluster do not. The motions of galaxies within a cluster are relatively unimportant for very distant galaxies that are moving away from us at speeds of thou-

FIGURE 21.4. *The Hubble Law.*
This early diagram prepared by E. Hubble and M. Humason shows the relationship between galaxy velocity and distance. Over the years the slope of the relation has been altered, with the addition of data for more galaxies and with improvements in the measurement of distances, but the general appearance of this figure is the same today.

FIGURE 21.3. *Velocity of Expansion as a Function of Distance.* Above is a row of galaxies, and below is the same row one billion years later, when the distances between adjacent galaxies have doubled from 1 to 2 Mpc. From the viewpoint of an observer in *any* galaxy in the row, the recession velocity of its nearest neighbor is 1 Mpc/billion years; of its next nearest neighbor, 2 Mpc/billion years; of the next, 3 Mpc/billion years; and so on. The velocity is proportional to the distance in all cases, for any pair of galaxies.

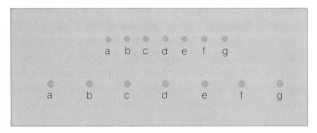

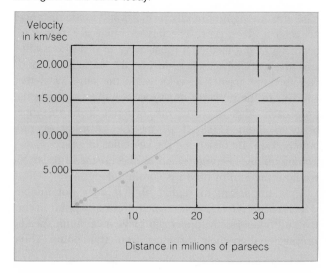

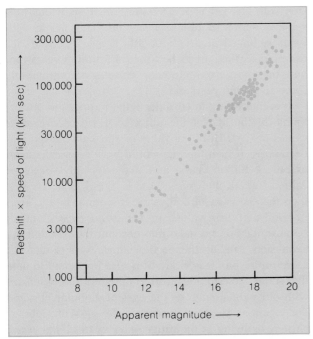

FIGURE 21.5. *A Modern Version of the Hubble Law.*
Often apparent magnitude instead of distance is used on the
horizontal axis since the two quantities are related, particularly if
the diagram is limited to galaxies of the same type, as this one
is. At lower left the small rectangle indicates the extent of the
relationship as Hubble first discovered it; today many more
galaxies, much dimmer, have been included.

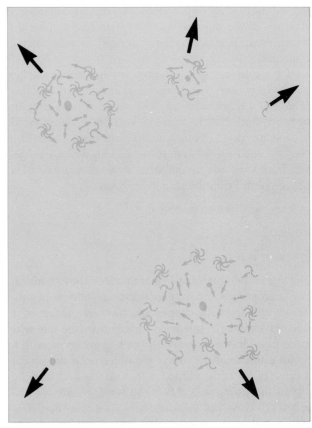

FIGURE 21.6. *Local Motions.*
Although there is a systematic overall expansion of the universe,
individual galaxies within clusters and even clusters within
superclusters have random individual motions. Hence within the
Local Group the galaxies are not uniformly receding from each
other.

sands of km/sec, but for nearby ones, the orbital motion can rival or exceed the motion caused by the
expansion of the universe. The Andromeda Nebula, for
example, is actually moving *toward* the Milky Way at
a speed of about 100 km/sec, whereas at its distance of
700 kpc from us, the universal expansion should give
it a velocity away from us of about 40 km/sec.

Hubble's Constant and the Age of the Universe

The relation discovered by Hubble can be written in
simple mathematical form:

$$v = Hd,$$

where v is a galaxy's velocity of recession and d is the
galaxy's distance in megaparsecs. The Hubble constant, H, is given in units of km/sec/Mpc.

The value of H is difficult to establish. It is obtained
by collecting as large a body of data as possible on
galactic velocities and distances, and then deducing the
value of H that best represents the relationship between
distance and velocity. Hubble did this first, finding
$H = 500$ km/sec/Mpc, which means that for every megaparsec of distance, the velocity increases by 500 km/
sec. The standard candles were not very well established in Hubble's day, when it was still news that nebulae were distant galaxies, and his value of H turned
out to be an overestimate. The best modern values for
H are between 50 and 100 km/sec/Mpc. Until very recently, a consensus had been developing that the best
value is 55 km/sec/Mpc. A new distance determination
for a number of galaxies has, however, led to a value
close to 90 km/sec/Mpc. The precise value of H has ex-

tremely important implications in terms of our understanding of the universe and its expansion, and an intense research effort is being devoted to refining the estimates.

If the universe is expanding, it follows that all the matter in it used to be closer together than it is today (Fig. 21.7). If we carry this logic to its obvious conclusion, we find that the universe was once concentrated in a single point, from which it has been expanding ever since. From the rate of expansion, we can calculate how long ago the galaxies were all together in a single point. From the simple expression

Time = distance/velocity,

we find

age = d/Hd = 1/H.

The age of the universe is equal to 1/H, if the expansion has been proceeding at a constant rate since it began. As we will learn in chapter 23, this assumption of a constant expansion rate is not strictly true (the rate was more rapid in the beginning), but it provides us with a useful estimate of the age of the universe. If we accept a value of 55 km/sec/Mpc for H, then we find that the age of the universe is 5.6×10^{17} sec = 1.8×10^{10} years. (To carry out this calculation, we must first convert the value of a megaparsec into kilometers.)

A simple calculation based on the observed expansion rate has led to a profound conclusion: our universe is about 18 billion years old. This fits in with what we have learned about galactic ages. The sun, for example, is thought to be about 4.5 billion years old, and the most aged globular clusters are apparently some 16 billion years old.

It is interesting to consider what happens to our estimate of the age of the universe if we choose other values of H. If Hubble's value of 500 km/sec/Mpc had been correct, then the age would have been calculated as only 2 billion years, and if the recently suggested value of about 90 km/sec/Mpc is found to be correct, then the age is 11 billion years. It is important to keep in mind that these values are somewhat overestimated, because they do not take into account the fact that the expansion rate at the earliest times was somewhat more rapid than it is today. It is also important to note that some age estimates lead to apparent discrepancies with other data, such as the ages of globular clusters. For example, we have discussed the ages of globular clusters as being in the range of 14 to 16 billion years, and now we find that a recent estimate of the value of H leads to an age of the universe that is less than 14 billion years. Clearly something is wrong with either the estimated ages of globular cluster or this value of H. Some astronomers have used this discrepancy to argue that H cannot be as large as 90 km/sec/kpc. Whatever the correct ages are, this situation serves to illustrate how uncertain some of the most important measurements in astronomy can be.

There are many other important questions about the universe that depend on the value of H, which is why so much effort is being expended to define it as accurately as possible. One of the principal reasons for development of the *Space Telescope* is to probe the most distant galaxies, to determine their distances and velocities so that the information can be used to help us find the correct value of H.

FIGURE 21.7. *Backtracking the Expansion.*
If the universe began as a single point, then the present rate of expansion provides information on how long ago the expansion began. Although there is considerable uncertainty in calculations, owing to the fact that the expansion rate has probably not been constant, astronomers arrive at an estimated age for the universe of between 15 and 20 billion years.

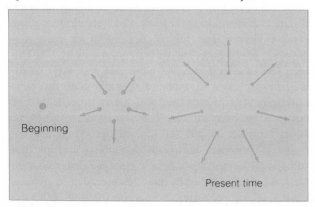

Redshifts as Yardsticks

The expansion of the universe has a very practical advantage for astronomers concerned with the properties of distant galaxies: it provides a means of determining the distances to these galaxies (Fig. 21.8). If we know the value of H, we can find the distance to a galaxy

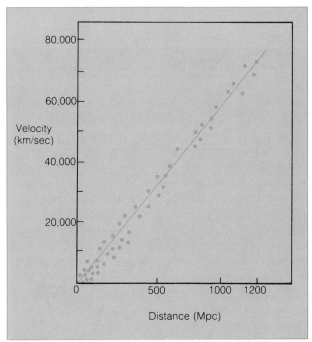

FIGURE 21.8. *Finding Distances from the Hubble Law.*
This diagram shows how the distance to a galaxy can be
deduced directly from the Hubble law, if the redshift, hence the
velocity, is measured.

simply by measuring its velocity as indicated by the
Doppler shift of its spectral lines. The distance is
given by

$$d = v/H.$$

Thus, if we adopt $H = 55$ km/sec/Mpc, then a galaxy
found to be receding at 5,500 km/sec, for example, has
the distance

$$d = 5,500/55 = 100 \text{ Mpc}.$$

This technique can be applied to the most distant
and faint galaxies, where it is nearly impossible to observe any standard candles. It is important to remember, however, that the use of Hubble's relation to find
distances is only as accurate as our value of H, and the
value that we use is derived in turn from distance determinations that depend on standard candles. Furthermore, because the expansion rate of the universe has
not been constant, for very distant galaxies we need to
know what the rate was in the past, in order to accurately determine the distance.

A Cosmic Artifact: The Microwave Background

Since Hubble's discovery of universal expansion, many
astronomers and physicists have explored its significance, both for what it may tell us about the future and
for the information it provides about the past. A number of scientists, beginning in the 1940s with George
Gamow and his associates, have carried out extensive
studies of the nature of the early universe. It was realized that if all the matter were originally compressed
into a small volume, it must have been very hot, for the
same reason that a gas heats up when compressed. The
fiery conditions inferred for the early stages of universal
expansion have led to the term **big bang**, which is the
commonly accepted title for all theories which hold
that the origin of the universe began with an expansion
from a single point.

Gamow's primary interest was in analyzing nuclear
reactions and element production during the big bang
(the modern understanding of this is described in
chapter 23), but in the course of his work he came to
another important realization: the universe must have
been filled with radiation when it was highly compressed and very hot, and this radiation should still be
with us. At first, when the temperature was in the billions of degrees, γ-ray radiation dominated, but later,
as the universe cooled, X rays and then ultraviolet light
filled the universe. During the first million years or so
after the expansion started, the radiation was constantly being absorbed and reemitted (that is, scattered) by the matter in the universe, but eventually
(when the temperature had dropped to around 4,000 K),
protons and electrons combined to form hydrogen atoms, and from that time on there was little interaction
between the matter and the radiation. The matter, in
due course, organized itself into galaxies and stars,
whereas the radiation has simply continued to cool
as the universe has continued to expand (Fig. 21.9).
The intensity and spectrum of the radiation are dependent only on the temperature of the universe, and
are described by the laws of thermal radiation (see
chapter 3).

Gamow and others made rough calculations which
showed that the temperature of the radiation filling the
universe is very low indeed, about 5 K. More recently,
R. H. Dicke and colleagues independently carried out
calculations yielding similar results. Using Wien's law,

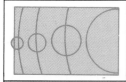

Astronomical Insight 21.2

THE DISTANCE PYRAMID

Having reached the ultimate distance scales, we can pause to look back at the progression of steps that got us here. As noted in the text, the use of Hubble's relation to estimate distances to the farthest galaxies depends critically on the value of H, which can be determined only from knowledge of the distances to a representative sample of galaxies. Those distances in turn depend on various standard candles, themselves calibrated through the use of other methods, such as the Cepheid variable period-luminosity relation, which are applicable only to the most nearby galaxies. Our knowledge of these close extragalactic distances rests on the known distances to objects within the Milky Way, such as star clusters containing variable stars, which are used to calibrate the period-luminosity relation. The distances to these clusters, in their turn, are known from more local techniques, such as spectroscopic parallaxes and main-sequence fitting, and these ultimately depend on the distances to the most nearby stars, derived from trigonometric parallaxes.

Thus the entire progression of distance scales, all the way out to the known limits of the universe, depends on our knowledge of the distance from the earth to the sun, the basis of trigonometric-parallax measurements. At every step of the way outward,

our ability to measure distances rests on the previous step. If we revise our measurement of local distances, we must accordingly alter all our estimates of the larger scales, which will affect our perception of the universe as a whole. This elaborate and complex interdependence of distance-determination methods is known as the **distance pyramid,** and it is sometimes depicted in graphic form, as presented here.

A very important step in the distance pyramid is represented by the Hyades cluster, a galactic star cluster some 43 parsecs away. The distance to this cluster is determined from the **moving-cluster method,** in which a combination of Doppler-shift measurements and stellar motion measurements reveals the true direction of cluster motion. A comparison of angular motion and true velocity then tells us the distance to the cluster. Because much of the rest of the distance pyramid depends on the application of this technique to the Hyades cluster, great care has been taken in the analysis, and every so often it is redone. Whenever the distance to the Hyades cluster is altered slightly, the impact spreads as astronomers judge how this affects other distance scales in the universe.

Distance	Key Object	Method
10^{-6} pc	Sun	↓ Radar, planetary motions
10^{-5}		
10^{-4}		
10^{-3}		| Trigonometric parallax
10^{-2}		
10^{-1}		
1	α Centauri	↓
10^{1}		↓ Moving-cluster method
10^{2}	Hyades cluster	
10^{3}		
10^{4}	Limits of Milky Way	↓ Main-sequence fitting
10^{5}	Magellanic Clouds	
10^{6}	Andromeda galaxy	| Cepheid variables
10^{7}		↓ ↓ Brightest stars
10^{8}	Virgo cluster	
10^{9}		
10^{10}		↓ Brightest galaxies

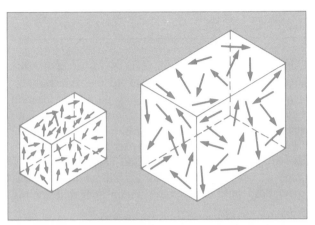

FIGURE 21.9. *The Cooling of the Universal Radiation.*
At left is a box representing the early universe, filled with intense gamma-ray radiation. At right is the same box, greatly expanded. The radiation is still there, but its wavelength has increased, and the temperature represented by its thermal spectrum has decreased.

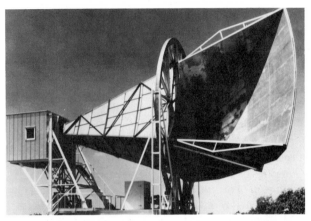

FIGURE 21.10. *The Radio Receiver that Discovered the 3° Background Radiation.*
This is the antenna at Bell Laboratories with which Penzias and Wilson found a persistent noise that turned out to be the remnant radiation from the big bang.

we can easily calculate the most intense wavelength of such radiation: if the temperature is 5 K, then the value of λ_{max} is 0.06 cm, in the microwave part of the radio spectrum.

Gamow's principal interest was in element production during the big bang, rather than in the remnant radiation. Furthermore, in the late 1940s the technology necessary to detect the radiation was years away from being ready. Hence no attempt was made in Gamow's time to search for it, and the radiation was forgotten until Dicke's work. By the mid-1960s, the technology was available, and Dicke and his colleagues set out to build a radio telescope capable of detecting radiation at the appropriate wavelength.

Because the expected signal would be very weak, and because the earth's atmosphere blocks out much of the microwave radiation that enters from space, the development of such a device was difficult. In 1965, while Dicke's group was still at work on the task, a pair of radio astronomers named Arno Penzias and Robert Wilson, working at Bell Laboratories (only a few miles away from Dicke's group at Princeton University), were puzzled by a persistent static in a new radio receiver (Fig. 21.10). Try as they might, they could not eliminate this noise, and they were forced to conclude that it was of cosmic origin.

The implication soon became clear: Penzias and Wilson had discovered the universal radiation that the theorists had predicted should exist. This was one of the most significant astronomical observations ever

made, for it provided a direct link to the origins of the universe. More than a piece of circumstantial evidence, this was a real artifact, the kind of hard evidence that carries weight in a court of law.

The Crucial Question of the Spectrum

There remained questions about the interpretation of the radiation detected by Penzias and Wilson, however, and doubt in the minds of some who did not accept the big-bang theory of the origin of the universe. If the radiation is really the remnant of a primeval fireball, then it is expected to fulfill certain conditions, and intensive efforts were made to see whether it does.

One important prediction is that the spectrum should be that of a simple glowing object whose emission of radiation is caused only by its temperature (that is, should be a thermal spectrum, as explained in chapter 3). If, on the other hand, the source of radiation were something other than the remnant of the initial universe, such as distant galaxies or intergalactic gas, it would have a different spectrum. Thus the shape of the spectrum, its intensity as a function of wavelength, was a very important factor in assessing the origin of the radiation.

Some of the early results were confusing, owing to unforeseen complication, but the situation seems to have been resolved. The spectrum does indeed follow the shape expected (Fig. 21.11), and the radiation is considered to be a relic of the big bang. The most difficult part of the task of measuring the spectrum was observing the intensity at short wavelengths, around 1 millimeter or less, where the earth's atmosphere is especially impenetrable. but this was finally accomplished with high-altitude balloon payloads.

The result of all the effort expended in determing the spectrum is that the temperature of the radiation is 2.7 K, and its peak emission occurs at a wavelength of 1.1 millimeter. The radiation is commonly referred to as the **microwave background,** or the **3° background radiation.**

The care that went into establishing the true nature of the background radiation reflects the importance of what it has to tell us about the universe. As we will learn in chapter 23, there are alternatives to the big-bang theory, and those who support these other theories have sought different explanations for the microwave radiation. Its close adherence to the spectrum expected of the radiation from the cosmic fireball envisioned in the big-bang theory has presented grave difficulties for these alternative points of view.

FIGURE 21.11. *The Measured Background Spectrum.*
This shows the spectrum of the cosmic background radiation, plotted as a function of frequency (a wavelength scale is given at the top as well). The solid line represents a thermal spectrum for a temperature of just under 3 K.

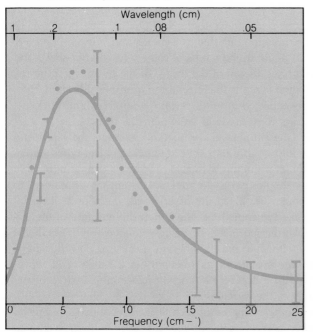

Isotropy and Daily Variations

If the background radiation is a remnant of the big bang, then it is also expected to fill the universe uniformly, with no preference for any particular location or direction. A medium or radiation field that has no preferred orientation, but instead looks the same in all directions, is said to be **isotropic**. Some of the early observations of the 3° background radiation were designed to test its isotropy, for its failure to meet this expectation could imply some origin other than the big bang.

A test of isotropy, in simple terms, is a measurement of the radiation's intensity in different directions, to see whether it varies. From the time it was discovered, the microwave background showed a high degree of isotropy, as expected. The issue has been pressed, however, for a variety of reasons. One alternative explanation of the radiation, for example, is that it arises from a vast number of individual objects, such as very distant galaxies, so closely spaced in the sky that their combined emission appears to come uniformly from all over. To test this idea, a group of astronomers have attempted to see whether the radiation is patchy on very fine scales, as it would be if it came from a number of point sources. So far no evidence of any clumpiness has been found, and the big-bang origin for the radiation has not been threatened.

There is a kind of subtle nonuniformity that is expected of the microwave radiation, even if it is the remnant of the big bang. Because of the Doppler shift, the intensity of the radiation as seen by a moving observer will vary with direction, depending on whether the observer is looking toward or away from the direction of motion. If the observer looks ahead, toward the direction of motion, then a blueshift occurs, the peak of the spectrum being shifted slightly toward shorter wavelengths (Fig. 21.12). In terms of temperature, this means that the radiation looks a little hotter when viewed in this direction. If the observer looks the other way, there is a slight redshift, and the measured temperature is cooler.

The earth is moving in its orbit about the sun; the sun is orbiting the galaxy; and the galaxy is moving along its own path through the Local Group. All of these motions combined represent a velocity of the earth with respect to the frame of reference established by the background radiation, a velocity that should produce slight differences in the radiation temperature

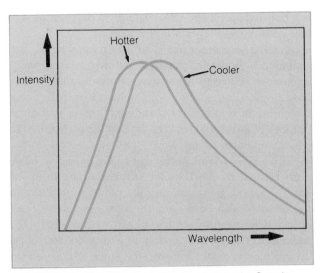

FIGURE 21.12. *The Effect of a Doppler Shift on the Cosmic Background Radiation.*
If the observer on the earth has a velocity with respect to the radiation, this will shift the peak of the spectrum a little, and affect the temperature that the observer deduces. Motion of the earth results in a daily cycle of tiny fluctuations in the observed temperature, known as the 24-hour anisotropy.

FIGURE 21.13. *The Cosmic Background Explorer.*
This satellite, to be launched in the late 1980s, will make the most precise and extensive measurements yet of the microwave background radiation.

when it is viewed from different directions. Because of the earth's rotation, if we simply point our radio telescope straight up, we should alternately see high and low temperatures, as our telescope points toward and then twelve hours later, away from, the direction of motion. This means that there should be a daily variation cycle in the radiation, referred to as the **24-hour anisotropy** (anisotropic is the term for a nonuniform medium).

The actual change in temperature from one direction to the other is very small, and was not successfully measured until very recently. The observed temperature difference from the forward to the rearward direction is only a few thousandths of a degree, and very sophisticated technology was required in order for this to be measured. The data show that the earth is moving with respect to the background radiation at a speed of about 350 km/sec. When the sun's known orbital velocity about the galaxy is taken into account, this implies that the galaxy itself is moving at 520 km/sec with respect to the background radiation. This motion must be a combination of the motion of the galaxy in its orbit within the Local Group, and possibly the motion of the Local Group within the supercluster to which it belongs.

The use of the 3° background radiation as a tool for

deducing the local motions with respect to the radiation is predicated on the assumption that the radiation is fundamentally isotropic. At present we have no reason to doubt this is true. The possibility remains, however, that the radiation is not isotropic; that the universe is somehow lopsided. If it is, we will have much difficulty measuring its asymmetry, because of the complex motions of the earth, our observing platform.

Astronomers are planning future observations to improve on their measurements of the 3° background radiation. A satellite called the *Cosmic Background Explorer* (Fig. 21.13) is due to be launched in the late 1980s, and will provide, from its vantage point above the earth's atmosphere, the best data yet on the spectrum and isotropy of the background radiation.

PERSPECTIVE

We have learned now that the universe, like all the subordinate objects within it, is a dynamic, evolving entity. One of the grandest stories in astronomy has been the unfolding of the concept of universal expansion. An idea forced on astronomers by the evidence of the redshifts, it has led directly to the big-bang picture, which in turn tells us the age of the universe, explains the origin of some of the elements, and is verified through the presence of the microwave background radiation.

Before we finish the story, by studying modern theories of the universe and their implications for its future, we must take a closer look at some of the objects that inhabit the universe. There is a breed of strange and bizarre entities whose very nature seems to be linked to the evolution of the universe, and which therefore will provide us with a few additional bits of evidence for our final discussion.

SUMMARY

1. It was discovered early in this century that galaxies tend to have redshifted spectra, which implies that other galaxies are moving away from us.

2. Hubble discovered in 1929 that the velocity of recession is proportional to distance, demonstrating that the universe is expanding.

3. The rate of universal expansion, expressed in terms of Hubble's constant H, is thought to be between 50 and 100 km/sec/Mpc.

4. The fact that the universe is expanding implies that it originated from a single point, and the rate of expansion, if constant, tells us that it began some 10 to 20 billion years ago. Thus the universe may be 10 to 20 billion years old.

5. The relationship between velocity and distance allows the distance to a galaxy to be estimated from its velocity, through the Doppler effect. This method of determining distance extends to the farthest limits of the observable universe.

6. The fact that the universe originated from a single point implies that it was very hot and dense initially, and this in turn implies that the early universe was filled with radiation. As the expansion has proceeded, this radiation has been transformed into microwave radiation, with a thermal spectrum corresponding to a temperature of about 3 K. The cosmic background radiation was discovered in 1965.

7. To distinguish a primordial origin for the background radiation from other possible origins (such as numerous galaxies spread throughout the universe), we must measure the spectrum to see whether it truly is a thermal spectrum. Observations to date are consistent with the assumption that it is.

8. Another important test is to determine whether the radiation is isotropic, or whether it might be unevenly intense in different directions. Careful observations have revealed no evidence of patchiness or unevenness which might imply that the radiation arises in a large number of individual sources, or that the early universe was nonuniform in any way.

9. There is a 24-hour anisotropy in the radiation, however, created by the Doppler effect resulting from the earth's motion. By comparing the radiation intensity in the forward and backward directions, we have been able to determine the earth's velocity with respect to the radiation. The earth's motion is a combination of its orbital motion about the sun, the sun's orbital motion about the galaxy, and the galaxy's motion within the Local Group.

REVIEW QUESTIONS

1. From the information on the Doppler shift in chapter 3, determine the speed of recession of a galaxy whose ionized calcium line (assume the rest wavelength is 3933 Å) is observed at a wavelength of 3936.93 Å.

2. Suppose three galaxies are situated initially so that the distance from galaxy A to galaxy C is twice the distance from A to B. After a period of time, the distance between galaxies A and B has doubled. How does the distance between A and C now compare with that between A and B?

3. To help understand how the age of the universe is determined from the rate of expansion, determine the age of a supernova remnant whose outermost portions are expanding away from the center at a rate of 1,000 km/sec, if its radius is 10 parsecs. (You will have to convert parsecs to kilometers to do this. Express your answer in years.)

4. If the past expansion of the universe was more rapid than the present expansion, the correct age of the universe is less than the value calculated by assuming that H has been constant at the present value. Suppose the correct present value of H is 55 km/sec/Mpc, but

the average value over the entire history of the universe has been 70 km/sec/Mpc. Using this value, calculate the age of the universe.

5. Using information from chapter 20, summarize the kinds of observations needed in order to determine the value of H, the Hubble constant.

6. Suppose the position of the ionized calcium line is observed in the spectra of three galaxies. The rest wavelength of this line is 3,936 Å, and in the three galaxies the line is observed at wavelengths of 3,936 Å, 4,028 Å, and 3,942 Å. Rank the three galaxies in order of increasing distance. (As an additional exercise, you may want to calculate the distances to the galaxies, but you need not do that to simply rank them.)

7. If the Hubble constant is $H = 55$ km/sec/Mpc, calculate the distance to a galaxy whose ionized calcium line is observed at a wavelength of 4,028 Å, instead of the rest wavelength of 3,933 Å.

8. What was the wavelength of maximum emission for the universal radiation field when hydrogen atoms formed from protons and electrons, when the temperature was 4,000 K?

9. Explain why it is necessary, when attempting to observe the cosmic background radiation, to cool the telescope and instruments to a low temperature.

10. Discuss the possible ways in which the cosmic background radiation could be anisotropic (apart from the 24-hour anisotropy), and the implications if it is.

ADDITIONAL READINGS

Abell, G. 1978. Cosmology—the origin and evolution of the universe. *Mercury* 7(3):45.

Barrow, J. D., and Silk, J. I. 1980. The structure of the early universe. *Scientific American* 242(2):118.

Layzer, D. 1975. The arrow of time. *Scientific American* 233(6):56.

Muller, R. A. 1978. The cosmic background radiation and the new aether drift. *Scientific American* 238(5):64.

Shu, F. H. 1982. *The physical universe.* San Francisco: W. H. Freeman.

Silk, J. 1980. *The big bang: the creation and evolution of the universe.* San Francisco: W. H. Freeman.

Weinberg, S. 1977. *The first three minutes.* New York: Basic Books.

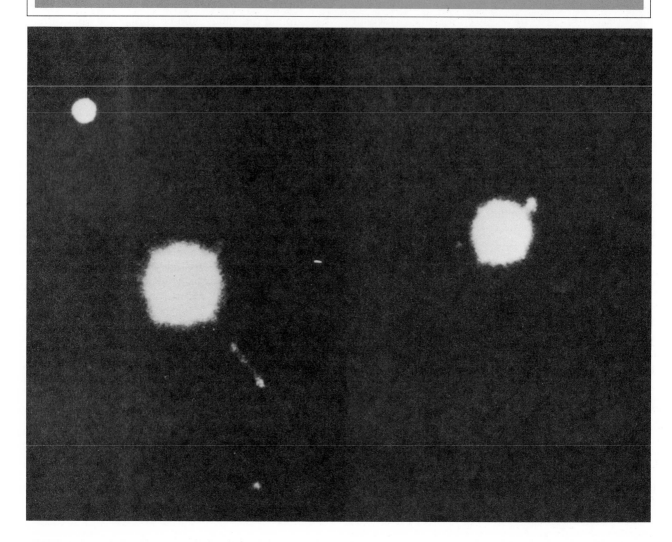

Learning Goals

22

In discussing galaxies in the preceding chapters, we have overlooked a variety of objects, some of them galaxies and some possibly not, that have unusual traits. As in many other situations in astronomy, the so-called peculiar objects, once understood, will have quite a bit to tell us about the more normal ones.

So that we could fully appreciate the bizarre nature of some of these astronomical oddities, it was best to delay their introduction until this point, where we have essentially completed our survey of the universe. We know about the expansion and the big bang, and we are just in the process of tying it all together. In this chapter we will uncover a number of vital clues in the cosmic puzzle.

The Radio Galaxies

Most of the first astronomical sources of radio emission to be discovered, other than the sun, were galaxies. The majority of these galaxies, when examined opti-

FIGURE 22.1 *Centaurus A.*
This is a giant elliptical radio galaxy, showing a dense lane of interstellar gas and dust across the central region. Many radio galaxies are giant ellipticals with peculiarities in visual appearance.

cally, turned out to be large ellipticals, often with some unusual-appearing structure (Fig. 22.1). The first of these objects to be detected, and one of the brightest, is called Cygnus A (its name is based on a preliminary cataloguing system in which the ranking radio sources in constellations are listed alphabetically). This galaxy has a strange double appearance, and astronomers eventually found, after sufficient refinement of radio-observing techniques (see the discussion of radio inter-ferometry in chapter 3), that the radio emission comes from two locations on opposite sides of the visible galaxy and well separated from it (Figs. 22.2 and 22.3).

This double-lobed structure is a common feature of **radio galaxies,** those with unusually strong radio emission. Most, if not all, ordinary galaxies emit radio radiation, but the term *radio galaxy* is applied only to the special cases where the radio intensity is many times greater than the norm. The core of the Milky Way is a strong radio source as viewed from the earth, but is not in a league with the true radio galaxies.

Although the double-lobed structure is standard among elliptical radio galaxies, the visual appearance varies quite a bit. We have already noted the double appearance of Cygnus A. Another bright source, the giant elliptical Centaurus A (Fig. 22.1), appears to have a dense band of interstellar matter bisecting it, and it has not one, but two pairs of radio lobes, one much farther out from the visible galaxy than the other. Other

FIGURE 22.2 *A Radio Image of a Radio Galaxy.*
Here the lighter areas represent maximum radio brightness. The galaxy as seen in visible wavelengths does not even appear here; this image is completely dominated by the double side lobes.

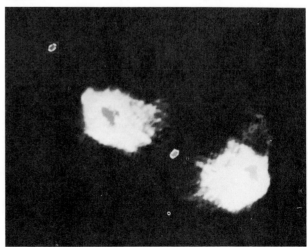

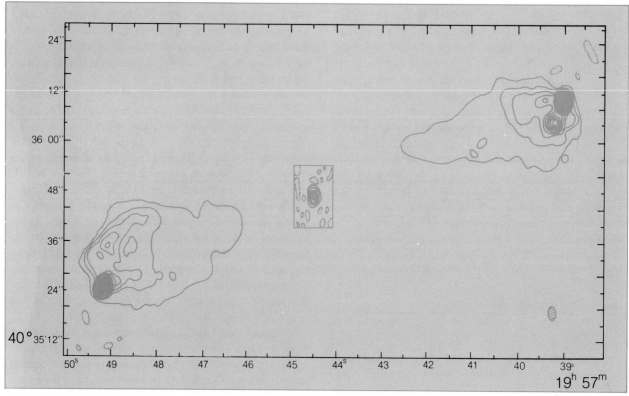

FIGURE 22.3 *A Map of a Radio Galaxy.* The intensity contours here represent radio emission from the two side lobes of a radio galaxy, Cygnus A. The blurred image drawn in at the center illustrates the size of the visible galaxy compared to gigantic side lobes.

giant elliptical radio galaxies have other kinds of strange appearances. One of the most famous is M87, also known as Virgo A, which has a jet protruding from one side that is aligned with one of the radio lobes (Fig. 22.4). Careful examination shows that this jet actually consists of a series of blobs that appear to have been ejected in sequence from the core of the galaxy.

The size scale of the radio galaxies can be enormous. In some cases the radio lobes or jets extend as far as a million parsecs from the central galaxy. Recall that the diameter of a large galaxy is only one-tenth of this distance, and that the Andromeda galaxy is less than a million parsecs from our position in the Milky Way.

The first radio galaxies were detected in the 1940s, and by the 1950s studies of the radio spectra of these objects were in progress. It was discovered that they are emitting by the synchrotron process (mentioned in chapter 7), which requires a strong magnetic field and

FIGURE 22.4 *A Giant Elliptical Radio Galaxy with a Jet.* This is M87, also known as Virgo A, showing its remarkable linear jet of hot gas. This is a short-exposure photograph, specially processed to maximize visibility of the jet and its lumpy structure. On a longer-exposure photograph, this galaxy looks like a normal elliptical, the light from the jet being drowned out by the intensity of light from the galaxy.

a supply of rapidly moving electrons. The electrons are forced to follow spiral paths around the magnetic-field lines, and as they do so, they emit radiation over a broad range of wavelengths. The characteristic signature of synchrotron radiation, in contrast with the thermal radiation from hot objects such as stars, is a sloped spectrum with no strong peak at any particular wavelength. The radiation from a synchrotron source is also polarized. Whenever an object is found to be emitting by the synchrotron process, astronomers immediately conlude that some highly energetic activity is taking place, producing the fast-moving electrons that are required to create the emission.

In some cases, it appears that a different process is responsible for the radio emission, but one that also requires a supply of rapidly moving electrons. If there is ordinary thermal radio emission from a galaxy, then the presence of electrons moving near the speed of light can alter the spectrum of the radio emission, as energy is transferred from the electrons to the radiation. This process is called **inverse Compton scattering,** and it has the same implication as synchrotron radiation: there must be a source of vast amounts of energy, in order to produce the rapidly moving electrons that are necessary to create the radiation.

The source of this energy in radio galaxies is a mystery, although some interesting speculation has been fueled by certain characteristics of the galaxies. The double-lobed structure indicates that two clouds of fast-moving electrons are located on either side of the visible galaxy, and this leads us naturally to the impression that the clouds of hot gas may have been ejected from the galactic core, perhaps by an explosive event that expelled material symmetrically in opposite directions. This impression is heightened by the many cases where jets are seen protruding from the core (Figs. 22.5, 22.6, and 22.7), and by instances where more than one pair of lobes, aligned in the same direction but with one farther out than the other, are seen.

One method envisioned by astronomers for producing prodigious amounts of energy from a small volume such as the core of a galaxy is to have a giant black hole there, surrounded by an accretion disk, and this has naturally been a suggested source for the energy in the radio galaxies. The general idea is that superheated

FIGURE 22.5 *A Radio Jet and Inner Lobes in Centaurus A.*
This is a radio image of the galaxy shown in Fig. 22.1, in the same orientation. It shows a pair of radio lobes close to the galaxy (there is more structure much farther out as well), and a section of a jet that is aligned with the lobes.

FIGURE 22.6 *Dual Radio Jets.*
While visible jets are rare, radio images often reveal jet-like structures. Here is a radio image of a galaxy with a pair of jets emanating from opposite sides, aligned with the normal double radio-emitting lobes. It appears that jets such as these are responsible for the formation of the radio lobes.

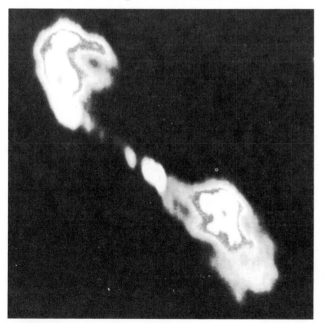

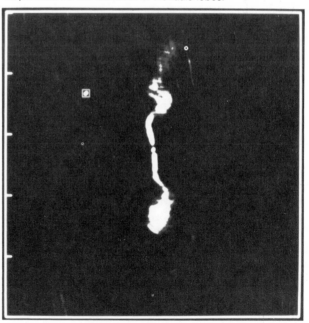

 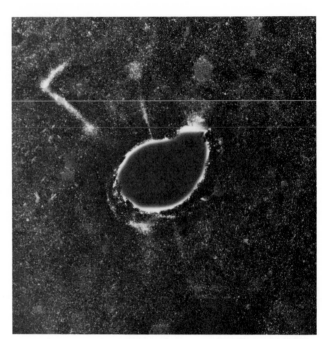

FIGURE 22.7 *Jets in an Unusual Galaxy*. At left is a normal barred spiral galaxy, a type of object not normally found to have jets from its center. At right, however, is a computer-enhanced image of an object called 3C120, which may be the nucleus of a spiral galaxy, and which clearly has two linear jets extending through it's center.

gas that has been compressed by the immense gravitational field of the black hole escapes along the rotation axes, and is channeled into two opposing jets. It is probably better to view this process not as explosive, but as a more or less steady expulsion of gas. It has been suggested that *all* radio galaxies have jets; that this is the only mechanism for forming the double radio lobes. The fact that the jets are not *observed* in all radio galaxies is thought to be an effect of how the jets are aligned with respect to our line of sight. The emission from rapidly moving particles is confined to a narrow angle in the forward direction; therefore, an observer to the side of a jet will see little of its radiation.

The theory that massive black holes are the energy sources in radio galaxies has received some observational support. Recently it was reported that evidence for such a central, massive object has been found in the heart of M87, the radio galaxy with the well-known visible jet. Measurements of the velocity dispersion near the center of the galaxy led astronomers to deduce that a large amount of mass must be confined in a very small volume at the core, and a black hole seemed the best explanation. Subsequent observations, however, have failed to confirm the reportedly high velocities of

stars in the inner portions of the galaxy, and the case for a black hole in M87 has been weakened. Further efforts to find direct evidence of supermassive black holes are being made.

Seyfert Galaxies and Explosive Nuclei

Although giant ellipticals are the strongest radio emitters among galaxies, they are by no means the only ones with evidence of energetic phenomena in their cores. Some spiral galaxies also display such behavior. We learned in chapter 17 that even the Milky Way is not immune; there is evidence of a compact, massive object existing at its center. There are other spiral galaxies with much more pronounced violence in their nuclei. These galaxies as a class are called **Seyfert galaxies** (Fig. 22.8), after the astronomer who discovered and catalogued many of them in the 1930s.

Seyfert galaxies have the appearance of ordinary spirals, except that their nuclei are unusually bright and

FIGURE 22.8 *A Seyfert Galaxy.*
This photo of NGC 4151, a well-studied example, shows the enormous intensity of light from the nucleus compared to the rest of the galaxy.

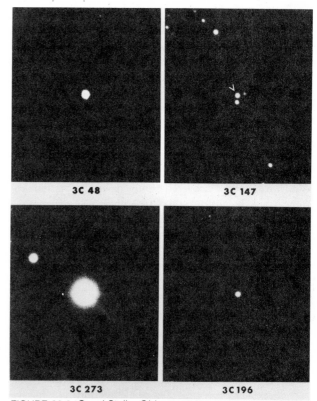

FIGURE 22.9 *Quasi-Stellar Objects.*
These star-like objects are quasars, whose spectra are quite unlike those of normal stars. As explained in the text, these objects are probably more distant, and therefore more luminous, than any normal galaxies.

blue in color, in contrast with the red color of most normal spiral-galaxy nuclei. About 10 percent of them are radio emitters, with spectra indicating that the synchrotron process is at work. The radio emission usually comes from the nucleus, rather than from double lobes. The spectrum of the visible light from a Seyfert nucleus typically shows emission lines, something completely out of character for normal galaxies. The emission lines, formed in an ionized gas, are sometimes very broad, indicating velocities of several thousand kilometers per second in the gas that produces the emission. Evidently the cores of these galaxies are in extreme turmoil, with hot gas swirling about and tremendous amounts of energy being generated.

Few spirals show such effects, and it is not clear whether the Seyfert behavior is a phase they all pass through at some time, or whether only a few act up in this manner. Later in this chapter we will discuss evidence supporting the first of these possibilities.

The Discovery of Quasars

In 1960, spectra were obtained of two starlike, bluish-colored objects that had been found to be sources of radio emission (Fig. 22.9). No radio stars were known, and astronomers became very interested in the two ob-

jects, called 3C48 and 3C273 (their designations in a catalogue of radio sources that had recently been compiled by the radio observatory of Cambridge University, in England).

The spectra of the two objects had a completely unprecedented appearance: they contained several strong emission lines at totally unrecognizable wavelengths. The lines defied identification, and the objects became known as **quasars** (short for quasi-stellar radio sources), or **quasi-stellar objects** (sometimes simply **QSOs**) because of their starlike appearance but nonstellar spectra.

The mysterious spectra of 3C48 and 3C273 continued to confound astronomers until finally, in 1963, Maarten Schmidt of the Hale Observatories found the answer to the puzzle. Schmidt realized that the spacing of some of the most prominent emission lines in 3C273 coincided perfectly with the separations of the bright lines of hydrogen, except that the entire pattern was displaced by hundreds of angstroms. The hydrogen line whose rest wavelength is 4,861 Å, for example, was found at a wavelength of 5,639 Å, and the line that is normally located at 4,340 Å was shifted to 5,034 Å. This indicated a redshift of 16 percent; that is, the object emitting these lines must be moving away from the earth at 16 percent of the speed of light, or 48,000 km/sec! In the case of 3C48, the shift was even greater, about 37 percent, corresponding to a velocity of 111,000 km/sec.

Since the early 1960s, hundreds of additional quasars have been discovered. Only a small fraction are radio sources, and in other ways they may differ from the first two quasars, but they invariably have highly redshifted emission lines, and in very many cases they also have weak absorption lines. In a number of quasars the redshift is so huge that spectral lines whose rest wavelengths are in the ultraviolet portion of the spectrum are shifted all the way into the visible region. The grand champion today is a quasar with a redshift of 378 percent, so that the strongest of all the hydrogen lines, whose rest wavelength is 1,216 Å, is shifted all the way to 4,596 Å. This quasar is *not* traveling away from us at 3.78 times the speed of light, however; for very large velocities, the Doppler-Shift formula has to be modified in accordance with relativistic effects (see appendix 13). The velocity of this quasar is 92 percent of the speed of light, still an enormous speed.

The Origin of the Redshifts

In order to understand the physical properties of quasars, we must first discover the reason for the high redshifts. We have already tacitly adopted the most obvious explanation, that they are a result of the Doppler effect in objects that are moving very rapidly, but we still must ascertain the nature of the motions. Furthermore, one alternative explanation has been suggested that has nothing to do with motions.

One consequence of Einstein's theory of general relativity, verified by experiment, is that light can be redshifted by a gravitational field. Photons struggling to escape an intense field lose some of their energy in the process, and as this happens, their wavelengths are shifted toward the red portion of the spectrum. We learned about this concept in chapter 16, when we dis-

cussed the behavior of light near black holes—black holes have such strong gravitational fields that no light can escape at all. For awhile it was considered possible that quasars are stationary objects sufficiently massive and compact to have large gravitational redshifts.

This suggestion has now largely been ruled out, for two reasons. One is that there is no known way for an object to be compressed enough to produce such strong gravitational redshifts without falling in on itself and becoming a black hole. If enough matter is squeezed into such a small volume that its gravity produces redshifts as large as those in quasars, no known force could prevent this object from collapsing further. A neutron star does not have as large a gravitational redshift as those found in quasars, and we already know that the only possible object with a stronger gravitational field than that of a neutron star is a black hole.

The second objection is that even if such a massive yet compact object could exist, its spectral lines would be very much broader than those observed in quasar spectra. The high pressure would distort electron energy levels so that the spectral lines would be smeared out (as in a white dwarf, but more extreme), and light emitted from slightly different levels in the object would have different gravitational redshifts, again causing spectral lines to be broadened.

For both of these reasons, we are forced to accept the Doppler-shift explanation for the redshifts. The problem then is to somehow explain how such large velocities can arise. One possibility is that the quasars are relatively nearby (by intergalactic standards), and are simply moving at very high speeds away from us, perhaps as the result of some explosive event (Fig. 22.10). This "local" explanation requires some care: to accept it, we must explain why no quasars have ever been found to have blueshifts. In other words, if quasars are nearby objects moving very rapidly, it is difficult to understand why none happen to be approaching us, but are instead all receding. We could argue that they originated in a nearby explosion (at the galactic center, perhaps) a long time ago, so any that happened to be aimed toward us have had time enough to pass by, and are now receding. There are serious difficulties with this picture, though, primarily in the amount of energy that would be required to get all these objects moving at the observed velocities, an amount that would dwarf the total light output of the galaxy over its entire lifetime.

THE REDSHIFT CONTROVERSY

In the text we presented a standard set of arguments demonstrating that quasar redshifts are cosmological because of the expansion of the universe, and implying that the quasars must be very distant objects. Although this viewpoint is accepted by most astronomers, it is not universally adopted, and there are those who favor other explanations of the redshifts.

The evidence cited by the opponents of the cosmological interpretation consists primarily of instances where quasars are found apparently associated with galaxies or clusters of galaxies, but do not have the same redshift as the galaxies. It is argued in these cases that the quasar is physically associated with the galaxy or cluster of galaxies, and is therefore at the same distance, so its redshift cannot be cosmological.

The evidence favoring these arguments can be quite striking. When a quasar is found within a cluster of galaxies, there is a natural tendency to think of it as physically a member of the group. It can be argued from statistical grounds that the chances of accidental alignments between galaxies and quasars are so low that the observed associations between these objects cannot be coincidental. There are even cases where long-exposure photographs appear to show gaseous filaments connecting a galaxy with a quasar that has a different redshift. This would seem to prove that the redshift cannot be cosmological.

This radical view of quasars leaves many important questions open. The most difficult of these is how to explain the redshifts, if they really are not cosmological, for the arguments cited in the text against gravitational and local Doppler redshifts are still valid. No satisfactory explanation of the quasar redshifts has been offered by the opponents of the cosmological viewpoint.

The counterarguments center on statistical calculations. One point is that the calculations showing that the quasar-galaxy associations have low probability of occurring by chance are made after the fact.

For example, after a few such associations were found, arguments were made that the associations were very unlikely to have been chance alignments of objects randomly distributed over the sky. This is somewhat akin to arguing that in a card game the chances of being dealt a certain combination of cards are very low. This is true before the deal, but meaningless after the fact. Thus many scientists claim it is incorrect to argue that chance associations of quasars and galaxies are improbable. More to the point, it is argued, quasars that are associated with galaxies are more likely to be discovered than those which are not; therefore, it is natural to find a disproportionately high number of quasars that happen to be aligned with galaxies in the sky. This is an example of a **selection effect** (described in the text), because the process by which quasars are found is not perfectly random and thorough; the attention of astronomers is naturally concentrated on regions where there are many galaxies, so those regions are more thoroughly searched for quasars.

Probably the most telling argument against the noncosmological redshifts for quasars arises from the fact (noted in the text) that there are a number of quasars embedded in galaxies or within clusters of galaxies that share the same redshift. Recent observations show, in fact, that *every* quasar that is sufficiently nearby for an associated galaxy to be detected is indeed embedded within a galaxy that has the same redshift as the quasar. There has not yet been an exception to this, and the data amount to virtual proof that quasars are the cores of galaxies and that their redshifts are cosmological.

The controversy is by no means over, however. Adherents of the noncosmological interpretation of the redshifts continue to find remarkable cases of apparent galaxy-quasar connections where the redshifts do not match. Perhaps observations made in the near future, with the *Space Telescope* or other planned instruments, will provide unequivocal evidence supporting one side or the other.

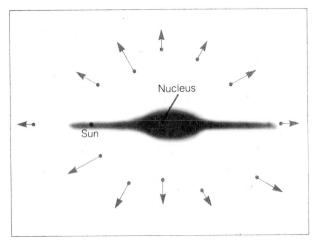

FIGURE 22.10 *A "Local" Hypothesis For Quasar Redshifts.*
If quasars were very rapidly-moving objects ejected from our
galactic nucleus some time ago, they could all be receding
from our position in the disk by now. This would explain the fact
that only redshifts are found, but there is no known mechanism
for providing the energy required to produce such great
speeds.

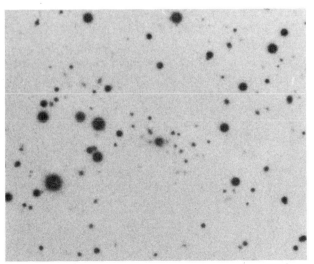

FIGURE 22.11 *A Quasar in a Cluster of Galaxies.*
The image indicated at the center of this very long-exposure
phtotograph is a quasar. The much fainter objects around it are
galaxies, forming a cluster of which the quasar is apparently a
member. The galaxies and the quasar have the same redshift,
indicating that they lie together at the same distance.

We are left with one other alternative, one that still poses the problems, but perhaps is more acceptable than the others. Let us consider the possibility that the quasars are very distant objects, moving away from us with the expansion of the universe. In this view they are said to be at "cosmological" distances, meaning that they obey Hubble's relation between distance and velocity, just as galaxies do. We find, if we adopt this assumption, that 3C273 is some 870 Mpc away, and 3C48 is more than 2,000 Mpc distant. (These distances are based on the assumption that the rate of expansion of the universe has been constant, something that is probably not true; this will be discussed in chapter 23.)

The best evidence that the quasars are at cosmological distances is the fact that some have been found in clusters of galaxies, with the same redshift as the galaxies (Fig. 22.11). This was an extremely difficult observation to make, because at the distances where the quasars exist, the much fainter galaxies are very hard to see, even with the largest telescopes.

Supporting evidence for the cosmological-redshift interpretation is provided by the **BL Lac objects.** Named for the prototype, an object called BL Lacertae that was once thought to be a variable star, these are elliptical galaxies with very bright central cores. The nucleus of a BL Lac object displays many of the properties of a quasar, with radio synchrotron emission, variable brightness, and enormous luminosity. The surrounding galaxy shows a normal absorption-line spectrum with a redshift that is consistent with the galaxy's apparent distance and which is undoubtedly comological. Generally, BL Lac objects have relatively small redshifts (by quasar standards), and they apparently are simply nearby quasars. The implications of the fact that they are embedded in galaxies will be discussed later in the chapter.

Distances of hundreds or thousands of megaparsecs, inferred for the quasars with very large redshifts, have important implications. One is that the light that we receive from quasars has traveled on its way to us for many billions of years. When we look at quasars, we are looking far into the history of the universe, and we must keep this in mind as we attempt to interpret them. The fact that quasars are seen only at high redshifts, meaning great distances, indicates that they were more common in the early days of the universe than they are now. This will prove to be a very important clue to their true nature.

It has recently been reported that there is a limit to quasar redshifts; that is, no quasars can be found with a redshift greater than this limit. The implication is fantastic, for it means that we are looking back to a time before the universe had organized itself into such objects as galaxies and quasars.

Another important implication of the great distances to quasars is that they are the most luminous objects known. Their apparent magnitudes (the brightest are thirteenth-magnitude objects) combined with their tremendous distances, imply that their luminosities are far greater than even the brightest galaxies, by factors of hundreds or even thousands. As we will see, explaining this astounding energy output is one of the central problems of modern astronomy.

The Properties of Quasars

Over 1,500 quasars have been catalogued, and more are being discovered continually. Many are being studied with care, primarily by means of spectroscopic observations. Within the last few years, it has even become possible to observe quasars at ultraviolet and X-ray wavelengths, with the use of satellite observatories such as the *International Ultraviolet Explorer* and the *Einstein Observatory*. The faintness of the quasars makes all these observations difficult, but their importance makes the effort worthwhile. The result of all the intensive work being done on quasars is that a great deal is now known about their external properties, even though their origin and especially their source of energy remain unsettled.

The blue color that characterized the first quasars discovered is a general property of all quasars (except for extremely redshifted ones, where the blue light has been shifted all the way into the red part of the spectrum), but the radio emission is not. Only about 10 percent are radio sources, contrary to the early impression that all are. This type of misunderstanding is known as a **selection effect,** because the process of selecting quasars was based at first on a certain assumption about their characteristics. For awhile, the only method used to look for quasars was to search for objects with radio emission, so naturally all the quasars that were found were radio sources. The radio emission, and usually the continuous radiation of visible light as well, show the characteristic synchrotron spectrum.

Apparently *all* quasars emit X rays (again, by the synchrotron process), according to the data collected by the *Einstein Observatory*, the most sensitive X-ray satellite yet launched. Hence X-ray observations may be the most reliable technique for finding new quasars.

In some cases, photographs of quasars reveal evidence of structure, instead of a single point of light. The most notable structure is seen in 3C273, which played such a key role in the initial discovery of quasars. This quasar shows a linear jet extending from one side (Fig. 22.12), closely resembling the one emanating from the giant radio galaxy M87. Perhaps this means that similar processes are occurring in these two rather different objects.

Many quasars vary in brightness, usually over times of several days to months or years (although in one case, variations were seen over just a few hours). This variability is very important, for it provides information on the size of the region in the quasar that is emitting the light. This region cannot be any larger than the distance light travels in the time over which the intensity varies. There is no way an entire object that might be hundreds or thousands of light-years across can vary in brightness in a time of a month or so. The light travel-time across the object would guarantee that we would see changes only over times of hundreds or

FIGURE 22.12 *A Quasar With a Jet.*
This is 3C273, one of the first two quasars discovered. This visible-light photograph reveals a linear jet very much like those seen in many radio galaxies. The radio structure of quasars is usually double-lobed, also similar to radio galaxies.

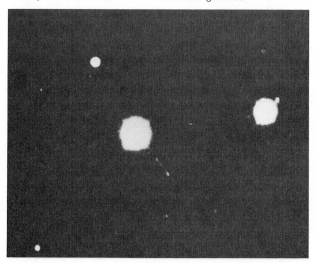

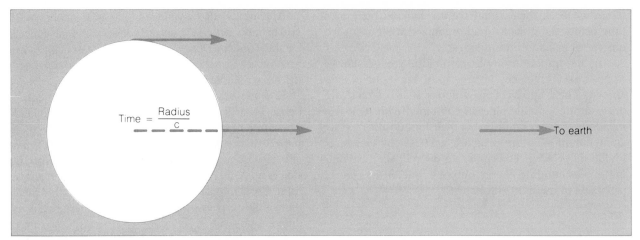

FIGURE 22.13 *The Implication of Time Variability for the Size of the Emitting Object.* This illustrates why an object cannot appear to vary in less time than it takes for light to travel across it. As a simple case, this spherical object is assumed to instantaneously change its luminosity. At earth, we observe the first hint of this change when light from the nearest part of the object reaches us, but we continue to see the brightness changing gradually as light from more distant portions reaches us.

thousands of years, as the light from different parts of the object reached us (Fig. 22.13). Therefore the observed short-term variations in quasars tell us that the fantastic energy emitted by these objects is produced within a volume no more than a light-month, or about 0.03 parsecs, in diameter. This obviously places stringent limitations on the nature of the emitting object.

The spectra of quasars have a great deal of detail (Fig. 22.14). All quasars have emission lines, generally of common elements such as hydrogen, helium, and often carbon, nitrogen, and oxygen. (Some of the latter elements have strong lines only in the ultraviolet, and therefore are best observed in cases where the redshift is sufficiently large to move these lines into the visible portion of the spectrum.) In many cases, the emission lines are very broad, showing that the gas that forms them has internal motions of thousands of kilometers per second. The degree of ionization tells us that the gas is subjected to an intense radiation field, which continually ionizes the gas by the absorption of energetic photons.

Many quasars have absorption lines in addition to the emission features. Strangely enough, the absorption lines are usually at a different redshift (usually smaller) than the emission lines, indicating that the gas that creates the absorption is not moving away from us as rapidly as the gas that produces the emission. Further confounding the issue is the fact that many quasars have multiple absorption redshifts; that is, they have several

distinct sets of absorption lines, each with its own redshift, and each therefore representing a distinct velocity. This shows that there are several absorbing clouds in our line of sight. There are two possible origins for these clouds (Fig. 22.15). One is that the quasar itself, which is moving at the high speed indicated by the red-

FIGURE 22.14 *A Schematic Quasar Spectrum.*
All quasars apparently have emission lines, usually of great breadth. Many also have absorption lines, usually at different (mostly smaller) redshifts than the emission lines. The absorption lines, discussed later in this chapter, are formed in material that is not receding from us as rapidly as the quasar.

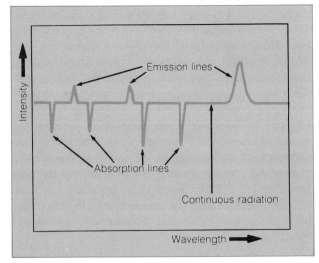

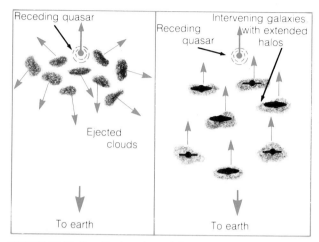

FIGURE 22.15 *Two Explanations of Quasar Absorption Lines.* In one scenario (left), the absorption lines are formed by clouds of gas ejected from the quasar. Clouds ejected directly towards the earth have lower velocities of recession than the quasar itself. In the other, more likely view (right), the absorption lines are formed in the extended halos of galaxies that happen to lie between us and the quasar. The galaxies are closer to us than the quasar, and therefore have lower velocities of recession and smaller redshifts.

shift of the emission lines, is ejecting clouds of gas, some of which happen to be aimed toward the earth. These clouds therefore have lower velocities of recession than the quasar, so the redshift of the absorption lines they form is less than the redshift of the emission lines formed in the quasar.

The alternative view, which is favored by most astronomers, is that the absorption lines arise in the gaseous halos of galaxies that happen to lie between us and the more distant quasar. Each galaxy has a recession velocity determined by the expansion of the universe, and because the intervening galaxies are not as far away as the quasars, they are not moving as fast and therefore have lower redshifts. If this interpretation is correct, then analysis of the absorption lines should provide important information on the nature of the extensive halos thought to surround many galaxies.

In some cases the absorption spectra of quasars show only lines created by hydrogen, and none created by heavier elements. These lines apparently arise in gas that has not been enriched by stellar processing, and it is thought that this gas might comprise intergalactic clouds that have never formed into galaxies, so that no stars have formed in them. These may be material remnants of the big bang; if so, they can tell us much about

the exact amount of nuclear processing that occurred in the early stages of the expansion of the universe.

Galaxies in Infancy?

A variety of theories, some of them rather fanciful, have been proposed to explain the quasars. We will limit discussion to the hypothesis that has gradually become widely accepted, but we will keep in mind that there are other suggestions.

The prevailing interpretation of the nature of quasars was inspired by the fact that quasars existed only in the long-ago past, by their association with the BL Lac objects (which clearly are galaxies), and by their resemblance to the nuclei of Seyfert galaxies. The statement that quasars existed only in the past is based on the fact that all are very far away; the light-travel time ensures that we are seeing things only as they were billions of years ago, not as they are today. The resemblance of quasars to Seyfert nuclei is striking: both are blue; both are radio sources in about 10 percent of the cases; both vary on similar time scales; and both have very similar emission-line spectra. Seyferts lack the complex absorption lines often seen in the spectra of quasars, but this would be expected if the quasar absorption lines are formed in the halos of intervening galaxies, since Seyfert galaxies are not so far away that there are likely to be many other galaxies along their lines of sight. The nuclei of Seyferts differ from quasars also in the amount of energy they emit, being considerably less luminous.

The picture that is developing is that quasars are very young galaxies, with some sort of youthful activity taking place in their centers. In this view, the Seyfert galaxies are descendants of quasars, still showing activity in their nuclei, but with diminished intensity. If we carry this idea a step further, we are led to the suggestion that normal galaxies like the Milky Way are later stages of the same phenomenon; recall the mildly energetic activity (by quasar standards) in the core of our galaxy.

The notion that quasars may be infant galaxies is supported by some rather direct evidence. As noted earlier, some quasars have been found associated with clusters of galaxies, showing that they can be physically located in the same region of space at the same time.

Astronomical Insight 22.2

THE DOUBLE QUASAR AND A NEW CONFUSION LIMIT

In chapter 18, in our discussion of radio emission lines produced by molecules in interstellar clouds, we introduced the concept of a **confusion limit.** This is a point beyond which, owing to the complexity of the universe, it is simply impossible to learn anything more by further observation. It is a fundamental limit on how deeply we can probe.

So far in the development of astronomy, the limitations have all been imposed by technological shortcomings, by the earth's atmosphere, or by other factors that can in principle be overcome. Now, however, we find the first ominous signs of a confusion limit that could prevent us from ever extending our view to the limits of the universe.

Recently a pair of quasars was discovered, very near each other in the sky, with properties so nearly identical that it has been concluded that they are in fact two images of the same quasar. If so, then astronomers have found the first example of a **gravitational lens,** something predicted to be a possibility on the basis of Einstein's theory of general relativity. Since a gravitational field can bend light rays, it is possible for such a field to act as a lens. Calculations have shown that the gravitational field of a galaxy can bend the light from a distant object around it. If the galaxy were perfectly aligned with the more distant object, then we would see either a single-point image, as if the galaxy were not there (this would happen only if we were precisely at the

focal point of the lens), or more likely, a circular ring-shaped image. In the more probable event that the galaxy is slightly off-center between us and the distant light source, we will see two or more distinct images, on either side of the intervening galaxy. The double quasar is thought to be such a case, the two images reaching us by coming around opposite sides of the gravitational lens formed by a galaxy between us and the quasar. Careful photographic studies have revealed what appears to be a galaxy between the two images of the quasar, which lends support to this interpretation.

If we are really seeing a gravitational lens in this case, it follows that if we extend our observations to more and more distant quasars, the likelihood of encountering other intervening galaxies will increase, so that our view is more and more likely to be distorted by gravitational lenses (a triple quasar has recently been found that might be another example). If we look far enough away, it will be like looking through a shower-stall glass, and everything in the distance will be distorted. At that point we will have reached a fundamental limit; it will be impossible to unambiguously study the farthest frontiers of the universe, because we will not be able to see clearly. We will not know the true locations of the objects we see, nor will we be able to say how many there are, because of the multiplicity of images.

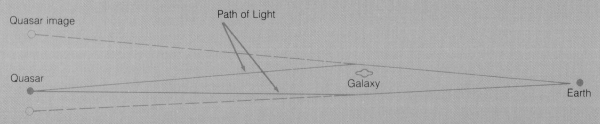

This diagram illustrates how an intervening galaxy can act as a gravitational lens, forming two images of a distant quasar.

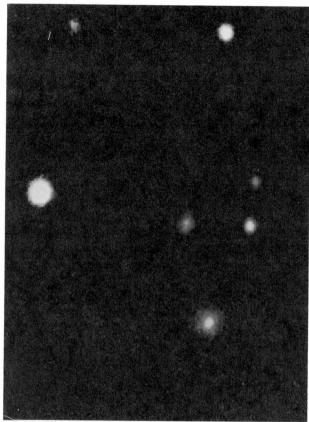

FIGURE 22.16 *A Quasar Embedded in a Galaxy.*
This photo, obtained with an electronic detector, contains several star images (top, sides), a galaxy (slightly elongated object just to the right of center), and a quasar (lower right). The quasar image is extended, just as the galaxy image is, and the fuzzy region surrounding it is indeed a galaxy. In every case where a quasar is close enough to us that a galaxy would be bright enough to be detected, one has been, indicating that all quasars lie at the centers of galaxies.

In addition, there are a few examples where careful observation has revealed a fuzzy, dimly glowing region surrounding a quasar (Fig. 22.16). This is probably a galaxy, at the center of which the quasar is embedded. It is likely that all quasars are located in the nuclei of galaxies, which are so much fainter that they usually cannot be seen. A short-exposure photograph of a Seyfert galaxy shows only the nucleus, which is very much brighter than the surrounding galaxy (Fig. 22.17). It is clear that if we observed one of these galaxies from so far away that it was near the limit of detectability, all we would see would be a blue starlike object resembling a quasar.

Although we can make a strong case that quasars are very young galaxies, we are still far from understanding all their properties. The main mystery remaining is the source of the tremendous energy of quasars and active galactic nuclei. A number of possibilities have been raised, but so far none is certain.

One idea invokes the presence in quasars of **antimatter,** which combines with normal matter, converting all of the mass of both into energy. Every subatomic particle has an analogous **antiparticle,** similar in mass but opposite in other quantities such as electrical charge. Recall the positron (mentioned in chapter 14), which is the antiparticle of the electron. The matter of which we and our surroundings are made is "normal" matter, but we do not know how much antimatter might have been produced in the big bang, nor whether much of it exists today in other parts of the universe. If a large amount does exist, or did in the distant past, then it is possible that it combined with normal matter in dense cores of young galaxies, releasing vast amounts of energy in the process. The rate at which energy is produced is given by Einstein's famous equation $E = mc^2$. Since the matter and antimatter that combine are totally annihilated, this is the most efficient possible mechanism for producing energy, easily capable of accounting for the tremendous luminosities of quasars.

Another suggestion, made shortly after the quasars were discovered, is that their power arises in multiple supernova explosions. The idea is that in the dense inner regions of a young galaxy, many massive, short-lived stars could form and quickly (by astronomical standards) evolve to the point where they explode in supernovae. If these explosions were taking place at a sufficiently great rate, they could maintain a steady luminosity comparable to that of a quasar. The observed variability could thus be a result of random fluctuations in the frequency of supernovae. A mechanism of this sort could help explain how a galaxy such as ours could have undergone a rapid buildup of heavy elements early in its history, as evidence suggests was the case. The supernova rate in the galactic core would decrease as the galaxy aged and the gas was gradually dispersed, slowing the rate of star formation.

A third possibility and the one most widely accepted, is that quasars are powered by massive black holes (Fig. 22.18). In the chaotic early days of collapse, when a galaxy was just forming out of a condensing cloud of primordial gas, a great deal of material might

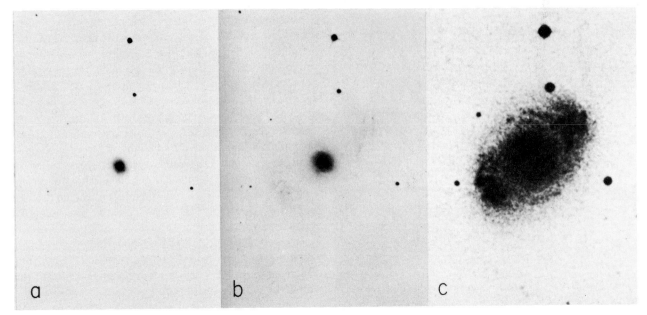

FIGURE 22.17 *Three Exposures of a Seyfert Galaxy*. At left is a short-exposure photograph of the Seyfert galaxy NGC 4151, in which only the nucleus is seen, resembling the image of a star. The center and right-hand images are longer exposures of the same object, revealing more of its outer, fainter structure. This sequence illustrates how a quasar, which is even more luminous than a Seyfert galaxy nucleus, can appear as a star-like image even though it is embedded in a galaxy.

have collected at the center. If many stars formed there, frequent collisions could have caused them to settle in more and more tightly, until they coalesced to form a black hole. The gravitational influence of such a massive object would have then stirred up the surroundings, creating the high electron velocities required to fuel the synchrotron emission. Infalling matter would

FIGURE 22.18 *Geometrical Model for a Quasar, Seyfert Galaxy, or a Radio Galaxy*. All of these objects have certain features in common that fit the picture shown here. A central object (most likely a supermassive black hole with an accretion disk) ejects opposing jets of energetic charged particles. These jets produce syndhrotron radiation, and build up double radio-emitting lobes on either side of the central object. These lobes extend well beyond the confines of the galaxy in which the central black hole is embedded.

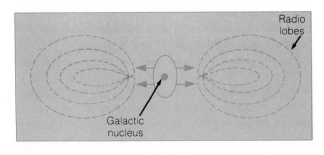

have formed an accretion disk around the black hole, from which X rays would be emitted, in analogy with stellar black holes and neutron stars in binary systems. In this picture the source of energy in quasars is the same as in radio galaxies, described at the beginning of the chapter. Even the existence of jets of hot gas is the same; such jets have been observed emanating from several quasars, as mentioned earlier.

Whatever the origin and power source of the quasars, we will certainly learn a lot about the nature of matter and the early history of the universe when we are able to answer all the questions about quasars. The *Space Telescope*, with its broad wavelength coverage and great sensitivity, will provide invaluable information on these fundamentally important objects.

PERSPECTIVE

In this chapter we have learned of new and wondrous things, the mighty radio galaxies and the enigmatic quasars. Along the way we have gained several important bits of information about the universe itself. It is time to tackle the fundamental question of the origin and fate of the cosmos.

ably explanation is that a quasar has a massive black hole at its core; this can produce enough energy (from the infall of matter) to account for the quasar's great luminosity as well as its time variability. By inference, similar objects may exist in the nuclei of galaxies such as Seyfert galaxies, radio galaxies, and the Milky Way.

SUMMARY

1. Radio galaxies emit vast amounts of energy in the radio portion of the spectrum. These galaxies, often giant ellipticals, commonly show structural peculiarities. The radio emission is nonthermal (usually synchrotron radiation), and in most cases is produced from two lobes on opposite sides of the visible galaxy.

2. The source of energy in a radio galaxy is unknown, but appears to be concentrated at the core.

3. Seyfert galaxies are spiral galaxies with compact, bright blue nuclei that produce emission lines characteristic of high temperatures and rapid motions.

4. In the early 1960s, several point sources of radio emission that looked like blue stars were found, and they were therefore called quasi-stellar objects, or quasars.

5. Quasars have emission-line spectra with very large redshifts, corresponding to velocities that are a significant fraction of the speed of light. If these redshifts are cosmological, then quasars are on the frontier of the universe, are extremely luminous, and are seen as they were billions of years ago.

6. Quasars, which are compact, blue objects, are sometimes radio sources and they usually emit X rays; in some cases they vary over times of only a few days. This implies that their tremendous energy output arises in a volume only a few light-days across.

7. Many quasars have absorption lines at a variety of redshifts (always smaller than the redshift of the emission lines), which may be created by matter being ejected from the quasars, or more likely, by intervening galactic halos.

8. The most satisfactory explanation of quasars is that they are very young galaxies. This interpretation is supported by the fact that they are only seen at great distances (that is, they only existed long ago), and by their resemblance to the nuclei of Seyfert galaxies. In several cases, photographs have revealed galaxies surrounding quasars, which confirms this suggestion.

9. The source of the energy that powers quasars is a premier mystery of modern astronomy. The most prob-

REVIEW QUESTIONS

1. Summarize the differences between radio galaxies and the radio emission from the Milky Way.

2. In radio galaxies with more than one pair of radio-emitting lobes on either side, the alignment between the inner and outer lobes is generally quite precise. What does this tell us about the source of lobes?

3. Suppose that in the spectrum of a quasar, emission lines are found at wavelengths of 2,189 Å, 2,790 Å, and 5,040 Å. If these are identified as the Lyman-alpha line of hydrogen (rest wavelength 1,216 Å), the three-times ionized carbon line (rest wavelength 1,550 Å), and the strong line of ionized magnesium (2,800 Å), what is the redshift of this quasar? How far away is it, if Hubble's constant has the value $H = 55$ km/sec/Mpc? (Note: You must use the relativistic redshift equation, See Appendix 13.) If the apparent magnitude of this quasar is $m = +16$, what is its absolute magnitude?

4. Explain, in your own words and using your own sketch, why the time scale of variations in light from a source such as a quasar puts a limit on the diameter of the emitting region.

5. List the similarities and differences between quasars and Seyfert galaxies.

6. Why are quasar absorption lines *always* observed at smaller redshifts than the emission lines? Would this necessarily be true of the redshifts if the emission lines were not cosmological?

7. From what you have learned about quasars in this chapter and about the *Space Telescope* in chapter 3, describe some important observations of quasars that might be made with the *Space Telescope*.

8. If quasars are young galaxies, how does this help demonstrate the difficulty of using faraway galaxies as standard candles in estimating large distances in the universe?

9. Explain why nearly all the first quasars detected

are radio sources, but only about 10 percent of *all* quasars are.

10. Explain why the material in intergalactic clouds may potentially provide clues about conditions very early in the life of the universe.

ADDITIONAL READINGS

Blandford, R. D., Begelman, M. C., and Rees, M. J. 1982. Cosmic jets. *Scientific American* 246(5):124.

Chaffee, F. H. 1980. The discovery of a gravitational lens. *Scientific American* 243(5):60.

Disney, M. J., and Veron, P. 1977. BL Lac objects. *Scientific American* 237(2):32.

Feigelson, E. D., and Schreier, E. J. 1983. The x-ray jets of Centaurus A and M87. *Sky and Telescope* 65(1):6.

Hamilton, D., Keel, W., and Nixon, J. F. 1978. Variable galactic nuclei. *Sky and Telescope* 55(5):372.

Hazard, C., and Mitton, S., eds. 1979. *Active galactic nuclei.* Cambridge, England: Cambridge University Press.

Kaufmann, W. 1978. Exploding galaxies and supermassive black holes. *Mercury* 6(5):78.

Lawrence, J. 1980. Gravitational lenses and the double quasar. *Mercury* 9(3):66.

Margon, B. 1983. The origin of the cosmic x-ray background. *Scientific American* 248(1):104.

Osmer, P. S. 1982. Quasars as probes of the distant and early universe. *Scientific American* 246(2):126.

Shipman, H. L. 1976. *Black holes, quasars, and the universe.* Boston: Houghton-Mifflin.

Silk, J. 1980. *The big bang: the creation and evolution of the universe.* San Francisco: W. H. Freeman.

Strom, R. G., Miley, G. K., and Oort, J. 1975. Giant radio galaxies. *Scientific American* 233(2):26.

Wyckoff, S., and Wehinger, P. 1981. Are quasars luminous nuclei of galaxies? *Sky and Telescope* 61(3):200.

COSMOLOGY: PAST, PRESENT, AND FUTURE OF THE UNIVERSE

23

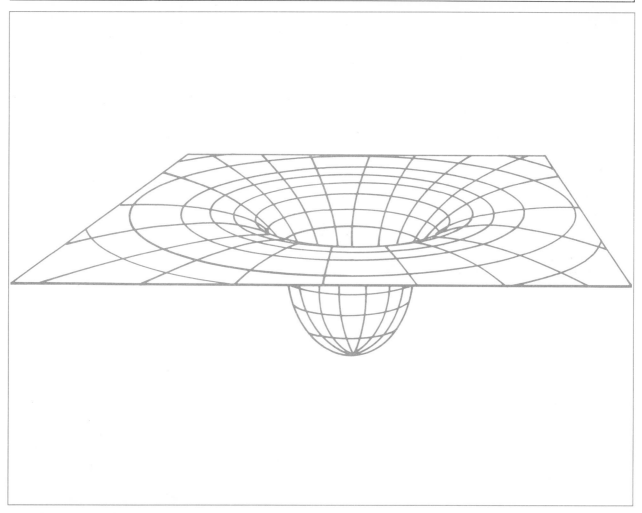

Learning Goals

THE NATURE OF COSMOLOGY
Underlying Assumptions
    Homogeneity and Isotropy: The Cosmological
    Principle
    Observational Tests
THE THEORY OF GENERAL RELATIVITY
The Equivalence of Gravity and Acceleration
Mathematical Description of the Universe
Three Possible Geometries
THE KEY QUESTION: AN OPEN OR CLOSED
UNIVERSE?
TOTAL MASS CONTENT
The Critical Density

Mass in Visible Form
The Question of the Missing Matter
THE DECELERATION OF THE EXPANSION
Looking Back in Time
The Abundance of Deuterium
THE HISTORY OF THE UNIVERSE
Radiation Dominance
Early Element Production
Decoupling of Matter and Radiation
Galaxy Formation
SPECULATIONS ON THE FUTURE

23 Given the time and opportunity, humans have, through the ages, devoted themselves to speculation on the grandest scale of all. Poets, philosophers, and theologians have approached the question of the origin and future of the universe in a countless variety of ways. So have astronomers, with the exception that a somewhat restrictive set of rules is followed: the answers that are accepted must not violate known laws of physics. By retaining this requirement, scientists attempt to approach the problem in an objective, verifiable manner.

There are difficulties in maintaining this idealized posture, however, and we shall try to clearly point them out. Because the universe in which we live is, as far as we know, unique, we have no opportunity to check our hypotheses through comparison with other examples. Furthermore, no matter how thoroughly and rigorously we trace the evolution of the universe by application of known physical laws, there will always be fundamental questions that are beyond the scope of physics. As a result, even the most careful and objective scientists reach a point where they have to make certain unverifiable assumptions, and at that point they become philosophers or theologians. In this chapter we restrict ourselves to questions that in principle have objective, verifiable answers.

Technically, the study of the universe as it now appears is **cosmology** which is really what this book has been all about. The study of the origin of the universe is **cosmogony,** and this word applies to the big-bang theory as well as to the earlier theories on the origin of the solar system, once thought to be the entire universe. In practice, the general subject of the nature of the universe and its evolution is lumped under the heading of cosmology, and so it is on a pursuit of this subject that we now embark.

Underlying Assumptions

To even begin to study the universe as a whole, we need to make certain assumptions. These can in principle be tested, although it is not clear that there is a practical way to do so. Therefore in making these assumptions, the astronomer is on the verge of acting as philosopher.

A central rule traditionally set forth by cosmologists is that the universe must look the same at all points within it. This does not mean that the appearance of the heavens should be literally identical everywhere, but it means that the general structure, the density and distribution of galaxies and clusters of galaxies, should be constant (Fig. 23.1). This assumption states that the universe is **homogeneous.**

A related but slightly different assumption has to do with the appearance of the universe when viewed in different directions. The assumed property in this case is **isotropy;** that is, the universe must look the same to an observer, no matter in which direction the telescope is pointed. We have already encountered this concept in connection with the cosmic microwave background radiation. Again, the assumption of isotropy does not imply identical constellations of galaxies and stars in all directions, just comparable ones.

Both of these assumptions are thought to apply only on the largest scales, larger than any obvious groupings in the universe, such as clusters or superclusters of galaxies. It should also be pointed out here, although it is not discussed until later in the chapter, that both as-

FIGURE 23.1 *The Cosmological Principle.*
The universe is assumed to be homogeneous and isotropic, meaning that it looks the same to all observers in all directions. Here is a segment of the universe, filled with galaxies. Their distribution, while not identical from place to place, is similar throughout.

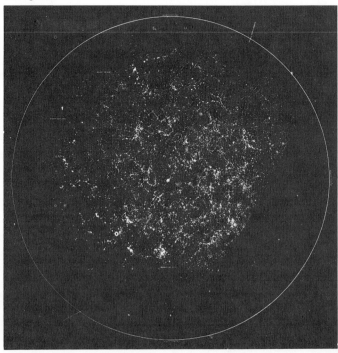

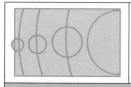

Astronomical Insight 23.1

THE MYSTERY OF THE NIGHTTIME SKY

A very simple question, first asked more than two centuries ago and discussed at length in the early 1800s by the German astronomer W. Olbers, leads to profound consequences. The question is why the sky is dark at night.

In an infinite universe filled with stars, every line of sight, regardless of direction, should intersect a stellar surface. The nighttime sky should therefore be uniformly bright, not dark, as the star images in the sky would literally overlap each other. The fact that this is not so presents a paradox, one worth pondering.

One suggested solution is that interstellar extinction can diminish the light from distant stars enough to make the sky appear black. Careful consideration shows, however, that this is not so. Light that is absorbed by dust in space causes the grains to heat up. If the universe were really filled with starlight as suggested by Olbers's paradox, then the grains would become so hot that they would glow, and we would still have a uniformly bright nighttime sky.

One potential solution to Olbers's paradox is to realize that the universe is not really infinite, at least from a practical point of view. At sufficiently great distances we expect to see no galaxies or stars, because we will be looking back to a time before the universe began. Thus, even though the universe has no edge, there is a horizon beyond which stars and galaxies do not exist. Therefore, this argument goes, all lines of sight do not have to intersect stellar surfaces; many reach the darkness beyond the horizon of the universe, and we have a dark nighttime sky.

Interestingly, it now appears that this horizon has been reached by observations. As noted in chapter 21, very recent data show that there appears to be a limit on quasar redshifts, implying that no quasars exist beyond a certain distance. Hence we may have already reached the point where we are looking back to a time before quasars had formed.

Given that the universe began in a very hot phase filled with energetic radiation, it seems, however, that if we look far enough into the past, we should see the brightness of the primeval fireball. We must therefore look for another solution to the dilemma.

The ultimate explanation of Olbers's paradox has to do with the expansion of the universe. The redshift caused by the expansion, if we look far enough away, becomes so great that the light never gets here in visible form. It becomes redshifted to extremely large wavelengths, losing nearly all its energy.

It is interesting that this simple question, if analyzed fully, could have led to the conclusion that the universe has a finite age or is expanding, a full century before these things were learned from other observational and theoretical developments.

sumptions have been questioned by some modern theories of the structure of the universe. In some current theories, it is even thought surprising that the universe should be as homogeneous as it is. Hence the assumption of homogeneity and isotropy should now be viewed more as an observation that must be explained, rather than as a postulate of cosmological theory.

The statement that the universe is homogeneous and isotropic is often referred to as the **Cosmological Principle.** It is often stated in terms of how the universe looks to observers; that is, the universe looks the same to all observers everywhere. It is difficult to verify the Cosmological Principle. We can test the assumption of isotropy by looking in all directions from earth, and indeed we do not find any deviations. On the other hand, we cannot test the homogeneity of the universe by traveling to various other locations to see whether things look any different. What we can do is count very dim, distant galaxies, and observe whether their densities appear to be any greater or less in some regions than in others, but even this is complicated by the fact that we are looking back in time to an era when the universe was more compressed and dense than it is now. The 3° background radiation gives us another tool for testing both homogeneity and isotropy, and it also appears to satisfy the Cosmological Principle. The best we can say for now is that the available evidence supports this principle, but does not completely rule out the possibility of subtle anisotropies or inhomogeneities.

Einstein's Relativity: Mathematical Description of the Universe

The simple act of making assumptions about the general nature of the universe by itself tells us very little about the past or future of the universe. It serves to elucidate certain properties of the universe as it is today, but to tie this into a quantitative description, to develop a theory capable of making predictions that can be tested, requires a mathematical framework. In developing such a mechanism for describing the universe, the astronomer seeks to reduce the universe to a set of equations and then to solve them, much as a stellar-structure theorist studies the structure and evolution of stars by constructing numerical models.

The most powerful mathematical tool for describing the universe was developed by Albert Einstein (Fig. 23.2). His theory of general relativity represents the properties of matter and its relationship to gravitational fields. Within the context of his theory, Einstein developed a set of relations, called the **field equations** (Fig. 23.3), that express in mathematical terms the interaction of matter, radiation, and gravitational forces in the universe. Although alternatives to general relativity have been developed and their consequences explored, most research in cosmology today involves finding solutions to Einstein's field equations, and testing these solutions with observational data.

The basic premise of general relativity is that acceleration caused by a gravitational field is indistinguishable from acceleration caused by a changing rate or direction of motion. One way to visualize this is to imagine we are inside a compartment with no windows (Fig. 23.4). If this compartment is on the earth's surface, our weight feels normal because of the earth's gravitational attraction. If, however, we are in space and the compartment is being accelerated at a rate equivalent to one earth gravity, our weight will feel normal here, as well. There is no experimental way for

FIGURE 23.2 *Albert Einstein.*
Among his many great contributions was the development of a mathematical formalism to describe the interaction of gravity, matter, and energy in the universe. This framework, general relativity, has withstood all observational and experimental tests applied to it so far.

us to tell the difference between the two situations, short of opening the door and looking out.

One consequence of the equivalence of gravity and acceleration is that an object passing near a source of gravitational pull (that is, any other object having mass) undergoes acceleration, and therefore follows a curved path. In a universe containing matter, this means that all trajectories of moving objects are

FIGURE 23.3 *The Field Equations.*
Here, in Einstein's own handwriting, are the equations that describe the physical state of the universe. To a large extent, the science of cosmology involves finding solutions to these equations.

EINSTEIN: Einheitliche Feldtheorie von Gravitation und Elektrizität   415

Jnabhängig von diesem affinen Zusammenhang führen wir eine kontravariante Tensordichte $\mathfrak{g}^{\mu\nu}$ ein, deren Symmetrieeigenschaften wir ebenfalls offen lassen. Aus beiden bilden wir die skalare Dichte

$$\mathfrak{H} = \mathfrak{g}^{\mu\nu} R_{\mu\nu}. \tag{3}$$

und postulieren, daß sämtliche Variationen des Integrals

$$\mathfrak{J} = \int \mathfrak{H} dx_1 dx_2 dx_3 dx_4$$

nach den $\mathfrak{g}^{\mu\nu}$ und $\Gamma^{\tau}_{\mu\nu}$ als unabhängigen (an den Grenzen nicht varierten) Variabeln verschwinden.

Die Variation nach den $\mathfrak{g}^{\mu\nu}$ liefert die 16 Gleichen

$$R_{\mu\nu} = 0, \tag{4}$$

die Variation nach den $\Gamma^{\tau}_{\mu\nu}$ zunächst die 64 Gleichungen

$$\frac{\partial \mathfrak{g}^{\mu\nu}}{\partial x_\alpha} + \mathfrak{g}^{\sigma\nu} \Gamma^{\mu}_{\sigma\alpha} + \mathfrak{g}^{\mu\sigma} \Gamma^{\nu}_{\alpha\sigma} - \delta^{\nu}_{\alpha} \left(\frac{\partial \mathfrak{g}^{\mu\beta}}{\partial x_\beta} + \mathfrak{g}^{\mu\beta} \Gamma^{\sigma}_{\beta\sigma} \right) - \mathfrak{g}^{\mu\nu} \Gamma^{\sigma}_{\alpha\sigma} = 0. \tag{5}$$

Wir wollen nun einige Betrachtungen anstellen, die uns die Gleichungen (5) durch einfachere zu ersetzen gestatten. Verjüngen wir die linke Seite von (5) nach den Indizes ν, α bzw. μ, α, so erhalten wir die Gleichungen

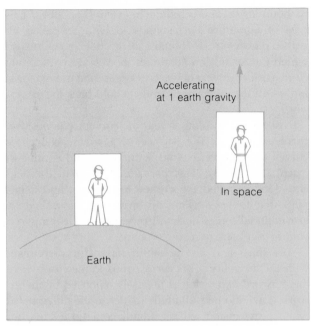

FIGURE 23.4 *General Relativity.*
The person in the enclosed room has no experimental or intuitive means of distinguishing whether he is motionless on the surface of the earth, or in space, accelerating at a rate equivilent to one earth gravity. The implication of this is that acceleration due to motion and that due to gravity are equivilent, and that in turn implies that space is curved in the presence of a gravitational field.

curved, and it is often said that space itself is curved. The degree of curvature is especially high close to massive objects (Fig. 23.5), but there is also an overall curvature of the universe, owing to its total mass content. The solutions to the field equations specify, among

other things, the degree and type of curvature. We will return to this point shortly.

Einstein's solution to the equations, developed in 1917, had a serious flaw, in his view: it did not allow for a static, nonexpanding universe. In what he later admitted was the biggest mistake he ever made, Einstein added an arbitrary force called the **cosmological constant** to the field equations, solely for the purpose of allowing the universe to be stationary, neither expanding nor contracting.

Others developed different solutions, always by making certain assumptions about the universe. Some of these assumptions were necessary in order to simplify the field equations so that they could be solved. For example, W. de Sitter in 1917 developed a solution that corresponded to an empty universe, one with no matter in it.

By the 1920s, solutions for an expanding universe had been found, primarily by the Soviet physicist A. Friedmann and later by the Belgian G. LeMaitre, who went so far as to propose a hot, dense state as the origin from which the universe has been expanding ever since. This was the true beginning of the big-bang idea, and it was developed some three years before Hubble's observational discovery of the universal expansion. Of course, once it was found that the universe is not static, but is actually in a dynamic state, the need for Einstein's cosmological constant diappeared. Nevertheless, modern cosmologists usually include it in the field equations, but its value is assumed to be zero.

The general-relativistic field equations allow for three possibilities regarding the curvature of the uni-

FIGURE 23.5 *Curvature of Space Near a Massive Star.* This curved surface represents the shape of space very close to a massive star. The best way to envision what it means to say that space is curved in this way is to imagine photons of light as marbles rolling on a surface of this shape.

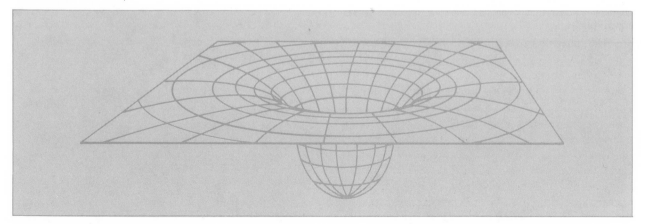

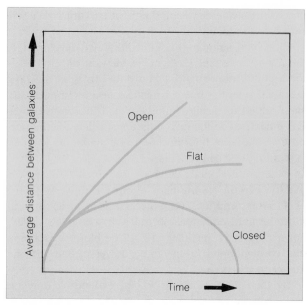

FIGURE 23.6 *The Three Possible Fates of the Universe.*
This diagram shows the manner in which the average distance between galaxies will change with time for the open, flat, and closed universe. In the first case, galaxies will continue to separate forever, although the rate of separation will slow. In the second case, the rate of separation will, in an infinite time, slow to a halt, but will not reverse. In the third case, the galaxies eventually begin to approach each other, and the universe returns to a single point.

verse, and three possible futures (Fig. 23.6). A central question of modern studies of cosmology has to do with deciding which possibility is correct. One is referred to as negative curvature, and an analogy to this is a saddle-shaped surface, which is curved everywhere, has

FIGURE 23.7 *A Saddle Surface.*
This is a representation of the geometry of an open universe, one that has negative curvature, is infinite in extent, and has no boundaries.

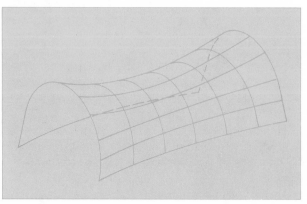

no boundaries, and is infinite in extent (Fig. 23.7). This type of curvature corresponds to what is called an **open universe,** a solution to the field equations in which the expansion continues forever, never stopping (it does slow down, however, because the gravitational pull of all the matter in it tends to hold back the expansion).

A second possibility is that the curvature is positive, corresponding to the surface of a sphere, which is curved everywhere, has no boundaries, and is finite in extent (Fig. 23.8). This possible solution to the field equations is called the **closed universe,** and it implies that the expansion will eventually be halted by gravitational forces and will reverse itself, leading to a contraction back to a single point.

The third and last possibility is a **flat universe,** with no curvature. This corresponds to the case where the outward expansion is precisely balanced by the inward gravitational pull of the matter in the universe, so that the expansion will eventually come to a stop, but will not reverse itself. This balanced, static state requires a perfect coincidence between the energy of expansion and the inward gravitational pull of the combined total mass of the universe, and therefore may seem very unlikely. We do not know what set of rules governed the amount of matter the universe was born

FIGURE 23.8 *The Surface of a Sphere.*
This surface represents a closed universe; one that has positive curvature, has a finite extent, but has no boundaries.

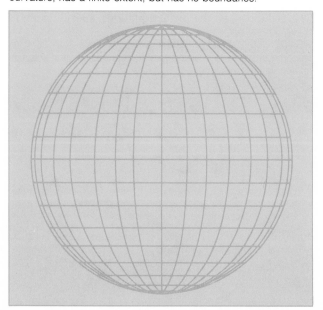

with, however, so there is no logical reason to rule out this possibility. Most astronomers generally pose the question in terms of whether the universe is open or closed, without specifically mentioning the third possibility, that the universe is flat. As we will see in the next section, if a perfect balance has been achieved, it will be very difficult to determine this from the observational evidence, which has large uncertainties.

Open or Closed: The Observational Evidence

Substantial intellectual and technological resources are devoted today to determining the fate of the universe. Theorists are at work developing and refining the solutions to the field equations, or seeking alternatives to general relativity that might provide equally or more valid representations. Observers are busily attempting to test the theoretical possibilities by finding situations where competing theories should lead to different observational consequences. This is a difficult job, because most of the differences will show up only on the largest scales, and therefore to detect them requires observations of the farthest reaches of the universe.

Within the context of general relativity, the premier question is whether the universe is open or closed. There are two general observational approaches to answering this question: determine whether there is enough matter in the universe to produce sufficient gravitational attraction to close it; or measure the rate of deceleration of the expansion, to see whether it is slowing rapidly enough to eventually stop and reverse itself.

Total Mass Content

The total mass in the universe is the quantity that determines whether gravity will halt the expansion or not. The field equations express the mass content of the universe in terms of the density; the amount of mass per cubic centimeter. This is convenient for observers, because it is obviously simpler to measure the density in our vicinity than to try to observe the total mass everywhere in the universe. The field equations can be solved for the value of the density that would produce an exact balance between expansion and gravitational

attraction; that is, the density corresponding to a flat universe. If the actual density is greater than this **critical density,** than there is sufficient mass to close the universe, and the expansion will stop. The critical density depends on the value of Hubble's constant H; for the value near 55 km/sec/Mpc, it is calculated to be roughly 6×10^{-29} grams/cm^3, or about 3 protons per cubic meter, a very low value by earthly standards.

The most straightforward way to measure the density of the universe is to simply count galaxies in some randomly selected volume of space, add up their masses, and divide the total by the volume. Care must be taken to choose a very large sample volume, so that clumpiness owing to clusters of galaxies is not important. This technique yields very low values for the density, only a few percent of the critical density. It may seem that we have already answered the question, and that the universe is open, but this method may overlook substantial quantities of mass.

One clue to this arises from the determination of cluster mass based on the velocity dispersion of the galaxies in the cluster (see chapter 20). These measurements of mass, for reasons still not entirely clear, always yield much higher values than those based on the estimated total mass of the visible galaxies in the cluster. The disparity can be as great as a factor of ten or more. Thus if the larger values are correct, the average density of the universe comes closer to the critical value. Even the larger values based on velocity-dispersion measurements, however, fall short of the amount needed to close the universe. If the mass density is to exceed the critical value, there must be large quantities of matter in some form that has not yet been detected (Fig. 23.9). The hidden matter cannot be inside clusters of galaxies, because its presence would have been detected by the velocity-dispersion measurements.

If there is a lot of invisible matter in the universe, it could take several different forms. One possibility that is obvious to us by now is that there may be many black holes in space between clusters. The astrophysicist Stephen Hawking hypothesized the existence of countless numbers of "mini" black holes in intergalactic space, formed during the early stages of the big bang. These would have very small masses, much smaller than even the mass of the earth, but would be so numerous that they could easily exceed the critical density. Current observational evidence argues against the existence of such objects in great numbers, however.

Another possibility is that neutrinos, the elusive sub-

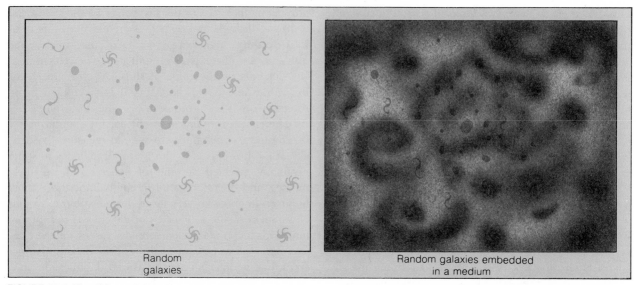

FIGURE 23.9 *The "Missing" Mass.* Are the isolated galaxies that we see all that there is in the universe? Or are they merely points that happen to glow, embedded in a universal sea of unknown substance, whose mass density overwhelms that represented by the galaxies?

atomic particles produced in nuclear reactions, have mass. Recall (from chapter 11) that these particles permeate space, freely traveling through matter and vacuum alike, but that standard theory says they have no mass. A recent controversial experiment indicates that this last assertion may not be correct; that they may contain miniscule quantities of mass after all. If so, they are sufficiently plentiful to provide more than the critical density, thereby closing the universe. Further experiments designed to determine whether the neutrino has mass have been performed, all of them so far being more consistent with the standard notion that these particles are massless. Thus, at present we cannot say whether neutrinos provide the matter density required to close the universe.

For now we are forced to conclude from the available observational evidence that the mass density of the universe is less than the critical density, and therefore that the universe is open.

The Deceleration of the Expansion

The second approach to answering the question of whether the universe is open or closed is perhaps being pursued more vigorously today. The objective is to compare the present expansion rate with what it was early in the history of the big bang, to see how much slowing, or deceleration, has occurred (Fig. 23.10). The expansion has certainly slowed in any case; the question is how much. If it has only decelerated a little, then we infer that it is not going to slow down enough to stop and reverse itself, but if it has already decelerated a lot, then we conclude that the expansion is coming to a halt, and that the universe is closed.

To establish the deceleration rate requires knowing both the present expansion rate and the expansion rate at a time long ago, shortly after the big bang started. Neither quantity is easy to determine: we already learned that the value of the Hubble constant H, which tells us the present expansion rate, is quite uncertain. To measure the expansion rate that occurred early in the history of the universe is even more difficult, for it involves determining the distances and velocities of very faraway galaxies, so distant that we see them as they were when the universe was young. Such objects are now at the very limits of detectability, and it is hoped that the *Space Telescope* will push the frontier back far enough to permit observations of velocities in the early days of the expansion.

We still must rely on standard candles to establish the distance scale, and for such distant objects this procedure becomes even more uncertain than usual. When we look so far back in time, we are seeing galaxies that are much younger than those near us, and which

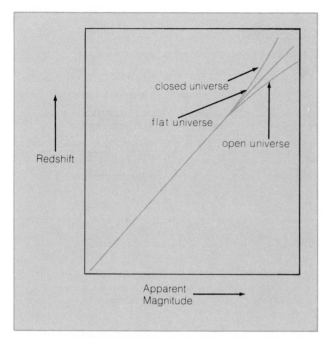

FIGURE 23.10 *The Effect of Deceleration in the Hubble Diagram*. This shows the relationship between velocity of recession and distance for the three possible cases: closed (upper curve), flat (middle), or open (lower). Present data seem to favor the open universe, although there is substantial uncertainty, largely because of problems in applying standard candle techniques to galaxies so far away that they are being seen as they were at a young age. The distinction is also made difficult by the subtlety of the contrast between the shapes of the curves.

therefore may have quite different properties than mature galaxies. For example, their content of stellar populations may be rather different than in nearby galaxies, their brightest stars and nebulae may be different, and even total galactic luminosities may be different. It may not be valid to assume that the brightest galaxy in a cluster of galaxies has the same absolute magnitude as in more nearby clusters.

The present evidence on deceleration is that the universe has not slowed very much, and therefore is not on its way to stopping and beginning to contract. As we have seen, however, the observational uncertainties are great, and this conclusion is not very firm.

An indirect technique for measuring the deceleration has recently been employed, and avoids many of the difficulties of observing extremely distant galaxies. Some light elements were created in the early stages of the expansion, and the amounts that were produced depend on the fraction of the universal energy that was in the form of mass rather than radiation at the time of element formation. The primordial mass density is in turn related to the present-day deceleration of the expansion, so if the primordial density is deduced from the abundances of elements that were created, we can infer the deceleration rate. The strategy, therefore, is to measure the abundance of some element that was produced in the big bang, and from that derive the deceleration.

The most abundant element (other than hydrogen) produced in the big bang is helium, so if we could measure how much helium was created, this would be a good indicator of the deceleration. The problem with this element, however, is that it is also produced in stellar interiors, so we have difficulty determining how much of what we see in the universe today is really left over from the big bang.

Better candidates are ^{7}Li, an isotope of the element lithium, and deuterium, the form of hydrogen that has one proton and one neutron in the nucleus. These species were also produced in the big bang, and, as far as we know, are not made in any other way. To date, most efforts have been concentrated on measuring the present-day abundance of deuterium, although in principle ^{7}Li can be measured as well. Although deuterium can be destroyed by nuclear processing in stars, it is not produced in that way (even if it were, it would not survive the high temperatures of stellar interiors without undergoing further reactions). The present deuterium abundance in the universe should therefore represent an upper limit on the quantity created in the big bang. Direct measurements of the amount of deuterium in space became possible in the 1970s with the launch of the *Copernicus* satellite, which made ultraviolet spectroscopic measurements and was able to observe absorption lines of interstellar deuterium atoms (Fig. 23.11). The abundance of deuterium that was found is sufficiently high that it implies a low density, in turn pointing to a small amount of deceleration. Thus this test, in agreement with others, indicates that the universe is open. There is some uncertainty, though, because the *Copernicus* data seem to indicate an uneven distribution of interstellar deuterium throughout the galaxy, and this would not be expected if the deuterium really formed exclusively in the big bang. A clumpy distribution of deuterium in the galaxy could indicate that some of it is somehow produced by stars, in which case its abundance is not a strict test of deceleration. A recent measurement of the ^{7}Li abundance

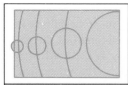

Astronomical Insight 23.2

THE INFLATIONARY UNIVERSE

While many of us have been preoccupied with the effects of inflation on the economy, cosmologists have begun to speak in terms of a different kind of inflation, one having to do with the early universe. The big-bang cosmology is widely accepted, but there remain some nagging problems. One is that it is thought to be very unlikely for a universe starting from a singularity as envisioned in the big bang to be as symmetric as the observed universe. It is unlikely that the universe should appear as homogeneous and isotropic as observations tell us it is. As we learned in the text, a great deal of effort has been put into testing this property of the universe, because any departures from homogeneity would provide information on imbalances or asymmetries in the early epochs of the expansion of the universe.

A new kind of expansion model has been developed in which the homogeneity of the universe is easily understood. Calculations have shown that at a certain point very early in the history of the universe, when the temperature was around 10^{27} K, conditions were right for small regions to separate themselves from the rest of the universe and then to expand very rapidly to much larger sizes. A reasonable analogy would be the creation of bubbles in a liquid, with the bubbles then growing much larger almost instantaneously. In this cosmological model, each "bubble" becomes a universe in its own right, with no possibilty of communication with other bubbles.

The major advantage of this so-called **inflationary model** is that a universe born of a tiny cell in the early expansion would be expected to remain symmetric and homogeneous as it grew. The reasons for this are somewhat abstract, but have to do with the fact that the various portions of the tiny region that was to expand into the present universe were so close together that they were in a form of equilibrium, with little or no variation in physical conditions being possible. In a larger region, such as in the standard big-bang cosmology, no such equilibrium would have existed, and there would be

FIGURE 23.11 *An Ultraviolet Absorption Line of Interstellar Deuterium.* The abundance of deuterium in space is determined from the analysis of absorption lines it forms in the spectra of background stars. Here we see a weak deuterium line, close to a strong absorption feature due to normal hydrogen. The spectrum shown here was obtained with the *Copernicus* satellite.

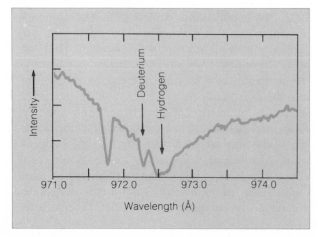

in the galaxy is consistent with the deuterium results; that is, the observed quantity of ^{7}Li is also large enough to imply that the universe is open.

It is worth noting that all the observational techniques that have been tried so far to determine whether the universe is open or closed have pointed toward a fairly close balance. That is, the data do not point to a universe that is closed or open by a wide margin; whatever the correct answer is, we can say that the universe is close to being flat. This is very significant, for such a close balance is not necessarily expected in the big-bang theory, and must be regarded in this theory as a coincidence. There are other models, however, such as the very recent **inflationary theory,** in which a flat or nearly flat universe is expected (see Astronomical Insight 23.2).

no reason to expect uniformity throughout the resulting universe.

Another advantage of the inflationary model is that in such a universe, it would be expected that the expansion would go on forever, but would continue to slow, approaching a stationary state; a flat universe is the natural result of rapid expansion from a tiny bubble as envisioned in this model. As we learned in the text, observations tell us that our universe is nearly flat (although perhaps not quite). If it were not so closely balanced, we would not have so much difficulty determining whether the universe is open or closed.

A third promising feature of the inflationary model is its agreement with the observational data on another kind of particle, one that has yet to be unambiguously detected. The standard big-bang theory predicts the formation in the early universe of a large number of **magnetic monopoles,** sub atomic particles with single magnetic poles, instead of the usual pair of opposing poles. Such particles would be very difficult to detect, the best technique being to look for tiny electrical currents they would briefly create in wires they passed near. So far, only one possible detection of a magnetic monopole has been reported, whereas the big-bang theory leads to the expectation that these particles should not be so rare. The inflationary theory holds that only a very few magnetic monopoles were created in each bubble that formed before expanding into a universe. Therefore, the lack of many observed monopoles is more consistent with the inflationary model than with the big-bang theory.

The homogeneity of a bubble in the inflationary universe is an advantage in explaining some observations, as we have seen, but is also a disadvantage. It is very difficult for us to understand how the matter ever became lumpy enough to form galaxies and clusters of galaxies in such a uniform universe. In the text we spoke of possible turbulence and other mechanisms having to do with concentrations of matter in the early universe, but we have trouble understanding how such concentrations would ever have formed in terms of the inflationary model.

The inflationary model, first suggested in 1981, is currently under active discussion among cosmologists. There will continue to be refinements to it, as more of the observational constraints are confronted. It will be very interesting to see whether it withstands the test of continued scrutiny.

The History of Everything

We are now at the forefront of modern cosmology, having outlined the present state of knowledge, and having pointed out the basis for current and planned observational tests. At this point it is useful to review the development of the universe, highlighting some of the significant events. The very fact that we can do this, that we can say with any degree of certainty what conditions were like at the beginning and at points along the way, is a triumph of modern science.

It is impossible to physically describe the universe as it was at the precise moment that the expansion began; it is physical nonsense to deal with infinitely high temperature and density. It is possible, however, to calculate the conditions that were present immediately after the expansion started, and at any later time. Many of the most interesting events in the early history of the universe, and the ones currently under the most active investigation, took place at very early times, before even a ten-thousandth of a second had passed. Under the conditions of density and temperature that existed then, matter and the forces that act on it were quite different from anything we can experience, even in the most advanced laboratory experiments. Even the familiar sub-atomic particles such as protons, neutrons, and electrons could not exist, but were replaced by *their* constituent particles.

Current particle-physics theory holds that the most fundamental particle is something called a **quark.** Modern theory also provides a basis for believing that all four of the fundamental forces in nature (see Astronomical Insight 2.2) are really manifestations of the same phenomenon. So far it has been possible to show that at least three of the fundamental forces (the electromagnetic force, and the so-called strong and weak nuclear forces) are manifestations of the same basic interaction. Some aspects of the theory have been confirmed by laboratory experiments. The theoretical framework connecting the three forces is called **grand-unifed theory** (**GUT** for short). This theory says that in the very first moments following the beginning of the expansion, until 10^{-35} seconds had passed, the universe contained only quarks and related particles and radiation, and the only forces operating in it were gravity and the unified force that was later to become recognizable as the electromagnetic, strong, and weak forces. It is suspected that at some yet earlier time, even gravity was combined with the others, and that the universe was very simple, consisting essentially of several kinds of elementary particles and one kind of force. Today a great deal of theoretical and experimental research is being devoted to developing the ultimate unified field theory, the one in which all forces are shown to be the same, only having different manifestations under conditions of low density and temperature (under what we view as "normal" conditions). This is the problem to which Albert Einstein devoted most of his later years without success, but now great progress is being made.

Let us now jump forward in time to an epoch in the early universe when matter was beginning to take on more familiar forms (Figs. 23.12 and 23.13), and the four fundamental forces were already acting as four distinct forces. At 0.01 seconds after the beginning of the expansion, the temperature was perhaps 100 billion degrees (10^{11} K), and electrons and positrons began to appear. The temperature dropped to 10 billion degrees (10^{10} K) by 1.09 seconds after the start, and by then protons and neutrons were appearing. At this point, most of the energy of the universe was still in the form of radiation. Within several seconds, conditions became better suited for nuclear reactions to take place efficiently, and the most active stage of element creation began. The principal products, in addition to helium and deuterium, were tritium (another form of hydrogen, with one proton and two neutrons in the nucleus), lithium (three protons and four neutrons), and to a minor extent, beryllium (four protons and five neutrons). Nearly all the available neutrons combined with protons to form helium nuclei, a process that was com-

FIGURE 23.12 *Element Formation in the Big Bang.* This diagram illustrates the relative rates of formation of various light elements by nuclear reactions during the early stages of the big bang expansion. The production rates are shown as functions of time and of temperature.

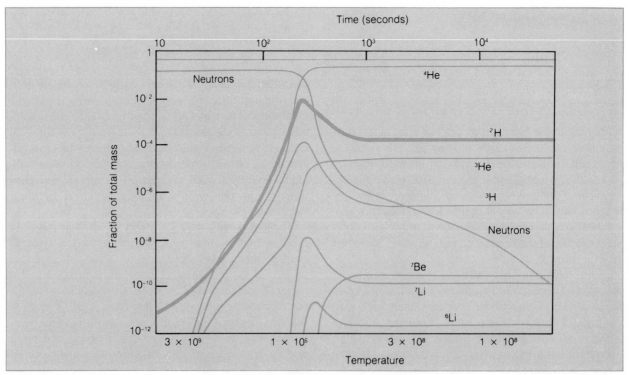

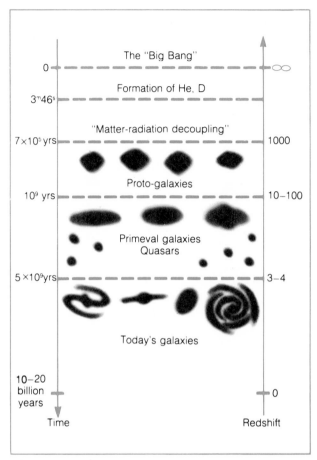

FIGURE 23.13 *The Evolution of the Universe.*
This diagram shows, as a function of time since the start of the big bang, significant stages in the evolution of the universe. It also shows the types of objects that existed at each stage, and the redshift at which they are (or would be) observed at the present time.

plete within four minutes after the beginning of the expansion. At this point, with some 22 to 28 percent of the mass in the form of helium, the reactions were essentially over, except for some production of lithium and beryllium during the next half hour.

The expansion and cooling continued, but nothing significant happened for a long time after the nuclear reactions stopped. Eventually the density of the radiation had become reduced enough that its energy was less than that contained in the mass; that is, the energy derived from $E = mc^2$ became greater than that contained in the radiation. At that point, it is said, the uni-

verse became matter-dominated rather than radiation-dominated.

Matter and radiation continued to interact, however, because free electrons scatter photons of light very efficiently. The strong interplay between matter and radiation finally ended nearly a million years after the start of the big bang, when the temperature became low enough to allow electrons and protons to combine into hydrogen atoms. The atoms still absorbed the reemitted light, but much less effectively, because they could do so only at a few specific wavelengths. From this time on, matter and radiation went separate ways. The radiation simply continued to cool as the universe expanded, reaching its present temperature of 2.7 K some 10 to 20 billion years after the expansion began.

Sometime in the first billion years or so, the matter in the expanding universe became clumpy, and fragmented into clouds and groups of clouds that eventually collapsed to form galaxies and clusters of galaxies. As we noted in chapter 20, the cause of the clumpiness and fragmentation is still unknown. Once galaxies began to form, the subsequent evolution of matter followed steps which have been described in the preceding chapters.

Because there is so much uncertainty about how the initial stages of fragmentation took place, there is a great deal of interest in probing far enough into the past to allow us to get data on this process. To do so will be another of the goals of the *Space Telescope*.

What Next?

To discuss the future of the universe is obviously a speculative venture. We cannot even answer with absolute certainty the basic question of whether the universe is open or closed. There have been attempts to calculate future conditions in the universe, in analogy to the theoretical work on past conditions, discussed in the last section. In the case of the future of the universe, however, the uncertainties are much greater, and the following descriptions should be regarded as *very* speculative.

If the universe is open, then it will have no definite end; it will just gradually run down. The radiation background will continue to decline in temperature, approaching absolute zero. As stellar processing continues in galaxies, the fraction of matter that is in the form of heavy elements will continue to grow, and the supply of hydrogen, the basic nuclear fuel, will diminish. It is predicted that all the hydrogen should be gone by about 10^{14} years after the birth of the universe, so the universe in which stars dominate has now lived approximately one ten-thousandth of its lifetime. The recycling process between stars and the interstellar medium will continue until this time, but gradually matter will become locked up in black holes, neutron stars, and white dwarfs. Dead and dying stars will continue to interact gravitationally, eventually colliding often enough in their wanderings that all planets will be lost (at about 10^{17} years) and galaxies will dissipate as their constituent stars are lost to intergalactic space (by about 10^{18} years). Further speculation shows that a new physical process will take over at later times. The new GUT theory tells us that the proton, a basically stable particle, may disintegrate in a very low-probability process that occurs on the average once in 10^{32} years for a given proton. When the universe reaches an age of about 10^{20} years, enough protons will begin to evaporate here and there that the energy produced will keep the remnant stars heated, although only to the modest temperature of perhaps 100 K. When the universal age is 10^{32} years, most protons will have decayed, and the universe will consist largely of free positrons, electrons, black holes, and radiation (the extremely cold remnant of the big bang). The final stage that has been foreseen occurs when the universe reaches an age of 10^{100} years, when sufficient time has passed for all black holes to evaporate, and nothing is left but a sea of positrons, electrons, and radiation (theory says that black holes can disintegrate with a very low probability, meaning that if we wait long enough, eventually they will do so).

If the universe is closed, then someday, perhaps some 50 billion years from now, the expansion will stop, and will be replaced by contraction. The deterioration just described will still take place, but will be arrested when the universe once again becomes hot and compressed, entering a new singularity. According to some views, purely conjectural and without possibility of verification, such a contraction would be followed by a new big bang, and the universe would be reborn. This concept of an oscillating universe, pleasing to the minds of many, will not be fulfilled unless the present weight of the evidence favoring an open universe is found to be in error.

PERSPECTIVE

Our story is essentially complete. We have discussed the universe and all the various objects and structures within it, and we have described what is known of the evolution and fate of the universe.

We have, however, omitted consideration of what is perhaps the most important ingredient of all: life. Although much of what we might say about this is beyond the scope of astronomy, it is appropriate to assess what can be learned from objective scientific examination.

SUMMARY

1. In cosmological studies, astronomers usually adopt the Cosmological Principle, which states that the universe is both homogeneous and isotropic. Existing observational data tend to support this assumption.

2. Einstein's theory of general relativity, which describes gravitation and its equivalence to acceleration, is used to mathematically describe the universe as a whole. In the context of this theory, there are field equations that represent the interaction of matter, radiation, and energy in the universe. Solutions to the field equations amount to definitions of the properties of the universe.

3. Modern cosmologists consider three general solutions to the field equations. These solutions correspond to a closed universe (positive curvature), an open universe (negative curvature), and a flat universe (no curvature).

4. A major question in modern astrophysics is whether the universe is closed (expansion eventually to be reversed) or open (expansion never to stop).

5. There are two ways to test whether the universe is open or closed: (1) ascertain whether there is enough mass in the universe to gravitationally halt the expansion; or (2) determine the rate of deceleration of the expansion, to see whether it is slowing enough to eventually stop.

6. The total mass content is measured in terms of the average density of the universe; the matter that is visi-

ble in the form of galaxies is not sufficient to close the universe. Various suggestions have been made concerning other forms in which the necessary mass could exist.

7. The deceleration is measured in two ways: (1) by comparing past and present expansion rates, through observations of very distant galaxies; and (2) by inferring the early expansion rate from the present-day abundances of elements that formed only during the big bang.

8. Both the total mass content that is observed and the inferred deceleration of the expansion indicate that the universe is open. This conclusion is not universally accepted, and observational tests are continuing.

9. The early stages of the universal expansion, up to the time when matter and radiation decoupled, can be described quite precisely and with certainty by modern physics. Following the initial moment of infinite density and temperature, the first atomic nuclei formed just over a second later, and all the early element production was finished within a few minutes. Matter and radiation decoupled almost a million years later; it is not so well understood how the universe subsequently organized itself into stars and galaxies.

10. The future of the universe appears to have two possibilities. If it is open, it will gradually become cold and disorganized. If it is closed, it will eventually contract, perhaps to a new beginning in another big bang.

forms a planetary nebula. The shell of gas, depending on the mass of the star and the initial velocity of ejection, will either escape completely or fall back onto the star. What determines this, and how is it analogous to the question of whether the universe is open or closed?

5. Why do astronomers not simply measure the density of matter within the well-observed solar neighborhood, within a few hundred parsecs of the sun, in determining whether there is enough mass to close the universe? What do you think the result would be if this local density were used? (You may want to review chapter 18.)

6. How do we know that the existence of massive black holes in the nuclei of all galaxies would not be sufficient to close the universe?

7. Summarize the difficulties in measuring very distant galaxies in order to infer the early expansion rate of the universe.

8. Why is deuterium rather than helium a better species to use in inferring the early expansion rate of the universe? Why would either be better than iron, for this purpose?

9. What would it tell us about the early expansion of the universe if it were discovered that substantial quantities of elements such as carbon, nitrogen, and oxygen had been formed in the big bang? Would this imply that the universe is open or closed?

10. Summarize the ways in which the *Space Telescope* should help answer cosmological questions raised in this chapter.

REVIEW QUESTIONS

1. In chapter 20 we discussed the fact that galaxies often are grouped into clusters. In this chapter we asserted that the universe appears to be homogeneous; that it has a uniform distribution of matter. Explain how galaxies can be grouped into clusters, and at the same time the universe can be homogeneous.

2. Recall, from chapter 21, that there is a 24-hour anisotropy in the 3° background radiation. Why does this not violate the assumption adopted here that the universe is isotropic?

3. Are Einstein's field equations more like Kepler's laws of planetary motion, or are they more akin to Newton's derivation of Kepler's laws?

4. Suppose a star ejects a spherical shell of gas as it

ADDITIONAL READINGS

Barrow, J. D., and Silk, J. 1980. The structure of the early universe. *Scientific American* 242(4):118.

Dicus, D. A.; Letaw, J. R.; Teplitz, D. C.; and Teplitz, V. L. 1983. The future of the universe. *Scientific American* 248(3):90.

Ferris, T. 1977. *The red limit: the search for the edge of the universe.* New York: William Morrow.

Field, G. B. 1982. The hidden mass in galaxies. *Mercury* 11(3):74.

Gaillard, M. K. 1982. Toward a unified picture of elementary particle interactions. *American Scientist* 70(5):506.

Page, D. N., and McKee, M. R. 1983. The future of the universe. *Mercury* 12(1):17.

Penzias, A. A. 1978. The riddle of cosmic deuterium. *American Scientist* 66:291.

Schramm, D. N. 1974. The age of the elements. *Scientific American* 230(1):69.

Silbar, M. L. 1982. Neutrinos: rulers of the universe? *Griffith Observer* 46(1):9.

Trefil, J. S. 1978. Einstein's theory of general relativity is put to the test. *Smithsonian* 11(1):74.

———.1983. How the universe began. *Smithsonian* 14(2):32.

———.1983. How the universe will end. *Smithsonian* 14(3):72.

Trefil, J. S. 1983. *The moment of creation.* New York: Charles Scribner's Sons.

Tucker, W., and Tucker, K. 1982. A question of galaxies. *Mercury* 11(5):151.

Weinberg, S. 1977. *The first three minutes.* New York: Basic Books.

Weisskopf, V. F. 1983. The origin of the universe. *American Scientist* 71(5):473.

Martin Rees

Professor Rees is one of the world's foremost authorities on theoretical studies of galaxy formation and of the highly energetic and still mysterious quasars and peculiar galaxies. Most of the modern ideas about the energy sources of these objects owe some debt to his work on accretion disks and their interaction with surrounding gas. One of his most noted achievements was the theoretical prediction that double radio sources such as radio galaxies should have jets of fast-moving, superheated gas emanating from their centers, creating the double radio-emitting lobes. Today sensitive radio observations have revealed the widespread existence of these jets. Professor Rees is well known for his ability to develop, from simple physical principles and scaling laws, new understandings of complex phenomena. For several years director of the Institute of Astronomy at Cambridge University, he is now Plumian Professor of Astronomy. Here are some of his thoughts on another area of interest to him, cosmology and the roles of consciousness and of fundamental physical

quantities in determining the nature of the universe.

According to the big-bang theory, about fifteen billion years ago all the material in the universe constituted an exceedingly compressed and hot gas—hotter than the centers of stars. The intense radiation in this compressed primordial fireball, though cooled and diluted by the expansion (the wavelengths being stretched and redshifted), could still be around pervading the whole universe. The background radiation that Penzias and Wilson discovered is a kind of "echo" of the explosion that initiated the universal expansion. The microwave photons now reaching us have been propagating uninterruptedly since before galaxies formed, maybe since the universe was a thousand times more compressed than today and at a temperature of several thousand degrees.

If we boldly ask what the universe would have been like when it was a few seconds old, we find that the fireball temperature would then be measured in billions of degrees. The nuclear reactions that occurred as it cooled through this temperature range can be calculated in detail, and reveal that the material emerging from the fireball would be about 75 percent hydrogen and 25 percent helium. This is gratifying, because the theory of synthesis of elements in stars and supernovae, which works so well for carbon, iron, et cetera, was always hard pressed to explain why there was so much helium, and why the helium was so uniform in its abundance. The attribution of helium formation to the big bang

thus solved a long-standing problem in nucleogenesis, and gave cosmologists confidence in extrapolating right back to the first few seconds of the universe's history.

Can we extrapolate back still further? As the inferred temperatures (or energies) become higher, our knowledge of the relevent physics becomes less confident. However, our universe may owe some key properties—its scale, and the proportions of matter and radiation that fill it—to processes occurring when the thermal energies were 10^{24} electron volts (meaning that individual particles had enormous kinetic energies, about 10^{12} ergs per particle) and the age of the universe was about 10^{-35} seconds. This is at least the hope of physicists working on grand-unified theories.

None of our ideas about the early universe are too firmly established. The hot big-bang concept has more than fashion to recommend it, and offers a consistent story about the origins of matter and radiation in our universe: it is certainly more plausible than any equally specific alternative. But it is hypothesis and not dogma; our present satisfaction may prove as illusory as that of a Ptolemaic astronomer who has just discovered another epicycle. Cosmologists must sometimes be chided for being "often in error but never in doubt." We must not be bemused by a self-consistent theory into closing our eyes to conflicting data, which could surface some day.

Having inferred something, tentatively at least, about how our expanding universe began, cosmolo-

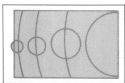
gists have set the context for asking questions about its future and eventual fate. Will the universe expand forever, and the galaxies fade and disperse? Or, on the other hand, will it recollapse? Will we all share the fate of an experimenter who ventured into a black hole?

We ask the universe which course it will follow, by measuring its deceleration or by comparing its mass content with the critical density that would cause it to recollapse. We do not yet have an unequivocal answer. Let us therefore consider both options for the long-term future of our universe.

What will happen if the universe recollapses? The redshifts of distant galaxies would be replaced by blueshifts, and galaxies would crowd together again. Space is already becoming more and more punctured as isolated regions—dead stars and galactic nuclei—collapse to black holes. But this would then be just a precursor of a universal squeeze to a big crunch that engulfs everything. Galaxies would merge; stars would move faster (just as atoms of a gas move faster), as the universe became compressed; stars would eventually be destroyed because the night sky was hotter than their centers. The final outcome would be a fireball like that which initiated the universe's expansion, though it would be somewhat more lumpy and unsynchronized. The earliest this could happen would be fifty billion years from now, so the breathing space is at least ten times the remaining lifetime of the sun.

But what about the other case,

when there is not enough gravitating stuff ever to halt the universe's expansion? The universe then has time to run down to a final heat death. If an astronomer had to answer the question "what is happening in the universe?" in just one sentence, he might say this: "Gravitational binding energy is being released, as stars, galaxies, and clusters progressively contract, this inexorable trend being delayed by rotation, nuclear energy, and the sheer scale of astronomical systems, which makes things happen slowly and staves off gravity's final victory." But if the universe expands indefinitely there *will* be enough time for all stars, all galaxies, to attain a terminal equilibrium. Stars all die, galaxies experience dynamical evolution, black holes grow. If protons do not live forever, then all ordinary stars will eventually decay, leaving only black holes. These too will eventually decay by evaporation. If protons did last forever, then the final heat death would be spun out over even longer periods. The time it would take before neutron stars tunneled into black holes (via quantum effects) is so enormous that, if written out in full, it would have as many zeros as the number of atoms in the observable universe!

Will our descendants have the potentiality to survive an infinite future? Or will they fry in the big crunch a few times 10^{10} years hence? These two outcomes are very different. We will need to compile a more complete inventory of what is in the universe, by observing all wavebands, and searching for black holes, before

we can pronounce a more reliable long-range forecast for the next hundred billion years.

But the initial conditions that could have led to anything like our present universe are actually *very restrictive,* compared to the range of possibilities that might have been set up. We know that our universe is still expanding after 10^{10} years. Had it recollapsed sooner, there would have been no time for stars to evolve. If it collapsed after less than a million years, it would have remained opaque and close to thermal equilibrium—and thermodynamic disequilibrium is a necessity for any kind of life. The expansion rate cannot however be too fast. Otherwise no gravitationally bound systems—neither stars nor galaxies—could have condensed from the expanding background (this is equivalent to the statement that the present density is not orders of magnitude below the critical density). There is therefore a sense in which the dynamics of the early universe must have been *finely tuned.* In Newtonian terms, the fractional difference between the initial potential and the kinetic energies of any spherical region must have been very small. The universe is close to a balance between being open and closed.

This fine-tuning requirement stimulates further speculations. For instance, all of the features of the everyday world and the astronomical scene are essentially determined by a few basic physical laws and constants, such as the masses of the elementary particles, and the relative strengths of the basic forces that operate between them. In many cases a

rather delicate balance seems to prevail. For instance, if nuclear forces were slightly *stronger* than they actually are relative to electromagnetism, the diproton (a nucleus consisting of only two protons) would be stable, ordinary hydrogen would not exist, and stars would evolve quite differently. If nuclear forces were slightly *weaker,* no chemical elements other than hydrogen would be stable, and chemistry would be very dull indeed. The details of stellar nucleosynthesis are sensitive to some apparent accidents. For instance, there is a particular resonance in the carbon nucleus, whose precise level is crucial in allowing carbon and oxygen both to be produced nonexplosively in stars.

Perhaps the contingency, that we are here to ponder such matters, itself poses some constraints on what physical laws can be like. In other words, given that we know our cosmic environment permits observers to exist, maybe we should not take the Copernican principle too far: we would not feel justified in assigning ourselves a central position in the universe, but it may be equally unrealistic to deny that our situation can be privileged in any sense. This line of argument is sometimes called the "anthropic principle." We might imagine an ensemble of universes with different physical laws and different fundamental ratios. Most of the universes would be "stillborn," in the sense that the prevailing laws would not permit anything interesting to happen in them. But maybe in some of them complex structures can evolve,

and we would have achieved something if we could show that any such universe has to possess features that our actual universe does possess. We must of course not be too anthropomorphic, nor too restrictive in envisioning the requirements for the emergence of a conscious observer.

The anthropic principle obviously cannot provide a scientific explanation in the proper sense. At best it can offer a stopgap satisfaction of our curiosity regarding phenomena for which we cannot yet obtain a genuine physical explanation. We can note that the world would be very different, and perhaps even "uncognisable," if the relative strengths of the nuclear and electromagnetic interactions were somewhat altered, but we can still hope that a grand-unified theory will relate the strengths of these forces. It will then seem as unnatural to envision tinkering with their ratios as it has been, since the work of Maxwell, to vary independently the magnetic and electric forces and the speed of light.

The main conceptual inadequacy in present-day physics is that there is no theory which reconciles gravity with the quantum principle. Normally there is no overlap in the domain of relevance of these two effects. Quantum effects come in only on the microscopic scale, and gravity is important only on large scales where quantum fluctuations do not matter. But in the big bang (and in gravitational collapse) a proper theory unifying gravity with the other forces of nature is a necessary prerequisite for understanding

why things in the early universe were as they were.

The eventual status of the anthropic principle will depend on what the final unified theory is like, if indeed it exists. If this theory has a *statistical* element which could in principle lead to a universe cooling down with different values of the fundamental ratios, then the ensemble idea would be put on a serious footing. If, on the contrary, this final theory yields unique numbers for all the ratios, then it may be inconceivable for us to envision a universe with different fundamental constants. But it would then seem coincidental, or even providential, that the constants happen to lie in the narrowly restricted range that permitted the evolution of complexity and consiousness in the low-energy world we inhabit.

"The most incomprehensible thing about the universe is that it is comprehensible" is one of the best known (and indeed most hackneyed) of Einstein's sayings. He meant by this that the basic physical laws, which our brains are attuned to understand, have such broad scope that they offer a framework for interpreting not just the everyday world but even the behavior of the remote cosmos. The physicist Eugene Wigner described this as "the unreasonable effectiveness of mathematics in the physical sciences." The proverbial rational man who loses keys at night searches only under the street lamps not because that is necessarily where he dropped them, but because his quest is otherwise quite certain to fail. Cosmologists approach their subject in a similar way. They start by using

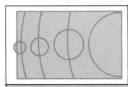

the physics that is validated locally, and making simplifying assumptions about symmetry, homogeneity, et cetera. There seems to be no reason why the universe *should* be so ordered that this permits any real progress—why the physics we study in the laboratory applies in quasars billions of light-years away, and in the early stages of the big bang. But unless there is a firm link with local physics, cosmology risks degeneration into ad hoc explanations on the level of "just so" stories. What does seem amazing is that this has led to some progress, that the universe is comprehensible. Questions about how the universe began and how it will end certainly cannot be answered, but can now at least be posed in a scientific spirit.

LIFE IN THE UNIVERSE

Introduction to Section VI

We finish our survey of the dynamic universe with a discussion of living organisms and the prospects for finding that we are not alone in the cosmos. To many, this is the most central question we can ask. In the sole chapter of this section, we examine the astronomer's attempt to answer it.

In discussing the question of life elsewhere in the universe, we begin by looking into the origins of life here on earth. In the process, we learn of biological evolution and the evidence that has been unearthed to tell us of our own beginnings. To do this, we will step outside the arena of astronomy, but we return to it in assessing whether the conditions that led to life on earth might exist on other planets in our solar system and in the galaxy at large. We discuss the probability that technological civilizations might be thriving here and there in the Milky Way.

We complete our brief treatment of life in the universe with a description of the strategy for searching for other civilizations. It seems most likely that radio communications will be our initial means of contact, and there is an interesting exercise in logic involved with choosing the wavelength at which to search for or send signals.

THE CHANCES OF COMPANIONSHIP

Learning Goals

We have attempted to answer all the fundamental questions about the physical universe that can be treated scientifically. Having done this, we know our place in the cosmos: we know something of the scale and of the universe, and we realize how insignificant our habitat is.

In this chapter we contemplate whether we as living creatures are unique in the universe, or whether even that distinction must be shared. Nothing that we have learned so far leads us to rule out the possibility that other life forms, some of them intelligent, exist. We believe that the earth and the other planets are a natural by-product of the formation of the sun, and we have evidence that some of the essential ingredients for life were present on the earth from the time it formed. Similar conditions must have been met countless times in the history of the universe, and will occur countless more times in the future.

Science cannot yet tell the full story of how life began, however, and we have not found any evidence that life actually does exist elsewhere. The mystery remains.

Life on Earth

We start by discussing the origins of the only life we know. Besides giving us some insight into the processes thought to have been at work on the earth, this will help us later, when we are speculating about whether the same processes have occurred elsewhere.

Before the time of Charles Darwin (Fig. 24.1) in the mid-1800s, the view was widely held that life could arise spontaneously from nonliving matter. Darwin's study of evolution, showing how species develop gradually as a result of environmental pressures, made such an idea seem improbable.

An alternative to the idea that life arose spontaneously was proposed in 1907 by the Swedish chemist S. Arrhenius, who suggested that life on earth was introduced billions of years ago from space, originally in the form of microscopic spores that floated through the cosmos, landing here and there to act as seeds for new biological systems. This idea, called the **panspermia** hypothesis, cannot be ruled out, but several arguments make it seem unlikely. It would take a very long time for such spores to permeate the galaxy, and their den-

sity in space would have to be very high in order for one or more of them to reach the earth by chance. More important, it seems very unlikely that the spores could survive the hazards of space, such as ultraviolet light and cosmic rays. Even if the panspermia concept is correct, the question of the ultimate origin of life remains, although it is transferred to some other location. In view of what is known today about the evolution of life and the early conditions on earth, scientists generally agree that life arose through natural processes occurring here, and was not introduced from some place else.

It is believed that the early atmosphere of the earth was composed chiefly of hydrogen (H_2) and hydrogen-bearing molecules such as ammonia (NH_3) and methane (CH_4), as well as water (H_2O). Therefore the first organisms must have developed in the presence of these ingredients.

In the 1950s, scientists performed experiments that involved reproducing the conditions of the early earth. The starting point of these experiments was to place water in containers filled with the type of atmosphere just described. Water was introduced because it is apparent that from very early times the earth had oceans, and because it is thought that life started in the oceans, where the liquid environment provided a medium in which complex chemical reactions could take place. Reactions occur much more slowly in solids and gases,

FIGURE 24.1 *Charles Darwin.*
The scientific inquiries by Darwin led to an understanding of evolution, one of the most profound concepts of human intellectual development.

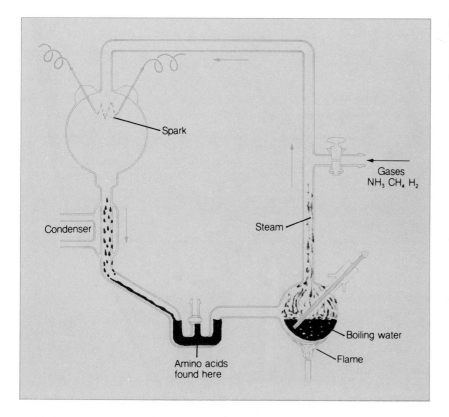

FIGURE 24.2 *Simulating the Early Earth.*
This is an apparatus constructed by Urey and Miller to reproduce conditions on the primitive earth, in hopes of learning how life forms could have developed.

in the former case because the atoms are not free to move about easily and interact, and in the latter case because the density of gas is low, and particles are relatively unlikely to encounter each other. Water is the most stable and abundant liquid that can form from the common elements thought to have been present when the earth was young.

The first of these experiments (Fig. 24.2) was carried out in 1953 by the American scientists H. Urey and S. Miller, who concocted a mixture of methane, ammonia, water, and hydrogen, and exposed it to electrical discharges, a possible source of energy on the primitive earth (ultraviolet light from the sun is another, but was more difficult to work with in the laboratory). After a week, the mixture turned a dark brown, and Urey and Miller analyzed its composition. What they found were large quantities of amino acids, complex molecules that form the basis of proteins, which are the fundamental substance of living matter. Other scientists later showed that exposing the mixture of elements to ultraviolet light produced the same results. These experiments demonstrated that at least some of the precursors of life probably existed in the primitive oceans

almost immediately after the earth had cooled enough to support liquid water. Other similar experiments produced more complex molecules, including sugars and larger fragments of proteins.

As noted in chapter 10, amino acids may have been present in the solar system even before the earth formed, because traces of them have been found in some meteorites (Fig. 24.3), and we know that meteorites are very old, representing the first solid material in the solar system. We also know that several kinds of complex molecules, including organic (carbon-bearing) molecules, exist inside dense interstellar clouds (see chapter 18), and we can speculate that amino acids formed in these regions. In order to have survived on the earth, these primordial amino acids would have had to reach our planet sometime *after* its molten period.

It is not so clear what direction things took, once amino acids and other organic molecules existed. Somehow these building blocks had to combine to form **ribonucleic acid** (RNA) and **deoxyribonucleic acid** (DNA; Fig. 24.4). These very complex molecules carry the genetic codes that allow living creatures to

FIGURE 24.3 *Primordial Amino Acids.*
This is a section of the Murchison meteorite, which fell in
Australia. Amino acids found in this carbonaceous chondrite
were undoubtedly present in it when it fell.

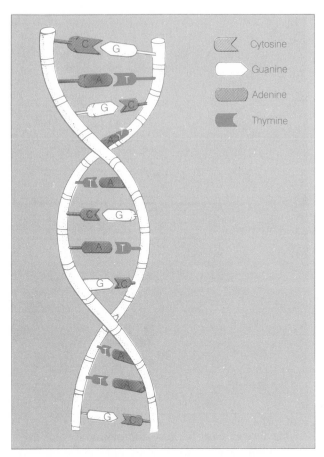

Cytosine
Guanine
Adenine
Thymine

FIGURE 24.4 *The DNA Molecule.*
This is a schematic diagram of a section of a DNA molecule.
DNA carries the genetic code that allows organisms to
reproduce themselves, and a critical question in understanding
the development of life on earth is to learn how DNA arose.

reproduce themselves. Experiments have successfully
produced molecules that are fragments of RNA and
DNA from conditions like those which prevailed on the
early earth, but not the complete forms required.
Maybe it is a simple matter of time; if such experiments
could be carried out for years or millennia, perhaps the
vital forms of these proteins would appear. This is one
of the areas of greatest uncertainty in our present
knowledge of how life began.

Fossil records tell us that the first microorganisms
appeared some 3 to 3.5 billion years ago, when the
earth was barely a billion years old (Fig. 24.5). The
evolution of more complex species seems to have followed naturally. At first the development was very
slow, only reaching the level of simple plants such as
algae a billion years later. Increasingly elaborate mul-
ticellular plant forms followed, and gradually altered
the earth's atmosphere by introducing free oxygen.
Meanwhile, the gases hydrogen and helium, light and
fast-moving enough to escape the earth's gravity, essentially disappeared. Nitrogen, always present from outgassing and volcanic activity, became more predominant through the decay of dead organisms. By about
one billion years ago, the earth's atmosphere had
reached its present composition.

The first broad proliferation of animal life occurred
about 600 million years ago, and the great reptiles
arose some 350 million years later. The dinosaurs died

out after about 200 million years, and mammals came
to dominance about 65 million years ago. Our primitive ancestors appeared only in the last 3 or 4 million
years. Once the development of intelligence had provided the ability to control the environment, the entire
world became our ecological home, and our physical
evolution essentially stopped. It remains to be seen
whether future ecological pressures will create future
evolution of the human species.

Could Life Develop Elsewhere?

The scenario just described, if at all accurate, seemingly should occur almost inevitably, given the proper
conditions. If this is so, then the question of whether

FIGURE 24.5 *A Fossil Microorganism.*
This is evidence for the fact that primitive life forms existed on the earth billions of years ago.

life exists elsewhere amounts to asking whether the conditions that existed on the primitive earth could have arisen elsewhere. (For a dissenting view, that life was still improbable, see Astronomical Insight 24.1.)

It is clear that no other planet in the solar system could have provided an environment exactly like that of the early earth. It therefore seems unlikely that earthlike organisms will be found anywhere else in the solar system (but recall the argument in chapter 7 that the atmosphere of Jupiter or the interior of its satellite Europa may come close to reproducing the conditions of the early earth). Mars is the only planet where life forms have been sought so far (Fig. 24.6), but even there the conditions are quite unlike those on earth. Given the billions of stars in the galaxy, however, and the vast number that are very similar to the sun, it seems highly probable that the proper conditions must have been reproduced many times in the history of the galaxy.

So far we have worked from the tacit assumption that life on other planets, if it exists, must be similar to life on earth, and we have considered only the question of whether there could be other earthlike environments. We may question the premise that life could only have developed in the form that we are familiar with, however. Here, obviously, we must indulge in speculation, having no examples of other types of life at hand for examination.

Life as we know it is based on carbon-bearing molecules, and it has been argued that only carbon has the capability of combining chemically in a sufficiently wide variety of ways with other common elements to produce the complexity of molecules thought to be necessary for life. After all, it is clear that the basic elements available to begin with must be the same every-

FIGURE 24.6 *Searching for Life in the Solar System.*
Mars was long thought to be the most likely home in the solar system for extraterrestrial life. Here we see a *Viking* lander in a simulated Martian environment (left), and an artist's concept of a planned sample return spacecraft, lifting off of the Martian surface with a cargo of soil and rocks (right). The *Viking* mission reached Mars in 1976, but found no evidence for life forms.

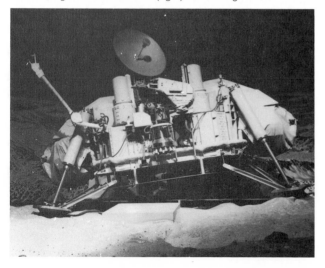

where, given the homogeneous composition of the universe. This may seem to rule out life forms based on anything other than carbon, but it has been pointed out that another common element, silicon, also has a very complex chemistry, and therefore might provide a basis for a radically different type of life. If so, we cannot begin to speculate on the conditions necessary for such life forms to arise.

Another assumption that might be subject to question is that water is a necessary medium. As noted earlier, it is the most abundant liquid that could form under the temperature and pressure conditions of the early earth, and it is thought that only a liquid medium could support the required level of chemical activity. Other liquids can exist under other conditions, however, and it is interesting to consider whether life forms of a wholly different type than we are familiar with might arise in oceans of strange composition. It is interesting to speculate, for example, about what goes on in the lakes of liquid methane that are thought to exist on Titan, the mysterious giant satellite of Saturn.

If we are satisfied that life probably has formed naturally in many places in the universe, we can address a related, and to many a more important, question: given the existence of life, how likely is it that intelligence will follow? Here we have no means of answering, except to reiterate that as far as we know, the evolution of our species on earth was the natural product of environmental pressures.

The Probability of Detection

In view of the limitations that prohibit faster-than-light travel, it is exceedingly unlikely that we will be able to visit other solar systems, seeking out life forms that may reside there. We will continue to explore our own system, so there is a reasonable chance that if life exists on any of the other planets of our sun, we will someday discover it. It seems, however, that our best hope of finding other intelligent races in the galaxy will be to make long-range contact with them, through radio or light signals. Since this requires both a transmitter and a receiver, we can hope to contact other civilizations only as advanced as ours, with the capability of constructing the necessary devices for interstellar communication.

A mathematical exercise in probabilities has been used for some years as a means of assessing, as objectively as possible, the chances of making contact with an extraterrestrial civilization. The aim is to separate the question into several distinct steps, each of which can be treated independently. The underlying assumption is that the number of technological civilizations in our galaxy today with the capability for interstellar communication is the product of the number of planets that exist with appropriate conditions, the probability that life arose on those planets, the probability that such life developed intelligence that gave rise to a technological civilization, and finally, the likelihood that the civilization was not destroyed, through evolution or catastrophe.

Mathematically, the so-called **Drake equation** (after Frank Drake, who has been its best-known advocate) is written:

$$N = R_* f_p n_e f_l f_i f_c L,$$

where N is the number of technological civilizations presently in existence, R_* is the number of stars of appropriate spectral type formed per year in the galaxy, f_p is the fraction of these stars that have planets, n_e is the number of earthlike planets per star, f_l is the fraction of these on which life arose, f_i is the fraction of those planets on which intelligence developed, f_c is the fraction of planets with intelligence on which a technological civilization evolved to the point where interstellar communications is possible, and, finally, L is the average lifetime of such a civilization.

By using this equation, we are able to isolate the factors that we can make educated guesses about from those we are more ignorant about. It is an interesting exercise to go through the terms of the equation one by one, to see what conclusions we reach under various assumptions. People who do this have to make sheer guesses for some of the terms, and the result is a variety of answers ranging from very optimistic to very pessimistic ones. We will adopt middle-of-the-road numbers for most of the unknown terms.

The first two factors, R_* and f_p, are quantities that in principle can be known with some certainty from observations. For the rate of star formation to be relevant in this exercise, stars must be of a spectral type similar to the sun's. A much cooler star would not have a temperate zone around it where a planet could have the moderate temperatures needed for life to begin (we are, throughout this exercise, limiting ourselves to life forms similar to our own). A very hot star would be too short lived; there would be insufficient time for life to develop on its planets, before the parent star blew up in a supernova explosion. Taking these considerations into account, some people estimate that up to ten suitable stars form in our galaxy per year. For the sake of discussion, we will be more cautious and adopt $R_* = 1$/year. Even this may be a little high for the present epoch in galactic history, but the star formation rate was surely much higher early in the lifetime of the galaxy, and low-mass stars of solar type which formed that long ago are still in their prime. Thus, an average formation rate of one sunlike star per year is probably reasonable.

From what we know of the formation of our solar system, it seems that the formation of planets is almost inevitable, except perhaps in double- or multiple-star systems. Let us assume that $f_p = 1$; that is, all stars of solar type have planets. Observations made with future instruments such as the *Space Telescope* may soon provide real information on this term.

The number of earthlike planets, n_e, is highly uncertain, and depends on how wide the zone is where the appropriate temperature conditions could exist. Recent studies show that this zone might be rather small; that the earth would not have been able to support life if its distance from the sun were only about 5 percent closer or farther than it is. Estimates of n_e vary from 10^{-6} to 1. Let us be moderate and assume $n_e = 0.1$; that in one out of ten planetary systems surrounding solar-type stars, there is a planet within the temperate zone where life can arise.

Now we get to the *really* speculative terms in the equation. We have no way of estimating how likely it is that life should begin, given the right conditions. It can be argued that because of the seeming naturalness of its development on earth, life would always begin if given the chance. Let us be optimistic here and agree with this, adopting $f_l = 1$.

Again, the chance of this life developing intelligence, and furthermore advancing to the point of being capable of interstellar communication, are complete unknowns. All we know is that in the only example that has been observed, both things happened. For the sake of argument, we therefore set both f_i and f_c equal to 1.

At this point it is instructive to put the values adopted so far into the equation. We find:

$$N = (1/\text{year})(1)(0.1)(1)(1)(1)L$$
$$= 0.1L$$

Having taken our chances and guessed at the values for all the other terms, we now face a critical question: how long can a technological civilization last? Ours has been advanced enough to send or receive interstellar radio signals for only about fifty years, and there is sufficient instability in our society to lead some pessimists to think we will not last many more decades. If we take this viewpoint and adopt fifty years as the average lifetime, then we find:

$$N = 5,$$

meaning that we expect the total number of technological civilizations present in the galaxy at any one moment to be very small, about 5. If this is correct, then the average distance between these outposts of civilization is nearly 20,000 light years (Fig. 24.7). The time it would take for communications to travel between civilizations would therefore be very much longer than their lifetimes, and we would have no hope of establishing a dialogue with anyone out there. If this estimate is correct, it is no surprise that we have not heard from anybody yet.

We can be more optimistic, though, and assume that technological civilization solves its internal problems and lives much longer than fifty years. Extremely optimistic people would argue that a civilization is immortal; that it colonizes other star systems besides its

FIGURE 24.7 *Possible Values of N.*
The number of technological civilizations in the galaxy may be very small (upper), in which case the average distance between them is very large; or N may be large, so that the distance between civilizations is relatively small. The chances for communication with alien races are much higher in the latter case, where the distances are only a few tens of light years.

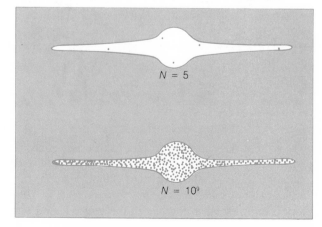

$N = 5$

$N = 10^9$

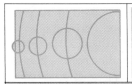

Astronomical Insight 24.1

THE CASE FOR A SMALL VALUE OF *N*

Although it is an interesting exercise to write down an equation for *N*, the number of technological civilizations in the galaxy, no two astronomers can agree on the values of the various terms.

One very pessimistic argument developed recently concludes that we are alone in the galaxy. This is based on the likely fact that if there were other civilizations, a significant fraction of them would have arisen long before ours. This is a straightforward consequence of the great age of the galaxy, some 5 or 10 billion years greater than the age of the earth. This argument postulates further that if technological civilizations, once begun, live a long time, then the galaxy must be inhabited by a number of very advanced races, much older and more mature than our own. There is little controversy about any of these assertions.

The next logical step in the argument, however, is hotly debatable. It says that any advanced civilization that has survived for millions or even billions of years will necessarily have done so by perfecting some form of interstellar travel, and colonizing other planets. Once this started, the argument goes, the spread of a given civilization throughout the galaxy would accelerate rapidly, and by this time in galactic history, no habitable planet such as the earth would remain uncolonized. The first successful galactic civilization, in this view, would quickly rise to complete dominance.

The only logical corollary, if the argument is accepted up to this point, is that no other civilizations exist, because if they did, the earth would have been colonized long ago. The absence of interstellar visitors on earth is taken as proof that there are no other civilizations in the galaxy.

Those who adhere to this line of reasoning believe that $N = 1$. Therefore at least one of the terms in Drake's equation must be very much smaller than the more optimistic values discussed in the text. One suggestion is that the temperate zone around a sun-like star where planets can have moderate conditions is really very much narrower than usually supposed, perhaps because of the more ready development of a Venus-like greenhouse effect than is normally thought to be the case. If the temperate zone is very small, then the term n_e, representing the number of habitable, earthlike planets per star, would be very much smaller than the value of 0.1 adopted in the text.

Another term that has recently been singled out is f_l, the fraction of earthlike planets on which life begins. As we learned in the text, somehow the amino acids that were present on the primitive earth had to arrange themselves into special combinations to produce the long chainlike protein molecules RNA and DNA. From a purely statistical point of view, the probability that the proper combination would come together by chance is very small. This apparently happened on the earth, but it may be so unlikely that it has not occurred anywhere else in the galaxy.

Whatever the correct values of the terms in the equation, the debate will rage on until our civilization reaches its life expectancy *L* and dies; or we discover another civilization; or we ourselves expand to colonize the galaxy and find no one else out there.

own, so that it is immune to any local crises such as planetary wars or suns expanding to become red giants. In that case, allowing a few billion years for the development of such civilizations, we can set $L = 10^{10}$ years (nearly equal to the age of the galaxy), and we find

$$N = 10^9,$$

in which case the average distance between civilizations is only about thirty light years (Fig. 24.7), coincidentally just a little less than the distance our own radio signals have traveled since the early days of radio and television (Fig. 24.8). If this estimate of N is correct, we should be hearing from somebody very soon.

We have presented two extreme views of the likelihood that other civilizations exist in our part of the galaxy. As we mentioned earlier, opinions among the scientists who seriously study this question vary throughout this large range. Those who favor the optimistic viewpoint advocate the idea that we should make deliberate attempts to seek out other civilizations.

The Strategy for Searching

The probability arguments just outlined are amusing and perhaps somewhat instructive, but obviously not very accurate. There are entirely too many unknowns in the equation for us to develop a reliable estimate of the chances for galactic companionship. Perhaps we

will not know for certain what the answer is until we make contact with another civilization.

The problem of developing an experiment to search for or send interstellar signals is that we do not know the ground rules. There is an infinite number of ways in which a distant civilization might choose to try to communicate, and it is impossible to search for them all. We must try to guess what the most probable technique would be.

In view of the power that is transmitted by radio signals and the relative lack of natural "noise" in the galaxy in that part of the spectrum, it has often been assumed that radio communications are most likely to succeed, although other techniques have been tried (Fig. 24.9). In the early 1960s, the U.S. National Radio Astronomy Observatory (Fig. 24.10) "listened" for transmissions from two nearby sunlike stars, tau Ceti and epsilon Eridani, without success. More recently, larger surveys of neighboring stars have been carried out, in the United States and in the Soviet Union, with similar results. A new search program was started in 1983, using a radio telescope belonging to Harvard University. This telescope is now entirely devoted to the search for extraterrestrial communications.

The wavelength chosen for most of these observa-

FIGURE 24.8 *Earth's Message to the Cosmos.*
As our entertainment and communications broadcasts travel out into space, they provide a history of our culture for anyone who may be receiving the signals. At the present time, the growing sphere that is filled with our broadcasts has a radius of over fifty light years.

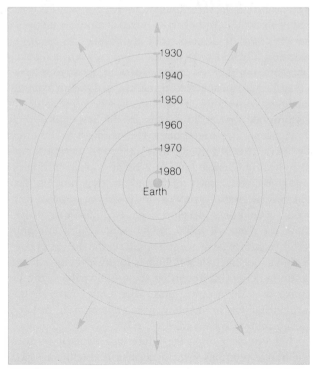

FIGURE 24.9 *Another Message From Earth.*
This recording of a message from earth is travelling beyond the solar system aboard the Voyager spacecraft. Only an advanced race of beings would have the technological skills necessary to learn how to listen to and decode the message of peace that it contains.

FIGURE 24.10 *A Radio Telescope Used in the Search For Life.*
This ia an antenna at the U.S. National Radio Astronomy
Observatory, in Green Bank, West Virginia, which has been
used in efforts to detect signals from alien civilizations. Most of
these attempts have been made at the 21-cm wavelength of
atomic hydrogen emission.

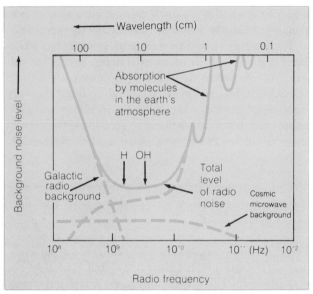

FIGURE 24.11 *The Galactic Noise Spectrum.*
This diagram shows the relative intensity of the background
noise in the galaxy as a function of wavelength. The quietest
portion of the spectrum is in the vicinity of the so-called "water
hole", between the natural emission wavelengths of H_2O and
OH molecules.

tions has been 21 centimeters, selected partly for practical reasons (radio telescopes designed for observing
at this wavelength already exist), and partly because it
has been hypothesized that if a civilization out there
wanted deliberately to send signals, it might choose to
do so at a wavelength that astronomers around the galaxy might be observing anyway, even if they were not
trying to find distant civilizations. Another wavelength
that has been considered but has been tried only in a
very limited search, lies in the microwave region, between the emission lines of interstellar water vapor and
OH. Dubbed the "water hole," this region has the advantages that it, too, might be a target of extraterrestrial
astronomers, and that it is in the quietest part of the
radio spectrum. (Various objects, ranging from supernova remnants to radio galaxies, create radio noise that
fills the galaxy, and the water hole lies in a region of
the spectrum where this noise is minimized; see Fig.
24.11.)

One pessimistic viewpoint is that we are all listen-
ing, and no one is sending. If that is the case, we can
hope to pick up only accidental emissions, such as entertainment broadcasts on radio or television, which
would be much weaker and more difficult to detect. It
has been estimated that at the present level of technology we could deliberately send radio signals that could
be received by a similarly advanced civilization up to
several thousand parsecs away, whereas our accidental

radio and television signals, broadcast indiscriminately
in all directions rather than being beamed at a specific
target, could be detected only at distances of one or two
parsecs. Therefore, it makes a big difference whether
someone out there is trying to send a message or not.

Although most thinking about methods of communication with other civilizations has been based on the
assumption that radio signals are the best choice, there
have been other suggestions. One is that a **laser** might
be used. This is a device that can emit a powerful, narrow beam of visible light at a single wavelength; lasers
today are being used for many purposes ranging from
microsurgery to weapons to telephone communications. The lasers developed so far, however, do not
have the range of radio transmissions, so at this point
most scientists favor radio signals as the means of communication. Laser technology is improving, but there is
a fundamental limitation on the use of visible light: the
universe is far noisier, more filled with confusing natural emissions, at visible wavelengths than in the radio
portion of the spectrum.

It is not clear what the future will bring, in terms of
deliberate searches for extraterrestrial intelligence. The

FIGURE 24.12 *Project Cyclops.*
The grandest plan seriously considered for the purpose of communicating with extraterrestrial civilizations, *Project Cyclops* would consist of 1000 to 2500 radio antennae, each 100 feet in diameter. This huge array, some 16 kilometers in extent, would be capable of detecting signals over a large portion of the radio spectrum, from planetary systems 1000 or more parsecs away.

most ambitious project so far is called *Project Cyclops* (Fig. 24.12), and it is probably several years or decades away from being implemented. The plan is to build a veritable forest of large radio antennae, along with a very sophisticated data-processing system, so that up to a million stars could be scanned for artificial emissions over a broad wavelength region in a few years' time. By searching over a wide range of wavelengths, the scientists planning this effort hope to eliminate much of the guesswork about which is the most likely portion of the spectrum to be used by alien civilizations. This also enhances the chances of detecting accidental emissions, which, if they are anything like our own radio and television signals, may be spread throughout much of the spectrum. The *Project Cyclops* antennae could also be used to send powerful radio signals, allowing the earth to broadcast its own beacon for others to find, and providing for the possibility of two-way communications.

The principal problem in implementing *Cyclops,* or any other large-scale effort, is the enormous cost. The version of *Project Cyclops* just outlined is currently estimated to cost some 10 billion dollars. Such a sum is rarely devoted to astronomy, and if it were, there would still be considerable debate over whether to spend it on this or on some other research program that might appear to have greater certainty of success. Presumably, astronomers in other solar systems would have to face the same questions, although we can hope that sufficiently advanced civilizations will have overcome the limitations of their planetary resources, and

will be willing and able to tackle such a massive task.

A less ambitious project, called the *Search for Extraterrestrial Life (SETI)* has been funded by NASA. The plan is to use existing radio dishes and specially developed, sophisticated computers to carry out a limited search of a number of stars. Like the more complex *Cyclops* plan, *SETI* will scan a number of wavelengths for possible signals, although will not cover so broad a spectrum nor such a great number of stars. For *SETI* to succeed in detecting another civilization, the value of N will have to be quite high. Although the scope of *SETI* is far short of that envisioned for *Cyclops,* it is nevertheless noteworthy that a federal government agency has taken the task seriously enough to fund a search for extraterrestrial life.

PERSPECTIVE

In this chapter we have introduced a bit of speculation, well-founded perhaps in the discussion of life on earth, but less so in the sections on the possibilities of life elsewhere.

We are now prepared to understand and appreciate new advances in astronomy. Having completed our study of the present knowledge of the universe, we will be able to put into perspective the major new discoveries that are sure to come as technology and theory continue to improve. The *Space Telescope,* the large optical telescopes now being designed, the comparable advances in radio, infrared, and X-ray techniques; all of these will inevitably lead to novel and unforeseen breakthroughs in our view of the universe. The next few decades will be exciting times for astronomy, and we can anticipate their coming with a sense of excitement and wonder.

signals at 21-cm and other wavelengths, with no success.

9. The chances for detecting or being detected by other civilizations depend strongly on whether or not deliberate attempts are made to send signals.

10. Future projects to search for and send signals have been planned, and will be sufficiently powerful to have a high probability of success if a large number of civilizations exist. However, such projects may not be funded and put into operation in the foreseeable future.

SUMMARY

1. Although there were early beliefs that life started on earth spontaneously or by primitive spores from space, most scientists today accept the theory that life began through natural, evolutionary processes.

2. Amino acids, fundamental components of living organisms, are formed readily in experiments designed to simulate early conditions on earth.

3. The steps that led to the development of the necessary forms of RNA and DNA are not yet fully understood, and probably occurred over a very long period of time.

4. Fossil evidence provides a record of the evolutionary steps leading from the first primitive life forms to modern civilization.

5. The conditions that prevailed on the early earth have probably been duplicated on other planets in the galaxy, though probably not on other planets in the solar system.

6. It is often assumed that only earthlike life could develop, because carbon is nearly unique in the complexity of its chemistry. However, there have been suggestions that at least one other element (silicon) may have the necessary properties.

7. Estimates of the number (N) of technological civilizations now in the galaxy can be made (with great uncertainty) based on what is known of the formation rate of sunlike stars, and what is guessed as the probability of such stars' having planets with the proper conditions, and the probability that life, leading to intelligence and technology, will develop on these planets. Estimates range from $N = 1$ to $N = 10^9$.

8. Some attempts have been made to search for radio

REVIEW QUESTIONS

1. Suppose life did start on earth as a result of a chance arrival of microscopic spores from space. If these spores travel through space with velocities typical of interstellar clouds, then their speeds might be about 20 km/sec. Calculate how long it would take a spore to travel at this speed from the nearest star (1.4 parsecs away) to earth. How long would it take for a spore to cross the galactic disk, a distance of 30,000 parsecs?

2. Summarize the evolution of the earth's atmosphere, as outlined in chapter 4. Does this provide a possible clue that should be looked for in assessing whether life exists on a planet that might be discovered orbiting a distant star?

3. We have noted here and in chapter 7 that some of the conditions thought necessary for the formation of life exist in the atmosphere of Jupiter. Some key conditions are probably absent, however. What are they?

4. Based on your answer to question 3, do you think life is likely to have formed inside of interstellar clouds? (Recall that rather complex organic molecules have been observed in dark clouds.)

5. Why would a planet only slightly closer to the sun than the earth probably not be able to support life? (Hint: Recall what you learned in chapter 5 about the origin of the contrasting conditions on the earth and on Venus.)

6. Repeat the calculation of N, the number of technological civilizations in the galaxy, with all parameters the same as adopted in the text, except that the probability of life beginning on an earthlike planet is only 10^{-4} rather than 1 (assume that the lifetime L is 10^{10} years).

7. What is N if all the values used in the text are adopted, except that L is set equal to 10,000 years?

8. Would you expect technological civilizations in the galaxy to be concentrated along the spiral arms, or distributed more or less uniformly throughout the disk? Explain your answer.

9. Would you expect life to be as likely to arise on planets orbiting Population II stars as on those orbiting Population I stars? Explain.

10. Explain why it makes such a big difference whether or not a deliberate attempt is made by a civilization to send radio signals; that is, explain why deliberate signals can be detected at much greater distances than accidental signals such as radio and television broadcasts.

ADDITIONAL READINGS

Abt, H. A. 1979. The companions of sun-like stars. *Scientific American* 236(4):96.

Ball, J. A. 1980. Extraterrestrial intelligence: where is everybody? *American Scientist* 68:656.

Barber, V. 1974. Theories of the chemical origin of life on earth. *Mercury* 3(5):20.

Goldsmith, D., ed. 1980. *The quest for extraterrestrial life.* Mill Valley, Ca.: University Science Books.

O'Neill, G. K. 1974. The colonization of space. *Physics Today* 27(9):32.

Papagiannis, M. D. 1982. The search for extraterrestrial civilizations—a new approach. *Griffith Observer* 11(1):112.

Pelligrino, C. R. 1979. Organic clues in carbonaceous meteorites. *Sky and Telescope* 57(4):330.

Pollard, W. G. 1979. The prevalence of earth-like planets. *American Scientist* 67:653.

Rood, R. T., and Trefil, J. S. 1981. *Are we alone?* New York: Scribners.

Sagan, C., and Drake, F. 1975. The search for extraterrestrial intelligence. *Scientific American* 232(5):80.

Sagan, C., and Shklovskii, I. S. 1966. *Intelligent life in the universe.* San Francisco: Holden-Day.

Tipler, F. J. 1982. The most advanced civilization in the galaxy is ours. *Mercury* 11(1):5.

Wetherill, C., and Sullivan, W. T. 1979. Eavesdropping on the earth. *Mercury* 8(2):23.

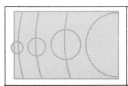

Guest Editorial

OUR FUTURE ON EARTH AND IN SPACE

Gerard K. O'Neill

A physicist by profession, Gerard K. O'Neill has become a leader in a variety of areas of modern and future technology. His main research area is high-energy particle physics, and one of his best-known contributions in that area was the development of the storage-ring technique for generating high-energy particle collisions. This technique is now widely used throughout the world. Dr. O'Neill is a professor in the Physics Department at Princeton University, and with a colleague has published a graduate-level textbook on particle physics. During the 1970s, Dr. O'Neill developed an interest in the humanization of space; in 1977 he published The High Frontier, *and in 1981 another book followed, entitled* 2081: A Hopeful View of the Human Future. *In these and other writings, Dr. O'Neill develops both philosophical and technical bases for the colonization of space, and his work has become widely recognized. His suggestion that space stations be built at the Lagrange points in the lunar orbit (see the discussion of Trojan asteroids in chapter 10 of this book)*

has received support from such groups as the L-5 Society, the L-5 point being the position 60 degrees behind the moon in its orbit. He is furthering the case for space colonization through a foundation called the Space Studies Institute, which supports research in this area. Finally, as an example of his versatility, Dr. O'Neill has recently founded a company, Geostar, Inc., to market his patented system for satellite positioning and message exchange.

We have a responsibility beyond mere curiosity to learn as much about the future as we can, because we must choose those actions which will insure not only the survival of humanity, but an improvement in its condition. Before we take action to try to solve such problems, and before we seek to influence the actions of other individuals or governments, we must understand the realistic possibilities. Fighting for the "right" cause can give us a sense of virtue, but we'll gain nothing by advocating actions that have no realistic chance of happening.

There is in my mind no question of the priority of expanding the race beyond the confines of earth; if there is a single, objectively identifiable purpose of our being, it is that. The real question is not whether it should be done, but how. The resources required to send colonists into space are enormous; how do we convince ourselves to pay the price? We must do it by making the effort pay for itself, by making it a profitable venture, in the economic and the political sense.

To assess the chances that this will succeed, we must evaluate the prevailing human condition and the ground rules which will take us into space. We must try to guess what it is about our society that will endure, and what will not.

I made my own value judgments years ago on two of the most basic questions that involve the future: I place a higher value on the freedom of the individual than on material wealth or the absence of risk, and after freedom, not before, I put the search for peace.

It seems wisest to guess that the political world of the future will still be fragmented into nations, and that nations will still be heavily armed. I do not see any *political* idea that has a realistic possibility of improving that situation. My stress of "political" is to remind us that there may be technological developments that will alter international confrontations in a fundamental way. We can hope, in any case, that the largest nations will continue to avoid direct warfare with each other.

It is also safe to assume that the most enduring institutions and characteristics of society will continue. Great universities will survive, in some form, as they have for centuries. Though governments will be overthrown and both the names and the boundaries of nations will shift, the same languages will still be spoken in the same geographical areas.

We'd predict further that technology will continue to be apolitical as it is today. Indeed, one of the remarkable features of modern society is that the universality of the

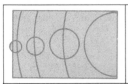

laws of nature forces different nations to develop almost identical designs for aircraft, automobiles, and all other technical artifacts, even though the same nations may be violently at odds with each other on political, religious, or ideological issues.

I have for several years advocated the colonization of space, pointing out not only its technical feasibility, but also its economic rationale. We have seen that manned space stations can be orbited and operated over long periods. That is a first step toward the day when more complex colonies might be built, housing not only a limited crew of astronauts, but also a community of workers whose task would be to establish a space-borne economy. Given the starting tools, such a task force could begin making use of resources in space. Recently a successful working model of a "mass-driver" was demonstrated. This is an electromagnetic machine for accelerating materials. A mass-driver on the moon could accelerate lunar materials at low cost to a high-orbital industrial facility. There the materials could be separated into pure metals, silicon, and oxygen. Research funded by the Space Studies Institute has shown how that separation can be carried out.

We are beginning to see some steps taken toward the goals outlined here. The Space Shuttle, providing for the first time the capability of frequent "commuter" flights into earth orbit, is now well on its way. Before long profit-making industries, given the availability of commercial space on the Shuttle and the possibility of privately owned vehicles in the near future, may begin to undertake commercial exploration of space. Already major airline companies are vying to see whose colors will fly with the first commercial passengers into earth orbit.

Turning now to the human experience of travel, spaceflight a century from now will be a far more luxurious experience than air travel today. Despite the best efforts of airlines, jet travel leaves the average passenger cramped in a tiny space for hours. Flight across the sea of space may have its hardships, but at least it will restore the elbowroom, the leisure, and the time for human interaction that the older generation tells us was so wonderful about the age of ocean liners, qualities that are preserved today only in cruise ships. Although spaceships will have to be light in weight, they can be large and comfortable in volume.

While we can't be sure which of several alternate solutions to each technical problem of the spaceships will prevail, the question of their crews is more predictable. A ship could easily be controlled throughout its entire voyage by an on-board computer, assisted only by occasional radioed commands. Indeed, that's the way our planetary spacecraft, *Mariner, Voyager, Viking*, and the rest, operate already. There would be some point in having a human crew to make simple repairs, but even those tasks could be automated. Yet I do expect that a passenger vessel will carry a human crew. The reason is the need for a well-defined authority figure to take command and prevent catastrophic conflicts of the kind that might otherwise occur among passengers thrown together in confined quarters, during voyages of several months' duration. The history of arctic exploration and our everyday experience in air and water transport make it clear that the captain's responsibility can't be shifted to a computer. We must keep in mind that the gadgets of our future will be very different from those we have now, but that people will be little different from ourselves.

Our race will inhabit the solar system; we know how to do it now. What remains is to look beyond the near future, to speculate about whether and how we might look farther outward, after colonization of the solar system.

While spaceships driven by mass-drivers and powered by solar-cell arrays are a sure thing, and laser-powered ion-drive ships are a very safe bet, interstellar craft of nearly the speed of light are much more speculative. Yet we already have a fair idea how to build them, and in guessing they'll exist eventually, I'm mindful that the prophets of technology have generally erred in the direction of timidity.

I am compelled to conclude that technological change will continue, and that we should take advantage of the possibilities it will open for finding new solutions to previously insoluble problems; that the facts don't warrant throwing away the freedoms we have worked so hard to enlarge and preserve; and finally, that the future is potentially even more exciting than the past, and that we shall meet it with courage and a spirit of adventure.

Appendix 1. Physical and Mathematical Constants

CONSTANT	SYMBOL	VALUE
Speed of light	c	2.9979250×10^{10} cm/sec
Gravitation constant	G	6.670×10^{-8} dyn cm^2/gm
Planck constant	h	6.62620×10^{-27} erg/sec
Electron mass	m_e	9.10956×10^{-28} gm
Proton mass	m_p	1.672661×10^{-24} gm
Stefan-Boltzmann constant	σ	5.66596×10^{-5} erg/cm^{2}deg^{4}sec
Wien constant	W	0.289789 cm deg
Boltzmann constant	k	1.38062×10^{-16} erg/deg
Astronomical unit	AU	1.495979×10^8 km
Parsec	pc	3.085678×10^{13} km = 3.261633 light years
Light year	ly	9.460530×10^{12} km
Solar mass	$M_\odot$	1.9891×10^{33} gm
Solar radius	$R_\odot$	6.9600×10^{10} cm
Solar luminosity	$L_\odot$	3.827×10^{33} erg/sec
Earth mass	$M_\oplus$	5.9742×10^{27} gm
Earth radius	$R_\oplus$	6378.140 km
Tropical year (equinox to equinox)		365.241219878 days
Sidereal year (with respect to stars)		365.256366 days = 3.155815×10^7 sec

Appendix 2. The Elements and Their Abundances

ELEMENT	SYMBOL	ATOMIC NUMBER	ATOMIC WEIGHT[1]	ABUNDANCE[2]
Hydrogen	H	1	1.0080	1.00
Helium	He	2	4.0026	0.063
Lithium	Li	3	6.941	1.55×10^{-9}
Beryllium	Be	4	9.0122	1.41×10^{-11}
Boron	B	5	10.811	2.00×10^{-10}
Carbon	C	6	12.0111	0.000372
Nitrogen	N	7	14.0067	0.000115
Oxygen	O	8	15.9994	0.000676
Fluorine	F	9	18.9984	3.63×10^{-8}
Neon	Ne	10	20.179	3.72×10^{-5}
Sodium	Na	11	22.9898	1.74×10^{-6}
Magnesium	Mg	12	24.305	3.47×10^{-5}
Aluminum	Al	13	26.9815	2.51×10^{-6}
Silicon	Si	14	28.086	3.55×10^{-5}
Phosphorus	P	15	30.9738	3.16×10^{-7}
Sulfur	S	16	32.06	1.62×10^{-5}
Chlorine	Cl	17	35.453	2×10^{-7}
Argon	Ar	18	39.948	4.47×10^{-6}
Potassium	K	19	39.102	1.12×10^{-7}
Calcium	Ca	20	40.08	2.14×10^{-6}
Scandium	Sc	21	44.956	1.17×10^{-9}
Titanium	Ti	22	47.90	5.50×10^{-8}
Vanadium	V	23	50.9414	1.26×10^{-8}
Chromium	Cr	24	51.996	5.01×10^{-7}
Manganese	Mn	25	54.9380	2.63×10^{-7}
Iron	Fe	26	55.847	2.51×10^{-5}
Cobalt	Co	27	58.9332	3.16×10^{-8}
Nickel	Ni	28	58.71	1.91×10^{-6}
Copper	Cu	29	63.546	2.82×10^{-8}
Zinc	Zn	30	65.37	2.63×10^{-8}
Gallium	Ga	31	69.72	6.92×10^{-10}
Germanium	Ge	32	72.59	2.09×10^{-9}
Arsenic	As	33	74.9216	2×10^{-10}
Selenium	Se	34	78.96	3.16×10^{-9}
Bromine	Br	35	79.904	6.03×10^{-10}
Krypton	Kr	36	83.80	1.6×10^{-9}
Rubidium	Rb	37	85.4678	4.27×10^{-10}
Strontium	Sr	38	87.62	6.61×10^{-10}
Yttrium	Y	39	88.9059	4.17×10^{-11}
Zirconium	Zr	40	91.22	2.63×10^{-10}
Niobium	Nb	41	92.906	2.0×10^{-10}
Molybdenum	Mo	42	95.94	7.94×10^{-11}
Technetium	Tc	43	98.906	
Ruthenium	Ru	44	101.07	3.72×10^{-11}

[1]The atomic weight of an element is its mass in **atomic mass units.** An atomic mass unit is defined as one-twelfth of the mass of the most common isotope of carbon, and has the value 1.660531×10^{-24} grams. In general, the atomic weight of an element is approximately equal to the total number of protons and neutrons in its nucleus.

[2]The abundances are given in terms of the number of atoms of each element compared with hydrogen, and are based on the composition of the sun. For very rare elements, particularly those toward the end of the list, the abundances can be quite uncertain, and the values given should not be considered exact.

Appendix 2.—*Continued*

ELEMENT	SYMBOL	ATOMIC NUMBER	ATOMIC WEIGHT[1]	ABUNDANCE[2]
Rhodium	Rh	45	102.905	3.55×10^{-11}
Palladium	Pd	46	106.4	3.72×10^{-11}
Silver	Ag	47	107.868	4.68×10^{-12}
Cadmium	Cd	48	112.40	9.33×10^{-11}
Indium	In	49	114.82	5.13×10^{-11}
Tin	Sn	50	118.69	5.13×10^{-11}
Antimony	Sb	51	121.75	5.62×10^{-12}
Tellurium	Te	52	127.60	1×10^{-10}
Iodine	I	53	126.9045	4.07×10^{-11}
Xenon	Xe	54	131.30	1×10^{-10}
Cesium	Cs	55	132.905	1.26×10^{-11}
Barium	Ba	56	137.34	6.31×10^{-11}
Lanthanum	La	57	138.906	6.46×10^{-11}
Cerium	Ce	58	140.12	4.37×10^{-11}
Praseodymium	Pr	59	140.908	4.27×10^{-11}
Neodymium	Nd	60	144.24	6.61×10^{-11}
Promethium	Pm	61	146	
Samarium	Sm	62	150.4	4.57×10^{-11}
Europium	Eu	63	151.96	3.09×10^{-12}
Gadolinium	Gd	64	157.25	1.32×10^{-11}
Terbium	Tb	65	158.925	2.63×10^{-12}
Dysprosium	Dy	66	162.50	1.29×10^{-11}
Holmium	Ho	67	164.930	3.1×10^{-12}
Erbium	Er	68	167.26	5.75×10^{-12}
Thulium	Tm	69	168.934	2.69×10^{-12}
Ytterbium	Yb	70	170.04	6.46×10^{-12}
Lutetium	Lu	71	174.97	6.92×10^{-12}
Hafnium	Hf	72	178.49	6.3×10^{-12}
Tantalum	Ta	73	180.948	2×10^{-12}
Tungsten	W	74	183.85	3.72×10^{-10}
Rhenium	Re	75	186.2	1.8×10^{-12}
Osmium	Os	76	190.2	5.62×10^{-12}
Iridium	Ir	77	192.2	1.62×10^{-10}
Platinum	Pt	78	195.09	5.62×10^{-11}
Gold	Au	79	196.967	2.09×10^{-12}
Mercury	Hg	80	200.59	1×10^{-9}
Thallium	Tl	81	204.37	1.6×10^{-12}
Lead	Pb	82	207.19	7.41×10^{-11}
Bismuth	Bi	83	208.981	6.3×10^{-12}
Polonium	Po	84	210	—
Astatine	At	85	210	—
Radon	Rn	86	222	—
Francium	Fr	87	223	—
Radium	Ra	88	226.025	—
Actinium	Ac	89	227	—
Thorium	Th	90	232.038	6.61×10^{-12}
Protactinium	Pa	91	230.040	—

(Continued on next page)

Appendix 2.—*Continued*

ELEMENT	SYMBOL	ATOMIC NUMBER	ATOMIC WEIGHT[1]	ABUNDANCE[2]
Uranium	U	92	238.029	4.0×10^{-12}
Neptunium	Np	93	237.048	—
Plutonium	Pu	94	242	—
Americium	Am	95	242	—
Curium	Cm	96	245	—
Berkelium	Bk	97	248	—
Californium	Cf	98	252	—
Einsteinium	Es	99	253	—
Fermium	Fm	100	257	—
Mendelevium	Md	101	257	—
Nobelium	No	102	255	—
Lawrencium	Lr	103	256	—

Appendix 3. Temperature Scales

As mentioned in the text, at the most basic level temperature can be defined in terms of the motion of particles in a gas (or a solid or liquid as well). We all have an intuitive idea of what heat is, and we all are familiar with at least one scale for measuring temperature.

The most commonly used scales are somewhat arbitrarily defined, with zero points not representing any truly fundamental physical basis. The popular **Fahrenheit** scale, for example, has water freezing at a temperature of 32°F and boiling at 212°F. On this scale, absolute zero, the lowest possible temperature (where all molecular motions cease), is −459°F.

The **centigrade** (or **Celsius**) scale is perhaps more well-founded, although it is based on the freezing and boiling points of water, rather than the more fundamental absolute zero. In this system, the freezing point is defined as 0°C, and the boiling point is 100°C. This scale has the advantage that there are exactly 100° between the freezing and boiling points, rather than 180° as in the Fahrenheit system. To convert from Fahrenheit to centigrade, we must first subtract 32°, and then multiply the remainder by 100/180, or 5/9. For example, 50°F is equal to $5/9 \times (50 - 32) = 10°C$. To convert from centigrade to Fahrenheit, we first multiply by 9/5 and then add 32°. Thus, $-10°C = 9/5 \times (-10) + 32 = 14°F$. On the centigrade scale, absolute zero occurs at −273°C.

The temperature scale preferred by scientists is a modification of the centigrade system. In this system, named after its founder, the British physicist Lord Kelvin, the same degree is used as in the centigrade scale; that is, one degree is equal to one one-hundredth of the difference between the freezing and boiling points of water. The zero point is different from that of the centigrade scale, however; it is set equal to absolute zero. Hence on this scale water freezes at 273°K and boils at 373°K. Comfortable room temperature is around 300°K. In modern usage, the degree symbol (°) is dropped, and we speak simply of temperatures in units of Kelvins (K).

Appendix 4. Radiation Laws

In chapter 3 we described several laws that apply to continuous radiation from hot objects such as stars. These laws were discussed in general terms, and a few simple applications were explained. In this appendix, the same laws are given in more precise mathematical form, and their use in that form is illustrated.

Wien's Law

In general terms, Wien's law states that the wavelength of maximum emission from a glowing object is inversely proportional to its temperature. Mathematically, this may be written as:

$$\lambda_{max} \propto 1/T,$$

where λ_{max} is the wavelength of strongest emission, T is the surface temperature of the object, and $\propto$ is a special mathematical symbol meaning "is proportional to."

Experimentation can determine the **proportionality constant,** specifying the exact relationship between λ_{max} and T, and Wien did this, finding

$$\lambda_{max} = W/T,$$

where W has the value 0.29 if λ_{max} is measured in centimeters and T in degrees absolute. With this equation it is possible to calculate λ_{max}, given T, or vice versa. Thus if we measure the spectrum of a star and find that the star emits most strongly at a wavelength of $2,000$ Å $= 2 \times 10^{-5}$ cm, then we can solve for the temperature:

$$T = W/\lambda_{max} = 0.29/2 \times 10^{-5} = 14,500°K.$$

This is a relatively hot star, and would appear blue to the eye. Note that it was necessary to measure the spectrum in ultraviolet wavelengths in order to find λ_{max}.

When solving problems with Wien's law, we can always use the equation form, as we have just done. Often, however, it is more convenient to compare the properties of two objects by considering the ratio of the temperatures or of the wavelengths of maximum emission. This, in effect, is what we did in the text when we compared two objects of different temperatures in order to determine how their values compared. For ex-

ample, we said that if one object is twice as hot as another, its value of λ_{max} is half that of the other.

We can see how this ratio technique works by writing the equation for Wien's law separately for object 1 and object 2:

$$\lambda_{max_1} = W/T_1, \text{ and } \lambda_{max_2} = W/T_2.$$

Now we can divide one equation by the other:

$$\frac{\lambda_{max_1}}{\lambda_{max_2}} = \frac{(W/T_1)}{(W/T_2)}, \text{ or}$$

$$\lambda_{max_1}/\lambda_{max_2} = T_2/T_1.$$

The numerical factor W has canceled out, and we are left with a simple expression relating the values of λ_{max} and T for the two objects. Now we see that if $T_1 = 2T_2$ (object 1 is twice as hot as object 2), then

$$\lambda_{max_1}/\lambda_{max_2} = T_2/2T_2 = 1/2,$$

or λ_{max} for object 1 is one-half that for object 2. In this extremely simple example, it probably would have been easier to just work it out in our heads, but what if we have a case where one object is 3.368 times hotter than the other?

A great deal can be learned from making comparisons in this way. Astronomers often use the sun as the standard for comparison, expressing various quantities in terms of solar values. Comparisons are also useful when the numerical constants (such as Wien's constant in the previous examples) are not known. If the trick of comparing is kept in mind, it is often possible to quickly work out answers to astronomical questions, simply by carrying around in our heads a few numbers describing the sun.

The Stefan-Boltzmann Law

As discussed in the text, the energy emitted by a glowing object is proportion to T^4, where T is the surface temperature. This can be written mathematically as

$$E = \sigma T^4,$$

where σ stands for a proportionality constant that has the value 5.7×10^{-5}, if the centimeter-gram-second metric units are used.

Let us now consider how much surface area a star has. The total energy the star emits, called its **luminosity** and usually denoted L, is

$$L = \text{surface area} \times E$$
$$= 4\pi R^2 \sigma T^4,$$

where R is the radius of the star. This equation is called the Stefan-Boltzmann law.

As in other cases, we can use the law directly in this form, or we may choose to compare properties of stars by writing the equation separately for two stars and then dividing. If we do this, we find

$$\frac{L_1}{L_2} = \frac{R_1^2 T_1^4}{R_2^2 T_2^4},$$

The constant factors 4π and σ have canceled out.

As an example of how to use this expression, suppose we determine that a particular star has twice the temperature of the sun, but only one-half the radius, and we wish to know how this star's luminosity compares with that of the sun. If we designate the star as object 1, and the sun as object 2, then

$$L_1 = \left(\frac{1}{2}\right)^2 2^4 = 1/4 \times 16 = 4.$$

The star has four times the luminosity of the sun.

The Stefan-Boltzmann law is particularly useful because it relates three of the most important stellar properties to each other.

The Planck Function

The radiation laws described here and in the text are actually specific forms of a much more general law, discovered by the great German physicist Max Planck. Wien's law, the Stefan-Boltzmann law, and some others not mentioned in the text bear a similar relation to Planck's law as the laws of planetary motion discovered by Kepler do to Newton's mechanics. Kepler's laws were first discovered by observation, but with Newton's laws of motion it is possible to derive Kepler's

laws theoretically. In the same fashion, the radiation laws discussed so far were found experimentally, but can be derived mathematically from the much more general and powerful Planck's law.

Planck's law is usually referred to as the Planck function, a mathematical relationship between intensity and wavelength (or frequency) that describes the spectrum of any glowing object at a given temperature. The Planck function specifically applies only to objects that radiate solely because of their temperature, objects that do not reflect any light or have any spectral lines. The popular term for such an object is **blackbody,** and we often refer to radiation from a blackbody as **blackbody radiation,** or more commonly, as **thermal radiation,** the term used in the text. Stars are not perfect radiators, but to a close approximation they can be treated as such. Thus in the text we apply Wien's law and the Stefan-Boltzmann law to stars (and even planets in certain circumstances) without pointing out the fact that to do so is only approximately correct.

The form of the Planck function for the radiation intensity B as a function of wavelength is

$$B = 2hc^2\lambda^{-5}/(e^{hc/\lambda kT} - 1),$$

where h is the Planck constant, c is the speed of light, and k is the Boltzmann constant (the values of all three are tabulated in appendix 1), λ is the wavelength (in cm), and T is the temperature of the object (on the absolute scale). The symbol e represents the base of the natural logarithm, something not used elsewhere in the text; for the present purpose, this may be regarded simply as a mathematical constant with the value 2.718.

In terms of frequency ν rather than wavelength, the expression is

$$B = 2h\nu^3 c^{-2}/(e^{h\nu/kT} - 1).$$

Either expression may be used to calculate the spectrum of continuous radiation from a glowing object at a specific temperature. In practice, the Planck function is used in a wide assortment of theoretical calculations that call for knowledge of the intensity of radiation so that the effects of radiation on physical conditions such as ionization can be assessed.

Appendix 5. Major Telescopes of the World (Two Meters or Larger)*

OBSERVATORY	LOCATION	TELESCOPE
Special Astrophysical Observatory	Mount Pastukhov, USSR	6.0–m Bol'shoi Teleskop Azimutal'nyi
Palomar Observatory	Palomar Mountain, California	5.08–m George Ellery Hale Telescope
Smithsonian Observatory	Mount Hopkins, Arizona	4.5–m Multiple Mirror Telescope
[Royal Greenwich Observatory]	[Canary Islands]	[4.2–m William Herschel Telescope]
Cerro Tololo Inter-American Observatory	Cerro Tololo, Chile	4.0–m
Anglo-Australian Observatory	Siding Spring Mountain, Australia	3.9–m Anglo-Australian Telescope
Kitt Peak National Observatory	Kitt Peak, Arizona	3.8–m Nicholas U. Mayall Telescope
Royal Observatory Edinburgh	Mauna Kea, Hawaii	3.8–m United Kingdom Infrared Telescope
Canada-France-Hawaii Observatory	Mauna Kea, Hawaii	3.6–m Canada-France-Hawaii Telescope
European Southern Observatory	Cerro La Silla, Chile	3.57–m
Max Planck Institute (Heidelberg)	Calar Alto, Spain	3.5–m
[Italian National Astronomical Observatory]	[Italy, final site to be chosen]	[3.5–m]
[Iraqi National Observatory]	[Mount Korek]	[3.5–m]
Lick Observatory	Mount Hamilton, California	3.05–m C. Donald Shane Telescope
Mauna Kea Observatory	Mauna Kea, Hawaii	3.0–m NASA Infrared Telescope
McDonald Observatory	Mount Locke, Texas	2.7–m
Haute Provence Observatory	St. Michele, France	2.6–m
Crimean Astrophysical Observatory	Simferopol, Ukrainian SSR	2.6–m
Byurakan Astrophysical Observatory	Mount Aragatz, Armenian SSR	2.6–m
Mount Wilson and Las Campanas Observatories	Mount Wilson, California	2.5–m Hooker Telescope
Mount Wilson and Las Camapanas Observatories	Las Campanas, Chile	2.5–m Irenée du Pont Telescope
Royal Greenwich Observatory	Canary Islands	2.5–m Isaac Newton Telescope
Wyoming Infrared Observatory	Mount Jelm, Wyoming	2.3–m Wyoming Infrared Telescope
Steward Observatory	Kitt Peak, Arizona	2.3–m
Mauna Kea Observatory	Mauna Kea, Hawaii	2.2–m
Max Planck Institute (Heidelberg)	Calar Alto, Spain	2.2–m
Max Planck Institute	LaSilla, Chile	2.2–m
[Mexican National Observatory]	[San Pedro Martir, Baja (Mexico) California]	[2.16–m]
La Plata Observatory	La Plata, Argentina	2.15–m
Kitt Peak National Observatory	Kitt Peak, Arizona	2.1–m
McDonald Observatory	Mount Locke, Texas	2.1–m Otto Struve Telescope
Karl Schwarzschild Observatory	Tautenberg, East Germany	2.0–m (largest Schmidt telescope)

*Telescopes described in brackets are planned or under construction, but not yet in operation.

Appendix 6. Planetary and Satellite Data

Orbital Data for the Planets

Planet	Sidereal Period	Semimajor Axis	Orbital Eccentricity*	Inclination of Orbital Plane	Rotation Period	Tilt of Axis
Mercury	0.241 yr	0.387 AU	0.206	7°0′15″	58^d65	28°
Venus	0.615	0.723	0.007	3°23′4″	243	3°
Earth	1.000	1.000	0.017	0°0′0″	$23^h56^m04^s$	23°27′
Mars	1.881	1.524	0.093	1°51′0″	$24^h37^m22^s$	23°59′
Jupiter	11.86	5.203	0.048	1°18′17″	9^h50^m	3°5′
Saturn	29.46	9.555	0.056	2°29′33″	10^h14^m	26°44′
Uranus	84.01	19.218	0.047	0°46′23″	15^h36^m:	97°55′
Neptune	164.8	30.110	0.009	1°46′22″	18^h12^m	28°48′
Pluto	248.8	39.44	0.250	17°10′12″	6^d387	35°

*The eccentricity of an orbit is defined as the ratio of the distance between foci to the semimajor axis. In practice, it is related to the perihelion distance P and the semimajor axis a by $P = a(1 = e)$, where e is the eccentricity; and to the aphelion distance A by $A = a(1 + e)$.

Physical Data for the Planets

Planet	Mass*	Diameter*	Density	Surface Gravity	Escape Velocity	Temperature	Albedo
Mercury	0.0553	0.382	5.43 g/cc	0.38 g	4.3 kmb/sec	100–700 K	0.06
Venus	0.815	0.949	5.25	0.90	10.3	730	0.76
Earth	1.000	1.000	5.518	1.00	11.2	200–300	0.39
Mars	0.107	0.532	3.95	0.38	5.0	130–290	0.15
Jupiter	317.9	10.86	1.33	2.64	59.5	130	0.51
Saturn	95.2	8.89	0.69	1.13	35.6	95	0.50
Uranus	14.5	4.01	1.19	1.07	21.2	95	0.66
Neptune	17.2	3.96	1.66	1.08	23.6	50	0.62
Pluto	0.002:	0.16–.47	0.5–.9	0.024–.034	0.9–1.1	40	0.25

*The masses and diameters are given in units of the earth's mass and diameter, which are 5.976×10^{27} grams and 12,756 kilometers, respectively. For the giant planets, the diameter given is the mean diameter.

Satellites*

Planet	Satellite**	Semimajor Axis	Period	Diameter	Mass	Density
Earth	Moon	3.84×10^5	$27^d.322$	3,476 km	7.35×10^{25}	3.34 gm/cm^3
Mars	Phobos	9.38×10^3	0.319	27 × 21 × 18	9.6×10^{18}	2.0
	Deimos	2.35×10^4	1.262	15 × 12 × 10	2.0×10^{18}	1.9
Jupiter	Metis	1.280×10^5	0.295	40		
	Adrastea	1.290×10^5	0.298	25 × 20 × 15		
	Amalthea	1.82×10^5	0.498	270 × 170 × 150		
	Thebe	2.22×10^5	0.675	110 × ? × 90		
	Io	4.25×10^5	1.769	3,630	8.92×10^{22}	3.53

*Data primarily from Burns, J. and Morrison, D. (eds.) 1984, *Natural Satellites* (Tucson: University of Arizona Press), in press

**Satellite data given in brackets indicate bodies whose existence is not confirmed.

Satellites—*Continued*

Planet	Satellite**	Semimajor Axis	Period	Diameter	Mass	Density
Jupiter	Europa	6.76×10^5	3.551	3,138	4.87×10^{22}	3.03
	Ganymede	1.08×10^6	7.155	5,262	1.49×10^{23}	1.93
	Callisto	1.90×10^6	16.689	4,800	1.08×10^{23}	1.70
	Leda	1.11×10^7	240	10:		
	Himalia	1.15×10^7	251	180		
	Lysithia	1.17×10^7	260	20:		
	Elara	1.18×10^7	260	80		
	Ananke	2.08×10^7	617	20:		
	Carme	2.24×10^7	692	30:		
	Pasiphae	2.33×10^7	735	40:		
	Sinope	2.38×10^7	758	30:		
Saturn	Atlas	1.377×10^5	0.602	$38 \times ? \times 26$		
	1980S27	1.394×10^5	0.613	$140 \times 100 \times 75$		
	1980S26	1.417×10^5	0.629	$110 \times 85 \times 65$		
	Janus	1.514×10^5	0.694	$220 \times 190 \times 160$		
	Epimetheus	1.515×10^5	0.695	$140 \times 115 \times 100$		
	Mimas	1.855×10^5	0.942	392	3.75×10^{22}	1.19
	Enceladus	2.380×10^5	1.370	500	8.4×10^{22}	1.13
	Telesto	2.947×10^5	1.888	$30 \times 25 \times 15$		
	Calypso	2.947×10^5	1.888	$? \times 25 \times 20$		
	Tethys	2.947×10^5	1.888	1,060	7.55×10^{23}	1.20
	[No name I]	$[3.3 \times 10^5]$		$[15 - 20]$		
	Dione	3.774×10^5	2.739	1,120	1.052×10^{24}	1.43
	1980S6	3.781×10^5	2.739	$36 \times ? \times <30$		
	[No name II]	$[3.78 \times 10^5]$		$[15 - 20]$		
	[No name III]	$[4.70 \times 10^5]$		$[15 - 20]$		
	Rhea	5.271×10^5	4.518	1,530	2.49×10^{24}	1.33
	Titan	1.22×10^6	15.945	5,150	1.346×10^{26}	1.88
	Hyperion	1.48×10^6	21.277	$350 \times 235 \times 200$		
	Iapetus	3.56×10^6	79.331	1,460	1.88×10^{24}	1.16
	Phoebe	1.30×10^7	550.4	220		
Uranus	Miranda	1.30×10^5	1.414	400:		
	Ariel	1.92×10^5	2.520	1330		
	Umbriel	2.67×10^5	4.144	1110		
	Titania	4.38×10^5	8.706	1600		
	Oberon	5.86×10^5	13.46	1630		
Neptune	[No Name]	$[8 \times 10^4:]$	[0.5:]	[≤100:]		
	Triton	3.54×10^5	5.877	3500		
	Nereid	5.51×10^6	365.21	400:		
Pluto	Charon	1.97×10^4	6.387	1000:		

Appendix 7. The Fifty Brightest Stars

STAR			SPECTRAL TYPE	APPARENT MAGNITUDE	DISTANCE	POSITION (1980) RIGHT ASCENSION	DECLINATION
α	Eri	Archernar	B3 V	0.51	36 pc.	$01^h 37.^m 0$	$-57°20'$
α	UMi	Polaris	F8 Ib	1.99	208	02 12.5	+89 11
α	Per	Mirfak	F5 Ib	1.80	175	03 22.9	+49 47
α	Tau	Aldebaran	K5 III	0.86	21	04 34.8	+16 28
β	Ori	Rigel	B8 Ia	0.14	276	05 13.6	−08 13
α	Aur	Capella	G8 III	0.05	14	05 15.2	+45 59
γ	Ori	Bellatrix	B2 III	1.64	144	05 24.0	+06 20
β	Tau	Elnath	B7 III	1.65	92	05 25.0	+28 36
ε	Ori	Alnilam	B0 Ia	1.70	490	05 35.2	−01 13
ζ	Ori	Alnitak	O9.5 Ib	1.79	490	05 39.7	−01 57
α	Ori	Betelgeuse	M2 Iab	0.41	159	05 54.0	+07 24
β	Aur	Menkalinan	A2 V	1.86	27	05 58.0	+44 57
β	CMa		B1 II–III	1.96	230	06 21.8	−17 56
α	Car	Canopus	F0 Ib–II	−0.72	30	06 23.5	−52 41
γ	Gem	Alhena	A0 IV	1.93	32	06 36.6	+16 25
α	CMa	Sirius	A1 V	−1.47	2.7	06 44.2	−16 42
ε	CMa	Adhara	B2 II	1.48	209	06 57.8	−28 57
δ	CMa		F8 Ia	1.85	644	07 07.6	−26 22
α	Gem	Castor	A1 V	1.97	14	07 33.3	+31 56
α	CMi	Procyon	F5 IV–V	0.37	3.5	07 38.2	+05 17
β	Gem	Pollux	K0 III	1.16	11	07 44.1	+28 05
γ	Vel		WC8	1.83	160	08 08.9	−47 18
ε	Car	Avior	K3 III?	1.90	104	08 22.1	−59 26
δ	Vel		A2 V	1.95	23	08 44.2	−54 38
β	Car	Miaplacidus	A1 III	1.67	26	09 13.0	−69 38
α	Hya	Alphard	K4 III	1.98	29	09 26.6	−08 35
α	Leo	Regulus	B7 V	1.36	26	10 07.3	+12 04
γ	Leo		K0 III	1.99	28	10 18.8	+19 57
α	UMa	Dubhe	K0 III	1.81	32	11 02.5	+61 52
α	Cru A ⎫	Acrux	B0.5 IV	1.39	114	12 15.4	−62 59
α	Cru B ⎭		B1 V	1.86	114	12 15.4	−62 59
γ	Cru	Gacrux	M4 III	1.69	67	12 30.1	−57 00
β	Cru		B0.5 III	1.28	150	12 46.6	−59 35
ε	UMa	Alioth	A0p	1.79	21	1253.2	+56 04

Appendix 7.—*Continued*

STAR		SPECTRAL TYPE	APPARENT MAGNITUDE	DISTANCE	POSITION (1980) RIGHT ASCENSION	DECLINATION
α Vir	Spica	B1 V	0.91	67	13 24.1	− 11 03
η UMa	Alkaid	B3 V	1.87	64	13 46.8	+ 49 25
β Cen	Hadar	B1 III	0.63	150	14 02.4	− 60 16
α Boo	Arcturus	K2 III	− 0.06	11	14 14.8	+ 19 17
α Cen A	Rigil Kentaurus	G2 V	0.01	1.3	14 38.4	− 60 46
α Cen B		K4 V	1.40	1.3	14 38.4	− 60 46
α Sco	Antares	M1 Ib	0.92	160	16 28.2	− 26 23
α TrA	Atria	K2 Ib	1.93	25	16 46.5	− 68 60
λ Sco	Shaula	B1 V	1.60	95	17 32.3	− 37 05
θ Sco		F0 Ib	1.86	199	17 35.9	− 42 59
ε Sgr	Kaus Australis	B9.5 III	1.81	38	18 22.9	− 34 24
α Lyr	Vega	A0 V	0.04	8	18 36.2	+ 38 46
α Aql	Altair	A7 IV–V	0.77	5	19 49.8	+ 08 49
α Pav	Peacock	B2.5 V	1.95	95	20 24.1	− 56 48
α Cyg	Deneb	A2 Ia	1.26	491	20 40.7	+ 45 12
α Gru	Al Na'ir	B7 IV	1.76	20	22 06.9	− 47 04
α PsA	Fomalhaut	A3 V	1.15	7	22 56.5	− 29 44

Appendix 8. The Constellations

NAME	GENITIVE	ABBREVIATION	POSITION RIGHT ASCENSION	DECLINATION
Andromeda	Andromedae	And	01^h	+40°
Antlia	Antliae	Ant	10	−35
Apus	Apodis	Aps	16	−75
Aquarius	Aquarii	Aqr	23	−15
Aquila	Aquilae	Aql	20	+05
Ara	Arae	Ara	17	−55
Aries	Arietis	Ari	03	+20
Auriga	Aurigae	Aur	06	+40
Boötes	Bootis	Boo	15	+30
Caelum	Caeli	Cae	05	−40
Camelopardalis	Camelopardalis	Cam	06	−70
Cancer	Cancri	Cnc	09	+20
Canes Venatici	Canum Venaticorum	CVn	13	+40
Canis Major	Canis Majoris	CMa	07	−20
Canis Minor	Canis Minoris	CMi	08	+05
Capricornus	Capricorni	Cap	21	−20
Carina	Carinae	Car	09	−60
Cassiopeia	Cassiopeiae	Cas	01	+60
Centaurus	Centauri	Cen	13	−50
Cepheus	Cephei	Cep	22	+70
Cetus	Ceti	Cet	02	−10
Chamaeleon	Chamaeleonis	Cha	11	−80
Circinis	Circini	Cir	15	−60
Columba	Columbae	Col	06	−35
Coma Berenices	Comae Berenices	Com	13	+20
Corona Australis	Coronae Australis	CrA	19	−40
Corona Borealis	Coronae Borealis	CrB	16	+30
Corvus	Corvi	Crv	12	−20
Crater	Crateris	Crt	11	−15
Crux	Crucis	Cru	12	−60
Cygnus	Cygni	Cyg	21	+40
Delphinus	Delphini	Del	21	+10
Dorado	Doradus	Dor	05	−65
Draco	Draconis	Dra	17	+65
Equuleus	Equulei	Equ	21	+10
Eridanus	Eridani	Eri	03	−20
Fornax	Fornacis	For	03	−30
Gemini	Geminorum	Gem	07	+20
Grus	Gruis	Gru	22	−45
Hercules	Herculis	Her	17	+30
Horologium	Horologii	Hor	03	−60
Hydra	Hydrae	Hya	10	−20
Hydrus	Hydri	Hyi	02	−75
Indus	Indi	Ind	21	−55
Lacerta	Lacertae	Lac	22	+45
Leo	Leonis	Leo	11	+15
Leo minor	Leonis Minoris	LMi	10	+35
Lepus	Leporis	Lep	06	−20
Libra	Librae	Lib	15	−15

Appendix 8.—*Continued*

NAME	GENITIVE	ABBREVIATION	POSITION RIGHT ASCENSION	DECLINATION
Lupus	Lupi	Lup	15	−45
Lynx	Lincis	Lyn	08	+45
Lyra	Lyrae	Lyr	19	+40
Mensa	Mensae	Men	05	−80
Microscopium	Microscopii	Mic	21	−35
Monoceros	Monocerotis	Mon	07	−05
Musca	Muscae	Mus	12	−70
Norma	Normae	Nor	16	−50
Octans	Octantis	Oct	22	−85
Ophiuchus	Ophiuchi	Oph	17	00
Orion	Orionis	Ori	05	+05
Pavo	Pavonis	Pav	20	−65
Pegasus	Pegasi	Peg	22	+20
Perseus	Persei	Per	03	+45
Phoenix	Phoenicis	Phe	01	−50
Pictor	Pictoris	Pic	06	−55
Pisces	Piscium	Psc	01	+15
Piscis Austrinus	Piscis Austrini	PsA	22	−30
Puppis	Puppis	Pup	08	−40
Pyxis	Pyxidis	Pyx	09	−30
Reticulum	Reticuli	Ret	04	−60
Sagitta	Sagittae	Sge	20	+10
Sagittarius	Sagittarii	Sgr	19	−25
Scorpius	Scorpii	Sco	17	−40
Sculptor	Sculptoris	Scl	00	−30
Scutum	Scuti	Sct	19	−10
Serpens	Serpentis	Ser	17	00
Sextans	Sextantis	Sex	10	00
Taurus	Tauri	Tau	04	+15
Telescopium	Telescopii	Tel	19	−50
Trianulum	Trianguli	Tri	02	+30
Triangulum Australe	Trianguli Australi	TrA	16	−65
Tucana	Tucanae	Tuc	00	−65
Ursa Major	Ursae Majoris	UMa	11	+50
Ursa Minor	Ursae Minoris	UMi	15	+70
Vela	Velorum	Vel	09	−50
Virgo	Virginis	Vir	13	00
Volans	Volantis	Vol	08	−70
Vulpecula	Vulpeculae	Vul	20	+25

Appendix 9. Mathematical Treatment of Stellar Magnitudes

Logarithmic Representation

In the text, magnitudes were discussed in terms of the brightness ratios between stars of different magnitudes. We generally avoided talking about two stars that differ by a fraction of a magnitude, because in such cases it is not simple to calculate the brightness ratio corresponding to the magnitude differences. If star 1 is 0.5 magnitudes brighter than star 2, for example, what is the brightness ratio? Or, if star 1 is a factor of 48.76 fainter than star 2, what is the difference in magnitudes?

Astronomers use an exact mathematical relationship between magnitude differences and brightness ratios, written as

$$m_1 - m_2 = 2.5 \log (b_2/b_1),$$

where m_1 and m_2 are the magnitudes of two stars, and b_2/b_1 is the ratio of their brightnesses. Log (b_2/b_1) represents the logarithm of this ratio; a logarithm is the power to which 10 must be raised to yield this ratio. Thus, if $b_2/b_1 = 100$, then $\log (b_2/b_1) = \log (10^2) = 2$, because 10 must be raised to the second power to yield 100. The magnitude difference is $2.5 \log (100) = 2.5 \times 2 = 5$. Similarly, if $b_2/b_1 = 0.001$, then $\log (b_2/b_1) = \log (0.001) = \log (10^{-3}) = -3$, and in this case the magnitude difference is $2.5 \times -3 = -7.5$ (the minus sign indicates that star 1 is brighter than star 2 in this example).

The method works equally well in cases where the power of 10 is not a whole number, as in $b_2/b_1 = 48.76$. Here $\log (b_2/b_1) = \log (48.76) = 1.69$ (we can usually find this by consulting logarithm tables or by using a scientific calculator). In this example, the magnitude difference is $2.5 \log (b_2/b_1) = 2.5 \log (48.76) = 2.5 \times 1.69 = 4.23$, so star 1 is 4.23 magnitudes fainter than star 2.

The equation can be used in other ways as well; solving for b_2/b_1 yields

$$b_2/b_1 = 10^{(m_1 - m_2)/2.5}$$
$$= 10^{0.4(m_1 - m_2)}.$$

Thus, if we know that the magnitudes of two stars differ by $m_1 - m_2$, then we multiply this difference by 0.4 and raise 10 to the power $0.4(m_1 - m_2)$, again using a calculator or tables, to get the brightness ratio b_2/b_1. As

a simple example, suppose $m_1 - m_2 = 5$; then $0.4(m_1 - m_2) = 0.4 \times 5 = 2$, and $10^{0.4(m_1 - m_2)} = 10^2 = 100$, as we knew it should. As a more complex example, consider the stars Betelgeuse (magnitude $+0.41$) and Deneb (magnitude $+1.26$). From the equation just cited, we see that Betelgeuse is $10^{0.4(1.26 - 0.41)} = 10^{0.34} = 2.19$ times brighter than Deneb.

Although in most cases we can follow the discussions of magnitudes and brightness ratios in the text without using the logarithmic representation, it is nevertheless useful to be familiar with this exact mathematical technique.

The Distance Modulus

Whenever we know both the apparent and absolute magnitudes of a star, a comparison of the two will give its distance. In the text we learned how to make this calculation, by following these steps:

1. Convert the difference m–M between apparent and absolute magnitiudes into a brightness ratio; that is, a numerical factor indicating how much brighter or fainter the star would appear at 10 parsecs distance than at its actual distance;

2. Using the inverse-square law, determine the change in distance required to produce this change in brightness;

3. Multiply this distance factor by 10 parsecs to find the distance to the star.

To do calculations mentally in this way can be laborious, especially in cases where the magnitude difference does not neatly correspond to a simple numerical factor, as it did in the examples given in the text. Hence astronomers use a mathematical equation expressing the relationship between distance and the distance modulus m–m. This equation, which works equally well for all cases, is

$$d = 10^{1 + .2(m-M)},$$

where d is the distance in parsecs to a star whose apparent magnitude is m and whose absolute magnitude is M.

In a simple example, where $m = 9$ and $M = -6$, we have

$$d = 10^{1+.2(15)}$$
$$= 10^{1+3}$$
$$= 10^4 \text{ parsecs.}$$

Now let us try a more complex case. Suppose the star is an M2 main-sequence star, so that $M = 13$, as found from the H-R diagram. The star's apparent magnitude is $m = 16$. Our equation tells us that

$$d = 10^{1+.2(16-13)}$$
$$= 10^{1+.6}$$
$$= 10^{1.6}$$
$$= 39.8 \text{ parsecs.}$$

Of course, we must use a slide rule, calculator, or mathematical tables, but this method is still relatively straightforward compared with steps outlined earlier, which involve mental calculations.

The Effect of Extinction on the Distance Modulus

When we discussed distance-determination techniques (in chapter 13), we ignored the effects of interstellar extinction. In any method that depends on the apparent brightness of a star, however, extinction can be important, particularly for very distant stars. Because extinction makes stars appear fainter than they otherwise would, our tendency is to overestimate distances if no allowance is made for it.

Recall that in the spectroscopic-parallax technique, we use the distance modulus $m–M$ to find the distance to a star, from the equation given earlier:

$$d = 10^{1+.2(m–M)},$$

where m is the apparent magnitude, M is the absolute magnitude, and d is the distance in parsecs to the star. To correct this equation for extinction, we add a term A_V, which refers to the extinction (in magnitudes) in visual light. Thus, if the extinction toward a particular star makes that star appear 2 magnitudes fainter than it otherwise would, then $A_V = 2$. If we insert this into the equation, we find:

$$d = 10^{1+.2(m–M–A_V)}.$$

Let us consider a simple example, a star whose apparent magnitude is $m = 12.4$ and whose absolute magnitude is $M = 2.4$. First, let us calculate the star's distance if extinction is ignored:

$$d = 10^{1+.2(12.4-2.4)} = 10^3 = 1,000 \text{ parsecs.}$$

Suppose we determine that the extinction in the direction of this star amounts to 1 magnitude. Now the distance is

$$d = 10^{1+.2(12.4-2.4-1)} = 10^{2.8} = 631 \text{ parsecs.}$$

One magnitude is a modest amount of extinction, yet by neglecting it, we overestimated the distance to this star by almost 60 percent. We can see from this example that extinction can have a drastic effect on distance estimates.

It is worthwhile to add a note about how the extinction A_V is determined. In chapter 18 we referred to the measurement of interstellar reddening by comparing the observed $B–V$ color index with what it would be if the star suffered no extinction. In other words, it is possible to determine how much redder, in terms of the $B–V$ color index, a star appears because of extinction. To carry that step further, astronomers define a **color excess** called $E(B–V)$, which is the difference between the observed and intrinsic values $B–V$:

$$E(B–V) = (B–V)_{observed} - (B–V)_{intrinsic}$$

Studies of the variation of interstellar extinction with wavelength show that the extinction at the visual wavelength is approximately equal to three times the color excess; that is:

$$A_V = 3\ E(B–V).$$

Hence determination of excess reddening leads to an estimate of A_V, which in turn can be used in the modified equation for finding the distance.

Appendix 10. The Proton-Proton Chain and the CNO Cycle

In the text we did not explain the details of the reactions that convert hydrogen into helium in stellar cores, although they are quite simple. To do so here, we will use the notation of the nuclear physicist. This is basically a shorthand in symbols. For example, a helium nucleus, containing two protons and two neutrons, is designated $^2_2$He, the subscript indicating the **atomic number** (the number of protons) and the superscript the **atomic weight** (the total number of protons and neutrons). Similarly, a hydrogen nucleus is $^1_1$H, and deuterium, a form of hydrogen with an extra neutron in the nucleus, is $^2_1$H. A special symbol (ν) is used for the **neutrino,** a massless subatomic particle emitted in some reactions, and e^+ indicates a **positron,** which is equivalent to an electron, but with positive electrical charge. The symbol γ indicates a gamma ray, a very short-wavelength photon of light emitted in some reactions.

Using this system, we can now spell out the proton chain:

$$^1_1\text{H} + {}^1_1\text{H} \rightarrow {}^2_1\text{H} + e^+ + \nu$$

$$^2_1\text{H} + {}^1_1\text{H} \rightarrow {}^3_2\text{He} + \gamma.$$

The $^3_2$He particle is a form of helium, but not the common form. Once we have this particle, it will combine with another:

$$^3_2\text{He} + {}^3_2\text{He} \rightarrow {}^4_2\text{He} + {}^1_1\text{H} + {}^1_1\text{H}.$$

We end up with a normal helium nucleus. A total of six hydrogen nuclei went into the reaction (remember, the first two steps had to occur twice, in order to produce two $^3_2$He particles for the final reaction), and there were two left at the end, so the net result is the conversion of four hydrogen nuclei into one helium nucleus.

The CNO cycle, which dominates at higher temperatures, is more complex, involving not only carbon, but also nitrogen and oxygen. Each of these elements has more than one form, with differing numbers of neutrons. Some of these **isotopes** are unstable, and spontaneously emit positrons, decaying into other species in the process. Here is the CNO cycle:

$$^{12}_6\text{C} + {}^1_1\text{H} \rightarrow {}^{13}_7\text{N} + \gamma$$

$$^{13}_7\text{N} \rightarrow {}^{13}_6\text{C} + e^+ + \nu$$

$$^{13}_6\text{C} + {}^1_1\text{H} \rightarrow {}^{14}_7\text{N} + \gamma$$

$$^{14}_7\text{N} + {}^1_1\text{H} \rightarrow {}^{15}_8\text{O} + \gamma$$

$$^{15}_8\text{O} \rightarrow {}^{15}_7\text{N} + e^+ + \nu$$

$$^{15}_7\text{N} + {}^1_1\text{H} \rightarrow {}^{12}_6\text{C} + {}^4_2\text{He}.$$

Here we end up with a helium nucleus and a carbon nucleus, although the particles going into the reaction were four hydrogen nuclei and a carbon nucleus. Along the way three isotopes of nitrogen and one of oxygen were created and then converted into something else, leaving neither element at the end. As in the proton-proton chain, the net result is the conversion of four hydrogen nuclei into one helium nucleus.

Appendix 11. Detected Interstellar Molecules*

NUMBER OF ATOMS	SYMBOL	NAME
2	H_2	Molecular hydrogen
	C_2	Diatomic carbon
	CH	Methylidyne
	CH^+	Methylidyne ion
	CN	Cyanogen
	CO	Carbon monoxide
	CO^+	Carbon monoxide ion
	CS	Carbon monosulfide
	OH	Hydroxyl
	NO	Nitric oxide
	NS	Nitrogen sulfide
	SiO	Silicon monoxide
	SiS	Silicon sulfide
	SO	Sulfur monoxide
3	C_2H	Ethynyl
	HCN	Hydrogen cyanide
	HNC	Hydrogen isocyanide
	NCO	Formyl
	HCO^+	Formyl ion
	N_2H^+	Protonated nitrogen
	HNO	Nitroxyl
	H_2O	Water
	HCS^+	Thioformyl ion
	H_2S	Hydrogen sulfide
	OCS	Carbonyl sulfide
	SO_2	Sulfur dioxide
4	H_2CO	Formaldehyde
	NH_3	Ammonia
	HNCO	Isocyanic acid
	HOCN**	Cyanic acid
	HNCS	Isothiocyanic acid
	C_3N	Cyanoethynyl
	H_2CS	Thioformaldehyde
5	$[CH_4]$	[Methane]
	C_4H	Butadynyl
	HCO_2H	Formic acid
	CH_2CO	Ketene
	HC_3N	Cyanoacetylene
	NH_2CN	Cyanamide
	CH_2NH	Methanimine
6	CH_3OH	Methanol
	CH_3CN	Methyl cyanide
	CH_3SH	Methyl mercaptan
	NH_2CHO	Formamide

*This list does not include isotopic variations—molecules that are identical except that one or more atoms are in rare isotopic forms, such as deuterium in place of hydrogen, or ^{13}C instead of the much more common ^{14}C. Bracketed [] species have been tentatively identified, but not yet confirmed.

**An alternative possible identification for this molecule is $HOCO^+$.

(Continued on next page)

Appendix 11.—*Continued*

NUMBER OF ATOMS	SYMBOL	NAME
7	CH_2CHCN	Vinyl cyanide
	$[CH_2CH_2O]$	[Ethylene oxide]
	CH_3C_2H	Methylacetylene
	CH_3CHO	Acetaldehyde
	CH_3NH_2	Methylamine
	HC_5N	Cyanodiacetylene
8	$HCOOCH_3$	Methyl formate
9	CH_3CH_3O	Dimethyl ether
	CH_3CH_2CN	Ethyl cyanide
	CH_3CH_2OH	Ethanol
	HC_7N	Cyano-hexa-tri-yne
10	$[NH_2CH_2COOH]$	[Glycene]
11	HC_9N	Cyano-octa-tetra-yne
13	$HC_{11}N$	Cyano-deca-penta-yne

Appendix 12. Galaxies of the Local Group

GALAXY*	TYPE**	ABSOLUTE MAGNITUDE	POSITION (1950) RIGHT ASCENSION	DECLINATION
M31 (Andromeda)	Sb	−21.1	$00^h40^m.0$	$+41°00'$
Milky Way	Sbc	−20.5	17 42.5	−28 59
M33 = NGC 598	Sc	−18.9	01 31.1	+30 24
Large Magellanic Cloud	Irr	−18.5	05 24	−69 50
IC 10	Irr	−17.6	00 17.6	+59 02
Small Magnellanic Cloud	Irr	−16.8	00 51	−73 10
M32 = NGC 221	E2	−26.4	00 40.0	+40 36
NGC 205	E6	−16.4	00 37.6	+41 25
NGC 6822	Irr	−15.7	19 42.1	−14 53
NGC 185	Dwarf E	−15.2	00 36.1	+48 04
NGC 147	Dwarf E	−14.9	00 30.4	+48 14
IC 1613	Irr	−14.8	01 02.3	+01 51
WLM	Irr	−14.7	23 59.4	−15 44
Fornax	Dwarf sph	−13.6	02 37.5	−34 44
Leo A	Irr	−13.6	09 56.5	+30 59
IC 5152	Irr	−13.5	21 59.6	−51 32
Pegasus	Irr	−13.4	23 26.1	+14 28
Sculptor	Dwarf sph	−11.7	00 57.5	−33 58
And I	Dwarf sph	−11	00 42.8	+37 46
And II	Dwarf sph	−11	01 13.6	+33 11
And II	Dwarf sph	−11	00 32.7	+36 14
Aquarius	Irr	−11	20 44.1	−13 02
Leo I	Dwarf sph	−11	10 05.8	+12 33
Sagittarius	Irr	−10	19 27.1	−17 47
Leo II	Dwarf sph	−9.4	11 10.8	+22 26
Ursa Minor	Dwarf sph	−8.8	15 08.2	+67 18
Draco	Dwarf sph	−8.6	17 19.4	+57 58
Carina	Dwarf sph		06 40.4	−50 55
Pisces	Irr	−8.5	00 01.2	+21 37

*Galaxy names are derived from a variety of sources, including several catalogues (designating galaxies as M, NGC, and IC, for example), and colloquial names bestowed by the discoverer. Many in this list are simply named after the constellation where they are found.

**The galaxy types listed here are described in the text, except for the *Dwarf sph* designation, which stands for *dwarf spheroidal* and refers to dwarf galaxies that do not easily fit into the category of dwarf ellipticals. Note that the absolute magnitudes of some of the dwarf spheroidals are comparable to those of the brightest individual stars in our galaxy.

Appendix 13. The Relativistic Doppler Effect

In the text we made the fantastic assertion that the largest redshift observed so far in any quasar is 378 percent; that is, the wavelengths of the spectral lines have been shifted toward the red part of the spectrum by a factor of 3.78.

Recall from the discussion of the Doppler effect in chapter 3 that the speed v of an object is related to the shift in wavelength ($\Delta\lambda$) and to the rest wavelength (λ), by

$$v = (\Delta\lambda/\lambda)c$$

where c is the speed of light. A redshift of 378 percent means that $(\Delta\lambda/\lambda) = 3.78$, and this appears to imply that the quasar is receding at more than $3\frac{1}{2}$ times the speed of light! The laws of physics tell us this is not possible.

Fortunately, Einstein's theory of relativity had already provided a way out of this dilemma, long before the quasars were discovered. The simple form of the Doppler realtion that we have used in actually only an approximation of the more complete form developed by Einstein, which takes into account time-dilation effects that occur when objects travel at speeds near that of light. The correct formula is:

$$\frac{\Delta\lambda}{\lambda} = \sqrt{\frac{1 + v/c}{1 - v/c}} - 1$$

which leads to the following solution for the velocity:

$$v = c\left[\frac{(z + 1)^2 - 1}{(z + 1)^2 + 1}\right]$$

where $z = \Delta\lambda/\lambda$ and is the most common form in which quasar redshifts are expressed.

If we use this expression to find the velocity of the most rapid quasar, which has z equal to 3.78, we get

$$v = 0.92\ c.$$

This quasar is moving at 92 percent of the speed of light, a very high velocity indeed, but not in violation of the known laws of physics.

GLOSSARY

Absolute magnitude   The magnitude a star would have if it were precisely ten parsecs away from the sun.

Absolute zero   The temperature at which all molecular or atomic motion stops, equal to $-273°C$ or $-459°F$.

Absorption line   A wavelength at which light is absorbed, producing a dark feature in the spectrum.

Acceleration   Any change in the state of rest or motion of a body; either a change in speed or direction.

Accretion disk   A rotating disk of gas surrounding a compact object (such as a neutron star or black hole), formed by material falling in ward.

Albedo   The fraction of incident light that is reflected from a surface such as that of a planet.

Alpha-capture reaction   A nuclear fusion reaction in which an alpha particle merges with an atomic nucleus. A typical example is the formation of ^{16}O by the fusion of an alpha particle with ^{12}C.

Alpha particle   A nucleus of ordinary helium, containing two protons and two neutrons.

Amino acid   A complex organic molecule of the type that forms proteins. Amino acids are fundamental constituents of all living matter.

Andromeda galaxy   The large spiral galaxy, located some 700,000 parsecs from the sun; the most distant object visible to the unaided eye.

Angstrom (Å)   The unit generally used in measuring wavelengths of visible and ultraviolet light; one angstrom is equal to 10^{-8} centimeters.

Angular diameter   The diameter of an object as seen in the sky, measured in units of angle.

Angular momentum   A measure of the mass, radius, and rotational velocity of a rotating or orbiting body. In the simple case of an object in circular orbit, the angular momentum is equal to the mass of the object times its distance from the center of the orbit times its orbital speed.

Annual motions   Motions in the sky caused by the earth's orbital motion about the sun. These include the seasonal variations of the sun's latitude, and the sun's motion through the zodiac.

Annular eclipse   A solar eclipse that occurs when the moon is near its greatest distance from the earth, so that its angular diameter is slightly smaller than that of the sun, and a ring, or annulus, of the sun's disk is visible surrounding the disk of the moon.

Anticyclone   A rotating wind system around a high-pressure area. On the earth, an anticycline rotates clockwise in the northern hemisphere, and counterclockwise in the southern hemisphere.

Antimatter   Matter composed of the antiparticles of ordinary matter. For each subatomic particle, there is an antiparticle that is its opposite in such properties as electrical charge, but its equivalent in mass. Matter and antimatter, if combined, annihilate each other, producing energy in the form of gamma rays according to the formula $E = mc^2$.

Aphelion   The point in the orbit of a solar-system object where it is farthest from the sun.

Apollo asteroid   An asteroid whose orbit brings it closer than 1 astronomical unit to the sun.

Asteroid   Any of the thousands of small, irregular bodies orbiting the sun, primarily in the asteriod belt, which is located between the orbits of Mars and Jupiter.

Asthenosphere   The deep portions of the earth's mantle, below the zone (the lithosphere) where convection currents are thought to operate. The term is also applied to similar zones in the interiors of the moon and other planets.

Astrology   The ancient belief that earthly affairs and human lives are influenced by the positions of the sun, moon, and planets with respect to the zodiac.

Astrometric binary   A double star recognized as such because the visible star or stars undergo periodic motion that is detected by astrometric measurements.

Astrometry   The science of accurately measuring stellar positions.

Astronomical unit (AU)   A unit of distance used in astronomy, equal to the average distance between the sun and the earth. One AU is equal to 1.4959787×10^8 kilometers.

Atomic number   The number of protons in the nucleus of an element. It is the atomic number that defines the identity of an element.

Atomic weight   The mass of an atomic nucleus in atomic mass units (one atomic mass unit is defined as the average mass of the protons and neutrons in a nucleus of ordinary carbon, ^{12}C). For most atoms, the atomic weight is approximately equal to the total number of protons and neutrons in the nucleus.

Autumnal equinox   The point where the sun crosses the celestial equator from north to south, around September 21. See also **equinox** and **vernal equinox.**

Barred spiral galaxy   A spiral galaxy whose nucleus has linear extensions on opposing sides, giving it a barlike shape. The spiral arms usually appear to emanate from the ends of the bar.

Basalt   An igneous silicate rock, common in regions formed by lava flows on the earth, the moon, and probably the other terrestrial planets.

Beta decay   A spontaneous nuclear reaction in which a neutron decays into a proton and an electron, with a neutrino being emitted also. The term has been generalized to mean any spontaneous reaction in which an electron and a neutrino (or their antimatter equivalents) are emitted.

Big bang   A term referring to any theory of cosmogony in which the universe began at a single point and was very hot initially, and has been expanding from that state since.

Binary star   A double-star system in which the two stars orbit a common center of mass.

Black hole   An object that has collapsed under its own gravitation to such a small radius that its gravitational force traps photons of light.

Bode's law   Also known as the Titius-Bode relation, this law is a simple numerical sequence that approximately represents the relative distances of the inner seven planets from the sun. Four is added to each number in the sequence 0, 3, 6, 12, . . . , and then each is divided by ten, resulting in the sequence 0.4, 0.7, 1.0, 1.6, . . . , which is approximately the sequence of planetary distances from the sun in astronomical units.

Bolide   An extremely bright meteor that explodes in the upper atmosphere.

Bolometric magnitude   A magnitude in which all wavelengths of light are included.

Breccia   Lunar rocks consisting of pebbles and soil fused together by meteorite impacts.

Burster   A sporadic source of intense X rays, probably consisting of a neutron star onto which new matter falls at irregular intervals.

Carbonaceous chondrite   A meteorite containing chondrules, with a high abundance of carbon and other volatile elements. Carbonaceous chondrites, thought to be very old, have apparently been unaltered since the formation of the solar system.

Cassegrain focus   A focal arrangement for a reflecting telescope, in which a convex secondary mirror reflects the image through a hole in the primary mirror to a focus at the bottom of the telescope tube. This arrangement is commonly used in situations where relatively lightweight instruments for analyzing the light are attached directly to the telescope.

Catastrophic theory   Any theory in which observed phenomena are attributed to sudden changes in conditions or to the intervention of an outside force or body.

Celestial equator   The imaginary circle formed by the intersection of the earth's equatorial plane with the celestial sphere. The celestial equator is the reference line for north-south (that is, declination) measurements in the standard equatorial coordinate system.

Celestial pole   The point on the celestial sphere directly overhead at either of the earth's poles.

Celestial sphere   The imaginary sphere formed by the sky. It is a convenient device for discussing and measuring the positions of astronomical objects.

Center of mass   In a binary-star system, or any system consisting of several objects, this is the point about which the mass is "balanced"; that is, the point which moves with a constant velocity through space, while the individual bodies in the system move about it.

Cepheid variable   A pulsating variable star, of a class named after the prototype δ Cephei. Cepheid variables obey a period-luminosity relationship, and are therefore useful as distance indicators. There are two classes of Cepheid variables, the so-called classical Cepheids, which belong to Population I, and the Population II Cepheids, also known as W Virginis stars.

Chondrite   A stony meteorite containing chondrules.

Chondrule   A spherical inclusion in certain meteorites, usually composed of silicates, and always of very great age.

Chromosphere   A thin layer of hot gas just outside the photosphere in the sun and other cool stars. The temperature in the chromosphere ranges from about 4,000 K at its inner edge to 10,000 or 20,000 K at its outer boundary. The chromosphere is characterized by the strong red emission line of hydrogen.

Closed universe   The possible state of the universe in which the expansion will eventually be reversed, and which is characterized by positive curvature, being finite in extent but having no boundaries.

CNO cycle   A nuclear-fusion-reaction sequence in which hydrogen nuclei are combined to form helium nuclei, and in which other nuclei, such as isotopes of carbon, oxygen, and nitrogen, appear as catalysts or by-products. The CNO cycle is dominant in the cores of stars located on the upper main sequence of the H-R diagram.

Color index   The difference B–V between the blue (B) and visual (V) magnitudes of a star. If B is less than V (that is, if the star is brighter in blue than in visual light), then the star has a negative color index, and is relatively hot. If B is greater than V, the color index is positive, and the star is relatively cool.

Coma   The extended, glowing region that surrounds the nucleus of a comet.

Comet   An interplanetary body, composed of loosely bound rocky and icy material, that forms a glowing head and extended tail when it enters the inner solar system.

Configuration   The position of a planet or the moon relative to the sun-earth line.

Confusion limit   A natural, fundamental limitation on the astronomer's ability to probe the universe, that cannot be overcome by technological improvement.

Conjunction   The alignment of two celestial bodies on the sky. In connection with the planets, a conjunction is the alignment of a planet with the sun. An inferior conjunction occurs when an inferior planet is directly between the sun and the earth, and a superior conjunction occurs when any planet is directly behind the sun as seen from the earth.

Constellation   A prominent pattern of bright stars, historically associated with mythological figures. In modern usage, each constellation incorporates a precisely defined region of the sky.

Continental drift   The slow motion of the continental masses over the surface of the earth, caused by the motions of the earth's tectonic plates, which in turn are probably caused by convection in the underlying asthenosphere.

Continuous radiation   Electromagnetic radiation that is emitted in a smooth distribution with wavelength, without spectral features such as emission and absorption lines.

Convection   The transport of energy by fluid motions occurring in gases, liquids, or semirigid material such as the earth's mantle. These motions are usually driven by the buoyancy of heated material, which tends to rise while cooler material descends.

Corona   The very hot, extended outer atmosphere of the sun and other cool main-sequence stars. The high temperature in the corona ($1 - 2 \times 10^6$ K) is probably caused by the dissipation of mechanical energy from the convective zone just below the photosphere.

Cosmic background radiation   The primordial radiation field that fills the universe. It was created in the form of gamma rays at the time of the big bang, but has since cooled so that today its temperature is 3 K and its peak wavelength is near 1.1 millimeters, in the microwave portion of the spectrum. Also known as the 3° background radiation.

Cosmic ray   A rapidly moving atomic nucleus from space. Some cosmic rays are produced in the sun, whereas others come from interstellar space and probably originate in supernova explosions.

Cosmogony   The study of the origins of the universe.

Cosmological constant   A term added to Einstein's field equations in order to allow for solutions in which the universe was static; that is, neither expanding nor contracting. Although the need for the term disappeared when it was discovered that the universe is expanding, the cosmological constant has been retained by modern cosmologists, but is usually assigned the value zero.

Cosmological Principle   The postulate, put forth by most cosmologists, that the universe is both homogeneous and isotropic; it is sometimes stated that the universe looks the same to all observers everywhere.

Cosmological redshift   A Doppler shift toward longer wavelengths that is caused by a galaxy's motion of recession, which in turn is caused by the expansion of the universe.

Cosmology   The study of the universe as a whole.

Coudé focus   A focal arrangement for a reflecting telescope, in which the image is reflected by a series of mirrors to a remote, fixed location where a massive, immovable instrument can be used to analyze it.

Cyclone   A rotating wind system, often associated with storms on the earth, about a low-pressure center. On the earth, cyclones rotate counterclockwise in the northern hemisphere, and clockwise in the southern hemisphere.

Declination   The coordinate in the equatorial system that measures positions in the north-south direction, with the celestial equator as the reference line. Declinations are measured in units of degrees, minutes, and seconds of arc.

Deferent   The large circle centered on or near the earth on which the epicycle for a given planet moved, in the geocentric theory of the solar system developed by ancient Greek astronomers such as Hipparchus and Ptolemy.

Degenerate gas   A gas in which either free electrons or free neutrons are as densely spaced as allowed by laws of quantum mechanics. Such a gas has extraordinarily high density, and its pressure is not dependent on temperature, as it is in an ordinary gas. Degenerate electron gas provides the pressure that supports white dwarfs against collapse, and degenerate neutron gas similarly supports neutron stars.

Deuterium   An isotope of hydrogen, containing in its nucleus one proton and one neutron.

Differential gravitational force   A gravitational force acting on an extended object, such that the portions of the object closer to the source of gravitation feel a stronger force than the portions farther away. Such a force, also known as a tidal force, acts to deform or disrupt the object, and is responsible for many phenomena, ranging from synchronous rotation of moons or double stars to planetary ring systems to the disruption of galaxies in clusters.

Differentiation   The sinking of relatively heavy elements into the core of a planet or other body. Differentiation can occur only in fluid bodies, so any planet that has undergone this process must have once been at least partially molten.

Distance modulus   The difference $m-M$ between the apparent and absolute magnitudes for a given star. This difference, which must be corrected for the effects of interstellar extinction, is a direct measure of the distance to the star.

Diurnal motion   Any motion related to the rotation of the earth. Diurnal motions include the daily risings and settings of all celestial objects.

Doppler effect   The shift in wavelength of light that is caused by relative motion between the source of light and the observer. The Doppler shift, $\Delta\lambda$, is defined as the difference between the observed and rest (laboratory) wavelengths for a given spectral line.

Dwarf elliptical galaxy   A member of a class of small spheroidal galaxies, similar to standard elliptical galaxies except for their small size and low luminosity. Dwarf galaxies are probably the most common in the universe, but cannot be detected at distances beyond the Local Group of galaxies.

Dwarf nova   A close binary-star system containing a white dwarf, in which material from the companion star falls onto the other at sporadic intervals, creating brief nuclear outbursts.

Eclipse   An occurrence in which one object is partially or totally blocked from view by another, or passes through the shadow of another.

Eclipsing binary   A double-star system in which one or both stars are periodically eclipsed by the other, as seen from earth. This situation can occur only when the orbital plane of the binary is viewed edge-on from the earth; or when the two stars are very close together.

Ecliptic   The plane of the earth's orbit about the sun, which is approximately the plane of the solar system as a whole. The apparent path of the sun across the sky is the projection of the ecliptic onto the celestial sphere.

Electromagnetic force   The force created by the interaction of electric and magnetic fields. The electromagnetic force can be either attractive or repulsive, and is important in countless situations in astrophysics.

Electromagnetic radiation   Waves consisting of alternating electric and magnetic fields. Depending on the wavelength, these waves may be known as gamma rays, X rays, ultraviolet radiation, visible light, infrared radiation, or radio radiation.

Electromagnetic spectrum   The entire array of electromagnetic radiation, arranged according to wavelength.

Electron   A tiny, negatively charged particle that orbits the nucleus of an atom. The charge is equal and opposite to that of a proton in the nucleus, and in a normal atom the number of electrons and protons is equal, so the overall electrical charge is zero. It is the electrons that emit and absorb electromagnetic radiation, by making transitions between fixed energy levels.

Ellipse   A geometrical shape such that the sum of the distances from any point on it to two fixed points called foci is constant. In any bound system where two objects orbit a common center of mass, their orbits are ellipses, with the center of mass at one focus.

Elliptical galaxy   One of a class of galaxies characterized by smooth spheroidal forms, few young stars, and little interstellar matter.

Emission line   A wavelength at which radiation is emitted, creating a bright line in the spectrum.

Emission nebula   A cloud of interstellar gas that glows by the light of emission lines. The source of excitation that causes the gas to emit may be radiation from a nearby star, or heating by any of a variety of mechanisms.

Endothermic reaction   Any reaction, nuclear or chemical, that requires more energy than is produced.

Energy   The ability to do work. Energy can be either kinetic, when it is a measure of the motion of an object, or potential, when it is stored but capable of being released into kinetic form.

Epicycle   A small circle on which a planet revolves, which in turn orbits another, distant body. Astronomers in ancient times used epicycles in theories of the solar system in order to devise a cosmology that had the earth at the center, but that also accurately accounted for the observed planetary motions.

Equatorial coordinates   The astronomical coordinate system in which positions are measured with respect to the celestial equator (in the north-south direction) and with respect to a fixed direction (in the east-west dimension). The coordinates used are declination (north-south, in units of angle) and right ascension (east-west, in units of time).

Equinox   Either of two points on the sky where the planes of the ecliptic and the earth's equator intersect. When the sun is at one of these two points, the lengths of night and day on the earth are equal. See also **autumnal equinox** and **vernal equinox.**

Erg A unit of energy that is equal to the kinetic energy of an object with a mass of two grams, moving at a speed of one centimeter per second. An erg is defined technically as the work required to raise a mass of one gram through a distance of one centimeter under a gravitational field equal to that at the earth's surface.

Escape velocity The velocity required for an object to escape the gravitational field of a body such as a planet. In a more technical sense, the escape velocity is the velocity at which the kinetic energy of the object equals its gravitational potential energy; if the object moves any faster, its kinetic energy exceeds its potential energy, and the object can escape the gravitational field.

Event horizon The "surface" of a black hole; the boundary of the region from within which no light can escape.

Evolutionary theory Any theory in which observed phenomena are thought to have arisen as a result of natural processes, requiring no outside intervention or sudden changes.

Excitation A state in which one or more electrons of an atom or ion are in energy levels above the lowest possible one.

Fluorescence The emission of light at a particular wavelength following excitation of the electron by absorption of light at another, shorter, wavelength.

Focus (1) The point at which light collected by a telescope is brought together to form an image; (2) one of two fixed points that define an ellipse (see also **ellipse**).

Force Any agent or phenomenon that produces acceleration of a mass.

Frequency The rate (in units of hertz, or cycles per second) at which electromagnetic waves pass a fixed point. The frequency, usually designated ν, is related to the wavelength λ and the speed of light c by $\nu = c/\lambda$.

Fusion reaction A nuclear reaction in which atomic nuclei combine to form more massive nuclei.

Galactic cluster A loose cluster of stars located in the disk or spiral arms of a galaxy.

Gamma ray A photon of electromagnetic radiation, whose wavelength is very short and whose energy is very high. Radiation whose wavelength is less than one angstrom is usually considered to be gamma-ray radiation.

Gegenshein The diffuse glowing spot (seen on the ecliptic opposite the sun's direction), created by sunlight reflected off of interplanetary dust.

Globular cluster A large, spherical cluster of stars located in the halo of the galaxy. These clusters, containing up to several hundred thousand members, are thought to be among the oldest objects in the galaxy.

Gram A unit of mass equal to the quantity of mass contained in one cubic centimeter of water.

Grand-unified theory (GUT) A theory in particle physics which postulates that the electromagnetic, weak interaction, and strong interaction forces are the same phenomenon, behaving differently under different physical conditions. It is expected that the gravitational force will eventually be included.

Granulation The spotty appearance of the solar surface (the photosphere) caused by convection in the layers just below.

Gravitational redshift A Doppler shift toward long wavelengths caused by the effect of a gravitational field on photons of light. Photons escaping a gravitational field lose energy to the field, which results in the redshift.

Greatest elongation The greatest angular distance from the sun that an inferior planet can reach, as seen from earth.

Greenhouse effect The trapping of heat near the surface of a planet, by atmospheric molecules (such as carbon dioxide) that absorb infrared radiation emitted by the surface.

GUT See **grand-unified theory.**

H II region A volume of ionized gas surrounding a hot star (see also **emission nebula**).

H-R diagram See **Hertsprung-Russell diagram.**

Half-life The time required for half of the nuclei of an unstable (radioactive) isotope to decay.

Halo (1) The extended outer portions of a galaxy such as the Milky Way. The halo is thought to contain a large fraction of the total mass of the galaxy, mostly in the form of dim stars and interstellar gas. (2) The extensive cloud of gas surrounding the head of a comet.

Helium flash A rapid burst of nuclear reactions in the degenerate core of a moderate-mass star in the hydrogen shell-burning phase. The flash occurs when the core temperature reaches a sufficiently high temperature to trigger the triple-alpha reaction.

Herbig-Haro objects Small emission nebulosities associated with regions of star formation. These objects are often found near T Tauri stars, and are thought to be powered by them, as beams of heated gas from the T Tauri stars strike surrounding interstellar material.

Hertz (Hz) A unit of frequency used in describing electromagnetic radiation; one hertz is equal to one cycle or wave per second.

Hertzsprung-Russell diagram A diagram on which stars are represented according to their absolute magnitudes (on the vertical axis) and spectral types (on the horizontal axis).

Because the physical properties of stars are interrelated, stars are not randomly located on such a diagram, but instead lie in well-defined regions according to their state of evolution. Very similar diagrams can be constructed that show luminosity instead of absolute magnitude, and temperature or color index in place of spectral type.

High-velocity star   A star whose velocity relative to the solar system is large. As a rule, high-velocity stars are Population II objects following orbital paths that are highly inclined to the plane of the galactic disk.

Homogeneous   Having the quality of being uniform in properties throughout. In astronomy, this term is often applied to the universe as a whole, which is postulated to be homogeneous.

Horizontal branch   A sequence of stars in the H-R diagram of a globular cluster, extending horizontally across the diagram to the left from the red-giant region. These stars are probably undergoing helium burning in their cores, by the triple-alpha reaction.

Hubble constant   The numerical factor, usually denoted H, that describes the rate of expansion of the universe. It is the proportionality constant in the Hubble law $v = Hd$, which relates the speed of recession of a galaxy (v) to its distance (d). The present value of H is not well known; estimates range from 55 to 90 km/sec/Mpc.

Hydrostatic equilibrium   The state of balance between gravitational and pressure forces that exists at all points inside any stable object such as a star or planet.

Igneous rock   A rock that was formed by cooling and hardening from a molten state.

Impact crater   A crater formed on the surface of a terrestrial planet or a satellite by the impact of a meteoroid or planetesimal.

Inertia   The tendency of an object to remain in its state of rest or uniform motion; this tendency is directly related to the mass of the object.

Inferior planet   A planet whose orbit lies closer to the sun than that of the earth; namely, Mercury or Venus.

Inflationary universe   A theoretical model in which the early universe formed small domains that then rapidly expanded, each such "bubble" forming a universe in its own right, with no possibility of communication with any other.

Infrared radiation   Electromagnetic radiation in the wavelength region just longer than that of visible light; that is, radiation whose wavelength lies roughly between 7,000 Å and 0.01 centimeter.

Interferometry   The use of interference phenomena in electromagnetic waves to precisely measure positions or to achieve gains in resolution. Interferometry in radio astronomy entails the use of two or more antennae to overcome the usually very coarse resolution of a single radio telescope; in visible-light observations, the goal is to eliminate the distorting effects of the earth's atmosphere.

Interstellar cloud   A region of relatively high density in the interstellar medium. Interstellar clouds have densities ranging between 1 and 10^6 particles per cubic centimeter, and in aggregate, contain most of the mass in interstellar space.

Interstellar extinction   The obscuration of starlight by interstellar dust. Light is scattered off of dust grains, so that a distant star appears dimmer than it otherwise would. The scattering process is most effective at short (blue) wavelengths; thus stars seen through interstellar dust appear reddened and dimmed.

Inverse-square law   In general, any law describing a force or other phenomenon that decreases in strength as the square of the distance from some central reference point. The term is often used by itself to mean the law stating that the intensity of light emitted by a source such as a star diminishes as the square of the distance from the source.

Ion   Any subatomic particle with a nonzero electrical charge. In standard practice, the term *ion* is usually applied only to positively charged particles such as atoms missing one or more electrons.

Ionization   Any process by which an electron or electrons are freed from an atom or ion. Generally, ionization occurs in two ways: by the absorption of a photon with sufficient energy; or by collision with another particle.

Ionosphere   The zone of the earth's upper atmosphere, between 80- and 500-km altitude, where charged subatomic particles (chiefly protons and electrons) are trapped by the earth's magnetic field. See also **Van Allen belts.**

Isotope   Any form of a given chemical element. Different isotopes of the same element have the same number of protons in their nuclei, but different numbers of neutrons.

Isotropic   Having the property of appearing the same in all directions. In astronomy, this term is often postulated to apply to the universe as a whole.

Kelvin   A unit of temperature, equal to one one-hundredth of the difference between the freezing and boiling points of water, and used in a scale whose zero point is absolute zero. A Kelvin is usually denoted by K.

Kiloparsec (Kpc)   A unit of distance equal to 1,000 parsecs.

Kinetic energy   The energy of motion. The kinetic energy of a moving object is equal to one-half times its mass times the square of its velocity.

Kirkwood's gaps   Narrow gaps in the asteroid belt created by orbital resonance with Jupiter.

Light-gathering power   The ability of a telescope to collect light from an astronomical source; the light-gathering power is directly related to the area of the primary mirror or lens.

Limb darkening   The dark region around the edge of the visible disk of the sun or a planet, caused by a decrease in temperature with height in the atmosphere.

Liquid metallic hydrogen   Hydrogen in a state of semirigidity that can exist only under conditions of extremely high pressure, as in the interiors of Jupiter and Saturn.

Lithosphere   The layer in the earth, moon, and terrestrial planets that includes the crust and the outer part of the mantle.

Local Group   The cluster of about thirty galaxies to which the Milky Way belongs.

Luminosity   The total energy emitted by an object per second; that is, the power of the object. For stars the luminosity is usually measured in units of ergs per second.

Luminosity class   One of several classes to which a star can be assigned on the basis of certain luminosity indicators in the star's spectrum. The classes range from I for supergiants to V for main-sequence stars (also known as dwarfs).

Lunar month   The synodic period of the moon, equal to 27 days 7 hours 43 minutes 11.5 seconds.

L-wave   A type of seismic wave that travels only over the surface of the earth.

Magellanic Clouds   The two irregular galaxies that are the nearest neighbors of the Milky Way; they are visible (to the unaided eye) in the southern hemisphere.

Magma   Molten rock from the earth's interior.

Magnetic braking   The slowing of the spin of a young star (such as the early sun) by magnetic forces exerted on the surrounding ionized gas.

Magnetic dynamo   A rotating internal zone inside the sun or a planet, thought to carry the electrical currents that create the solar or planetary magnetic field.

Magnetic monopole   A particle postulated to have been formed in the early universe, consisting of a single magnetic pole.

Magnetosphere   A region that surrounds a star or planet and that is permeated by the magnetic field of that body.

Magnitude   A measure of the brightness of a star, based on a system established by Hipparchus in which stars were ranked according to how bright they appeared to the unaided eye. In the modern system, a difference of 5 magnitudes corresponds exactly to a brightness ratio of 100, so that a star of a given magnitude has a brightness that is $100^{1/5} = 2.512$ times that of a star 1 magnitude fainter.

Main sequence   The strip in the H-R diagram (running from upper left to lower right) where most stars, those which are converting hydrogen to helium by nuclear reactions in their cores, are found.

Main-sequence fitting   A distance-determination technique in which an H-R diagram for a cluster of stars is compared with a standard H-R diagram to establish the absolute-magnitude scale for the cluster H-R diagram.

Main-sequence turn-off   In an H-R diagram for a cluster of stars, the point where the main sequence turns off toward the upper right. The main-sequence turn-off, showing which stars in the cluster have evolved to become red giants, is an indicator of the age of the cluster.

Mantle   The semirigid outer portion of the earth's interior, extending from roughly the midway point nearly to the surface, and consisting of the mesosphere (the lower portion) and the asthenosphere.

Mare (pl. _maria_)   Any of several extensive, smooth lowland areas on the surface of the moon or Mercury that were created by extensive lava flows early in the history of the solar system.

Mass   The quantity of matter contained in a body or object, measurable only by virtue of its inertia or the force it exerts when subjected to a gravitational field.

Mass-to-light ratio   The mass of a galaxy, in units of solar masses, divided by the galaxy's luminosity, in units of the sun's luminosity. The mass-to-light ratio is an indicator of the relative quantities of Population I and Population II stars in a galaxy.

Mean solar day   The average length of the solar day, as measured throughout the year; the mean solar day is precisely equal to twenty-four hours.

Megaparsec (Mpc)   A unit of distance equal to 10^6 parsecs.

Meridian   The great circle on the celestial sphere that passes through both poles and directly overhead; that is, the north-south line directly overhead.

Mesosphere   (1) The layer of the earth's atmosphere between roughly 50-and 80-km altitude, where the temperature decreases with height; (2) the layer below the asthenosphere in the earth's mantle.

Metamorphic rock   A rock formed by heat and pressure in the earth's interior.

Meteor   A bright streak or flash of light created when a meteoroid enters the earth's atmosphere from space.

Meteorite   The remnant of a meteoroid that survives a fall through the earth's atmosphere and reaches the ground.

Meteoroid   A small interplanetary body.

Meteor shower   A period during which meteors are seen with high frequency, occurring when the earth passes through a swarm of meteoroids.

Micrometeorite   A microscopically small meteorite.

Microwave background   See *cosmic background radiation* and *3° background radiation*.

Milky Way   Historically, the diffuse band of light stretching across the sky; our cross-sectional view of the disk of our galaxy. In modern usage, the term *Milky Way* refers to our galaxy as a whole.

Neutrino   A subatomic particle, without mass or electrical charge, that is emitted in certain nuclear reactions.

Neutron   A subatomic particle with no electrical charge and a mass nearly equal to that of the proton; neutrons and protons are the chief components of the atomic nucleus.

Neutron star   A very compact, dense stellar remnant whose interior consists entirely of neutrons, and which is supported against collapse by degenerate neutron gas pressure.

Newtonian focus   A focal arrangement for reflecting telescopes in which a flat mirror is used to reflect the image through a hole in the side of the telescope tube.

Nonthermal radiation   Radiation not caused solely by the temperature of an object. The term is most often applied to sources of continuous radiation such as synchrotron radiation.

Nova   A star that temporarily flares up in brightness, most likely as a result of nuclear reactions caused by the deposition of new nuclear fuel on the surface of a white dwarf in a binary system. See also *dwarf nova* and **recurrent nova**.

Nucleus   The central, dense concentration in an atom, comet, or galaxy.

OB association   A group of young stars whose luminosity is dominated by O and B stars.

Occam's razor   The principle that the simplest explanation of any natural phenomenon is most likely the correct one.

Oort cloud   The cloud of cometary bodies hypothesized to be orbiting the sun at a great distance, from which comets originate.

Open universe   A possible state of the universe. In this state, the expansion of the universe will never stop; it is characterized by negative curvature, being infinite in extent and having no boundaries.

Opposition   A planetary configuration in which a superior planet is positioned exactly in the opposite direction from the sun, as seen from earth.

Optical binary   A pair of stars that happen to appear near each other on the sky, but which are not in orbit; not a true binary.

Orbital resonance   A situation in which the periods of two orbiting bodies are simple multiples of each other, so that they are frequently aligned, and gravitational forces exerted by the outer body may move the inner body out of its original orbit. This is one mechanism thought responsible for creating gaps in the rings of Saturn, and for creating Kirkwood's gaps in the asteroid belt.

Organic molecule   Any of a large class of carbon-bearing molecules that are found in living matter.

Paleomagnetism   Vestigial traces or artifacts of ancient magnetic fields.

Parallax   Any apparent shift in position caused by an actual motion or displacement of the observer. See also **stellar parallax.**

Parsec (pc)   A unit of distance equal to the distance to a star whose stellar parallax is one second of arc. One parsec is equal to 206,265 astronomical units, 3.03×10^{13} kilometers, or 3.26 light-years.

Peculiar velocity   The deviation of a star's velocity from perfect circular motion about the galactic center.

Penumbra   (1) The light, outer portion of a shadow, such as the portion of the earth's shadow where the moon is not totally obscured during a lunar eclipse; (2) the light, outer portion of a sunspot.

Perihelion   The point in the orbit of any sun-orbiting body where it most closely approaches the sun.

Permafrost   A permanent layer of ice just below the surface of certain regions on earth and probably on Mars.

Photometer   A device, usually having a photoelectric cell, for measuring the brightnesses of astronomical objects.

Photon   A particle of light having wave properties but also acting as a discrete unit.

Photosphere   The visible surface layer of the sun and stars; the layer from which continuous radiation escapes and where absorption lines form.

Planck constant   The numerical factor h relating the frequency ν of a photon to its energy E in the expression $E = h\nu$. The Planck constant has the value $h = 6.62620 \times 10^{-27}$ erg $\cdot$ sec.

Planck function (also known as the Planck law)   The mathematical expression describing the continuous thermal spectrum of a glowing object. For a given temperature, the Planck function specifies the intensity of radiation as a function of either frequency or wavelength.

Planetary nebula   A cloud of glowing, ionized gas, usually taking the form of a hollow sphere or shell, ejected by a star in the late stages of its evolution.

Planetesimal   A small (diameter up to several hundred kilometers) solar-system body of the type that first condensed from the solar nebula. Planetesimals are thought to have been the principal bodies that combined to form the planets.

Plate tectonics   A general term referring to the motions of lithospheric plates over the surface of the earth or other terrestrial planets. See also **continental drift.**

Polarization   The preferential alignment of the magnetic and electric fields in electromagnetic radiation.

Population I   A class of stars with relatively high abundances of heavy elements. These stars are generally found in the disk and spiral arms of spiral galaxies, and are relatively young. The term *Population I* is also commonly applied to other components of galaxies associated with star formation, such as the interstellar material.

Population II   A class of stars with relatively low abundances of heavy elements. These stars are generally found in a spheroidal distribution about the galactic center and throughout the halo, and are relatively old.

Positron   A subatomic particle with the same mass as the electron, but with a positive electrical charge; the antiparticle of the electron.

Potential energy   Energy that is stored, and which may be converted into kinetic energy under certain circumstances. In astronomy, the most common form of potential energy is gravitational potential energy.

Precession   The slow shifting of star positions on the celestial sphere, caused by the 26,000-year periodic wobble of the earth's rotation axis.

Primary mirror   The principal light-gathering mirror in a reflecting telescope.

Prime focus   The focal arrangement in a reflecting telescope in which the image is allowed to form inside the telescope structure at the focal point of the primary mirror, so that no secondary mirror is needed.

Prograde motion   Orbital or spin motion in the forward or "normal" direction; in the solar system, this is counterclockwise as viewed looking down from above the north pole.

Proper motion   The motion of a star across the sky, usually measured in units of arcseconds per year.

Proton-proton chain   The sequence of nuclear reactions in which four hydrogen nuclei combine, through intermediate steps involving deuterium and ^{3}He, to form one helium nucleus. The proton-proton chain is responsible for energy production in the cores of stars on the lower main sequence of the H-R diagram.

Protostar   A star in the process of formation, specifically one that has entered the slow gravitational contraction phase.

Pulsar   A rapidly rotating neutron star that emits periodic pulses of electromagnetic radiation, probably by the emission of beams of radiation from the magnetic poles, which sweep across the sky as the star rotates.

P wave   A seismic wave that is a compressional, or density, wave. *P* waves can travel through both solid and liquid portions of the earth, and are the first to reach any location remote from an earthquake site.

QSO   See **quasi-stellar object.**

Quadrature   The configuration where a superior planet or the moon is 90 degrees away from the sun, as seen from the earth.

quantum   The amount of energy associated with a photon, equal to $h\nu$, where h is the Planck constant, and ν is the frequency. The quantum is the smallest amount of energy that can exist at a given frequency.

Quantum mechanics   The physics of atomic structure and the behavior of subatomic particles, based on the principle of the quantum.

Quark   The most fundamental of elementary particles; the basic particle of which all others are formed.

Quasar   See **quasi-stellar object.**

Quasi-stellar object   Any of a class of extragalactic objects characterized by emission lines with very large redshifts. The quasi-stellar objects are thought to lie at great distances, in which case they existed only at earlier times in the history of the universe; they may be young galaxies.

Radiation pressure   Pressure created by light hitting a surface.

Radiative transport   The transport of energy, inside a star or in other situations, by radiation.

Radioactive dating   A technique for estimating the age of material such as rock, based on the known initial isotopic composition and the known rate of radioactive decay for unstable isotopes originally present.

Radio galaxy   Any of a class of galaxies whose luminosity is greatest in radio wavelengths. Radio galaxies are usually elliptical galaxies, with synchrotron radiation emitted from one or more pairs of lobes located on opposite sides of the visible galaxy.

Ray   A bright streak of ejecta emanating from an impact crater, especially on the moon or on Mercury.

Recurrent nova   A star known to flare up in nova outbursts more than once. A recurrent nova is thought to be a binary system containing a white dwarf and a mass-losing star; the white dwarf sporadically flares up when material falls onto it from the companion.

Red giant   A star that has completed its core hydrogen-burning stage, and has begun hydrogen shell-burning, causing its outer layers to become very extended and cool.

Reflecting telescope   A telescope that brings light to a focus using mirrors.

Reflection nebula   An interstellar cloud containing dust that shines by light reflected from a nearby star.

Refracting telescope   A telescope that uses lenses to bring light to a focus.

Refractory   The property of being able to exist in solid form under conditions of very high temperature. Refractory elements are characterized by a high temperature of vaporization; they are the first to condense into solid form when a gas cools, as in the solar nebula.

Regolith   On the surface of the moon, the layer of debris created by the impact of meteorites; the lunar surface layer.

Resolution   In an image, the ability to separate closely spaced features; that is, the clarity or fineness of the image. In a spectrum, the ability to separate features that are close together in wavelength.

Rest wavelength   The wavelength of a spectral line as measured in a laboratory, when there is no relative motion between source and observer.

Retrograde motions   Orbital or spin motion in the opposite direction from prograde motion; in the solar system, retrograde motions are clockwise as seen from above the north pole.

Right ascension   The east-west coordinate in the equatorial coordinate system. The right ascension is measured in units of hours, minutes, and seconds to the east from a fixed direction on the sky, which itself is defined as the line of intersection of the ecliptic and the celestial equator.

Rille   A type of winding, sinuous valley commonly found on the moon.

Roche limit   The point near a massive body (such as a planet or star) inside of which the tidal forces acting on an orbiting body exceed the gravitational force holding it together. The location of the Roche limit depends on the size of the orbiting body.

RR Lyrae variable   A member of a class of pulsating variable stars named after the prototype star, RR Lyrae. These stars are blue-white giants with pulsational periods of less than one day, and are Population II objects found primarily in globular clusters.

Saros cycle   An 18-year, 11-day repeating pattern of solar and lunar eclipses caused by a combination of the tilt of the lunar orbit with respect to the ecliptic and the precession of the plane of the moon's orbit.

Scattering   The random reflection of photons by particles such as atoms or ions in a gas, or dust particles in interstellar space.

Schwarzschild radius   The radius within which an object has collapsed at the point when light can no longer escape the gravitational field, as the object becomes a black hole.

Secondary mirror   The second mirror in a reflecting telescope (after the primary mirror), usually either convex in shape, to reflect the image out of a hole in the bottom of the telescope to the cassegrain focus or to a series of flat mirrors in a coudé focus; or flat, to reflect the image out of the side of the telescope to the Newtonian focus.

Sedimentary rock   A rock formed by the deposition and hardening of layers of sediment, usually either underwater or in an area subject to flooding.

Seeing   The blurring and distortion of point sources of light such as stars, caused by turbulent motions in the earth's atmosphere.

Seismic wave   A wave created in a planetary or satellite interior, usually caused by an earthquake.

Selection effect   The tendency for a conclusion based on observations to be influenced by the method used to select the objects for observation. An example was the early belief that all quasars are radio sources, when the principal method used to discover quasars was to look for radio sources and then to see whether they had other properties associated with quasars.

Semimajor axis   One-half of the major, or long, axis of an ellipse.

Seyfert galaxy   Any of a class of spiral galaxies (first recognized by Carl Seyfert) having unusually bright, blue-colored nuclei.

Shear wave   A wave that consists of transverse motions; that is, motions perpendicular to the direction of wave travel.

Sidereal day   The rotation period of the earth with respect to the stars (or as seen by a distant observer), equal to 23 hours, 56 minutes, 4.091 seconds.

Sidereal period   The orbital or rotational period of any object with respect to the fixed stars, or as seen by a distant observer.

Solar day   The synodic rotation period of the earth with respect to the sun; that is, the length of time from one local noon, when the sun is on the meridian, to the next local noon.

Solar flare   An explosive outburst of ionized gas from the sun, usually accompanied by X-ray emission and the injection of large quantities of charged particles into the solar wind.

Solar motion   The deviation of the sun's velocity from perfect circular motion about the center of the galaxy; that is, the sun's peculiar velocity.

Solar nebula   The primordial gas and dust cloud from which the sun and planets condensed.

Solar wind   The stream of charged subatomic particles flowing steadily outward from the sun.

Solstice   The occasion when the sun, as viewed from the earth, reaches its farthest northern point (the summer solstice) or its farthest southern point (the winter solstice).

Spectrogram A photograph of a spectrum.

Spectrograph An instrument for recording the spectra of astronomical bodies or other sources of light.

Spectroscope An instrument allowing an observer to view the spectrum of a source of light.

Spectroscopic binary A binary system recognized as a binary because its spectral lines undergo periodic Doppler shifts as the orbital motions of the two stars cause them to move toward and away from the earth. If lines of only one star are seen, it is a single-lined spectroscopic binary; if lines of both stars are seen, it is a double-lined spectroscopic binary.

Spectroscopic parallax The technique of distance determination for stars in which the absolute magnitude is inferred from the H-R diagram and then compared with the observed apparent magnitude to yield the distance.

Spectroscopy The science of analyzing the spectra of stars or other sources of light.

Spectrum (pl. spectra) An arrangement of electromagnetic radiation according to wavelength.

Spectrum binary A binary system recognized as a binary because its spectrum contains lines of two stars of different spectral types.

Spin-orbit coupling A simple relationship between the orbital and spin periods of a satellite or planet, caused by tidal forces that have slowed the rate of rotation of the orbiting body. Synchronous rotation is the simplest and most common form of spin-orbit coupling.

Spiral density wave A spiral wave pattern in a rotating, thin disk, such as the rings of Saturn or the plane of a spiral galaxy like the Milky Way.

Spiral galaxy Any of a large class of galaxies exhibiting a disk with spiral arms.

Standard candle A general term for any astronomical object whose absolute magnitude can be inferred from its other observed characteristics, and which is therefore useful as a distance indicator.

Stefan-Boltzmann law The law of continuous radiation stating that for a spherical glowing object such as a star, the luminosity is proportional to the square of the radius and the fourth power of the temperature.

Stefan's law An experimentally derived law of continuous radiation stating that the energy emitted by a glowing body per square centimeter of surface area is proportional to the fourth power of the absolute temperature.

Stellar parallax The apparent annual shifting of position of a nearby star with respect to more distant background stars. The term *stellar parallax* is often assumed to mean the parallax angle, which is one-half of the total angular motion a star undergoes. See also **parallax** and **parsec.**

Stellar wind Any stream of gas flowing outward from a star, including the very rapid winds from hot, luminous stars; the intermediate-velocity, rarefied winds from stars like the sun; and the slow, dense winds from cool supergiant stars.

Stratosphere The layer of the earth's atmosphere between 10 and 50 kilometers in altitude, where the temperature increases with height.

Subduction The process in which one tectonic plate is submerged below another along a line where two plates collide. A subduction zone is usually characterized by a deep trench and an adjoining mountain range.

Supercluster A cluster of clusters of galaxies.

Supergiant A star in its late stages of evolution which is undergoing shell burning and is therefore very extended in size and very cool; an extremely large giant.

Supergranulation The pattern of large cells seen in the sun's chromosphere, when viewed in the light of the strong emission line of ionized hydrogen.

Superior planet Any planet whose orbit lies beyond the earth's orbit around the sun.

Supernova The explosive destruction of a massive star that occurs when all sources of nuclear fuel have been consumed, and the star collapses catastrophically.

S wave A type of seismic wave that is a transverse, or shear, wave, and which can travel only through rigid materials.

Synchronous rotation A situation in which the rotational and orbital periods of an orbiting body are equal, so that the same side is always facing the companion object.

Synchrotron radiation Continuous radiation produced by rapidly moving electrons traveling along lines of magnetic force.

Synodic period The orbital or rotational period of an object as seen by an observer on earth. For the moon or a planet, the synodic period is the interval between repetitions of the same phase or configuration.

Tetonic activity Geophysical processes involving motions of tectonic plates and associated volcanic and earthquake activity. See also **plate tectonics** and **continental drift.**

Thermal radiation Continuous radiation emitted by any object whose temperature is above absolute zero.

Thermal spectrum The spectrum of continuous radiation from a body that glows because it has a temperature above absolute zero; a thermal spectrum is one that is described mathematically by the Planck function.

Thermosphere The layer of the earth's atmosphere above the mesosphere, extending upward from a height of 80 kilometers, where the temperature rises with altitude.

3° background radiation The radiation field filling the universe, left over from the big bang. See also **cosmic background radiation.**

Tidal force A gravitational force that tends to stretch or distort an extended object. See **differential gravitational force.**

Total eclipse Any eclipse in which the eclipsed body is totally blocked from view or totally immersed in shadow.

Transit telescope A telescope designed to point straight overhead and accurately measure the times at which stars cross the meridian.

Transverse wave See **shear wave.**

Triple-alpha reaction A nuclear fusion reaction in which three helium nuclei (or alpha particles) combine to form a carbon nucleus.

Trojan asteroid Any of several asteroids orbiting the sun at stable positions in the orbit of Jupiter, either 60° ahead of the planet or 60° behind it.

Troposphere The lowest temperature zone in the earth's atmosphere, extending from the surface to a height of about 10 kilometers, in which the temperature decreases with altitude.

T Tauri star A young star still associated with the interstellar material from which it formed, typically exhibiting brightness variations and a stellar wind.

24-hour anisotropy The daily fluctuation in the observed temperature of the cosmic background radiation caused by a combination of the earth's motion with respect to the background and its rotation.

21-cm line The emission line of atomic hydrogen, whose wavelength is 21.11 centimeters, emitted when the spin of the electron with respect to that of the proton reverses itself. The 21-cm line is the most widely used and effective means of tracing the distribution of interstellar gas in the Milky Way and other galaxies.

Ultraviolet The portion of the electromagnetic spectrum between roughly 100 angstroms and 3,000 angstroms.

Umbra (1)The dark inner portion of a shadow, such as the part of the earth's shadow where the moon is in total eclipse during a lunar eclipse; (2) the dark central portion of a sunspot.

Van Allen belts Zones in the earth's magnetosphere where charged particles are confined by the earth's magnetic field. There are two main belts, one centered at an altitude of roughly 1.5 times the earth's radius, and the other between 4.5 and 6.0 times the earth's radius.

Velocity curve A plot showing the orbital velocity of stars in a spiral galaxy versus distance from the galactic center.

Velocity dispersion A measure of the average velocity of particles or stars in a group or cluster with random internal motions. In globular clusters and elliptical galaxies, the velocity dispersion can be used to infer the central mass.

Vernal equinox The occasion when the sun crosses the celestial equator from south to north, usually occurring around March 21. See also *autumnal equinox* and *equinox.*

Visual binary Any binary system in which both stars can be seen through a telescope or on photographs.

Volatile The property of being easily vaporized. Volatile elements stay in gaseous form except at very low temperatures; they did not condense into solid form during the formation of the solar system.

Wavelength The distance between wavecrests in any type of wave.

White dwarf The compact remnant of a low-mass star, supported against further gravitational collapse by the pressure of the degenerate electron gas that fills its interior.

Wien's law An experimentally discovered law applicable to thermal continuum radiation which states that the wavelength of maximum emission intensity is inversely proportional to the absolute temperature.

X ray A photon of electromagnetic radiation in the wavelength interval between about 1 angstrom and 100 angstroms.

Zeeman effect The broadening or splitting of spectral lines caused by the presence of a magnetic field in the gas where the lines are formed.

Zenith The point directly overhead.

Zero-age main sequence The main sequence in the H-R diagram formed by stars that have just begun their hydrogen-burning lifetimes, and have not yet converted any significant fraction of their core mass into helium. The zero-age main sequence forms the lower left boundary of the broader band representing the general main sequence.

Zodiac A band circling the celestial sphere along the ecliptic, broad enough to encompass the paths of all the planets visible to the naked eye; in some usages, the sequence of constellations lying along the ecliptic.

Zodiacal light A diffuse band of light visible along the ecliptic near sunrise and sunset, created by sunlight scattered off of interplanetary dust.

INDEX

oratory. **Color Plate 23.** *upper:* © 1980 Anglo-Australian Telescope Board; *lower left:* The National Radio Astronomy Observatory, operated by Associated Universities, Inc. under contract with the National Science Foundation; *lower right:* Prepared by the Royal Greenwich Observatory and reproduced with permission of the Science and Engineering Council. **Color Plate 24.** *upper:* Photograph obtained by P. A. Wehringer and S. Wyckoff with the University of Hawaii 2.2-m telescope on Mauna Kea; *lower left:* M. Malkan; *lower right:* Institute for Astronomy and Planetary-Geosciences Data Processing Facility, University of Hawaii.

Figure Credits

Chapter 1
Fig. 1.2 U.S. Naval Observatory photograph. **Fig. 1.3** © 1980 Anglo-Australian Telescope Board. **Fig. 1.14.** Courtesy High Altitude Observatory, National Center for Atmospheric Research; sponsored by the National Science Foundation. **Fig. 1.15.** NASA photograph. **Fig. 1.16.** NASA photograph. **Fig. 1.17.** Reprinted from C. Ronan, *The Astronomers,* 1964, New York: Hill and Wang, with permission. **Fig. 1.18.** The Granger Collection. **Fig. 1.19.** The Granger Collection. **Fig. 1.22.** The Bettmann Archive. **Fig. 1.24.** The Granger Collection.

Chapter 2
Fig. 2.1. City Museum, Torun, Poland; **Fig. 2.3.** The Bettman Archive. **Fig. 2.4.** The Bettman Archive. **Fig. 2.6.** The Bettman Archive. **Fig. 2.8.** The Granger Collection. **Fig. 2.9.** The Granger Collection. **Fig. 2.10.** Yerkes Observatory. **Fig. 2.11.** NASA photograph. **Fig. 2.12.** The Granger Collection. **Fig. 2.13.** The Granger Collection. **Fig. 2.16.** NASA photograph.

Chapter 3
Fig. 3.15. University of Wyoming. **Fig. 3.17.** Palomar Observatory, California Institute of Technology. **Fig. 3.18.** A. G. D. Phillip. **Fig. 3.19.** MMTO: Universtiy of Arizona and Smithsonian Institution. **Fig. 3.20.** MMTO: University of Arizona and Smithsonian Institution. **Fig. 3.21.** Courtesy Duncan Chesley, University of Hawaii. **Astronomical insight 3.2** (p. 59): Anglo-Australian Telescope Board. **Fig. 3.22.** Neils Bohr Laboratory. **Fig. 3.23.** The National Radio Astronomy Observatory, operated by Associated Universities, Inc., under contract with the National Science Foundation. **Fig. 3.24.** The National Radio Astronomy Observatory, operated by Associated Universities, Inc., under contract with the National Science Foundation. **Fig. 3.25.** NASA photograph. **Fig. 3.26.** NASA illustration. **Fig. 3.27.** W. C. Cash. **Fig. 3.28.** AURA, Inc., Kitt Peak National Observatory. **Fig. 3.29.** NASA illustration.

Chapter 4
Fig. 4.1. NASA photograph. **Fig. 4.4.** From F. K. Lutgens and E. J. Tarbuck, 1979, *The Atmosphere: An Introduction to Meteorology,* Fig. 7.3, p. 150. Englewood Cliffs: Prentice-Hall. @ Prentice-Hall. Reprinted with the permission of the publisher. **Fig. 4.5.** NASA photograph. **Fig. 4.12.** From F. S. Sawkins, C. G. Chase, D. C. Darby, and G. Rupp, 1978. *The Evolving Earth: A Text in Physical Geology,* Fig. 6.2, p. 163. New York; Macmillan. © Macmillan. Reprinted with the permission of the publisher. **Fig. 4.14.** Mount Wilson and Las Campanas Observatories, Carnegie Institution of Washington. **Fig. 4.15.** NASA photograph. **Fig. 4.16.** NASA photograph. **Fig.**

4.17. NASA photograph. **Fig. 4.18.** NASA photograph. **Fig. 4.19.** NASA photograph. **Fig. 4.20.** NASA photograph. **Fig. 4.21.** NASA photograph. **Fig. 4.22.** NASA photograph. **Fig. 4.23.** NASA photograph. **Fig. 4.24.** NASA photograph. **Fig. 4.25.** NASA photograph.

Chapter 5
Fig. 5.1. NASA photograph. **Fig. 5.2.** NASA photograph. **Fig. 5.4.** NASA photograph. **Fig. 5.7.** Laboratory for Atmospheric and Space Physics, University of Colorado, sponsored by NASA. **Fig. 5.8.** U.S. Geological Survey. **Fig. 5.9.** U.S. Geological Survey. **Fig. 5.10.** TASS from SOVFOTO. **Fig. 5.11.** TASS from SOVFOTO. **Fig. 5.12.** NASA photograph. **Fig. 5.16.** NASA photograph. **Fig. 5.17.** NASA photograph. **Fig. 5.18.** NASA photograph. **Fig. 5.20.** NASA photograph.

Chapter 6
Fig. 6.1. Mount Wilson and Las Campanas Observatories, Carnegie Institution of Washington. **Fig. 6.2.** Historical Pictures Services, Inc. **Fig. 6.3.** Mount Wilson and Las Campanas Observatories, Carnegie Institution of Washington. **Fig. 6.4.** NASA photograph. **Fig. 6.5.** NASA illustration. **Fig. 6.7.** NASA photograph. **Fig. 6.8.** NASA photograph. **Fig. 6.9.** NASA photograph. **Fig. 6.10.** NASA photograph. **Fig. 6.11.** NASA photograph. **Fig. 6.12.** NASA photograph. **Fig. 6.13.** NASA photograph. **Fig. 6.14.** NASA photograph. **Fig. 6.15.** NASA photograph. **Fig. 6.16.** NASA photograph. **Fig. 6.17.** NASA photograph. **Fig. 6.18** NASA photograph. **Fig. 6.19.** NASA photograph.

Chapter 7
Fig. 7.1. Mount Wilson and Las Campanas Observatories, Carnegie Institution of Washington. **Fig. 7.3.** NASA photograph. **Fig. 7.4.** NASA photograph. **Fig. 7.5.** NASA illustration. **Fig. 7.6.** NASA illustration. **Fig. 7.7.** NASA illustration. **Fig. 7.8.** NASA photograph. **Fig. 7.11.** NASA illustration. **Fig. 7.12.** NASA photograph. **Fig. 7.13.** NASA photograph. **Fig. 7.14.** NASA photograph. **Fig. 7.15.** NASA photograph. **Fig. 7.16.** NASA photograph. **Fig. 7.17.** NASA photograph. **Fig. 7.18.** NASA photograph. **Fig. 7.19.** NASA photograph. **Fig. 7.21.** NASA photograph. **Fig. 7.22.** NASA photograph.

Chapter 8
Fig. 8.1. Mount Wilson and Las Campanas Observatories, Carnegie Institution of Washington. **Fig. 8.2.** Yerkes Observatory. **Fig. 8.3.** NASA illustration. **Fig. 8.4.** NASA photograph. **Fig. 8.5.** Data from Beatty, J. K., O'Leary, B., and Chaikin, A., eds. 1981, *The New Solar System,* Cambridge, England: Cambridge University Press. **Fig. 8.6.** NASA photograph. **Fig. 8.7.** NASA photograph. **Fig. 8.9.** NASA photograph. **Fig. 8.10.** NASA photograph. **Fig. 8.11.** Data from Beatty, J. K., O'Leary, B., and Chaikin, A., eds. 1981, *The New Solar System,* Cambridge, England: Cambridge University Press; and from Owen, t. 1982, *Scientific American,* 246(2):98. **Fig. 8.12.** NASA photograph. **Fig. 8.13.** NASA photograph. **Fig. 8.14.** NASA photograph. **Fig. 8.15.** NASA photograph. **Fig. 8.16.** NASA photograph. **Fig. 8.17.** NASA photograph. **Fig. 8.18.** NASA photograph. **Fig. 8.19.** NASA photograph. **Fig. 8.20.** NASA photograph. **Fig. 8.21.** Laboratory for Atmospheric and Space Physics, University of Colorado, sponsored by NASA. **Fig. 8.22.** NASA photograph. **Fig. 8.23.** NASA photograph. **Fig. 8.24.** NASA photograph.

Chapter 9
Fig. 9.1. Project Stratoscope, Princeton University, supported by the National Science Foundation and NASA. **Fig. 9.3.** NASA photo-

graph, courtesy of H. Reitsema. **Fig. 9.4.** NASA photograph. **Fig. 9.5.** NASA illustration. **Fig. 9.6.** NASA photograph. **Fig. 9.8.** Lowell Observatory photograph. **Fig. 9.9.** Lowell Observatory photograph. **Fig. 9.10.** U.S. Naval Observatory photograph.

Chapter 10

Fig. 10.1. Eleanor F. Helin, Palomar Observatory. **Fig. 10.2.** NASA photograph. **Fig. 10.4.** The Granger Collection. **Fig. 10.6.** Palomar Observatory, California Institute of Technology. **Fig. 10.8.** Palomar Observatory, California Institute of Technology. **Fig. 10.10.** NASA photograph. **Fig. 10.11.** Fr. R. E. Royer. **Fig. 10.12.** New Mexico State University. **Fig. 10.13.** Yerkes Observatory. **Fig. 10.14.** Griffith Observatory, Ronald A. Oriti Collection. **Fig. 10.15.** Griffith Observatory, Ronald A. Oriti Collection. **Astronomical Insight 10.2** (p. 189). E. E. Barnard Observatory photograph by G. Emerson. **Fig. 10.16.** Griffith Observatory, Ronald A. Oriti Collection. **Fig. 10.17.** Griffith Observatory, Ronald A. Oriti Collection. **Fig. 10.18.** Griffith Observatory.**Fig. 10.19.** Yerkes Observatory. **Fig. 10.20.** Yerkes Observatory. **Fig. 10.21.** Photo by S. Suyama, courtesy of J. Weinberg. **Fig. 10.22.** NASA photograph, courtesy of D. E. Brownlee.

Chapter 11

Fig. 11.1. Mount Wilson and Las Campanas Observatories, Carnegie Institution of Washington. **Fig. 11.5.** Mount Wilson and Las Campanas Observatories, Carnegie Institution of Washington. **Fig. 11.6.** Project Stratoscope, Princeton University; sponsored by the National Science Foundation and NASA. **Fig. 11.7.** NASA photograph. **Fig. 11.8.** Sacramento Peak Observatory, operated by AURA, Inc., under contract with the National Science Foundation. **Fig. 11.9.** High Altitude Observatory, National Center for Atmospheric Research, sponsored by the National Science Foundation. **Fig. 11.10.** NASA photograph. **Fig. 11.11.** High Altitude Observatory, National Center for Atmospheric Research, sponsored by the National Science Foundation. **Fig. 11.12.** NASA photograph. **Fig. 11.13.** High Altitude Observatory, National Center for Atmospheric Research, sponsored by the National Science Foundation. **Fig. 11.15.** Prepared by the Royal Greenwich Observatory and reproduced with the permission of the Science and Engineering Research Council; courtesy of J. A. Eddy. **Fig. 11.16.** Palomar Observatory, California Institute of Technology. **Fig. 11.17.** NASA photograph. **Fig. 11.18.** J. A. Eddy.

Chapter 12

Fig. 12.7. From E. E. Barnard, 1927, *A Photographic Atlas of Selected Regions of the Milky Way*, Carnegie Institution of Washington, photographed at the Mount Wilson Observatory.

Chapter 13

Fig. 13.1. Historical Pictures Services, Inc. **Fig. 13.6.** From Abt, H., Meinel, A., Morgan, W. W., and Tapscott, R., 1968, *An Atlas of Low-Dispersion Grating Stellar Spectra*, Tucson: Kitt Peak National Observatory. **Fig. 13.7.** Harvard College Observatory. **Fig. 13.8.** Mount Wilson and Las Campanas Observatories, Carnegie Institution of Washington. **Astronomical Insight 13.2** (p. 239). Harvard College Observatory. **Fig. 13.17.** Princeton University. **Fig. 13.18.** Estate of Henry Norris Russel. Reprinted with permission. **Fig. 13.23.** D. L. Lambert.

Chapter 15

Fig. 15.1. From E. E. Barnard, 1927, *A Photographic Atlas of Selected Regions of the Milky Way*, Carnegie Institution of Washington, photographed at the Mount Wilson Observatory. **Fig. 15.2.** E. E. Barnard Observatory, photographed by G. Emerson. **Fig. 15.3.** Lick Observatory photograph. **Fig. 15.4.** Lick Observatory photograph. **Fig. 15.6.** P. J. Flower. **Fig. 15.8.** Lick Observatory photograph. **Fig. 15.9.** Lick Observatory photograph. **Fig. 15.10.** Photograph by B. J. Bok, made at the prime focus of the 4-m reflector at the Cerro Tololo Inter-American Observatory. **Fig. 15.11.** Photograph by B. J. Bok, made at the prime focus of the 4-m telescope at the Cerro Tololo Inter-American Observatory. **Fig. 15.13.** Lick Observatory photograph. **Fig. 15.17.** Data from I. Iben. **Fig. 15.24.** U.S. Naval Observatory photograph. **Fig. 15.25.** Data from I. Iben. **Fig. 15.29.** Data from E. P. J. van den Heuvel, 1976, *IAU Symposium 73, Structure and Evolution of Close Binary Systems*, ed. P. Eggleton, S. Mitton, and J. Whelan, Dordrecht: Reidel.

Chapter 16

Fig. 16.1. Palomar Observatory, California Institute of Technology. **Fig. 16.2.** G. A. Wegner. **Fig. 16.4.** E. E. Barnard Observatory, photograph by G. Emerson. **Fig. 16.5.** Palomar Observatory, California Institute of Technology. **Fig. 16.7.** Lick Observatory photograph. **Fig. 16.9.** Data from P. C. Joss. **Fig. 16.10.** Data from R. N. Manchester and J. H. Taylor, 1977, *Pulsars*, San Francisco: W. H. Freeman. **Fig. 16.13.** Harvard-Smithsonian Center for Astrophysics. **Fig. 16.17.** Harvard-Smithsonian Center for Astrophysics.

Chapter 17

Fig. 17.1. Mount Wilson and Las Campanas Observatories, Carnegie Institution of Washington. **Fig. 17.3.** @ 1980 Anglo-Australian Telescope Board. **Fig. 17.4.** Harvard College Observatory. **Fig. 17.7.** Harvard College Observatory. **Fig. 17.8.** Estate of H. Shapley. Reprinted with permission. **Fig. 17.12.** E. E. Barnard Observatory, photograph by G. Emerson. **Fig. 17.13.** The National Radio Astronomy Observatory, operated by Associated Universities, Inc., under contract with the National Science Foundation. **Fig. 17.15.** U.S. Naval Observatory photograph. **Fig. 17.17.** Harvard-Smithsonian Center for Astrophysics.

Chapter 18

Fig. 18.1. From E. E. Barnard, 1927, *A Photographic Atlas of Selected Regions of the Milky Way*, Carnegie Institution of Washington; photographed at the Mount Wilson Observatory. **Fig. 18.9.** From E. E. Barnard, 1927, *A Photographic Atlas of Selected Regions of the Milky Way*, Carnegie Institution of Washington; photographed at the Mount Wilson Observatory. **Fig. 18.11.** Lick Observatory photograph. **Fig. 18.12.** From E. E. Barnard, 1927, *A Photographic Atlas of Selected Region of the Milky Way*, Carnegie Institution of Washington; photographed at the Yerkes Observatory. **Fig. 18.14.** C. Heiles. **Fig. 18.15.** E. E. Barnard Observatory, photographed by G. Emerson. **Fig. 18.16.** NASA photograph. **Fig. 18.17.** *X-ray image:* Harvard-Smithsonian Center for Astrophysics; *optical image:* K. Kamper and S. van den Bergh; *radio image:* the National Radio Astronomy Observatory, operated by Associated Universities, Inc., under contract with the National Science Foundation. **Fig. 18.20.** Courtesy T. R. Gull and J. Heckathorn, photographed at the Kitt Peak National Observatory. **Fig. 18.21.** W. C. Cash

Chapter 19

Fig. 19.1. @ 1980 Anglo-Australian Telescope Board. **Fig. 19.6.** A. J. Kalnajs. **Fig. 19.8.** After T. M. Eneev, N. N. Kuzlov, and R. A. Sunyaev, 1973, *Astronomy and Astrophysics*, 22:41.

Chapter 20

Fig. 20.1. Lick Observatory. **Fig. 20.2.** Palomar Observatory, California Institute of Technology. **Fig. 20.3.** U.S. Naval Observatory photograph. **Fig. 20.4.** U.S. Naval Observatory photograph. **Fig. 20.6.** Mount Wilson and Las Campanas Observatories, Carnegie Institution of Washington. **Fig. 20.7.** @ 1980 Anglo-Australian Telescope Board. **Fig. 20.8.** Mount Wilson and Las Campanas Observatories, Carnegie Institution of Washington. **Fig. 20.10.** Palomar Observatory, California Institute of Technology. **Fig. 20.11.** Palomar Observatory, California Institute of Technology. **Fig. 20.12.** The National Radio Astronomy Observatory, operated by Associated Universities, Inc., under contract with the National Science Foundation. **Fig. 20.16.** Palomar Observatory, California Institute of Technology. **Fig. 20.17.** Mount Wilson and Las Campanas Observatories, Carnegie Institution of Washington. **Fig. 20.18.** H. Spinrad. **Fig. 20.19.** Photographed by P. W. Hodge at the Boyden Observatory. **Fig. 20.20.** Photographed by P. W. Hodge at the Lick Observatory. **Fig. 20.21.** A. N. Moffat. **Fig. 20.22.** NASA photograph. **Fig. 20.23.** Palomar Observatory, California Institute of Technology. **Fig. 20.25.** Photograph from the Palomar Observatory, California Institute of Technology; comptuer simulation courtesy A. Toomre and J. Toomre. **Fig. 20.26.** The Kitt Peak National Observatory. **Fig. 20.27.** F. Owen, National Radio Astronomy Observatory. **Fig. 20.28.** Harvard-Smithsonian Center for Astrophysics. **Fig. 20.29.** Illustration from *Sky and Telescope* magazine by Rob Hess, @ 1982 Sky Publishing Corporation, by permission; also with permission of R. B. Tully.

Chapter 21

Fig. 21.1. Palomar Observatory, California Institute of Technology. **Fig. 21.4.** Estate of E. Hubble and M. Humason. Reprinted with permission. **Fig. 21.5.** After J. Silk, 1980, *The Big Bang*, San Francisco: W. H. Freeman. **Fig. 21.10.** Bell Laboratories. **Fig. 21.11.** Data from D. P. Woody and P. L. Richards, 1981, *Astrophysical Journal*, 248:18. **Fig. 21.13.** NASA photograph.

Chapter 22

Figure 22.1. @ Association of Universities for Research in Astronomy; the Cerro Tololo Inter-American Observatory. **Fig. 22.2.** The National Radio Astronomy Observatory, operated by Associated Universities, Inc., under contract with the National Science Foundation. **Fig. 22.3.** Mullard Radio Astronomy Observatory, University of Cambridge. **Fig. 22.4.** Photograph by H. Arp, image processing by J. Lorre, Image Processing Laboratory, JPL. **Fig. 22.5.** National Radio Astronomy Observatory, operated by Associated Universities, Inc., under contract with the National Science Foundation. **Fig. 22.6.** The National Radio Astronomy Observatory, operated by Associated Universities, Inc., under contract with the National Science Foundation. **Fig. 22.7.** Photograph by H. Arp; image processing by J. Lorre, Image Processing Laboratory, JPL. **Fig. 22.8.** Palomar Observatory, California Institute of Technology. **Fig. 22.9.** Palomar Observatory, California Institute of Technology. **Fig. 22.11.** H. Spinrad. **Fig. 22.12.** The Kitt Peak National Observatory. **Fig. 22.16.** M. Malkan. **Fig. 22.17.** Photographs from the Mount Wilson Observatory, Carnegie Institution of Washington; composition by W. W. Morgan. **Fig. 22.18.** Adapted from J. Silk, 1981, *The Big Bang* (San Francisco: W. H. Freeman).

Chapter 23

Fig. 23.1. @ 1978, 1979 by the C-Evolution Quarterly, Box 428, Sausalito, CA. **Fig. 23.2.** The Granger Collection. **Fig. 23.3.** The Granger Collection. **Fig. 23.11.** J. B. Rogerson and D. G. York. **Fig. 23.12.** Data from R. V. Wagoner 1973, Astrophysical Journal, 179:343. **Fig. 23.13.** J. M. Shull.

Chapter 24

Fig. 24.1. The Granger Collection. **Fig. 24.2.** @ West Publishing Co. **Fig. 24.3.** Photo by John Fields, the Trustees, the Australian Museum. **Fig. 24.4.** @ West Publishing Co. **Fig. 24.5.** Photograph by Division of Photography, Field Museum of Natural History. **Fig. 24.6.** NASA photograph. **Fig. 24.9.** NASA photograph. **Fig. 24.10.** The National Radio Astronomy Observatory, operated by Associated Universitites, Inc., under contract with the National Science Foundation. Data from D. Goldsmith and T. Owen, 1980, *The Search for Life in The Universe;* **Fig. 24.11.** Menlo Park, Cal.: Benjamin Cummings. **Fig. 24.12.** NASA photograph.

†